The AutoCAD® 2007 Tutor for Engineering Graphics

The AutoCAD® 2007 Tutor for Engineering Graphics

ALAN J. KALAMEJA

Autodesk

Australia • Canada • Mexico • Singapore • Spain • United Kingdom • United States

The AutoCAD® 2007 Tutor for Engineering Graphics
Alan J. Kalameja

Autodesk Press Staff:

Vice President, Technology and Trades SBU:
David Garza

Director of Learning Solutions:
Sandy Clark

Senior Acquisitions Editor:
James DeVoe

Managing Editor:
Tricia Coia

Marketing Director:
Deborah S. Yarnell

Channel Manager:
Dennis Williams

Academic Marketing Manager:
Gayathri Baskaran

Production Director:
Patty Stephan

Senior Content Project Manager:
Stacy Masucci

Technology Project Manager:
Linda Verde

Editorial Assistant:
Niamh Matthews

COPYRIGHT © 2007, 2006 Thomson Delmar Learning. Thomson, the Star Logo, and Delmar Learning are trademarks used herein under license. Autodesk, AutoCAD, and the AutoCAD logo are registered trademarks of Autodesk. Delmar Learning uses "Autodesk Press" with permission from Autodesk for certain purposes.

Printed in the United States of America
1 2 3 4 5 XX 07 06

For more information contact
Thomson Delmar Learning
Executive Woods
5 Maxwell Drive, PO Box 8007,
Clifton Park, NY 12065-8007

Or find us on the World Wide Web at
www.delmarlearning.com

ALL RIGHTS RESERVED. No part of this work covered by the copyright hereon may be reproduced in any form or by any means—graphic, electronic, or mechanical, including photocopying, recording, taping, Web distribution, or information storage and retrieval systems—without the written permission of the publisher.

For permission to use material from the text or product, contact us by

Tel: (800) 730-2214
Fax: (800) 730-2215

www.thomsonrights.com

Library of Congress Cataloging-in-Publication Data
Kalameja, Alan J.
   The AutoCAD 2007 tutor for engineering graphics / Alan J. Kalameja.
      p. cm.
Includes bibliographical references and index.
ISBN 1-4180-4902-6 (pbk.)
   1. Engineering graphics.   2. AutoCAD.
I. Title.
T357.K35 2006
620'.0042028566—dc22
                                    2006020333

NOTICE TO THE READER

Publisher does not warrant or guarantee any of the products described herein or perform any independent analysis in connection with any of the product information contained herein. Publisher does not assume, and expressly disclaims, any obligation to obtain and include information other than that provided to it by the manufacturer.

The reader is expressly warned to consider and adopt all safety precautions that might be indicated by the activities herein and to avoid all potential hazards. By following the instructions contained herein, the reader willingly assumes all risks in connection with such instructions.

The publisher makes no representation or warranties of any kind, including but not limited to, the warranties of fitness for particular purpose or merchantability, nor are any such representations implied with respect to the material set forth herein, and the publisher takes no responsibility with respect to such material. The publisher shall not be liable for any special, consequential, or exemplary damages resulting, in whole or part, from the readers' use of, or reliance upon, this material.

CONTENTS

Introduction .... **xxiii**
    ONLINE COMPANION XXVIII
    ACKNOWLEDGMENTS XXVIII
    ABOUT THE AUTHOR XXVIII
    CONVENTIONS XXIX
    NOTES TO THE STUDENT AND INSTRUCTOR CONCERNING THE USE OF
        TUTORIAL EXERCISES XXIX
    SUPPLEMENTS XXIX

Chapter 1   Getting Started with AutoCAD **1**
    THE AUTOCAD INTRODUCTORY WORKSPACE SCREEN 1
    THE AUTOCAD CLASSIC WORKSPACE 2
    ADDITIONAL WORKSPACES 3
    THE STATUS BAR 4
        Additional Status Bar Buttons 7
    COMMUNICATING WITH AUTOCAD 7
        The Command Line 7
        Understanding the Command Prompt 7
        Dynamic Input 8
    METHODS OF CHOOSING COMMANDS 8
    PULL-DOWN MENUS 8
    TOOLBARS 9
    ACTIVATING TOOLBARS 9
    DOCKING TOOLBARS 10
    DASHBOARDS 11
        Dashboard Control Panels 11
    DIALOG BOXES AND ICON MENUS 11
    TOOL PALETTES 12
    RIGHT-CLICK SHORTCUT MENUS 13
    COMMAND ALIASES 14
    STARTING A NEW DRAWING 15
    OPENING AN EXISTING DRAWING 15
    BASIC DRAWING COMMANDS 16
    THE LINE COMMAND 17
        Dynamic Input and Lines 19
    THE DIRECT DISTANCE MODE FOR DRAWING LINES 20
    USING OBJECT SNAP FOR GREATER PRECISION 22

OBJECT SNAP MODES	25
Center (Cen)	25
Endpoint (End)	25
Extension (Ext)	25
From (Fro)	26
Insert (Ins)	27
Intersection (Int)	27
Extended Intersection (Int)	28
Midpoint (Mid)	28
Midpoint of Two Selected Points (M2P)	29
Nearest (Nea)	30
Node (Nod)	30
Parallel (Par)	30
Perpendicular (Per)	31
Quadrant (Qua)	31
Tangent (Tan)	32
CHOOSING RUNNING OBJECT SNAP	33
OBJECT SNAP TRACKING	35
POLAR TRACKING	36
SETTING A POLAR SNAP VALUE	39
SETTING A RELATIVE POLAR ANGLE	41
MORE ON OBJECT SNAP TRACKING MODE	42
USING TEMPORARY TRACKING POINTS	44
ALTERNATE METHODS USED FOR PRECISION DRAWING: CARTESIAN COORDINATES	46
ABSOLUTE COORDINATE MODE FOR DRAWING LINES	48
RELATIVE COORDINATE MODE FOR DRAWING LINES	49
POLAR COORDINATE MODE FOR DRAWING LINES	50
COMBINING COORDINATE MODES FOR DRAWING LINES	51
THE CIRCLE COMMAND	52
Circle by Radius Mode	53
Circle by Diameter Mode	53
Dynamic Input and Circles	54
USING THE PLINE COMMAND	54
Dynamic Input and Plines	56
ERASING OBJECTS	57
SAVING A DRAWING FILE	58
Save	59
Save As	59
EXITING AN AUTOCAD DRAWING SESSION	60
TUTORIAL EXERCISE: 01_GAGE BLOCK.DWG	**62**
TUTORIAL EXERCISE: 01_ANGLE BLOCK.DWG	**64**
TUTORIAL EXERCISE: 01_ANGLE PLATE.DWG	**66**
TUTORIAL EXERCISE: 01_PATTERN.DWG	**69**
TUTORIAL EXERCISE: 01_TEMPLATE.DWG	**73**
TUTORIAL EXERCISE: 01_POLYGON.DWG	**77**

Chapter 2 Drawing Setup and Organization ... 87

SETTING DRAWING UNITS ... 87
ENTERING ARCHITECTURAL VALUES FOR DRAWING LINES ... 88
SETTING THE LIMITS OF THE DRAWING ... 90
CALCULATING THE LIMITS OF THE DRAWING FOR MODEL SPACE ... 91
THE QUICK SETUP WIZARD ... 93
USING GRID IN A DRAWING ... 94
SETTING A SNAP VALUE ... 95
CONTROLLING SNAP AND GRID ... 96
    Grid Behavior Tools ... 97
CONTROLLING DYNAMIC INPUT ... 98
THE ALPHABET OF LINES ... 100
ORGANIZING A DRAWING THROUGH LAYERS ... 101
THE LAYER PROPERTIES MANAGER DIALOG BOX ... 102
CREATING NEW LAYERS ... 104
APPLYING CHANGES MADE TO LAYERS ... 106
ASSIGNING COLOR TO LAYERS ... 107
    Index Color Tab ... 107
    True Color Tab ... 108
    Color Books Tab ... 108
ASSIGNING LINETYPES TO LAYERS ... 108
ASSIGNING LINEWEIGHT TO LAYERS ... 110
THE LINEWEIGHT SETTINGS DIALOG BOX ... 111
THE LINETYPE MANAGER DIALOG BOX ... 111
THE LAYERS TOOLBAR ... 113
CONTROL OF LAYER PROPERTIES ... 113
THE PROPERTIES TOOLBAR ... 114
MAKING A LAYER CURRENT ... 114
THE STATUS OF LAYERS ... 115
USING THE LAYER PREVIOUS COMMAND ... 115
RIGHT-CLICK SUPPORT FOR LAYERS ... 116
CONTROLLING THE LINETYPE SCALE ... 116
ADVANCED LAYER TOOLS ... 118
    Layer Filters ... 118
    Layer Groups ... 118
    Layer States Manager ... 120
ADDITIONAL LAYER TOOLS ... 121
    Match Object's Layer (**LAYMCH**) ... 122
    Change to Current Layer (**LAYCUR**) ... 123
    Isolate Object's Layer (**LAYISO**) ... 124
    Copy Objects to New Layer (**COPYTOLAYER**) ... 125
    Layer Walk (**LAYWALK**) ... 126
    Freeze Object's Layer (**LAYFRZ**) ... 127
    Turn Object's Layer Off (**LAYOFF**) ... 127
    Lock Object's Layer (**LAYLCK**) ... 128
    Unlock Object's Layer (**LAYULK**) ... 128
    Merging Layers into One (**LAYMRG**) ... 129
    Deleting All Objects on a Layer (**LAYDEL**) ... 130

CREATING TEMPLATE FILES 131
    TUTORIAL EXERCISE: CREATING LAYERS USING THE LAYER PROPERTIES MANAGER DIALOG BOX 134

Chapter 3 AutoCAD Display and Basic Selection Operations 143
THE ZOOM COMMAND 143
ZOOMING IN REAL TIME 146
USING ZOOM-ALL 148
USING ZOOM-CENTER 148
USING ZOOM-EXTENTS 149
USING ZOOM-WINDOW 150
USING ZOOM-PREVIOUS 151
USING ZOOM-OBJECT 151
USING ZOOM-SCALE 152
USING ZOOM-IN 153
USING ZOOM-OUT 153
THE PAN COMMAND 153
SUPPORT FOR WHEEL MOUSE 155
CREATING NAMED VIEWS 156
    Changing the Boundary of a Named View 159
    Saving Layer Settings with Named Views 160
CREATING OBJECT SELECTION SETS 161
    Object Pre-Highlight 161
    Selecting Objects by Individual Picks 162
    Selecting Objects by Window 163
    Selecting Objects by Crossing Window 164
    Selecting Objects by a Fence 164
    Removing Objects from a Selection Set 165
    Selecting the Previous Selection Set 166
    Selecting Objects by a Crossing Polygon 166
CYCLING THROUGH OBJECTS 167
PRE-SELECTING OBJECTS 167
THE QSELECT COMMAND 168
    TUTORIAL EXERCISE: 03_SELECT OBJECTS.DWG 170

Chapter 4 Modifying Your Drawings 183
METHODS OF SELECTING MODIFY COMMANDS 183
LEVEL I MODIFY COMMANDS 184
MOVING OBJECTS 185
    Press and Drag Move 186
COPYING OBJECTS 187
    Press and Drag Copy 188
SCALING OBJECTS 189
    SCALE—Reference 190
ROTATING OBJECTS 191
    ROTATE—Reference 192

CREATING FILLETS AND ROUNDS ... 193
- Filleting by Radius ... 194
- Fillet as a Cornering Tool ... 195
- Performing Multiple Fillets ... 195
- Filleting Polylines ... 196
- No Trim Fillets ... 197
- Filleting Parallel Lines ... 198

CREATING CHAMFERS ... 198
- Chamfer by Equal Distances ... 199
- Chamfer by Angle ... 200
- Chamfering a Polyline ... 200
- Chamfering without Trimming ... 201
- Chamfer Project—Beam ... 202

OFFSETTING OBJECTS ... 203
- Offsetting Using a Through Point ... 203
- Offsetting by a Distance ... 203
- Performing Multiple Offsets ... 205
- Other Offset Options ... 205

TRIMMING OBJECTS ... 206
- Selecting Individual Cutting Edges ... 206
- Selecting All Objects as Cutting Edges ... 207
- Trimming by a Crossing Box ... 207
- Extending a Cutting Edge ... 208
- More about Individual Cutting Edges ... 209
- Trimming Exercise—Floor Plan ... 210

EXTENDING OBJECTS ... 211
- Selecting Individual Boundary Edges ... 211
- Extending Multiple Objects ... 212
- Extending a Boundary Edge ... 213
- Toggling from Extend to Trim ... 214
- Extend Exercise—Piping Diagram ... 214

BREAKING OBJECTS ... 215
- Breaking an Object ... 215
- Identifying a New First Break Point ... 216
- Breaking Circles ... 216
- Break at Point ... 217

LEVEL II MODIFY COMMANDS ... 218
CREATING ARRAYS ... 218
CREATING RECTANGULAR ARRAYS ... 218
- Creating Rectangular Patterns with Positive Offset ... 218
- Creating Rectangular Patterns with Negative Offsets ... 220

CREATING POLAR ARRAYS ... 222
MIRRORING OBJECTS ... 224
- Mirroring and Copying ... 224
- Mirroring by Flipping ... 225
- Mirroring Text ... 226
- More Information on Mirroring Text ... 226

Section	Page
STRETCHING OBJECTS	227
The Basics of Stretching	227
How Stretching Affects Dimensions	228
Stretching Using Multiple Crossing Windows	228
EDITING POLYLINES	229
Changing the Width of a Polyline	229
Joining Objects into a Single Polyline	230
Curve Generation	230
Generating Splines	230
Generating Fit Curves	231
Linetype Generation of Polylines	232
Offsetting Polyline Objects	232
Multiple Polyline Editing	233
EXPLODING OBJECTS	235
LENGTHENING OBJECTS	235
JOINING OBJECTS	236
UNDOING OPERATIONS	237
REDOING OPERATIONS	238
TUTORIAL EXERCISE: ANGLE.DWG	**239**
TUTORIAL EXERCISE: GASKET.DWG	**245**
TUTORIAL EXERCISE: TILE.DWG	**250**

Chapter 5 Performing Geometric Constructions ... 261

Section	Page
METHODS OF SELECTING OTHER DRAW COMMANDS	261
CONSTRUCTING ARCS	262
3 Points Arc Mode	263
Start, Center, End Mode	264
Continue Mode	264
CREATING A BOUNDARY	265
ADDITIONAL OPTIONS FOR CREATING CIRCLES	268
3 Points Circle Mode	268
2 Points Circle Mode	268
Constructing an Arc Tangent to Two Lines using CIRCLE-TTR	269
Constructing an Arc Tangent to a Line and Arc using CIRCLE-TTR	270
Constructing an Arc Tangent to Two Arcs using CIRCLE-TTR	272
Constructing an Arc Tangent to Two Arcs using CIRCLE-TTR	273
Constructing an Arc Tangent to Two Arcs using CIRCLE-TTR	274
Constructing a Line Tangent to Two Arcs or Circles	276
CIRCLE—Tan Tan Tan	277
QUADRANT VERSUS TANGENT OSNAP OPTION	278
CREATING FILLED-IN DOTS	279
CONSTRUCTING ELLIPTICAL SHAPES	281
CREATING MULTILINE OBJECTS	283
The MLINE Command	284
Creating a New Multiline Style	285
Editing Multilines	287
CREATING POINT OBJECTS	290
Setting a New Point Style	290

DIVIDING OBJECTS INTO EQUAL SPACES 291
MEASURING OBJECTS 293
CREATING POLYGONS 294
CREATING A RAY CONSTRUCTION LINE 296
CREATING RECTANGLE OBJECTS 296
    Rectangle by Picking Two Diagonal Points 296
    Changing Rectangle Properties 297
    Rectangle by Dimensions 298
    Rectangle by Area 298
CREATING A REVISION CLOUD 299
    Calligraphy Revision Clouds 301
CREATING SPLINES 302
MASKING TECHNIQUES WITH THE WIPEOUT COMMAND 303
CREATING CONSTRUCTION LINES WITH THE XLINE COMMAND 304
OGEE OR REVERSE CURVE CONSTRUCTION 306
    TUTORIAL EXERCISE: GEAR-ARM.DWG **308**
    TUTORIAL EXERCISE: PATTERN1.DWG **316**

Chapter 6 Working with Text, Fields, and Tables 329
AUTOCAD TEXT COMMANDS 329
ADDING MULTILINE TEXT 330
FORMATTING FRACTIONAL TEXT 332
TEXT JUSTIFICATION MODES 333
JUSTIFYING MULTILINE TEXT 333
INDENTING TEXT 334
FORMATTING WITH TABS 335
BULLETING AND NUMBERING TEXT 335
CHANGING THE MTEXT WIDTH 336
IMPORTING TEXT INTO YOUR DRAWING 337
BACKGROUND MASKING 338
MULTILINE TEXT SYMBOLS 338
CREATING SINGLE LINE TEXT 339
ADDITIONAL DYNAMIC TEXT APPLICATIONS 340
EDITING TEXT 342
GLOBALLY MODIFYING THE HEIGHT OF TEXT 344
GLOBALLY MODIFYING THE JUSTIFICATION OF TEXT 346
SPELL-CHECKING TEXT 347
CREATING DIFFERENT TEXT STYLES 349
    The Text Style Control Box 350
FIELDS 351
    Fields and Multiline Text 352
    Fields and Drawing Properties 353
CREATING TABLES 354
    Moving a Table 356
    Merging Cells in a Table 356
    Tables and Grips 356

  Tables and Fields 357
  Tables and Microsoft Excel 357
CREATING TABLE STYLES 358
 TUTORIAL EXERCISE: 06_PUMP_ASSEMBLY.DWG 361
 TUTORIAL EXERCISE: 06_TABLE.DWG 373
 TUTORIAL EXERCISE: 06_TABLE SUMMATIONS.DWG 378

Chapter 7 Object Grips and Changing the Properties of Objects 385
USING OBJECT GRIPS 385
OBJECT GRIP MODES 387
ACTIVATING THE GRIP SHORTCUT MENU 389
  Using the Grip—STRETCH Mode 390
  Using the Grip—SCALE Mode 390
  Using the Grip—MOVE/COPY Mode 391
  Using the Grip—MIRROR Mode 392
  Using the Grip—ROTATE Mode 393
  Using the Grip—Multiple ROTATE Mode 394
  Using Grip Offset Snap for Rotations 395
  Using Grip Offset Snap for Moving 396
  Grip Project—Lug.Dwg 398
MODIFYING THE PROPERTIES OF OBJECTS 399
USING THE QUICK SELECT DIALOG BOX 402
MODIFYING THE PROPERTIES OF INDIVIDUAL OBJECTS 405
USING THE PICKADD FEATURE OF THE PROPERTIES PALETTE 407
PERFORMING MATHEMATICAL CALCULATIONS 407
USING THE LAYER CONTROL BOX TO MODIFY OBJECT PROPERTIES 410
DOUBLE-CLICK EDIT ON ANY OBJECT 414
MATCHING THE PROPERTIES OF OBJECTS 415
  Matching Dimension Properties 417
  Matching Text Properties 418
  Matching Hatch Properties 419
  Matching Polyline Properties 420
  Matching Viewport Properties 420
  Match Properties and Tables 422
 TUTORIAL EXERCISE: 07_MODIFY-EX.DWG 424

Chapter 8 Shape Description/Multiview Projection 433
SHAPE DESCRIPTION 433
RELATIONSHIPS BETWEEN VIEWS 436
REVIEW OF LINETYPES AND CONVENTIONS 437
ONE-VIEW DRAWINGS 438
TWO-VIEW DRAWINGS 440
THREE-VIEW DRAWINGS 443
RUNOUTS 447
 TUTORIAL EXERCISE: ORTHOGRAPHIC BLOCK.DWG 448
 TUTORIAL EXERCISE: 08_SHIFTER.DWG 451

Chapter 9 Creating Section Views 473
SECTION VIEW BASICS 473
FULL SECTIONS 477
HALF SECTIONS 478
ASSEMBLY SECTIONS 479
ALIGNED SECTIONS 480
OFFSET SECTIONS 481
SECTIONING RIBS 482
BROKEN SECTIONS 482
REVOLVED SECTIONS 483
REMOVED SECTIONS 483
ISOMETRIC SECTIONS 484
ARCHITECTURAL SECTIONS 484
THE BHATCH COMMAND 485
AVAILABLE HATCH PATTERNS 487
SOLID FILL HATCH PATTERNS 489
GRADIENT PATTERNS 491
HATCH PATTERN SYMBOL MEANINGS 493
HATCHING ISLANDS 493
HATCH PATTERN SCALING 494
HATCH PATTERN ANGLE MANIPULATION 495
TRIMMING HATCH AND FILL PATTERNS 496
MODIFYING ASSOCIATIVE HATCHES 497
EDITING HATCH PATTERNS 499
ADVANCED HATCHING TECHNIQUES 500
    Gap Tolerance 502
PRECISION HATCH PATTERN PLACEMENT 502
INHERIT HATCHING PROPERTIES 503
EXCLUDE HATCH OBJECTS FROM OBJECT SNAPS 504
    TUTORIAL EXERCISE: 09_COUPLER.DWG 505
    TUTORIAL EXERCISE: 09_ELEVATION.DWG 511

Chapter 10 Auxiliary Views 525
AUXILIARY VIEW BASICS 525
CONSTRUCTING AN AUXILIARY VIEW 528
    TUTORIAL EXERCISE: 10_BRACKET.DWG 532

Chapter 11 Adding Dimensions to Your Drawing 549
DIMENSIONING BASICS 549
PLACEMENT OF DIMENSIONS 551
PLACING EXTENSION LINES 552
GROUPING DIMENSIONS 553
DIMENSIONING TO VISIBLE FEATURES 554
DIMENSIONING TO CENTERLINES 554
CONTINUE AND BASELINE DIMENSIONS 555

DIMENSIONING CYLINDERS 556
DIMENSIONING SYSTEMS 556
    Ordinate Dimensions 557
    Architectural Dimensions 558
REPETITIVE DIMENSIONS 558
METHODS OF CHOOSING DIMENSION COMMANDS 559
BASIC DIMENSION COMMANDS 561
LINEAR DIMENSIONS 561
ALIGNED DIMENSIONS 563
    Rotated Linear Dimensions 564
CONTINUE DIMENSIONS 565
BASELINE DIMENSIONS 566
THE QDIM COMMAND 567
    Continuous Mode 567
    Staggered Mode 568
    Baseline Mode 569
    Ordinate Mode 570
DIAMETER AND RADIUS DIMENSIONING 570
USING QDIM FOR RADIUS AND DIAMETER DIMENSIONS 572
ADDING JOGGED DIMENSIONS 573
ADDING ARC DIMENSIONS 573
LEADER LINES 574
THE QLEADER COMMAND 575
DIMENSIONING ANGLES 577
DIMENSIONING SLOTS 578
ADDING ORDINATE DIMENSIONS 579
EDITING DIMENSIONS 583
    Flipping Dimension Arrows 586
GEOMETRIC DIMENSIONING AND TOLERANCING (GDT) 587
    GDT Symbols 588
DIMENSION SYMBOLS 589
CHARACTER MAPPING FOR DIMENSION SYMBOLS 590
GRIPS AND DIMENSIONS 591
    TUTORIAL EXERCISE: 11_FIXTURE.DWG 595
    TUTORIAL EXERCISE: 11_DIMENSION VIEWS.DWG 604

Chapter 12 Managing Dimension Styles 611
THE DIMENSION STYLE MANAGER DIALOG BOX 611
THE LINES TAB 614
    Dimension Line Settings 614
    Extension Lines 615
    Extension Linetypes 617
    Fixed Length Extension Lines 617
THE SYMBOLS AND ARROWS TAB 618
    Arrowhead Settings 618
    Center Mark Settings 620
    Controlling Dimension Arc Length Symbol 620
    Controlling the Radius Dimension Jog Angle 621

THE TEXT TAB	621
Text Appearance	622
Text Placement	622
Vertical Text Placement	623
Horizontal Text Placement	624
Text Alignment	625
THE FIT TAB	626
Fit Options	626
Text Placement	629
Scale for Dimension Features	630
Fine-Tuning	631
THE PRIMARY UNITS TAB	631
Linear and Angular Dimension Units	632
Rounding Off Dimension Values	632
Applying a Dimension Prefix or Suffix	633
Measurement Scale	633
THE ALTERNATE UNITS TAB	634
Alternate Units	634
THE TOLERANCES TAB	635
Tolerance Format	636
CONTROLLING THE ASSOCIATIVITY OF DIMENSIONS	636
The DIMASSOC System Variable	636
Reassociating Dimensions	637
USING DIMENSION TYPES IN DIMENSION STYLES	639
OVERRIDING A DIMENSION STYLE	642
MODIFYING THE DIMENSION STYLE OF AN OBJECT	644
TUTORIAL EXERCISE: 12_DIMEX.DWG	**646**
TUTORIAL EXERCISE: 12_ARCHITECTURAL DIMENSION.DWG	**659**

Chapter 13   Analyzing 2D Drawings 673

USING INQUIRY COMMANDS	673
FINDING THE AREA OF AN ENCLOSED SHAPE	674
FINDING THE AREA OF AN ENCLOSED POLYLINE OR CIRCLE	675
FINDING THE AREA OF A SHAPE BY SUBTRACTION	678
USING FIELDS IN CALCULATIONS	680
THE DIST (DISTANCE) COMMAND	683
INTERPRETATION OF ANGLES USING THE DIST COMMAND	684
THE ID (IDENTIFY) COMMAND	684
THE LIST COMMAND	686
TUTORIAL EXERCISE: C-LEVER.DWG	**688**
CHECKING THE ACCURACY OF C-LEVER.DWG	693
SOLUTIONS TO THE QUESTIONS ON C-LEVER	694
TUTORIAL EXERCISE: 13_FIELD_CALC.DWG	**699**

Chapter 14   Working with Drawing Layouts 719

INTRODUCING DRAWING LAYOUTS	719
MODEL SPACE	719

TILED VIEWPORTS IN MODEL SPACE	720
CREATING VIEWPORTS IN MODEL MODE	721
MODEL SPACE AND PAPER SPACE	722
PAPER SPACE FEATURES	722
SETTING UP A PAGE	723
FLOATING MODEL SPACE	726
SCALING VIEWPORT IMAGES	727
CONTROLLING THE LIST OF SCALES	728
USING THE ZOOM XP OPTION	729
LOCKING VIEWPORTS	730
MAXIMIZING A VIEWPORT	731
Hiding Layout and Model Tabs	732
CREATING A PAPER SPACE LAYOUT	733
USING A WIZARD TO CREATE A LAYOUT	739
ARRANGING ARCHITECTURAL DRAWINGS IN PAPER SPACE	740
CREATING MULTIPLE DRAWING LAYOUTS	742
TRANSLATING DISTANCES BETWEEN SPACES	746
ASSOCIATIVE DIMENSIONS AND LAYOUTS	749
HATCH SCALING RELATIVE TO PAPER SPACE	751
NOTES TO THE STUDENT AND INSTRUCTOR	753
TUTORIAL EXERCISE: 14_HVAC.DWG	**754**

Chapter 15   Plotting Your Drawings 765

CONFIGURING A PLOTTER	765
PLOTTING FROM PAPER SPACE	770
ENHANCING YOUR PLOTS WITH LINEWEIGHTS	772
CREATING A COLOR-DEPENDENT PLOT STYLE TABLE	777
CONVERTING DRAWINGS TO DIFFERENT PLOT STYLES	784
PUBLISHING TO THE WEB	787

Chapter 16   Working with Blocks 793

WHAT ARE BLOCKS?	793
METHODS OF SELECTING BLOCK COMMANDS	795
CREATING A LOCAL BLOCK USING THE BLOCK COMMAND	795
CREATING A DRAWING FILE USING THE WBLOCK COMMAND	798
INSERTING BLOCKS	799
APPLICATIONS OF BLOCK INSERTIONS	801
TRIMMING TO BLOCK OBJECTS	802
EXTENDING TO BLOCK OBJECTS	803
EXPLODING BLOCKS	804
MANAGING UNUSED BLOCK DATA WITH THE PURGE DIALOG BOX	804
REDEFINING BLOCKS	806
BLOCKS AND THE DIVIDE COMMAND	808
BLOCKS AND THE MEASURE COMMAND	809
RENAMING BLOCKS	810
TABLES AND BLOCKS	811

ADDITIONAL TIPS FOR WORKING WITH BLOCKS	812
Create Blocks on Layer 0	812
Create Blocks Full Size If Applicable	812
Use Grid When Proportionality, but Not Scale, Is Important	813
USING DESIGNCENTER	813
DESIGNCENTER COMPONENTS	815
USING THE TREE VIEW	816
INSERTING BLOCKS THROUGH THE DESIGNCENTER	817
INSERTING BLOCKS USING THE TOOL PALETTE	819
Creating a New Tool Palette	822
Transparency Controls for Palettes	824
WORKING WITH MULTIPLE DRAWINGS	825
Opening Multiple Drawings	825
Working between Drawings	826
ADVANCED BLOCK TECHNIQUES—CREATING DYNAMIC BLOCKS	827
Working with Parameters and Actions	831
Working with Value Set Properties for Parameters	835
Base Parameter Applications	837
Alignment Parameter Applications	839
Multiple Alignment Point Applications	841
Scale and Angle Offset Applications	842
Centering Objects	845
Stretching a Block Equally in Opposite Directions	847
Array Applications	850
Working with Visibility States	853
Using Lookup Parameters and Tables	857
TUTORIAL EXERCISE: ELECTRICAL SCHEMATIC	**861**
TUTORIAL EXERCISE: 16_DYNAMIC_SWITCH_PLATE.DWG	**867**
TUTORIAL EXERCISE: 16_DYNAMIC_ELECTRICAL_LIST.DWG	**872**

Chapter 17 Working with Attributes … 881

ABOUT ATTRIBUTES	881
CREATING ATTRIBUTES THROUGH THE ATTRIBUTE DEFINITION DIALOG BOX	883
Attribute Tag	883
Attribute Prompt	884
Attribute Value	884
Attribute Modes	884
TUTORIAL EXERCISE: B TITLE BLOCK.DWG	**885**
FIELDS AND ATTRIBUTES	892
CONTROLLING THE DISPLAY OF ATTRIBUTES	893
Visibility Normal	893
Visibility On	893
Visibility Off	893
EDITING ATTRIBUTE PROPERTIES	894
CHANGING OR REPLACING VALUES	896
EDITING ATTRIBUTES GLOBALLY	896
TUTORIAL EXERCISE: 17_COMPUTERS.DWG	**900**

EXTRACTING ATTRIBUTES 903
CREATING TEMPLATE FILES 903
REDEFINING ATTRIBUTES 905
    Exploding a Block with Attributes 905
    Using the Properties Palette to Edit Tags 906
    Redefining the Block with Attributes 907
    TUTORIAL EXERCISE: 17_RESISTORS.DWG **908**
ENHANCED ATTRIBUTE EDIT AND EXTRACTION TOOLS 913
THE ENHANCED ATTRIBUTE EDITOR DIALOG BOX 914
THE BLOCK ATTRIBUTE MANAGER 915
    TUTORIAL EXERCISE: 17_BLOCK ATTRIB MGR.DWG **917**
THE ATTRIBUTE EXTRACTION WIZARD 919
    TUTORIAL EXERCISE: 17_CAD DEPT MODULES.DWG **920**
SYSTEM VARIABLES THAT CONTROL ATTRIBUTES 925
    ATTREQ 925
    ATTMODE 925
    ATTDIA 925

Chapter 18   Working with External References and Raster Image and DWF Files 927
COMPARING EXTERNAL REFERENCES AND BLOCKS 927
CHOOSING EXTERNAL REFERENCE COMMANDS 928
ATTACHING AN EXTERNAL REFERENCE 931
OVERLAYING AN EXTERNAL REFERENCE 935
THE XBIND COMMAND 938
IN-PLACE REFERENCE EDITING 940
BINDING AN EXTERNAL REFERENCE 943
USING LIST VIEW AND TREE VIEW 945
CLIPPING AN EXTERNAL REFERENCE 946
OTHER OPTIONS OF THE EXTERNAL REFERENCES PALETTE 948
    Unload 948
    Reload 948
    Detach 948
EXTERNAL REFERENCE NOTIFICATION TOOLS 949
USING ETRANSMIT 951
WORKING WITH RASTER IMAGES 951
USING THE DRAWORDER COMMAND 958
ATTACHING DWF FILES 960
    TUTORIAL EXERCISE: EXTERNAL REFERENCES **963**

Chapter 19   Creating Advanced Layouts 973
ARRANGING DIFFERENT VIEWS OF THE SAME DRAWING 973
    TUTORIAL EXERCISE: 19_ROOF PLAN.DWG **974**
CREATING A DETAIL PAGE IN LAYOUT MODE 981
ADDITIONAL VIEWPORT CREATION METHODS 987
MATCHING THE PROPERTIES OF VIEWPORTS 991

NOTES TO THE STUDENT AND INSTRUCTOR 993
    TUTORIAL EXERCISE: 19_ARCHITECTURAL DETAILS.DWG 994

Chapter 20   Solid Modeling Fundamentals 1003
3D MODEL TYPES 1003
    Wireframe Models 1003
    Surface Models 1004
    Solid Models 1004
SOLID MODELING BASICS 1005
CONSTRUCTING IN THE 3D MODELING WORKSPACE 1007
    The 3D Modeling Workspace 1008
    Using the Dashboard 1009
THE UCS COMMAND 1010
THE UCS–SPECIFY ORIGIN OF UCS AND ORIGIN OPTIONS 1013
THE UCS–3POINT OPTION 1014
THE UCS–X/Y/Z ROTATION OPTIONS 1015
THE UCS–OBJECT OPTION 1017
THE UCS–FACE OPTION 1017
THE UCS–VIEW OPTION 1018
USING DYNAMIC UCS MODE 1019
USING THE UCS DIALOG BOX 1022
THE UCS ICON DIALOG BOX 1024
THE APPEARANCE OF THE 3D COORDINATE SYSTEM ICON 1024
THE PLAN COMMAND 1025
VIEWING 3D MODELS 1026
VIEWING WITH FREE ORBIT 1027
VIEWING WITH CONSTRAINED ORBIT 1027
VIEWING WITH CONTINUOUS ORBIT 1028
    Additional Options of the Orbit Commands 1028
THE VIEW TOOLBAR 1028
VISUALIZING SOLID MODELS 1030
2D WIREFRAME 1031
3D WIREFRAME VISUAL STYLE 1031
HIDDEN VISUAL STYLE 1031
REALISTIC VISUAL STYLE 1032
CONCEPTUAL VISUAL STYLE 1033
CREATING A VISUAL STYLE 1033
    Visual Style Settings 1035
    Controlling Visual Style from the Dashboard 1037
SOLID MODELING COMMANDS 1040
CREATING A BOX 1042
CREATING A WEDGE 1043
CREATING A CYLINDER 1044
CREATING A CONE 1045
CREATING A SPHERE 1046
CREATING A TORUS 1047
CREATING A PYRAMID 1049

USING BOOLEAN OPERATIONS ON SOLID PRIMITIVES... 1050
CREATING SOLID UNIONS 1052
SUBTRACTING SOLIDS 1052
CREATING INTERSECTIONS 1053
    3D Applications of Unioning Solids 1054
    3D Applications of Moving Solids..1055
    3D Applications of Subtracting Solids 1056
UNDERSTANDING INTERSECTIONS 1057
CREATING SOLID EXTRUSIONS 1059
    Extruding an Existing Face 1061
CREATING REVOLVED SOLIDS 1062
CREATING A SOLID BY SWEEPING 1064
CREATING A SOLID BY LOFTING 1065
CREATING A HELIX 1068
HELIX APPLICATIONS 1070
CREATING POLYSOLIDS 1070
FILLETING SOLID MODELS 1072
CHAMFERING SOLID MODELS 1074
MODELING TECHNIQUES USING VIEWPORTS 1075
OBTAINING MASS PROPERTIES OF A SOLID MODEL 1076
SYSTEM VARIABLES USED FOR SOLID MODELS 1077
    The ISOLINES System Variable 1077
    The FACETRES System Variable 1078
    The DISPSILH System Variable 1079

    TUTORIAL EXERCISE: 20_BRACKET01.DWG **1080**
    TUTORIAL EXERCISE: COLLAR.DWG **1084**
    TUTORIAL EXERCISE: VACUUM ATTACHMEN.DWG **1089**

Chapter 21   Concept Modeling and Editing Solids 1107

CONCEPT MODELING 1107
    Dragging Basic Shapes 1107
    Using Dynamic UCS to Construct on Faces 1108
USING GRIPS TO MODIFY SOLID MODELS 1110
    Key Grip Locations 1110
    Grip Editing a Cone 1111
    Editing with Grips and Dynamic Input 1111
MANIPULATING SUBOBJECTS 1112
    Editing Subobjects of a Solid Model 1113
ADDING EDGES AND FACES TO A SOLID MODEL 1114
PRESSING AND PULLING BOUNDING AREAS 1115
USING PRESS AND PULL ON BLOCKS 1118
ADDITIONAL METHODS FOR EDITING SOLID MODELS 1122
USING THE 3DMOVE COMMAND 1123
THE 3DALIGN COMMAND 1124
THE 3DROTATE COMMAND 1126
THE MIRROR3D COMMAND 1130
THE 3DARRAY COMMAND 1131

DETECTING INTERFERENCES OF SOLID MODELS 1132
SECTIONING A SOLID 1133
SLICING SOLID MODELS 1136
    Slicing a Solid with a Surface 1137
    Terrain Modeling 1139
THE SOLIDEDIT COMMAND 1141
EXTRUDING (FACE EDITING) 1143
MOVING (FACE EDITING) 1144
ROTATING (FACE EDITING) 1145
OFFSETTING (FACE EDITING) 1146
TAPERING (FACE EDITING) 1147
DELETING (FACE EDITING) 1148
COPYING (FACE EDITING) 1149
IMPRINTING (BODY EDITING) 1150
SEPARATING SOLIDS (BODY EDITING) 1151
SHELLING (BODY EDITING) 1152
CLEANING (BODY EDITING) 1153
    TUTORIAL EXERCISE: 21_3DROTATE.DWG 1154
    TUTORIAL EXERCISE: 21_VALVE HEAD SECTION.DWG 1160

Chapter 22   Creating 2D Multiview Drawings from a Solid Model .. 1171
THE SOLVIEW AND SOLDRAW COMMANDS 1171
CREATING ORTHOGRAPHIC VIEWS 1175
CREATING AN AUXILIARY VIEW 1177
CREATING A SECTION VIEW 1179
CREATING AN ISOMETRIC VIEW 1182
EXTRACTING 2D VIEWS WITH FLATSHOT 1184
    TUTORIAL EXERCISE: 22_SOLID DIMENSION.DWG 1188

Chapter 23   Producing Renderings and Motion Studies 1197
AN INTRODUCTION TO RENDERINGS 1197
AN OVERVIEW OF PRODUCING RENDERINGS 1199
CREATING AND PLACING LIGHTS FOR RENDERING 1206
AN INTRODUCTION TO MATERIALS 1214
WORKING WITH MATERIALS 1217
USING MATERIAL TEMPLATES 1223
ASSIGNING MATERIALS BY LAYER 1227
APPLYING A BACKGROUND 1232
WALKING AND FLYING THROUGH A MODEL 1238
ANIMATING THE PATH OF A CAMERA 1240
    TUTORIAL EXERCISE: SUNLIGHT STUDY 1243

Index 1249

INTRODUCTION

Engineering graphics is the process of defining an object graphically before it is constructed and used by consumers. Previously, this process for producing a drawing involved the use of drawing aids such as pencils, ink pens, triangles, T-squares, and so forth to place an idea on paper before making changes and producing blue-line prints for distribution. The basic principles and concepts of producing engineering drawings have not changed, even when the computer is used as a tool.

This text uses the basics of engineering graphics to produce 2D drawings and 3D computer models using AutoCAD and a series of tutorial exercises that follow each chapter. Following the tutorials in most chapters, problems are provided to enhance your skills in producing engineering drawings. A brief description of each chapter follows:

CHAPTER 1 – GETTING STARTED WITH AUTOCAD

This first chapter introduces you to the following fundamental AutoCAD concepts: Screen elements and workspaces; use of function keys; opening an existing drawing file; using Dynamic Input for feedback when accessing AutoCAD commands; basic drawing techniques using the LINE, CIRCLE, and PLINE commands; understanding absolute, relative, and polar coordinates; using the Direct Distance mode for drawing lines; using all Object snap modes, and polar and object tracking techniques; using the ERASE command; and saving a drawing. Drawing tutorials follow at the end of this chapter.

CHAPTER 2 – DRAWING SETUP AND ORGANIZATION

This chapter introduces the concept of drawing in real-world units through the setting of drawing units and limits. The importance of organizing a drawing through layers is also discussed through the use of the Layer Properties Manager dialog box. Color, linetype, and lineweight are assigned to layers and applied to drawing objects. Advanced Layer tools such as isolating, filtering, and states, and how to create template files are also discussed in this chapter.

CHAPTER 3 – AUTOCAD DISPLAY AND BASIC SELECTION OPERATIONS

This chapter discusses the ability to magnify a drawing using numerous options of the ZOOM command. The PAN command is also discussed as a means of staying in a zoomed view and moving the display to a new location. Productive uses of real-time zooms and pans along with the effects a wheel mouse has on ZOOM and PAN are included. All object selection set modes are discussed, such as Window, Crossing, Fence, and All, to name a few. Finally, this chapter discusses the ability to save the image of your display and retrieve the saved image later through the View Manager dialog box.

CHAPTER 4 – MODIFYING YOUR DRAWINGS

This chapter is organized into two parts. The first part covers basic modification commands and includes the following: MOVE, COPY, SCALE, ROTATE, OFFSET, FILLET, CHAMFER, TRIM, EXTEND, and BREAK. The second part covers advanced methods of modifying drawings and includes ARRAY, MIRROR, STRETCH, PEDIT, EXPLODE, LENGTHEN, JOIN, UNDO, and REDO. Tutorial exercises follow at the end of this chapter as a means of reinforcing these important tools used in AutoCAD.

CHAPTER 5 – PERFORMING GEOMETRIC CONSTRUCTIONS

This chapter discusses how AutoCAD commands are used for constructing geometric shapes. The following drawing-related commands are included in this chapter: ARC, DONUT, ELLIPSE, MLINE, POINT, POLYGON, RAY, RECTANG, SPLINE, and XLINE. Tutorial exercises are provided at the end of this chapter.

CHAPTER 6 – WORKING WITH TEXT, FIELDS, AND TABLES

Use this chapter for placing text in your drawing. Various techniques for accomplishing this task include the use of the MTEXT and DTEXT commands. The creation of text styles and the ability to edit text once it is placed in a drawing are also included. A method of creating intelligent text, called Fields, is discussed in this chapter. Creating tables, table styles, and performing summations on tables are also covered here. Tutorial exercises are included at the end of this chapter.

CHAPTER 7 – OBJECT GRIPS AND CHANGING THE PROPERTIES OF OBJECTS

The topic of grips and how they are used to enhance the modification of a drawing is presented. The ability to modify objects through the Properties Palette is discussed in great detail. A tutorial exercise is included at the end of this chapter to reinforce the importance of changing the properties of objects.

CHAPTER 8 – SHAPE DESCRIPTION/MULTIVIEW PROJECTION

Describing shapes and producing multiview drawings using AutoCAD are the focus of this chapter. The basics of shape description are discussed, along with proper use of linetypes, fillets, rounds, chamfers, and runouts. Tutorial exercises are available at the end of this chapter and outline the steps used for creating a multiview drawing using AutoCAD.

CHAPTER 9 – CREATING SECTION VIEWS

The basics of section views are described in this chapter, including full, half, assembly, aligned, offset, broken, revolved, removed, and isometric sections. Hatching techniques through the use of the Hatch and Gradient dialog box are also discussed. The ability to apply a gradient hatch pattern is also discussed. Tutorial exercises that deal with the topic of section views follow at the end of the chapter.

CHAPTER 10 – AUXILIARY VIEWS

Producing auxiliary views using AutoCAD is discussed in this chapter. Items discussed include rotating the snap at an angle to project lines of sight perpendicular to a surface to be used in the preparation of the auxiliary view. A tutorial exercise is provided in this chapter.

CHAPTER 11 – ADDING DIMENSIONS TO YOUR DRAWING

Basic dimensioning rules are introduced in the beginning of this chapter before all AutoCAD dimensioning commands are discussed. Placing linear, diameter, and radius dimensions is then demonstrated. The powerful QDIM command allows you to place baseline, continuous, and other dimension groups in a single operation. A tutorial exercise is provided at the end of this chapter.

CHAPTER 12 – MANAGING DIMENSION STYLES

A thorough discussion of the use of the Dimension Styles Manager dialog box is included in this chapter. The ability to create, modify, and override dimension styles is discussed. A detailed tutorial exercise is provided at the end of this chapter.

CHAPTER 13 – ANALYZING 2D DRAWINGS

This chapter provides information on analyzing a drawing for accuracy purposes. The AREA, ID, LIST, and DIST commands are discussed in detail, in addition to how they are used to determine the accuracy of various objects in a drawing. A tutorial exercise follows that allows users to test their drawing accuracy.

CHAPTER 14 – WORKING WITH DRAWING LAYOUTS

This chapter deals with the creation of layouts before a drawing is plotted out. A layout takes the form of a sheet of paper and is referred to as Paper Space. A wizard to assist in the creation of layouts is also discussed. Once a layout of an object is created, scaling through the Viewports toolbar is discussed. The creation of numerous layouts for the same drawing is also introduced, including a means of freezing layers only in certain layouts. Various exercises are provided throughout this chapter to reinforce the importance of layouts.

CHAPTER 15 – PLOTTING YOUR DRAWINGS

Printing or plotting your drawings out is discussed in this chapter through a series of tutorial exercises. One tutorial demonstrates the use of the Add-A-Plotter wizard to configure a new plotter. Plotting from a layout is discussed through a tutorial. This includes the assignment of a sheet size. Tutorial exercises are also provided to create a color-dependent plot style. Plot styles allow you to control the appearance of your plot. Other tutorial exercises available in this chapter include plotting to the web.

CHAPTER 16 – WORKING WITH BLOCKS

This chapter covers the topic of creating blocks in AutoCAD. Creating local and global blocks such as doors, windows, and pipe symbols will be demonstrated. The Insert dialog box is discussed as a means of inserting blocks into drawings. The chapter continues by explaining the many uses of the DesignCenter. This feature allows the user to display a browser containing blocks, layers, and other named objects that can be dragged into the current drawing file. The use of tool palettes is also discussed as a means of dragging and dropping blocks and hatch patterns into your drawing. This chapter also discusses the ability to open numerous drawings through the Multiple Document Environment and transfer objects and properties between drawings. The creation of dynamic blocks, an advanced form of manipulating blocks, is also discussed, with numerous examples to try out. A tutorial exercise can be found at the end of this chapter.

CHAPTER 17 – WORKING WITH ATTRIBUTES

This chapter introduces the purpose for creating attributes in a drawing. A series of four commands step the user to a better understanding of attributes. The first command is ATTDEF and is used to define attributes. The ATTDISP command is used to control the display of attributes in a drawing. Once attributes are created and assigned to a block, they can be edited through the ATTEDIT command. Finally, attribute information can be extracted using the ATTEXT command or Attribute Extraction wizard. Extracted attributes can then be imported into such applications as Microsoft Excel and Access. Various tutorial exercises are provided throughout this chapter to help the user become better acquainted with this powerful feature of AutoCAD.

CHAPTER 18 – WORKING WITH EXTERNAL REFERENCES AND RASTER IMAGE AND DWF FILES

The chapter begins by discussing the use of External References in drawings. An external reference is a drawing file that can be attached to another drawing file. Once the referenced drawing file is edited or changed, these changes are automatically seen once the drawing containing the external reference is opened again. Performing in-place editing of external references is also demonstrated. Importing image files is also discussed and demonstrated in this chapter. Working with DWF overlay files is also discussed in this chapter. A tutorial exercise follows at the end of this chapter to let the user practice using external references.

CHAPTER 19 – CREATING ADVANCED LAYOUTS

This very important chapter is designed to utilize advanced techniques used in laying out a drawing before it is plotted. The ability to lay out a drawing consisting of various images at different scales is also discussed. The ability to create user-defined rectangular viewports through the Viewports Toolbar is demonstrated. The creation of non-rectangular viewports will also be demonstrated. A tutorial exercise follows to let the user practice this advanced layout technique.

CHAPTER 20 – SOLID MODELING FUNDAMENTALS

This chapter begins with a comparison between isometric, extruded, wireframe, surfaced, and solid model drawings. The chapter continues with a discussion on the use of the 3D Modeling workspace. Creating User Coordinate Systems and how they are positioned to construct objects in 3D is a key concept to master in this chapter. Creating User Coordinate Systems dynamically is also emphasized. The display of 3D images through the 3DORBIT command is discussed along with the creation of visual styles. Creating various solid primitives such as boxes, cones, and cylinders is discussed in addition to the ability to construct complex solid objects through the use of the Boolean operations of union, subtraction, and intersection. The chapter continues by discussing extruding, rotating, sweeping, and lofting operations for creating solid models in addition to filleting and chamfering solid models. Tutorial exercises follow at the end of this chapter.

CHAPTER 21 – CONCEPT MODELING AND EDITING SOLIDS

This chapter begins with a detailed study on how concept models can easily be created by dragging on grips located at key locations of a solid primitive. The ability to press and pull faces of a solid model and easily change its shape is also discussed. The 3DALIGN and 3DROTATE commands are discussed as a means of introducing the editing capabilities of AutoCAD on solid models. The process of cutting a section out of a solid model using a section plane is discussed. Modifications can also be made to a solid model through the use of the SOLIDEDIT command, which is discussed in this chapter. The ability to extrude existing faces, imprint objects, and create thin wall shells is demonstrated throughout this chapter. A tutorial exercise can be found at the end of this chapter.

CHAPTER 22 – CREATING 2D MULTIVIEW DRAWINGS FROM A SOLID MODEL

Once the solid model is created, the SOLVIEW command is used to lay out 2D views of the model, and the SOLDRAW command is used to draw the 2D views. Layers are automatically created to assist in the annotation of the drawing through the use of dimensions. The use of the FLATSHOT command is also explained as another means of projecting 2D geometry from a 3D model. A tutorial exercise is available at the end of this chapter, along with instructions on how to apply the techniques learned in this chapter to other solid models.

CHAPTER 23 – PRODUCING RENDERINGS AND MOTION STUDIES

This chapter introduces you to the uses and techniques of producing renderings from 3D models in AutoCAD. A brief overview of the rendering process is covered, along with detailed information about placing lights in your model, producing scenes from selected lighting arrangements, loading materials through the materials library supplied in AutoCAD, attaching materials to your 3D models, creating your own custom materials, applying a background to your rendered image, and experimenting with the use of motion path animations for creating walkthroughs of 3D models.

ONLINE COMPANION

This new edition contains a special Internet companion piece. The Online Companion is your link to AutoCAD on the Internet. Monthly updates include a command of the month, FAQs, and tutorials. You can find the Online Companion at:

> http://www.autodeskpress.com/resources/olcs/index.aspx

ACKNOWLEDGMENTS

I wish to thank the staff at Autodesk Press for their assistance with this document, especially John Fisher, Sandy Clark, and Stacy Masucci.

> I would also like to thank Kevin Lang, Engineering Design Graphics, Trident Technical College, Charleston, SC for performing the technical edit on the chapters that deal with the 3D functionality of AutoCAD.

The publisher and author would like to thank and acknowledge the many professionals who reviewed the manuscript to help us publish this AutoCAD text. A special acknowledgment is due to the following instructors, who reviewed the chapters in detail:

> Wen. M. Andrews, J. Sargeant Reynolds Community College, Richmond, VA
>
> Kirk Barnes, Ivy Tech State College, Bloomington, IN
>
> Paul Ellefson, Hennepin Technical College, Brooklyn Park, MN
>
> Rajit Gadh, University of Wisconsin, Madison, WI
>
> Gary J. Masciadrelli, Springfield Technical Community College, Springfield, MA

ABOUT THE AUTHOR

Alan J. Kalameja is the Department Head of Design and Construction at Trident Technical College, located in Charleston, South Carolina. He has been at the college for more than 25 years and has been using AutoCAD since 1984. He has authored the AutoCAD Tutor for Engineering Graphics in Release 10, 12, 14, and AutoCAD 2000, 2002, 2004, 2006, and 2007. He is also one of the principal authors involved in creating the Autodesk Inventor 11 Essentials PLUS book. All of his books are published by Thomson Delmar Learning.

CONVENTIONS

All tutorials in this publication use the following conventions in the instructions:

Whenever you are told to enter text, the text appears in boldface type. This may take the form of entering an AutoCAD command or entering such information as absolute, relative, or polar coordinates. You must follow these and all text inputs by pressing the ENTER key to execute the input. For example, to draw a line using the LINE command from point 3,1 to 8,2, the sequence would look like the following:

Command: **L** *(For LINE)*
Specify first point: **3,1**
Specify next point or [Undo]: **8,2**
Specify next point or [Undo]: *(Press* ENTER *to exit this command)*

Instructions for selecting objects are in italic type. When instructed to select an object, move the pickbox on the object to be selected and press the pick button on the mouse or digitizing puck.

If you enter the wrong command for a particular step, you may cancel the command by pressing the ESC key. This key is located in the upper left-hand corner of any standard keyboard.

Instructions in some tutorials are designed to enter all commands, options, coordinates, and so forth, from the keyboard. You may use the same commands by selecting them from the pull-down menu area, or from one of the floating toolbars.

Other tutorial exercises are provided with minimal instructions to test your ability to complete the exercise.

NOTES TO THE STUDENT AND INSTRUCTOR CONCERNING THE USE OF TUTORIAL EXERCISES

Various tutorial exercises have been designed throughout this book and can be found at the end of each chapter. The main purpose of each tutorial is to follow a series of steps toward the completion of a particular problem or object. Performing the tutorial will also prepare you to undertake the numerous drawing problems also found at the end of each chapter.

As you work on the tutorials, you should follow the steps very closely, taking care not to make a mistake. However, most individuals rush through the tutorials to get the correct solution in the quickest amount of time, only to forget the steps used to complete the tutorial. A typical comment made by many is "I completed the tutorial . . . but I don't understand what I did to get the correct solution."

It is highly recommended to both student and instructor that all tutorial exercises be performed two or even three times. Completing the tutorial the first time will give you the confidence that it can be done; however, you may not understand all the steps involved. Completing the tutorial a second time will allow you to focus on where certain operations are performed and why things behave the way they do. This still may not be enough. More complicated tutorial exercises may need to be performed a third time. This will allow you to anticipate each step and have a better idea what operation to perform in each step. Only then will you be comfortable and confident to attempt the many drawing problems that follow the tutorial exercises.

The CD-ROM in the back of the book contains AutoCAD drawing files for the Try It! exercises. To use drawing files, copy files to your hard drive, then remove their read-only attribute. Files cannot be used without AutoCAD. Files are located in the /Drawing Files/ directory.

SUPPLEMENTS

e.resource™--This is an educational resource that creates a truly electronic classroom. It is a CD-ROM containing tools and instructional resources that enrich your classroom and make your preparation time shorter. The elements of *e.resource* link directly to the text and tie together to provide a unified instructional system. Spend your time teaching, not preparing to teach.

ISBN 1-4180-5323-6

Features contained in *e.resource* include:

- **Instructors Guide:** This contains a comprehensive course syllabus and additional notes on each chapter.
- **Syllabus:** Lesson plans created by chapter. You have the option of using these lesson plans with your own course information.
- **Chapter Hints:** Objectives and teaching hints that provide the basis for a lecture outline that helps you to present concepts and material.
- **PowerPoint® Presentation:** These slides provide the basis for a lecture outline that helps you to present concepts and material. Key points and concepts can be graphically highlighted for student retention. There are more than 300 slides, covering every chapter in the text.
- **Exam View Computerized Test Bank:** More than 600 questions of varying levels of difficulty are provided in true/false and multiple-choice formats. Exams can be generated to assess student comprehension or questions can be made available to the student for self-evaluation.
- **Video and Animation Resources:** These AVI files graphically depict the execution of key concepts and commands in drafting, design, and AutoCAD and let you bring multimedia presentations into the classroom.

Spend your time teaching, not preparing to teach!

CHAPTER 1

Getting Started with AutoCAD

This chapter begins with an explanation of the components that make up a typical AutoCAD display screen. You will learn various methods of selecting commands; some from the pull-down menu, others from toolbars, and still others by entering the command from the keyboard. You will learn how to begin a new drawing. Once in a drawing, you will construct line segments with the LINE command in addition to drawing lines accurately with the Direct Distance mode, absolute coordinates, relative coordinates, and polar coordinates. Object Snap will be discussed as a way to construct all types of objects with greater precision. Additional drawing aids such as Object Snap Tracking and Polar Tracking will be discussed. Other basic drawing commands such as constructing circles and polylines will be shown. You will also be introduced to the ERASE command for removing drawing objects.

THE AUTOCAD INTRODUCTORY WORKSPACE SCREEN

Before creating a new drawing or opening an existing drawing, you must first choose a workspace to work in. Workspaces are considered task-oriented environments that use a default drawing template and even launch such items as toolbars and palettes depending on the workspace. AutoCAD gives you a number of pre-defined workspaces; namely, 3D Modeling and AutoCAD Classic.

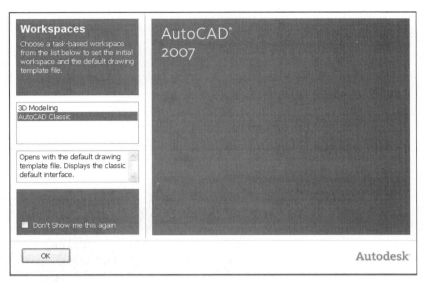

Figure 1-1

THE AUTOCAD CLASSIC WORKSPACE

Once you choose the AutoCAD Classic workspace, the drawing screen illustrated in the following image appears. Note: A number of additional items such as the Sheet Set Manager and default tool palette also load. After turning all of these extra items off, your screen should appear similar to the following image.

Figure 1-2

Use the following table to identify the items found in the AutoCAD Classic Workspace display screen.

In	Description
1	The workspace you are currently working in.
2	The AutoCAD file being edited is shown as [Drawing Name] at the top of the screen.
3	Pull-down menu area that contains various commands under such categories as File, Edit, View, etc.
4	The Layers toolbar is used to create, display, and control various layer states.
5	The Standard toolbar contains buttons for such commands as NEW and OPEN.
6	The Styles toolbar controls text, dimensions, and table styles.
7	The Properties toolbar is designed for manipulating color, linetypes, and line weights.
8	The Draw toolbar contains buttons for such commands as LINE, CIRCLE, ARC, and HATCH.
9	The Modify toolbar contains buttons for such commands as FILLET, CHAMFER, ERASE, and ARRAY.
10	The User Coordinate System icon is displayed in the lower corner of the drawing editor to alert you to the coordinate system currently in use.
11	The command line area is where AutoCAD prompts you for input depending on which command is currently in progress.
12	The status bar displays buttons used to turn off such items as Grid, Snap, and Osnap, to name a few.
13	Layouts switch you to the environment where you prepare your drawing to be plotted.
14	The graphics cursor is used for picking points on the screen or for selecting objects to edit.

ADDITIONAL WORKSPACES

Clicking on Tools in the pull-down menu and then clicking on Workspaces will display the available workspaces that you can choose from to work in, as shown in the following image. The presence of the checkmark alongside the AutoCAD Classic workspace signifies that it is current. In addition to using these pre-existing workspaces, you can also arrange your screen to your liking and save these screen changes as your own custom workspace.

Figure 1–3

 Note: In the previous image, a Default AutoCAD workspace may be present in the list. This workspace is automatically created if you are upgrading the AutoCAD 2007 software from a previous version of AutoCAD.

THE STATUS BAR

The status bar, illustrated in the following image, is used to toggle on or off the following modes: Coordinate Display, Snap, Grid, Ortho, Polar Snap, Object Snap, Object Tracking, Dynamic User Coordinate System (Dynamic UCS), Dynamic Input, Line Weight, and Model/Paper. Click the button once to turn the mode on or off. A button with a recessed appearance indicates that the mode is on. For example, click the OSNAP button once to turn the Object Snap on; click the OSNAP button again to turn the Object Snap off. Clicking the MODEL button enters the Page Setup mode.

Figure 1–4

The following table gives a brief description of each component located in the status bar:

Button	Tool	Description
2.8843, 3.4529	Coordinate Display	Toggles the coordinate display, located in the lower left corner of the status bar, on or off. When the coordinate display is off, the coordinates are updated when you pick an area of the screen with a mouse. When the coordinate display is on, the coordinates dynamically change with the current position of the cursor.
SNAP	SNAP	Toggle Snap mode on or off. The SNAP command forces the cursor to align with grid points. The current snap value can be modified and can be related to the spacing of the grid.
GRID	GRID	Toggles the display of the grid on or off. The actual grid spacing is set by the GRID command and not by this function key.
ORTHO	ORTHO	Toggles Ortho mode on or off. Use this key to force objects such as lines to be drawn horizontally or vertically.
POLAR	POLAR	Toggles the Polar Tracking on or off. Polar Tracking can force lines to be drawn at any angle, making it more versatile than Ortho mode. The Polar Tracking angles are set through a dialog box. Also, if you turn Polar Tracking on, Ortho mode is disabled, and vice versa.
OSNAP	OSNAP	Toggles the current Object Snap settings on or off. This will be discussed later in this chapter.
OTRACK	OTRACK	Toggles Object Snap Tracking on or off. This feature will also be discussed later in this chapter.
DUCS	DUCS	Toggles the Dynamic User Coordinate System on or off. This feature is used mainly for modeling in 3D.
DYN	DYN	Toggles Dynamic Input on or off. When it is turned on, your attention is directed to your cursor position as commands and options are executed. When it is turned off, all commands and options are accessed through the Command prompt at the bottom of the display screen.
LWT	LWT	Toggles Lineweight on or off. When turned off, no lineweights are displayed. When turned on, lineweights that have been assigned to layers are displayed in the drawing.
MODEL	MODEL	Toggles between Model mode and Paper mode (Layout).

Right-clicking one of the buttons displays the shortcut menu in the following image. Choose Settings to access various dialog boxes that control certain features associated with the button.

Figure 1–5

You can also access most tools located in the status bar through the function keys located at the top of any standard computer keyboard. The following table describes each function key.

Function Key	Definitions
F1	Displays AutoCAD Help Topics
F2	Toggle Text/Graphics Screen
F3	Object Snap settings On/Off
F4	Toggle Tablet Mode On/Off
F5	Toggle Isoplane Modes
F6	Toggle Coordinates On/Off
F7	Toggle Grid Mode On/Off
F8	Toggle Ortho Mode On/Off
F9	Toggle Snap Mode On/Off
F10	Toggle Polar Mode On/Off
F11	Toggle Object Snap Tracking On/Off
F12	Toggle Dynamic Input (DYN) On/Off

Note: Most of the function keys are similar in operation to the modes found in the status bar except for the following:

When you press F1, the AutoCAD Help Topics dialog box is displayed.

Pressing F2 takes you to the text screen consisting of a series of previous prompt sequences. This may be helpful for viewing the previous command sequence in text form.

Use F4 to toggle Tablet mode on or off. This mode is only activated when the digitizing tablet has been calibrated for the purpose of tracing a drawing into the computer.

Pressing F5 scrolls you through the three supported Isoplane modes used to construct isometric drawings (Right, Left, and Top).

ADDITIONAL STATUS BAR BUTTONS

Located at the far right end of the status bar are additional buttons used to manage the appearance of the AutoCAD display screen. The following table gives a brief description of these buttons.

Button	Tool	Description
	Communication Center	Activates a dialog box used for performing automatic software updates.
	Toolbar/Window Positions Toggle	Locks the position of all toolbars on the display screen.
	Status Bar Menu Controls	Activates a menu used for turning on or off certain status bar buttons.
	Clean Screen	Removes all toolbars from the screen, giving your display an enlarged appearance. Pick this button again to return the toolbars to the screen.

COMMUNICATING WITH AUTOCAD

THE COMMAND LINE

How productive the user becomes in using AutoCAD may depend on the user's degree of understanding of the command execution process with AutoCAD. One of the means of command execution is through the command prompt that is located at the bottom of the display screen. As a command is selected from a toolbar, AutoCAD prompts the user with a series of steps needed to complete this command. In the following image, the CIRCLE command is chosen as the command. The next series of sentences in the command line prompts the user to first specify or locate a center point for the circle. After this is accomplished, you are then prompted to specify the radius of the circle. These actions are also called the default prompt.

```
Command: CIRCLE
Specify center point for circle or [3P/2P/Ttr (tan tan radius)]:
Specify radius of circle or [Diameter]:
11.0702, 5.5776, 0.0000   SNAP GRID ORTHO POLAR OSNAP OTRACK DUCS DYN LWT MODEL
```

Figure 1–6

UNDERSTANDING THE COMMAND PROMPT

In the previous image of the command line, notice the string of CIRCLE command options displayed as the following:

[3P/2P/Ttr (tan tan radius)]

Items identified inside the square brackets are referred to as options. Typing in this option from the keyboard activates it.

DYNAMIC INPUT

Yet another more efficient means of command execution within AutoCAD is through the Dynamic input feature. This feature is activated by picking the DYN button located in the status bar at the bottom of your display screen. Whether a command is picked from the toolbar or pull-down menu or entered from the keyboard, you see immediate feedback at your cursor location. The following image illustrates how the CIRCLE command prompts display at the cursor location. As the cursor is moved around, the Specify center point for circle prompt also moves. Also notice that the current screen position is displayed. If the down directional arrow is typed on the keyboard, options of the CIRCLE command display. Typing the down arrow cycles through the available options in executing the CIRCLE command.

Figure 1–7

METHODS OF CHOOSING COMMANDS

AutoCAD provides numerous ways to enter or choose commands for constructing or editing drawings: pull-down menus, toolbars, dialog boxes, right-click shortcut menus, and command aliases for direct entry from the keyboard.

PULL-DOWN MENUS

Pull-down menus provide an easy way to access most AutoCAD commands. The pull-down menu area is located at the top of the main AutoCAD display screen. Here, various categories exist such as File, Edit, View, Insert, and so on. Clicking on one of these category headings will pull down a menu consisting of commands related to this heading. For example, clicking on the Format heading followed by Layer tools in the following image will pull down the menu containing extra commands used for managing layers in a drawing.

Figure 1-8

TOOLBARS

A toolbar, which represents an additional method for initiating commands, is illustrated as shown on the left in the following image. The toolbar in this example allows you access to most ZOOM commands. Other toolbars can be found by choosing Toolbars from the View pull-down menu. When the cursor rolls over a tool, a 3D border is displayed, along with a tooltip that explains the purpose of the command, as shown on the right in the following image.

Figure 1-9

ACTIVATING TOOLBARS

Many toolbars are available to assist the user in executing other types of commands. By default, six toolbars are already active or displayed in all drawings: the Draw, Layers, Modify, Properties, Standard, and Styles toolbars. To activate a different toolbar, move the cursor over the top of any command button and press the right mouse button. A shortcut menu appears that displays all toolbars, as shown in the following image. In this example, placing a check beside the Dimension name displays this toolbar.

Figure 1–10

DOCKING TOOLBARS

It is considered good practice to line the top or side edges of the display screen with toolbars. The method of moving toolbars to the sides of your screen is called docking. Press down on the toolbar title strip and slowly drag the toolbar to the screen edge until the toolbar appears to jump. Letting go of the mouse button will dock the toolbar to this side of the screen. Practice this by docking various toolbars to your screen.

 Tip: To prevent docking, press the CTRL key as you drag the toolbar. This will allow you to move the toolbar into the upper or lower portions of the display screen without the toolbar docking. Also, if a toolbar appears to disappear, it might actually be alongside or below toolbars that already exist. Turning off toolbars will assist in finding the missing one.

Figure 1–11

DASHBOARDS

A dashboard is similar to a tool palette; however it contains numerous categories of buttons that are based on a task-oriented workspace. To activate the dashboard, click on Palettes under the Tool pull-down menu area, as shown in the following image.

Figure 1–12

DASHBOARD CONTROL PANELS

The following image illustrates the appearance of the dashboard. Usually this special palette displays along the right side of your display screen. While the dashboard is primarily designed for creating, viewing, and rendering 3D models, a 2D control panel does exist. To activate this control panel, right-click in the blue dashboard title strip and pick Control panels from the menu. All valid control panels will display, as shown in the following image. If there is a check present, the control panel is present in the dashboard. To remove a control panel from the dashboard, remove the checkmark.

Figure 1–13

DIALOG BOXES AND ICON MENUS

Settings and other controls can be changed through dialog boxes. Illustrated on the left in the following image is the Drawing Units dialog box, which will be discussed in chapter 2. Illustrated on the right in the following image is the Hatch Pattern Palette. Palettes are similar to dialog boxes with the exception that palettes allow display of small images. The example shown here demonstrates how the user can pick hatch patterns more easily.

Certain dialog boxes can be increased in size by moving your cursor over their borders; this is true of the Hatch Pattern Palette dialog box. When two arrows appear, hold down the pick button of the mouse (usually the left button) and stretch the dialog box in that direction. If the cursor is moved to the corner of the dialog box, the box is stretched in two directions. These methods can also be used to make the dialog box smaller, although there is a default size for the dialog boxes, which limits smaller sizes. The dialog box cannot be stretched if no arrows appear when you move your cursor over the border of the dialog box; this is true of the Drawing Units dialog box as well.

Figure 1–14

TOOL PALETTES

Commands can also be accessed from a number of tool palettes. To launch the Tool Palette, click on the Tool Palettes Window button, which is located in the Standard toolbar, as shown in the following image. The Tool Palette is a long, narrow bar that consists of numerous tabs. Two tabs, namely, Modify and Draw, are illustrated below. Use these tabs to access the more popular drawing and modify commands. While this image shows two palettes, in reality only one will be present on your screen at any one time. Simply click a different tab to display the commands associated with the tab.

Figure 1-15

RIGHT-CLICK SHORTCUT MENUS

Many shortcut or cursor menus have been developed to assist with the rapid access to commands. Clicking the right mouse button activates these commands. The Default shortcut menu is illustrated on the left in the following image. It is displayed whenever you right-click in the drawing area and no command or selection set is in progress.

On the right in the following image is an example of the Edit shortcut menu. This shortcut consists of numerous editing and selection commands. This menu activates whenever you right-click in the display screen with an object or group of objects selected but no command is in progress.

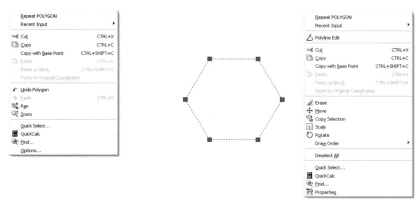

Figure 1-16

Right-clicking in the COMMAND prompt area of the display screen activates the shortcut menu, as shown on the left in the following image. This menu provides quick access to the Options dialog box, which is used to control various settings in AutoCAD. Also, a record of the six most recent commands is kept, which allows the user to select from this group of previously used commands.

Illustrated on the right in the following image is an example of a Command-Mode shortcut menu. When you enter a command and right-click, this menu will display options of the command. This menu supports a number of commands. In the figure, the 3P, 2P, and Ttr (tan tan radius) listings are all options of the CIRCLE command.

Figure 1–17

COMMAND ALIASES

Commands can be executed directly through keyboard entry. This practice is popular for users who are already familiar with the commands. However, users must know the exact command name, including its exact spelling. To assist with the entry of AutoCAD commands from the keyboard, numerous commands are available in shortened form, referred to as *aliases*. For example, instead of typing in LINE, all that is required is L. The letter E can be used for the ERASE command, and so on. These command aliases are listed throughout this book. The complete list of all command aliases can be found in the file ACAD.PGP, which is located in the Support directory. Once you are comfortable with the keyboard, command aliases provide a fast and efficient method of activating AutoCAD commands.

STARTING A NEW DRAWING

To begin a new drawing file, select the NEW command using one of the following methods:

>From the Standard toolbar
>
>From the pull-down menu (File > New)
>
>From the keyboard (New)

Command: **NEW**

Entering the NEW command displays the dialog box illustrated in the following image. This dialog box contains four buttons designed to open a drawing, start a drawing from scratch, start a drawing using a template, and use a wizard to make changes to drawing settings before creating.

Figure 1–18

OPENING AN EXISTING DRAWING

The OPEN command is used to edit a drawing that has already been created. Select this command from one of the following:

>From the Standard toolbar
>
>From the pull-down menu (File > Open)
>
>From the keyboard (Open)

When you select this command, a dialog box appears similar to the following image. Listed in the edit box area are all files that match the type shown at the bottom of the dialog box. Because the file type is .DWG, all drawing files supported by AutoCAD are listed. To choose a different folder, use standard Windows file management techniques by clicking in the Look in edit box. This will display all folders associated with the drive. Clicking the folder will display any drawing files contained in it.

Figure 1–19

BASIC DRAWING COMMANDS

The following sections discuss some basic techniques used in creating drawings. These include drawing lines, circles, and polylines, using object snap modes and tracking, and erasing objects. These basic drawing tools can easily be accessed using the Draw toolbar, as shown in the following image.

Figure 1–20

The following table gives a brief description of the LINE, CIRCLE, and PLINE commands:

Button	Tool	Key-In	Function
	Line	L	Draws individual or multiple line segments
	Circle	C	Constructs circles of specified radius or diameter
	Pline	PL	Used to construct a polyline, which is similar to a line except that all segments made with the PLINE command are considered a single object

THE LINE COMMAND

Use the LINE command to construct a line from one endpoint to the other. Choose this command from one of the following:

    From the Draw toolbar

    From the pull-down menu (Draw > Line)

    From the keyboard (L or Line)

As the first point of the line is marked, the rubber-band cursor is displayed along with the normal crosshairs to assist in locating where the next line segment will be drawn. The LINE command stays active until the user either executes the Close option or issues a null response by pressing ENTER at the prompt "To point."

Try It! – Create a new drawing from scratch. Study the following image and follow the command sequence for using the LINE command.

 Command: **L** *(For LINE)*
Specify first point: *(Pick a point at "A")*
Specify next point or [Undo]: *(Pick a point at "B")*
Specify next point or [Undo]: *(Pick a point at "C")*
Specify next point or [Close/Undo]: *(Pick a point at "D")*
Specify next point or [Close/Undo]: *(Pick a point at "E")*
Specify next point or [Close/Undo]: *(Pick a point at "F")*
Specify next point or [Close/Undo]: **C** *(To close the shape and exit the command)*

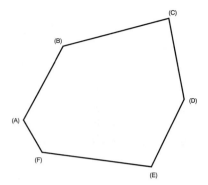

Figure 1–21

If a mistake is made in drawing a segment, as illustrated in the following image, the user can correct the error without exiting the LINE command. The built-in UNDO command within the LINE command removes the previously drawn line while still remaining in the LINE command. Refer to the following image and prompts to use the Undo option of the LINE command.

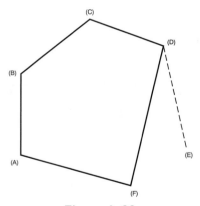

Figure 1–22

Command: **L** *(For LINE)*
Specify first point: *(Pick a point at "A")*
Specify next point or [Undo]: *(Pick a point at "B")*
Specify next point or [Undo]: *(Pick a point at "C")*
Specify next point or [Close/Undo]: *(Pick a point at "D")*
Specify next point or [Close/Undo]: *(Pick a point at "E")*
Specify next point or [Close/Undo]: **U** *(To undo or remove the segment from "D" to "E" and still remain in the LINE command)*

Specify next point or [Close/Undo]: *(Pick a point at "F")*
Specify next point or [Close/Undo]: **End** *(For Endpoint mode)*
of *(Select the endpoint of the line segment at "A")*
Specify next point or [Close/Undo]: *(Press ENTER to exit this command)*

Another option of the LINE command is the Continue option. The dashed line segment in the following image was the last segment drawn before the LINE command was exited. To pick up at the last point of a previously drawn line segment, type the LINE command and press ENTER. This will activate the Continue option of the LINE command.

Command: **L** *(For LINE)*
Specify first point: *(Press ENTER to activate Continue Mode)*
Specify next point or [Undo]: *(Pick a point at "B")*
Specify next point or [Undo]: *(Pick a point at "C")*
Specify next point or [Close/Undo]: End *(For Endpoint mode)*
of *(Select the endpoint of the vertical line segment at "A")*
Specify next point or [Close/Undo]: *(Press ENTER to exit this command)*

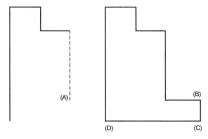

Figure 1–23

DYNAMIC INPUT AND LINES

With Dynamic Input turned on in the status bar, additional feedback can be obtained when drawing line segments. In addition to the command prompt and down arrow being displayed at your cursor location, a dynamic distance and angle are displayed to assist you in the construction of the line segment, as shown in the following image.

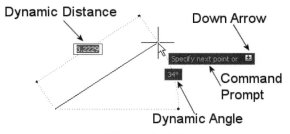

Figure 1–24

Command prompts for using Dynamic Input now appear in the drawing window next to the familiar AutoCAD cursor.

- When constructing line segments, dynamic dimensions in the form of a distance and an angle appear on the line. If the distance dimension is highlighted, entering a new value from your keyboard will change its value.
  - Pressing the TAB key will switch from the distance dimension to the angle dimension where you can change its value.
- By default in Dynamic Input, coordinates for the second point of a line are considered relative. In other words, you do not need to type the @ symbol in front of the coordinate. The @ symbol means "last point" and will be discussed in greater detail later in this chapter.
- You can still enter relative and polar coordinates as normal using the @ symbol if you desire. These older methods of coordinate entry will override the default dynamic input setting.
- The appearance of an arrow symbol in the Dynamic Input prompt area indicates that this command has options associated with it. To view these command options, press the down arrow on your keyboard. These options will display on your screen. Continue pressing the down arrow until you reach the desired command option and then press the ENTER key to select it.
- Dynamic Input can be toggled on or off in the status bar by clicking on the DYN button or by typing the F12 function key.

THE DIRECT DISTANCE MODE FOR DRAWING LINES

Another method is available for constructing lines, and it is called drawing by Direct Distance mode. In this method, the direction a line will be drawn in is guided by the location of the cursor. You enter a value, and the line is drawn at the specified distance at the angle specified by the cursor. This mode works especially well for drawing horizontal and vertical lines. The following image illustrates an example of how the Direct Distance mode is used.

Try It! – Create a new drawing from scratch. Turn Ortho mode on. Then use the following command sequence to construct the line segments using the Direct Distance mode of entry.

 Command: **L** *(For LINE)*
Specify first point: **2.00,2.00**
Specify next point or [Undo]: *(Move the cursor to the right and enter a value of 7.00 units)*

Specify next point or [Undo]: *(Move the cursor up and enter a value of 5.00 units)*
Specify next point or [Close/Undo]: *(Move the cursor to the left and enter a value of 4.00 units)*
Specify next point or [Close/Undo]: *(Move the cursor down and enter a value of 3.00 units)*
Specify next point or [Close/Undo]: *(Move the cursor to the left and enter a value of 2.00 units)*
Specify next point or [Close/Undo]: **C** *(To close the shape and exit the command)*

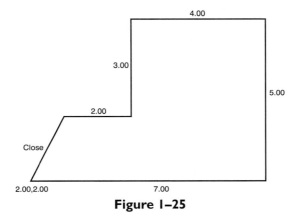

Figure 1–25

The following image shows another example of an object drawn with Direct Distance mode. Each angle was constructed from the location of the cursor. In this example, Ortho mode is turned off.

 Try It! – Create a new drawing from scratch. Be sure Ortho mode is turned off.

Then use the following command sequence to construct the line segments using the direct distance mode of entry.

Command: **L** *(For LINE)*

Specify first point: *(Pick a point at "A")*
Specify next point or [Undo]: *(Move the cursor and enter 3.00)*
Specify next point or [Undo]: *(Move the cursor and enter 2.00)*
Specify next point or [Close/Undo]: *(Move the cursor and enter 1.00)*
Specify next point or [Close/Undo]: *(Move the cursor and enter 4.00)*
Specify next point or [Close/Undo]: *(Move the cursor and enter 2.00)*
Specify next point or [Close/Undo]: *(Move the cursor and enter 1.00)*
Specify next point or [Close/Undo]: *(Move the cursor and enter 1.00)*
Specify next point or [Close/Undo]: **C** *(To close the shape and exit the command)*

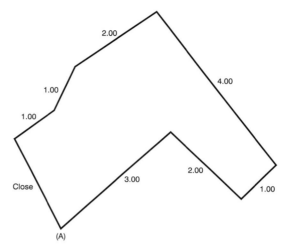

Figure 1–26

USING OBJECT SNAP FOR GREATER PRECISION

A major productivity tool that allows locking onto key locations of objects is Object Snap. The following image is an example of the construction of a vertical line connecting the endpoint of the fillet with the endpoint of the line at "A." The LINE command is entered and the Endpoint mode activated. When the cursor moves over a valid endpoint, an Object Snap symbol appears along with a tooltip indicating which OSNAP mode is currently being used. Another example of the use of Object Snap is in dimensioning applications where exact endpoints and intersections are needed.

 Try It! – Open the drawing file 01_Endpoint. Use the illustration in the following image and the command sequence below to draw a line segment from the endpoint of the arc to the endpoint of the line.

 Command: **L** *(For LINE)*
Specify first point: **End** *(For Endpoint mode)*
of *(Pick the endpoint of the fillet illustrated in the following image)*
Specify next point or [Undo]: **End** *(For Endpoint mode)*
of *(Pick the endpoint of the line at "A")*
Specify next point or [Undo]: *(Press ENTER to exit this command)*

Perform the same operation to the other side of this object using the Endpoint mode of OSNAP.

Figure 1-27

Object Snap modes can be selected in a number of different ways. Illustrated in the following image is the standard Object Snap toolbar. The following table gives a brief description of each Object Snap mode. In this table, notice the Key-In column. When the Object Snap modes are executed from keyboard input, only the first three letters are required.

Figure 1-28

The following table gives a brief description of each Object Snap mode:

Button	Tool	Key-In	Function
	Center	Cen	Snaps to the centers of circles and arcs
	Endpoint	End	Snaps to the endpoints of lines and arcs
	Extension	Ext	Creates a temporary extension line or arc when your cursor passes over the endpoint of objects; you can specify new points along the temporary line
	From	Fro	Snaps to a point at a specified distance and direction from a selected reference point
	Insert	Ins	Snaps to the insertion point of blocks and text
	Intersection	Int	Snaps to the intersections of objects
	Apparent Intersection	Int	Mainly used in creating 3D wireframe models; finds the intersection of points not located in the same plane

	Midpoint	Mid	Snaps to the midpoint of lines and arcs
	Midpoint Between 2 Points	M2P	Snaps to the middle of two selected points
	Nearest	Nea	Snaps to the nearest point found along any object
	Node	Nod	Snaps to point objects (including dimension definition points) and text objects (including multiline text and dimension text)
	None	Non	Disables object snap
	Osnap Settings	Osnap	Launches the Drafting Settings dialog box and activates the Object Snap tab
	Parallel	Par	Draws an object parallel to another object
	Perpendicular	Per	Snaps to a perpendicular location on an object
	Quadrant	Qua	Snaps to four key points located on a circle
	Tangent	Tan	Snaps to the tangent location of arcs and circles

The following image shows the Object Snap modes that can be activated when you hold down SHIFT or CTRL and press the right mouse button while within a command such as LINE or MOVE. This shortcut menu will appear wherever the cursor is currently positioned.

Figure 1–29

OBJECT SNAP MODES

 CENTER (CEN)

Use the Center mode to snap to the center of a circle or arc. To accomplish this, activate the mode by clicking the Center button and moving the cursor along the edge of the circle or arc similar to the following image. Notice the AutoSnap symbol appearing at the center of the circle or arc.

Figure 1–30

 ENDPOINT (END)

The Endpoint mode is one of the more popular Object Snap modes; it is helpful in snapping to the endpoints of lines or arcs. One application of Endpoint is during the dimensioning process, where exact distances are needed to produce the desired dimension. Activate this mode by clicking the Endpoint button, and then move the cursor along the edge of the object to snap to the endpoint. In the case of the line or arc shown in the following image, the cursor does not actually have to be positioned at the endpoint; favoring one end automatically snaps to the closest endpoint.

Figure 1–31

 EXTENSION (EXT)

When you select a line or an arc, the Extension mode creates a temporary path that extends from the object. Picking the end of the line at "A" shown in the following image acquires the angle of the line. A tooltip displays the current extension distance and angle. Picking the end of the arc at "B" acquires the radius of the arc and displays the current length in the tooltip.

Figure 1–32

 FROM (FRO)

Use the From mode along with a secondary Object Snap mode to establish a reference point and construct an offset from that point. Open the drawing file 01_Osnap From. In the following image, the circle needs to be drawn 1.50 units in the X and Y directions from point "A." The CIRCLE command is activated and the Object Snap From mode is used in combination with the Object Snap Intersection mode. The From option requires a base point. Identify the base point at the intersection of corner "A." The next prompt asks for an offset value; enter the relative coordinate value of @1.50,1.50 (this identifies a point 1.50 units in the positive X-direction and 1.50 units in the positive Y-direction). This completes the use of the From option and identifies the center of the circle at "B." Study the following command sequence to accomplish this operation:

 Try It! – Open the drawing file 01_Osnap From. Use the illustration and prompt sequence below for constructing a circle inside the shape with the aid of the Object Snap From mode.

Command: **C** *(For CIRCLE)*
Specify center point for circle or [3P/2P/Ttr (tan tan radius)]: **From**
Base point: **Int** *(For Intersection Mode)*
of *(Select the intersection at "A" in the following image)*
<Offset>: **@1.50,1.50**
Specify radius of circle or [Diameter]: **D** *(For Diameter)*
Specify diameter of circle: **1.25**

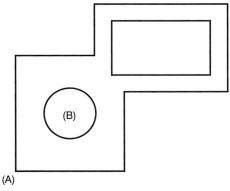

Figure 1–33

INSERT (INS)

The Insert mode snaps to the insertion point of an object. In the case of the text object in the following image, activating the Insert mode and positioning the cursor anywhere on the text snaps to its insertion point, in this case at the lower left corner of the text at "A." The other object illustrated in the following image is called a block. It appears to be constructed of numerous line objects; however, all objects that make up the block are considered to be a single object. Blocks can be inserted in a drawing. Typical types of blocks are symbols such as doors, windows, bolts, and so on—anything that is used several times in a drawing. In order for a block to be brought into a drawing, it needs an insertion point, or a point of reference. The Insert mode, when you position the cursor on a block, will snap to the insertion point of a block.

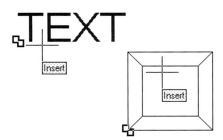

Figure 1–34

INTERSECTION (INT)

Another popular Object Snap mode is Intersection. Use this mode to snap to the intersection of two objects. Position the cursor anywhere near the intersection of two objects and the intersection symbol appears. See the following image.

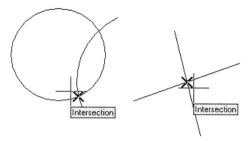

Figure 1–35

EXTENDED INTERSECTION (INT)

Another type of intersection snap is the Extended Intersection mode, which is used to snap to an intersection not considered obvious from the previous example. The same Object Snap Intersection button is for performing an extended intersection operation. The following image shows two lines that do not intersect. Activate the Extended Intersection mode and pick both lines. Notice the intersection symbol present where the two lines, if extended, would intersect.

Figure 1–36

 MIDPOINT (MID)

The Midpoint mode snaps to the midpoint of objects. Line and arc examples are shown in the following image. When activating the Midpoint mode, touch the object anywhere with some portion of the cursor; the midpoint symbol will appear at the exact midpoint of the object.

Getting Started with AutoCAD   29

Figure 1-37

MIDPOINT OF TWO SELECTED POINTS (M2P)

This Object Snap mode snaps to the midpoint of two selected points. To access this mode, type either M2P or MTP at the command prompt. While this Object Snap mode is not accessible through a toolbar button, it can be found by pressing Shift + Right Mouse Button to display the Object Snap menu, as shown in the following image.

The following command sequence and illustration in the following image show the construction of a circle at the midpoint of two selected points.

Command: **C** *(For CIRCLE)*
Specify center point for circle or [3P/2P/Ttr (tan tan radius)]: **M2P**
First point of mid: **End**
of *(Pick the endpoint at "A")*
Second point of mid: **End**
of *(Pick the endpoint at "B")*
Specify radius of circle or [Diameter]: **0.50**

Figure 1-38

NEAREST (NEA)

The Nearest mode snaps to the nearest point it finds on an object. Use this mode when a point on an object needs to be selected and an approximate location on the object is sufficient. The nearest point is calculated based on the closest distance from the intersection of the crosshairs perpendicular to the object or the shortest distance from the crosshairs to the object. In the following image, the appearance of the Nearest symbol helps to show where the point identified by this mode is actually located.

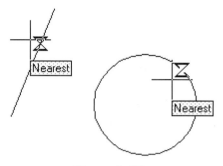

Figure 1–39

NODE (NOD)

The Node mode snaps to a node or point. Picking the point in the following image snaps to its center. The Nearest mode shown in the following image can also be used to snap to a point.

Figure 1–40

PARALLEL (PAR)

Use the Parallel mode to construct a line parallel to another line. In the following image, the LINE command is started and a beginning point of the line is picked. The Parallel mode is activated by selecting the Parallel icon then hovering the cursor over the existing line. The existing line is highlighted at "A" and the Parallel symbol appears. Finally, moving the cursor to the approximate position that makes the new line parallel to the one just selected allows the Parallel mode to construct a parallel line, the tracking path and the tooltip giving the current distance and angle.

Figure 1–41

⊥ PERPENDICULAR (PER)

The Perpendicular mode is helpful for snapping to an object normal (or perpendicular) from a previously identified point. The following image shows a line segment drawn perpendicular from the point at "A" to the inclined line "B." A 90° angle is formed with the perpendicular line segment and the inclined line "B." With this mode, the user can also construct lines perpendicular to circles.

Figure 1–42

◇ QUADRANT (QUA)

Circle quadrants are defined as points located at the 0°, 90°, 180°, and 270° positions of a circle, as in the following image. Using the Quadrant mode will snap to one of these four positions as the edge of a circle or arc is selected. In the example of the circle in the following image, the edge of the circle is selected by the cursor location. The closest quadrant to the cursor is selected.

Figure 1–43

TANGENT (TAN)

The Tangent mode is helpful in constructing lines tangent to other objects such as the circles in the following image. In this case, the Tangent mode is being used in conjunction with the LINE command. The point at "A" is anchored to the top of the circle using the Quadrant mode. When dragged to the next location, the line will appear to be tangent. Then, with Tangent mode activated and the location at "B" picked, the line will be tangent to the large circle near "B."

Figure 1–44

 Try It! – Open the drawing file 01_Tangent. Follow this command sequence for constructing a line segment tangent to two circles:

 Command: **L** *(For LINE)*

Specify first point: **Qua** *(For Quadrant mode)*
from *(Select the circle near "A")*
Specify next point or [Undo]: **Tan** *(For Tangent mode)*
to *(Select the circle near "B")*

 Try It! – Open the drawing file 01_Osnap. Various objects consisting of lines, circles, arcs, points, and blocks need to be connected up with line segments at their key locations. Use the prompt sequence and the following image for performing this operation.

Command: **L** *(For LINE)*

Specify first point: **End**
of *(Pick the endpoint at "A")*
Specify next point or [Undo]: **Nod**
of *(Pick the node at "B")*
Specify next point or [Undo]: **Tan**
to *(Pick the circle at "C")*

Specify next point or [Close/Undo]: **Int**
of *(Pick the intersection at "D")*
Specify next point or [Close/Undo]: **Int**
of *(Pick the line at "E")*
and *(Pick the horizontal line at "F")*
Specify next point or [Close/Undo]: **Qua**
of *(Pick the circle at "G")*
Specify next point or [Close/Undo]: **Cen**
of *(Pick the arc at "H")*
Specify next point or [Close/Undo]: **Mid**
of *(Pick the line at "J")*
Specify next point or [Close/Undo]: **Per**
to *(Pick the line at "K")*
Specify next point or [Close/Undo]: **Ins**
of *(Pick on the I-Beam symbol near "L")*
Specify next point or [Close/Undo]: **Nea**
to *(Pick the circle at "M")*
Specify next point or [Close/Undo]: *(Press ENTER to exit this command)*

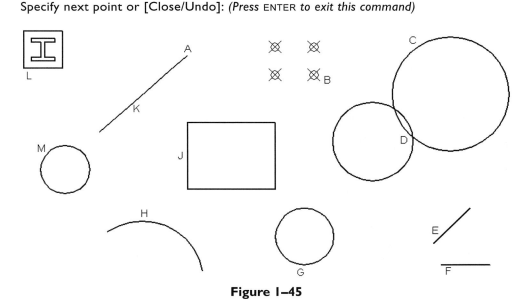

Figure 1–45

CHOOSING RUNNING OBJECT SNAP

So far, all Object Snap modes have been selected from the shortcut menu or entered at the keyboard. The problem with these methods is that if you're using a certain mode over a series of commands, you have to select the Object Snap mode every time. It is possible to make the Object Snap mode or modes continuously present through

Running Osnap. Right-click on the OSNAP button located in the status bar at the bottom of the drawing area, as shown in the following image. Selecting Settings from the shortcut menu will activate the Drafting Settings dialog box illustrated in the following image. By default, the Endpoint, Center, Intersection, and Extension modes are automatically selected. Whenever the cursor lands over an object supported by one of these four modes, a yellow symbol appears to alert the user to the mode. It is important to know that when changes are made inside this dialog box, the changes are applied to all drawing files. Even if a drawing session is terminated without saving, the changes made to the Object Snap modes are automatic. Other Object Snap modes can be selected by checking their appropriate boxes in the dialog box; removing the check disables the mode.

Figure 1–46

These Object Snap modes remain in effect during drawing until you click the OSNAP button illustrated in the status bar in the following image; this turns off the current running Object Snap modes. To reactivate the running Object Snap modes, click the OSNAP button again and the previous set of Object Snap modes will be back in effect. Object Snap can also be activated and reactivated by pressing the F3 function key.

Figure 1–47

OBJECT SNAP TRACKING

Tracking mode provides a quick and easy way to locate the center of the rectangle without drawing any construction geometry; an example is illustrated on the left in the following image. A hole needs to be displayed as a circle in the front view. The hole will be placed in the center of the rectangle identified by "A" and "B." When the user begins tracking and picks a point on the screen, AutoCAD forces the next point to conform to an orthogonal path that could extend vertically or horizontally from the first point. The first point may retain an X or Y value and the second point may retain separate X or Y values. Use the following command sequence to get an idea of the operation of Tracking mode. For best results, be sure Ortho mode is turned off.

 Try It! – Open the drawing file 01_Tracking. Follow the prompt sequence and illustrations below for completing this operation.

 Command: **C** *(For CIRCLE)*
Specify center point for circle or [3P/2P/Ttr (tan tan radius)]: **TK** *(For Tracking)*
First tracking point: **Mid**
of *(Pick the midpoint at "A" in the following image)*
Next point (Press enter to end tracking): **Mid**
of *(Pick the midpoint at "B" in the following image)*
Next point (Press enter to end tracking): *(Press* ENTER*)*
Specify radius of circle or [Diameter]: **D** *(For Diameter)*
Specify diameter of circle: **0.50**

The result is illustrated on the right in the following image, with the hole drawn in the center of the rectangular shape. Running OSNAP can also be set to Midpoint, OSNAP turned on, and OTRACK turned on to achieve the same results as in the command sequence above. This will be discussed later in this chapter.

 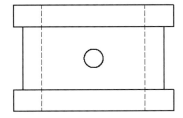

Figure 1–48

POLAR TRACKING

Previously in this chapter, the Direct Distance mode was highlighted as an efficient means of constructing orthogonal lines without the need for using one of the coordinate modes of entry. However, it did not lend itself to lines at angles other than 0°, 90°, 180°, and 270°. Polar Tracking has been designed to remedy this. This mode allows the cursor to follow a tracking path that is controlled by a preset angular increment. The POLAR button located at the bottom of the display in the Status area turns this mode on or off. Right-clicking on POLAR in the status bar at the bottom of the screen and picking Settings will display the Drafting Settings dialog box shown in the following image and the Polar Tracking tab.

Figure 1–49

A few general terms are defined before continuing:

Tracking Path—This is a temporary dotted line that can be considered a type of construction line. Your cursor will glide or track along this path (see the following image).

Tooltip—This displays the current cursor distance and angle away from the tracking point (see the following image).

Figure 1–50

 Tip: Both POLAR and ORTHO modes cannot be turned on at the same time. Once you turn POLAR on, ORTHO automatically turns off, and vice versa.

 Try It! – To see how the Polar Tracking mode functions, construct an object that consists of line segments at 10° angular increments. Create a new drawing starting from scratch. Then, set the angle increment through the Polar Tracking tab of the Drafting Settings dialog box, shown in the following image.

Figure 1–51

Start the LINE command, anchor a starting point at "A," and move the cursor to the upper right until the tooltip reads 20°. Enter a value of 2 units for the length of the line segment (see the illustration on the left in the following image).

Move the cursor up and to the left until the tooltip reads 110°, as shown on the right in the following image, and enter a value of 2 units. (This will form a 90° angle with the first line.)

Figure 1–52

Move the cursor until the tooltip reads 20°, as shown on the left in the following image, and enter a value of 1 unit.

Move the cursor until the tooltip reads 110°, as shown on the right in the following image, and enter a value of 1 unit.

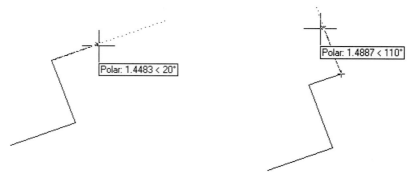

Figure 1–53

Move the cursor until the tooltip reads 200°, as shown on the left in the following image, and enter a value of 3 units.

Move the cursor to the endpoint, as shown on the right in the following image, or use the Close option of the LINE command to close the shape and exit the command.

Figure 1–54

SETTING A POLAR SNAP VALUE

An additional feature of using polar snap is illustrated in the following image. Clicking the Snap and Grid tab of the Drafting Settings dialog box displays the dialog box in the following image. Clicking the Polar Snap option found along the lower right corner of the dialog box allows the user to enter a polar distance. When SNAP and POLAR are both turned on and the cursor is moved to draw a line, not only will the angle be set but the cursor will also jump to the next increment set by the polar snap value.

Figure 1–55

Try It! – Open the drawing file 01_Polar. Set the polar angle to 30° and a polar snap distance to 0.50 unit increments. Be sure POLAR and SNAP are both turned on in your status bar and that all other modes are turned off. Begin constructing the object in the following image using the command prompt sequence below as a guide.

 Command: **L** *(For LINE)*
Specify first point: **7.00,4.00**
Specify next point or [Undo]: *(Move your cursor down until the tooltip reads Polar: 2.5000<270 and pick a point)*
Specify next point or [Undo]: *(Move your cursor right until the tooltip reads Polar: 1.5000<0 and pick a point)*
Specify next point or [Close/Undo]: *(Polar: 2.0000<30)*
Specify next point or [Close/Undo]: *(Polar: 2.0000<60)*
Specify next point or [Close/Undo]: *(Polar: 2.5000<90)*
Specify next point or [Close/Undo]: *(Polar: 3.0000<150)*
Specify next point or [Close/Undo]: *(Polar: 1.5000<180)*
Specify next point or [Close/Undo]: *(Polar: 2.5000<240)*
Specify next point or [Close/Undo]: *(Polar: 2.5000<120)*
Specify next point or [Close/Undo]: *(Polar: 1.5000<180)*
Specify next point or [Close/Undo]: *(Polar: 3.0000<210)*
Specify next point or [Close/Undo]: *(Polar: 2.5000<270)*
Specify next point or [Close/Undo]: *(Polar: 2.0000<300)*
Specify next point or [Close/Undo]: *(Polar: 2.0000<330)*
Specify next point or [Close/Undo]: *(Polar: 1.5000<0)*
Specify next point or [Close/Undo]: *(Polar: 2.5000<90)*
Specify next point or [Close/Undo]: **C** *(To close the shape)*

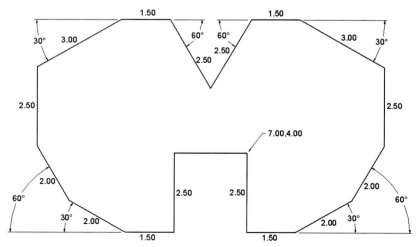

Figure 1–56

SETTING A RELATIVE POLAR ANGLE

An additional feature of using polar snap is illustrated in the following image. When activating the Polar tab of the Drafting Settings dialog box, located in the lower right corner are two settings that deal with the Polar Angle measurement; they are Absolute and Relative to last segment.

Absolute—This is the default setting when dealing with Polar Angle measurement. This setting controls all angle measurements based on the position of the current user coordinate system, the icon located in the lower left corner of all AutoCAD drawing screens.

Relative to last segment—When changing the Polar Angle measurement to Relative to last segment, the Polar Tracking angle is based on the last line segment drawn.

Figure 1–57

To get a better understanding of the two different Polar Angle measurement settings, study the following image. The illustration on the left is an example of the Absolute setting. The 150° angle was drawn from point (A) to point (B). This angle is derived from the absolute position of angle 0 (zero) set by default to the 3 o'clock position as defined in the Drawing Units dialog box. In the illustration on the right, the same line segment is drawn. However, this time the Relative to last segment setting is used. Notice how the 120° angle is calculated. The angle is based on the last line segment, not on an angle calculated in relation to 0 (zero). This is the reason for the parallel line segment at (C).

Figure 1–58

MORE ON OBJECT SNAP TRACKING MODE

Object Snap Tracking works in conjunction with Object Snap. Before you can track from an Object Snap point, you must first set an Object Snap mode or modes from the Object Snap tab of the Drafting Settings dialog box. Object Snap Tracking can be toggled on or off with the OTRACK button located in the status bar in the following image.

| SNAP | GRID | ORTHO | POLAR | OSNAP | OTRACK | DUCS | DYN | LWT | MODEL |

Figure 1-59

The advantage of using Object Snap Tracking is in the ability to choose or acquire points to be used for construction purposes. Acquired points are temporarily selected by hovering the cursor over the point versus selecting with the mouse. Care must be taken when acquiring points that the points are in fact not picked. They are used only for construction purposes. For example, two line segments need to be added to the object shown on the left in the following image to form a rectangle. Here is how to perform this operation using Polar Tracking.

 Try It! – Open the drawing file 01_Otrack Lines. Notice in the status bar that POLAR, OSNAP, and OTRACK are all turned on. Be sure that Running Osnap is set to Endpoint mode. Enter the line command and pick a starting point for the line at "A." Then, move the cursor directly to the left until the tooltip reads 180°. The starting point for the next line segment is considered acquired (see the illustration on the right in the following image).

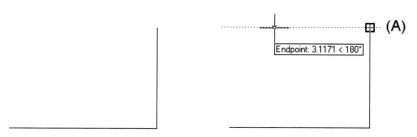

Figure 1-60

Rather than enter the length of this line segment, move your cursor over the top of the corner at "B" to acquire this point (be careful not to pick the point here). Then move your cursor up until the tooltip reads 90° (see the following image).

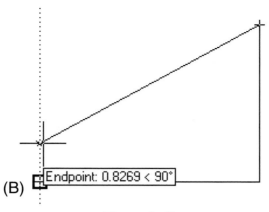

Figure 1–61

Move your cursor up until the tooltip now reads angles of 90° and 180°. Also notice the two tracking paths intersecting at the point of the two acquired points. Picking this point at "C" will construct the horizontal line segment (see the illustration on the left in the following image). Finally, slide your cursor to the endpoint at "D" to complete the rectangle (see the illustration on the right in the following image).

Figure 1–62

Tip: Pausing the cursor over an existing acquired point a second time will remove the tracking point from the object.

Try It! – Open the drawing file 01_Otrack Pipes. Set Running Osnap mode to Midpoint. Set the polar angle to 90°. Be sure POLAR, OSNAP, and OTRACK are all turned on. Enter the line command and connect all fittings with lines illustrated in the following image.

Figure 1-63

USING TEMPORARY TRACKING POINTS

A very powerful construction tool includes the ability to use an extension path along with the Temporary Tracking Point tool to construct objects under difficult situations, as illustrated in the following image.

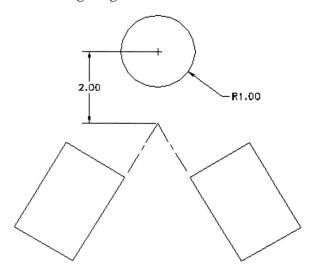

Figure 1-64

An extension path is similar to a tracking path except that it is present when the Object Snap Extension mode is activated. To construct the circle in relation to the two inclined rectangles, follow the next series of steps.

Try It! – Open the drawing file 01_Temporary Point. Set Running Osnap to Endpoint, Intersection, Center, and Extension. Check to see that OSNAP and OTRACK are turned on and all other modes are turned off. Activate the circle command; this will prompt you to specify the center point for the circle. Move the cursor over the corner of the right rectangle at "A" to acquire this point. Move the cursor up and to the left, making sure the tooltip lists the Extension mode (see the following image).

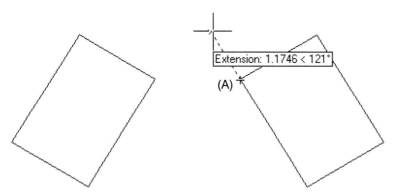

Figure 1–65

With the point acquired at "A," move the cursor over the corner of the left rectangle at "B" and acquire this point. Move the cursor up and to the right, making sure the tooltip lists the Extension mode, as shown in the following image.

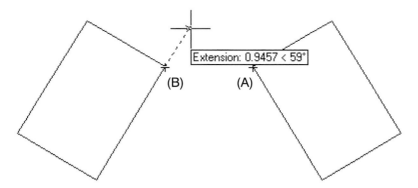

Figure 1–66

Move the cursor vertically until both acquired points intersect as shown in the following image. The center of the circle is located 2 units above this intersection. Click the Object Snap Temporary Tracking button and pick this intersection.

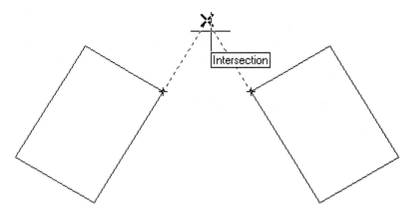

Figure 1–67

Next, move the cursor directly above the temporary tracking point, as shown on the left in the following image. The tooltip should read 90°. Entering a value of 2 units identifies the center of the circle.

The completed construction operation is illustrated on the right in the following image.

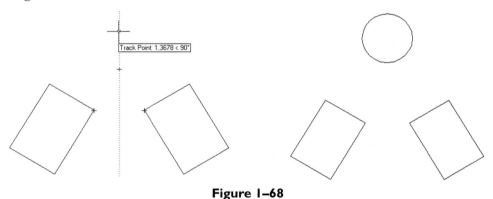

Figure 1–68

ALTERNATE METHODS USED FOR PRECISION DRAWING: CARTESIAN COORDINATES

Before drawing precision geometry such as lines and circles, it is essential to have an understanding of coordinate systems. The Cartesian or rectangular coordinate system is a system constructed of an orthogonal axis intersecting at an origin that creates four quadrants, allowing location of any point by specifying the coordinates. A coordinate is made up of a horizontal and vertical pair of numbers identified as X and Y. The coordinates are then plotted on a type of graph or chart. An example of a rectangular coordinate system is shown in the following image. The coordinates of the origin are 0,0. From the origin, all positive directions move up and to the right. All negative directions move down and to the left.

The coordinate axes are divided into four quadrants that are labeled I, II, III, and IV, as shown in the following image. In Quadrant I, all X and Y values are positive. Quadrant II has a negative X value and positive Y value. Quadrant III has negative values for X and Y. Quadrant IV has positive X values and negative Y values.

 Note: When you begin a drawing in AutoCAD, the screen display reflects Quadrant I of the Cartesian coordinate system, as shown in the following image. The origin 0,0 is located in the lower left corner of the drawing screen.

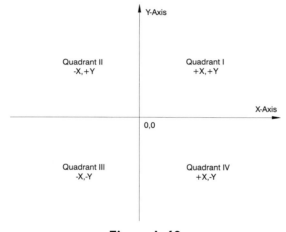

Figure 1–69

For each set of (X,Y) coordinates, X values represent distances from the origin horizontally to the right if positive and horizontally to the left if negative. Y values represent distances from the origin vertically up if positive and vertically down if negative. The following image shows a series of coordinates plotted on the number lines. One coordinate is identified in each quadrant to show the positive and negative values. As an example, coordinate 3,2 in Quadrant I represents a point 3 units to the right and 2 units vertically up from the origin. The coordinate −5,3 in Quadrant II represents a point 5 units to the left and 3 units vertically up from the origin. Coordinate −2,−2 in Quadrant III represents a point 2 units to the left and vertically down from the origin. Lastly, coordinate 2,−4 in Quadrant IV represents a point 2 units to the right and 4 units vertically down from the origin.

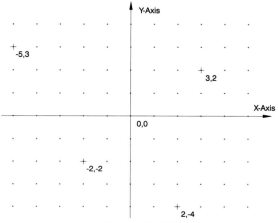

Figure 1-70

ABSOLUTE COORDINATE MODE FOR DRAWING LINES

When drawing geometry such as lines, the user must use a method of entering precise distances, especially when accuracy is important. This is the main purpose of using coordinates. The simplest and most elementary form of coordinate values is absolute coordinates. Absolute coordinates conform to the following format:

$$X,Y$$

One problem with using absolute coordinates is that all coordinate values refer back to the origin 0,0. This origin on the AutoCAD screen is usually located in the lower left corner when a new drawing is created. The origin will remain in this corner unless it is altered with the LIMITS command.

 Try It! – Create a new drawing file starting from scratch. For this exercise, turn off Dynamic Input by clicking on the DYN button located in the status bar. Then use the line command prompts below and the following image to construct the shape.

Command: **L** *(For LINE)*
Specify first point: **2,2** *(at "A")*
Specify next point or [Undo]: **2,7** *(at "B")*
Specify next point or [Undo]: **5,7** *(at "C")*
Specify next point or [Close/Undo]: **7,4** *(at "D")*
Specify next point or [Close/Undo]: **10,4** *(at "E")*
Specify next point or [Close/Undo]: **10,2** *(at "F")*
Specify next point or [Close/Undo]: **C** *(To close the shape and exit the command)*

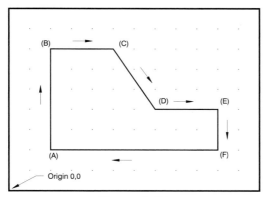

Figure 1–71

As shown in the previous example, all points on the object make reference to the origin at 0,0. Even though absolute coordinates are useful in starting lines, there are more efficient ways to construct lines and draw objects.

RELATIVE COORDINATE MODE FOR DRAWING LINES

With absolute coordinates, the horizontal and vertical distance from the origin at 0,0 must be kept track of at all times in order for the correct coordinate to be entered. With complicated objects, this is difficult to accomplish, and as a result, the wrong coordinate may be entered. It is possible to reset the last coordinate to become a new origin or 0,0 point. The new point would be relative to the previous point, and for this reason, this point is called a relative coordinate. The format is as follows:

$$@X,Y$$

In this format, we use the same X and Y values with one exception: the At symbol or @ resets the previous point to 0,0 and makes entering coordinates less confusing.

 Try It! – Create a new drawing file starting from scratch. Use the LINE command prompts below and the following image to construct the shape.

Command: **L** *(For LINE)*
Specify first point: **2,2** *(at "A")*
Specify next point or [Undo]: **@0,4** *(to "B")*
Specify next point or [Undo]: **@4,2** *(to "C")*
Specify next point or [Close/Undo]: **@3,0** *(to "D")*
Specify next point or [Close/Undo]: **@3,-4** *(to "E")*
Specify next point or [Close/Undo]: **@-3,-2** *(to "F")*
Specify next point or [Close/Undo]: **@-7,0** *(back to "A")*
Specify next point or [Close/Undo]: *(Press ENTER to exit this command)*

Tip: In each command, the @ symbol resets the previous point to 0,0.

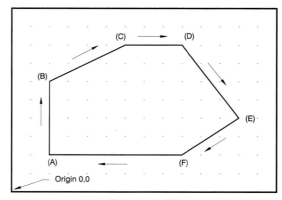

Figure 1-72

POLAR COORDINATE MODE FOR DRAWING LINES

Another popular method of entering coordinates is the Polar Coordinate mode. The format is as follows:

@Distance<Direction

As the preceding format implies, the Polar Coordinate mode requires a known distance and a direction. The @ symbol resets the previous point to 0,0. The direction is preceded by the < symbol, which reads the next number as a polar or angular direction. The following image illustrates the directions supported by the Polar Coordinate mode.

Figure 1-73

Try It! – Create a new drawing file starting from scratch. Use the LINE command prompts below and the following image to construct the shape.

Command: **L** *(For LINE)*
Specify first point: **3,2** *(at "A")*
Specify next point or [Undo]: **@8<0** *(to "B")*
Specify next point or [Undo]: **@5<90** *(to "C")*
Specify next point or [Close/Undo]: **@5<180** *(to "D")*
Specify next point or [Close/Undo]: **@4<270** *(to "E")*
Specify next point or [Close/Undo]: **@2<180** *(to "F")*
Specify next point or [Close/Undo]: **@2<90** *(to "G")*
Specify next point or [Close/Undo]: **@1<180** *(to "H")*
Specify next point or [Close/Undo]: **@3<270** *(to close back to "A")*
Specify next point or [Close/Undo]: *(Press ENTER to exit this command)*

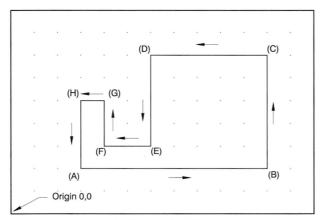

Figure 1-74

COMBINING COORDINATE MODES FOR DRAWING LINES

So far, the preceding pages concentrated on using each example of coordinate modes (absolute, relative, and polar) separately to create geometry. While the examples focused on each individual mode, it is important to note that maximum productivity is usually obtained through use of a combination of modes during a drawing session. It is fairly common to use one, two, or three coordinate modes in combination with one another. In the following image, the drawing starts with an absolute coordinate, changes to a polar coordinate, and changes again to a relative coordinate. The user should develop proficiency in each mode in order to be most productive.

Try It! – Create a new drawing file starting from scratch. Use the LINE command prompts below and the following image to construct the shape.

Command: **L** *(For LINE)*
Specify first point: **2,2** *(at "A")*
Specify next point or [Undo]: **@3<90** *(to "B")*
Specify next point or [Undo]: **@2,2** *(to "C")*
Specify next point or [Close/Undo]: **@6<0** *(to "D")*
Specify next point or [Close/Undo]: **@5<270** *(to "E")*
Specify next point or [Close/Undo]: **@3<180** *(to "F")*
Specify next point or [Close/Undo]: **@3<90** *(to "G")*
Specify next point or [Close/Undo]: **@2<180** *(to "H")*
Specify next point or [Close/Undo]: **@-3,-3** *(back to "A")*
Specify next point or [Close/Undo]: *(Press ENTER to exit this command)*

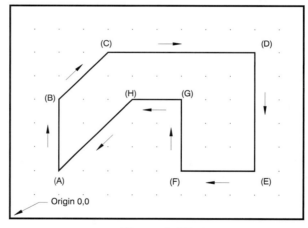

Figure 1–75

THE CIRCLE COMMAND

The CIRCLE command constructs circles of various radii or diameter. This command can be selected from any of the following:

    From the Draw toolbar

    From the pull-down menu (Draw > Circle)

    From the keyboard (C or Circle)

Choosing Circle from the Draw pull-down menu displays the cascading menu shown in the following image. All supported methods of constructing circles are displayed in the list. Circles may be constructed by radius or diameter. This command also supports circles defined by three points and construction of a circle tangent to other objects in the drawing. These last two modes will be discussed in chapter 5, "Performing Geometric Constructions."

Figure 1-76

CIRCLE BY RADIUS MODE

Use the CIRCLE command and the Radius mode to construct a circle by a radius value that you specify. After selecting a center point for the circle, the user is prompted to enter a radius for the desired circle.

 Try It! – Create a new drawing file starting from scratch. Use the CIRCLE command prompts below and the illustration on the left in the following image to construct a circle by radius.

 Command: **C** *(For CIRCLE)*
Specify center point for circle or [3P/2P/Ttr *(tan tan radius)*]: *(Mark the center at "A")*
Specify radius of circle or [Diameter]: **1.50**

CIRCLE BY DIAMETER MODE

Use the CIRCLE command and the Diameter mode to construct a circle by a diameter value that you specify. After selecting a center point for the circle, you are prompted to enter a diameter for the desired circle.

 Try It! – Create a new drawing file starting from scratch. Use the CIRCLE command prompts below and the illustration on the right in the following image to construct a circle by diameter.

Command: **C** *(For CIRCLE)*
Specify center point for circle or [3P/2P/Ttr (tan tan radius)]: *(Mark the center at "A")*
Specify radius of circle or [Diameter]: **D** *(For Diameter)*
Specify diameter of circle: **3.00**

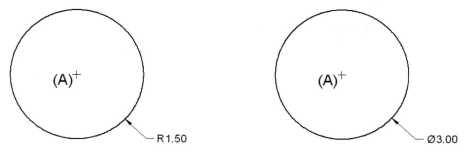

Figure 1–77

DYNAMIC INPUT AND CIRCLES

When using Dynamic Input mode for constructing circles, notice the appearance of a dynamic radius readout when you drag your cursor as shown in the following image.

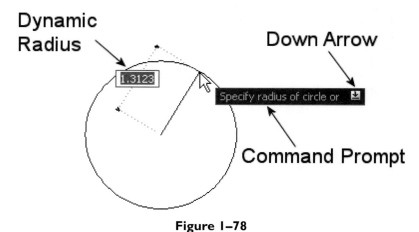

Figure 1–78

USING THE PLINE COMMAND

Polylines are similar to individual line segments except that a polyline can consist of numerous segments and still be considered a single object. Width can also be assigned to a polyline, unlike regular line segments; this makes polylines perfect for drawing borders and title blocks. Polylines can be constructed by selecting any of the following:

> From the Draw toolbar
>
> From the pull-down menu (Draw > Polyline)
>
> From the keyboard (PL or Pline)

Study the following images and both command sequences that follow to use the PLINE command.

 Try It! – Create a new drawing file starting from scratch. Follow the command prompt sequence and illustration below to construct the polyline.

 Command: **PL** *(For PLINE)*
Specify start point: *(Pick a point at "A" in the following image)*
Current line-width is 0.0000
Specify next point or [Arc/Close/Halfwidth/Length/Undo/Width]: **W** *(For Width)*
Specify starting width <0.0000>: **0.10**
Specify ending width <0.1000>: *(Press ENTER to accept the default)*
Specify next point or [Arc/Close/Halfwidth/Length/Undo/Width]: *(Pick a point at "B")*
Specify next point or [Arc/Close/Halfwidth/Length/Undo/Width]: *(Pick a point at "C")*
Specify next point or [Arc/Close/Halfwidth/Length/Undo/Width]: *(Pick a point at "D")*
Specify next point or [Arc/Close/Halfwidth/Length/Undo/Width]: *(Pick a point at "E")*
Specify next point or [Arc/Close/Halfwidth/Length/Undo/Width]: *(Press ENTER to exit this command)*

Figure 1–79

Try It! – Create a new drawing file starting from scratch. Follow the command prompt sequence and illustration below to construct the polyline object. The Direct Distance mode of entry could also be used to construct this object.

Command: **PL** *(For PLINE)*
Specify start point: 2,2.5 Current line-width is 0.0000
Specify next point or [Arc/Close/Halfwidth/Length/Undo/Width]: **@8<0** *(To "B")*
Specify next point or [Arc/Close/Halfwidth/Length/Undo/Width]: **@1<90** *(To "C")*
Specify next point or [Arc/Close/Halfwidth/Length/Undo/Width]: **@1<0** *(To "D")*
Specify next point or [Arc/Close/Halfwidth/Length/Undo/Width]: **@2<90** *(To "E")*
Specify next point or [Arc/Close/Halfwidth/Length/Undo/Width]: **@2<180** *(To "F")*
Specify next point or [Arc/Close/Halfwidth/Length/Undo/Width]: **@2<270** *(To "G")*
Specify next point or [Arc/Close/Halfwidth/Length/Undo/Width]: **@6<180** *(To "H")*
Specify next point or [Arc/Close/Halfwidth/Length/Undo/Width]: **@2<90** *(To "I")*
Specify next point or [Arc/Close/Halfwidth/Length/Undo/Width]: **@2<180** *(To "J")*
Specify next point or [Arc/Close/Halfwidth/Length/Undo/Width]: **@2<270** *(To "K")*
Specify next point or [Arc/Close/Halfwidth/Length/Undo/Width]: **@1<0** *(To "L")*
Specify next point or [Arc/Close/Halfwidth/Length/Undo/Width]: **C** *(To close the shape and exit the PLINE command)*

Figure 1–80

DYNAMIC INPUT AND PLINES

Using Dynamic Input for polylines is similar to lines. As you move your cursor while in the PLINE command, you can observe the appearance of your pline through the dynamic distance and angle features, as shown in the following image. As with all dynamic input modes, typing the down arrow on your keyboard will display options for the PLINE command that you can cycle through.

Figure 1–81

 ERASING OBJECTS

Throughout the design process, as objects such as lines and circles are placed in a drawing, changes in the design will require the removal of objects. The ERASE command will delete objects from the database. The ERASE command is selected from any of the following:

    From the Modify toolbar

    From the pull-down menu (Modify > Erase)

    From the keyboard (E or Erase)

On the left in the following image, line segments "A" and "B" need to be removed in order for a new line to be constructed, closing the shape. Two ways of erasing these lines will be introduced here.

When first entering the ERASE command, you are prompted to "Select objects:" to erase. Notice that your cursor changes in appearance from crosshairs to a pick box, as shown on the right in the following image. Move the pick box over the object to be selected and pick this item. Notice that it will be highlighted as a dashed object to signify it is now selected. At this point, the "Select objects:" prompt appears again. Additional objects may be selected at this point. Once all the objects are selected, pressing ENTER will perform the erase operation.

    Command: **E** *(For ERASE)*
Select objects: *(Pick line "A" as shown on the right in the following image)*
Select objects: *(Press* ENTER *to perform the erase operation)*

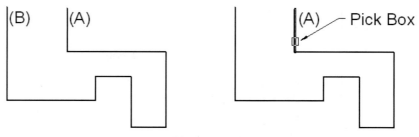

Figure 1–82

Another method of erasing is illustrated on the left in the following image. Instead of using the ERASE command, highlight the line by selecting it with the cursor crosshairs from the "command:" prompt. Square boxes also appear at the endpoints and midpoints of the line. With this line segment highlighted, press DELETE on the keyboard, resulting in removal of the line from the drawing.

With both line segments erased, a new line is constructed from the endpoint at "A" to the endpoint at "B," as shown on the right in the following image.

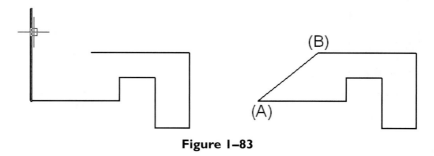

Figure 1–83

SAVING A DRAWING FILE

You can save drawings using the QSAVE, SAVE, and SAVEAS commands. The QSAVE command can be selected from the following:

    From the Standard toolbar

    From the pull-down menu (File > Save)

    From the keyboard (Save)

The SAVEAS command can be selected from any of the following:

    From the pull-down menu (File > Save As)

    From the keyboard (Saveas)

Two of these three commands are found on the File menu, as shown on the left in the following image.

Figure 1-84

SAVE

Selecting Save from the File pull-down menu, as shown in the previous illustration, activates the QSAVE command, which stands for Quick Save. If a drawing file has never been saved and this command is selected, the dialog box shown on the right in the previous image is displayed. Once a drawing file has been initially saved, selecting this command performs an automatic save and no longer displays the Save Drawing As dialog box.

SAVE AS

Using the SAVEAS command always displays the dialog box as shown on the right in the previous image. Simply click the Save button or press ENTER to save the drawing under the current name, displayed in the edit box. This command is more popular for saving the current drawing under an entirely different name. Simply enter the new name in place of the highlighted name in the edit box. Once a drawing is given a new name through this command, it also becomes the new current drawing file.

The ability to exchange drawings with past releases of AutoCAD is still important to many industry users. When the Files of type edit box is selected in the following image, a drop-down list appears. Use this list to save a drawing file in AutoCAD 2004, 2000, and even R14 formats. The user can also save a drawing file as a Drawing Standard (.dws), a Drawing Template (.dwt), and a Drawing Interchange Format (.dxf). The Drawing Interchange Format is especially useful with opening up an AutoCAD drawing in a competitive CAD system.

Figure 1-85

EXITING AN AUTOCAD DRAWING SESSION

It is considered good practice to properly exit any drawing session. One way of exiting is by choosing the Exit option from the File menu, as shown on the left in the following image. You can also use the CLOSE command to end the current AutoCAD drawing session.

Whenever an AutoCAD drawing session is exited, a built-in safeguard gives a second chance to save the drawing, especially if changes were made and a Save was not performed. You may be confronted with three options, illustrated in the AutoCAD alert dialog box shown on the right in the following image. By default, the Yes button is highlighted.

Figure 1-86

If changes were made to a drawing but no Save was performed, the user can now save them by clicking the Yes button before exiting the drawing. Changes to the drawing will be saved and the software exits back to the operating system.

If changes were made but the user does not want to save them, clicking the No button is appropriate. Changes to the drawing will not be saved and the software exits back to the operating system.

If changes are made to the drawing and the Exit option is chosen mistakenly, choosing the Cancel button cancels the Exit option and returns the user to the current drawing.

TUTORIAL EXERCISE: 01_GAGE BLOCK.DWG

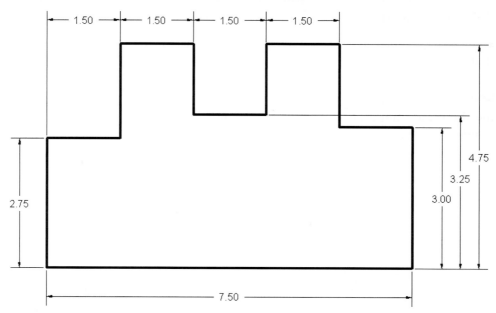

Figure 1-87

Purpose
This tutorial is designed to allow you to construct a one-view drawing of the Gage Block using Polar Tracking and Direct Distance mode.

System Settings
Use the current default settings for the limits of this drawing, (0,0) for the lower left corner and (12,9) for the upper right corner.

Suggested Commands
Open the drawing file called 01_Gage Block. The LINE command will be used entirely for this tutorial in addition to the Polar Tracking and Direct Distance modes. Running Object Snap should already be set to the following modes: Endpoint, Center, Intersection, and Extension.

STEP 1

Begin this tutorial by first turning Polar Tracking on. This can be accomplished by clicking on Polar in the status bar located at the bottom of your display screen. Then activate the LINE command and follow the next series of prompt sequences to complete this object.

Command: **L** (For LINE)

Specify first point: **1,1**
Specify next point or [Undo]: Move cursor to the right and type **7.5**
Specify next point or [Undo]: Move cursor up and type **3**
Specify next point or [Close/Undo]: Move cursor to the left and type **1.5**
Specify next point or [Close/Undo]: Move cursor up and type **1.75**
Specify next point or [Close/Undo]: Move cursor to the left and type **1.5**
Specify next point or [Close/Undo]: Move cursor down and type **1.5**
Specify next point or [Close/Undo]: Move cursor to the left and type **1.5**
Specify next point or [Close/Undo]: Move cursor up and type **1.5**
Specify next point or [Close/Undo]: Move cursor to the left and type **1.5**
Specify next point or [Close/Undo]: Move cursor down and type **2**
Specify next point or [Close/Undo]: Move cursor to the left and type **1.5**
Specify next point or [Close/Undo]: **C** (To close the shape)

TUTORIAL EXERCISE: 01_ANGLE BLOCK.DWG

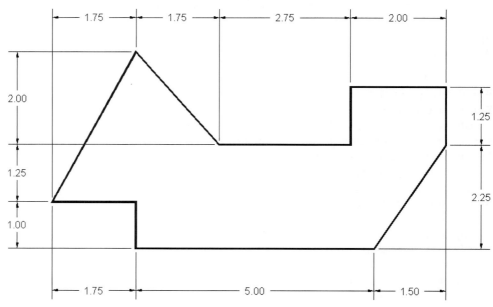

Figure 1-88

This tutorial is designed to allow you to construct a one-view drawing of the Angle Block using a combination of Polar Tracking, Direct Distance mode, and relative coordinates.

System Settings
Use the current default settings for the limits of this drawing, (0,0) for the lower left corner and (12,9) for the upper right corner.

Suggested Commands
Open the drawing file called 01_Angle Block. The LINE command will be used entirely for this tutorial in addition to the Polar Tracking and Direct Distance modes. Running Object Snap should already be set to the following modes: Endpoint, Center, Intersection, and Extension.

STEP 1

Begin this tutorial by first checking that Polar Tracking is turned on. Then activate the LINE command and follow the next series of prompt sequences to complete this object.

Command: **L** (*For LINE*)

Specify first point: **3,1**
Specify next point or [Undo]: *Move cursor to the right and type* **5**
Specify next point or [Undo]: **@1.5,2.25**
Specify next point or [Close/Undo]: *Move cursor up and type* **1.25**
Specify next point or [Close/Undo]: *Move cursor to the left and type* **2**
Specify next point or [Close/Undo]: *Move cursor down and type* **1.25**
Specify next point or [Close/Undo]: *Move cursor to the left and type* **2.75**
Specify next point or [Close/Undo]: **@-1.75,2**
Specify next point or [Close/Undo]: **@-1.75,-3.25**
Specify next point or [Close/Undo]: *Move cursor to the right and type* **1.75**
Specify next point or [Close/Undo]: **C** (*To close the shape*)

TUTORIAL EXERCISE: 01_ANGLE PLATE.DWG

AB=5.00
BC=2.50
CD=3.00
DE=3.00
EF=3.75
FG=3.50
GH=1.75
HI=6.50
IJ=2.50

Figure 1-89

This tutorial is designed to allow you to construct a one-view drawing of the Angle Plate using Polar Tracking set to relative mode and Direct Distance mode.

System Settings
Use the current default settings for the limits of this drawing, (0,0) for the lower left corner and (12,9) for the upper right corner.

Suggested Commands
Open the drawing file called 01_Angle Plate. The LINE command will be used entirely for this tutorial in addition to the Polar Tracking and Direct Distance modes. Polar Tracking will need to be set to a new incremental angle of 15°. Also, Polar Tracking will need to be set to relative mode. Running Object Snap should already be set to the following modes: Endpoint, Center, Intersection, and Extension.

STEP 1

Right-click on POLAR in the status bar at the bottom of the display screen and pick Settings from the menu. When the Drafting Settings dialog box appears, click on the Polar Tracking tab. While in this tab, set the Incremental Angle to 15° under Polar Angle Settings area. Then set the Polar Angle measurement to Relative to last segment, as shown in the following image. Click the OK button to save the settings and exit the Drafting Settings dialog box.

Figure 1-90

STEP 2

After setting the angle increment and relative mode under the Polar Tracking tab of the Drafting Settings dialog box, activate the LINE command and follow the next series of prompt sequences to complete this object.

Command: **L** (For LINE)

Specify first point: **3,1**

Specify next point or [Undo]: *(Move your cursor to the right until the polar tooltip reads 0° and enter 5)*

Specify next point or [Undo]: *(Move your cursor until the polar tooltip reads 45° and enter 2.5)*

Specify next point or [Close/Undo]: *(Move your cursor until the polar tooltip reads 330° and enter 3)*

Specify next point or [Close/Undo]: *(Move your cursor until the polar tooltip reads 60° and enter 3)*

Specify next point or [Close/Undo]: *(Move your cursor until the polar tooltip reads 75° and enter 3.75)*

Specify next point or [Close/Undo]: *(Move your cursor until the polar tooltip reads 75° and enter 3.5)*

Specify next point or [Close/Undo]: *(Move your cursor until the polar tooltip reads 30° and enter 1.75)*

Specify next point or [Close/Undo]: *(Move your cursor until the polar tooltip reads 270° and enter 6.5)*

Specify next point or [Close/Undo]: *(Move your cursor until the polar tooltip reads 120° and enter 2.5)*

Specify next point or [Close/Undo]: **C** (To close the shape)

STEP 3

When finished with this problem, change the Increment angle under Polar Angle Settings back to 90° and the Polar Angle measurement back to Absolute, as shown in the following image.

Figure 1-91

TUTORIAL EXERCISE: 01_PATTERN.DWG

Figure 1-92

Purpose
This tutorial is designed to allow you to construct a one-view drawing of the pattern using Polar Tracking techniques.

System Settings
Use the current default settings for the limits of this drawing, (0,0) for the lower left corner and (12,9) for the upper right corner.

Layers
Create the following layer with the format:

Name   Color   Linetype

Object   Green   Continuous

Suggested Commands
Open the drawing file called 01_Pattern. The LINE command will be used entirely for this tutorial in addition to the Polar Tracking mode. Running Object Snap should already be set to the following modes: Endpoint, Center, Intersection, and Extension. Dynamic Input has been turned off for this exercise.

STEP 1

Open the drawing file 01_Pattern. Activate the Drafting Settings dialog box, click on the Polar Tracking tab, and change the Increment angle setting to 10° (see the following image). Verify that POLAR, OSNAP, and OTRACK are all turned on.

Figure 1–93

STEP 2

Activate the LINE command, select a starting point, move your cursor to the right, and enter a value of 4 units (see the following image). Notice that your line is green because the current layer, Object, has been assigned the green color.

 Command: **L** *(For LINE)*
Specify first point: **3,1**
Specify next point or [Undo]: *(Move your cursor to the right and enter 4)*

Figure 1–94

STEP 3

While still in the LINE command, move your cursor directly up, and enter a value of 3 units (see the following image).

Specify next point or [Undo]: *(Move your cursor up and enter 3)*

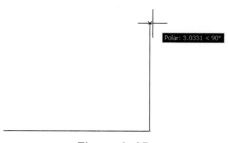

Figure 1–95

STEP 4

While still in the LINE command, move your cursor up and to the right until the tooltip reads 10°, and enter a value of 2 units (see the following image).

Specify next point or [Close/Undo]: *(Move your cursor up and to the right at a 10° angle and enter 2)*

Figure 1–96

STEP 5

While still in the LINE command, move your cursor up and to the left until the tooltip reads 100°, and enter a value of 2 units (see the following image).

Specify next point or [Close/Undo]:
  (Move your cursor up and to the left at a 100° angle and enter 2)

Figure 1–97

STEP 6

While still in the LINE command, first acquire the point at "A." Then move your cursor below and to the left until the Polar value in the tooltip reads 190°, and pick the point at "B" (see the following image).

Specify next point or [Close/Undo]:
  (Acquire the point at "A" and pick the new point at "B")

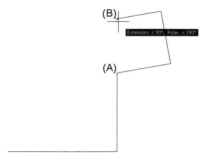

Figure 1–98

STEP 7

While still in the LINE command, move your cursor directly up, and enter a value of 1 unit (see the following image).

Specify next point or [Close/Undo]:
  (Move your cursor up and enter 1)

Figure 1–99

STEP 8

While still in the LINE command, first acquire the point at "C." Then move your cursor to the left until the tooltip reads Polar: < 180°, and pick the point at "D" (see the following image).

Specify next point or [Close/Undo]:
  (Acquire the point at "C" and pick the new point at "D")

Figure 1–100

STEP 9

While still in the LINE command, complete the object by closing the shape (see the following image).

Specify next point or [Close/Undo]: **C** *(To close the shape and exit the LINE command)*

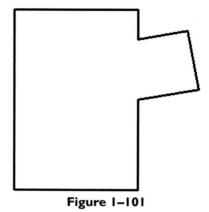

Figure 1–101

TUTORIAL EXERCISE: 01_TEMPLATE.DWG

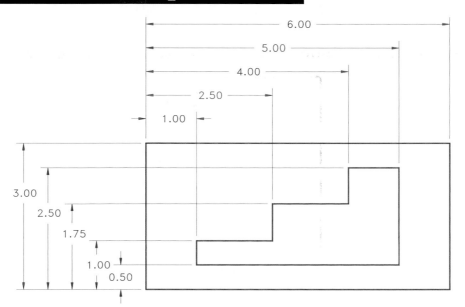

Figure 1–102

Purpose
This tutorial is designed to allow you to construct a one-view drawing of the template using Relative Coordinate mode in combination with the Direct Distance mode. The Direct Distance mode can also be used to perform this exercise.

System Settings
Use the current default settings for the limits of this drawing, (0,0) for the lower left corner and (12,9) for the upper right corner.

Layers
The following layer has already been created:

Name	Color	Linetype
Object	Green	Continuous

Suggested Commands
The LINE command will be used entirely for this tutorial, in addition to a combination of coordinate systems. The ERASE command could be used (however, using this command will force the user to exit out of the LINE command), although a more elaborate method of correcting mistakes while using the LINE command is to execute the Undo option. This option allows the user to delete (or undo) previously drawn lines without having to exit the LINE command. The Object Snap From mode will also be used to construct lines from a point of reference. The coordinate mode of entry and the Direct Distance mode will be used throughout this tutorial exercise.

STEP 1

Open the drawing file 01_Template.Dwg. Then use the LINE command to draw the outer perimeter of the box. One method of constructing the box is to use an absolute coordinate point followed by polar coordinates. Use the following command sequence and image as guides for performing this task.

 Command: **L** *(For LINE)*
[Draw>Line]
Specify first point: **2,2**
Specify next point or [Undo]: **@6<0**
Specify next point or [Undo]: **@3<90**
Specify next point or [Close/Undo]: **@6<180**
Specify next point or [Close/Undo]: **C** *(To close the shape)*

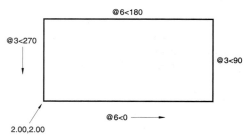

Figure 1–103

An alternate method is to use the Direct Distance mode. Because the box consists of horizontal and vertical lines, Ortho mode is first turned on; this will force all movements to be in the horizontal or vertical direction. To construct a line segment, move the cursor in the direction in which the line is to be drawn and enter the exact value of the line. The line is drawn at the designated distance in the current direction of the cursor. Repeat this procedure for the other lines that make up the box, as shown in the following image.

 Command: **ORTHO**

[Button is found on the bottom of the AutoCAD screen; pressed in is on]
ON/OFF <OFF>: **ON**

 Command: **L** *(For LINE)*
[Draw>Line]
Specify first point: **2,2**
Specify next point or [Undo]: *(Move the cursor to the right and enter a value of 6.00 units)*
Specify next point or [Undo]: *(Move the cursor up and enter a value of 3.00 units)*
Specify next point or [Close/Undo]: *(Move the cursor to the left and enter a value of 6.00 units)*
Specify next point or [Close/Undo]: **C** *(To close the shape)*

Figure 1–104

STEP 2

The next step is to draw the stair step outline of the template using the LINE command again. However, we first need to identify the starting point of the template.

Absolute coordinates could be calculated, but in more complex objects, this would be difficult. A more efficient method is to use the Object Snap From mode along with the Object Snap to Intersection mode to start the line relative to another point. Both Object Snap selections are found on the Object Snap toolbar. Use the following command sequence and image as guides for performing this operation.

   Command: L (For LINE)
[Draw>Line]

   Specify first point: From

   Base point: Int
of *(Pick the intersection at "A")*
<Offset>: **@1.00,0.50**

The relative coordinate offset value begins a new line a distance of 1.00 units in the X direction and 0.50 units in the Y direction.

Continue with the LINE command to construct the stair step outline shown in the following image. Use the Direct Distance mode to accomplish this task. In this example, Direct Distance mode is a good choice to use, especially since all lines are either horizontal or vertical. Use the following command sequence to construct the object with this alternate method.

Specify next point or [Undo]: *(Move the cursor to the right and enter a value of 4.00 units)*
Specify next point or [Undo]: *(Move the cursor up and enter a value of 2.00 units)*
Specify next point or [Close/Undo]: *(Move the cursor to the left and enter a value of 1.00 units)*
Specify next point or [Close/Undo]: *(Move the cursor down and enter a value of 0.75 units)*
Specify next point or [Close/Undo]: *(Move the cursor to the left and enter a value of 1.50 units)*
Specify next point or [Close/Undo]: *(Move the cursor down and enter a value of 0.75 units)*
Specify next point or [Close/Undo]: *(Move the cursor to the left and enter a value of 1.50 units)*
Specify next point or [Close/Undo]: **C** *(To close the shape)*

Figure 1–105

Figure 1–106

STEP 3

Dimensions may be added at a later time upon the request of your instructor. The completed problem is shown in the following image.

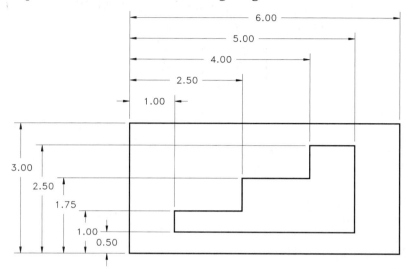

Figure 1–107

TUTORIAL EXERCISE: 01_POLYGON.DWG

Figure 1-108

Purpose
This tutorial is designed to allow you to construct a polygon using existing geometry as a guide. Polar Tracking and Object Snap Tracking will be used to accomplish this task. This drawing is available on the CD under the file name 01_Polygon.Dwg.

System Settings
No special system settings need to be made to this drawing file.

Layers
The drawing file 01_Polygon.Dwg has the following layers already created:

Name	Color	Linetype
Center	Yellow	Center
Dimension	Yellow	Continuous
Object	Green	Continuous
Defpoints	White	Continuous

Suggested Commands
The POLYGON command is used to start this tutorial exercise. To identify the center of the polygon, Polar Tracking will be used along with the Osnap Extension mode and the Temporary Tracking Point tool.

STEP 1

Open the drawing file 01_Polygon.Dwg and activate the POLYGON command. Polar tracking and Object Snap tracking will be used to identify the center of the polygon (see the following image).

Command: **POL** *(For POLYGON)*
Enter number of sides <4>: **6**

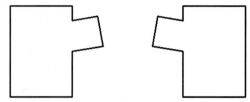

Figure 1–109

STEP 2

At the prompt "Specify center of polygon or [Edge]:" begin locating the center of the polygon by acquiring the endpoint at "A" in the following image. This creates a tracking path at the same angle as the line selected. Move your cursor above and to the right as illustrated in this figure.

Specify center of polygon or [Edge]:
   (Acquire the point at "A")

Figure 1–110

STEP 3

While still at the prompt "Specify center of polygon or [Edge]:" acquire the endpoint at "B" in the following image. This creates another tracking path at the same angle as the line selected. Move your cursor above and to the left as illustrated in the following image.

(Acquire the point at "B" in the following image)

Figure 1–111

STEP 4

Now move both temporary tracking points to meet at their intersection, as shown in the following image, but do not pick a point at this location. The polygon needs to be constructed 3 units above this point. Click the Temporary Tracking Point Osnap mode and then pick a point at the intersection of "C." Notice the following prompt sequence that appears:

Specify temporary OTRACK point: *(Pick a point at the intersection at "C")*

Figure 1–112

STEP 5

Move your cursor directly above this tracking point, as shown in the following image, and enter a value of 3 units. This will locate the center of the polygon 3 units directly above the last tracking point. Complete the construction by following the remaining POLYGON command prompts:

*(Move your cursor up and enter **3**)*
Enter an option [Inscribed in circle/
   Circumscribed about circle] <I>:
   (Press ENTER *to accept this default value)*
Specify radius of circle: **1.50**

The completed construction task is illustrated in the following image.

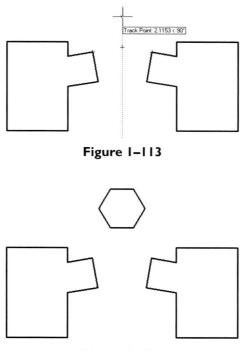

Figure 1-113

Figure 1-114

PROBLEMS FOR CHAPTER 1

Directions for Problems 1–1 and 1–2

Supply the appropriate absolute, relative, and/or polar coordinates for these figures in the table that follows each object.

PROBLEM 1-1

	Absolute	Relative	Polar
From Pt (1)	.50,.50	.50,.50	.50,.50
To Pt (2)	___ , ___	___ , ___	___ , ___
To Pt (3)	___ , ___	___ , ___	___ , ___
To Pt (4)	___ , ___	___ , ___	___ , ___
To Pt (5)	___ , ___	___ , ___	___ , ___
To Pt (6)	___ , ___	___ , ___	___ , ___
To Pt (7)	___ , ___	___ , ___	___ , ___
To Pt (8)	___ , ___	___ , ___	___ , ___
To Pt (9)	___ , ___	___ , ___	___ , ___
To Pt (10)	___ , ___	___ , ___	___ , ___
To Pt (11)	___ , ___	___ , ___	___ , ___
To Pt (12)	___ , ___	___ , ___	___ , ___
To Pt (13)	___ , ___	___ , ___	___ , ___
To Pt (14)	___ , ___	___ , ___	___ , ___
To Pt (15)	___ , ___	___ , ___	___ , ___
To Pt (16)	___ , ___	___ , ___	___ , ___
To Pt (17)	___ , ___	___ , ___	___ , ___
To Pt	Enter	Enter	Enter

PROBLEM 1-2

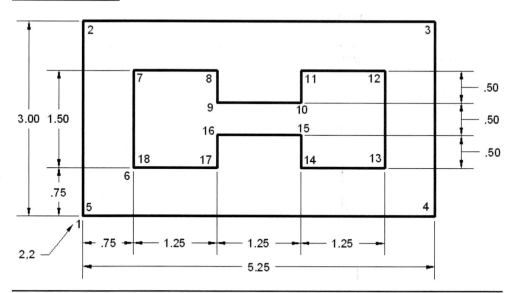

	Absolute	Relative	Polar
From Pt (1)	2.00,2.00	2.00,2.00	2.00,2.00
To Pt (2)	___,___	___,___	___,___
To Pt (3)	___,___	___,___	___,___
To Pt (4)	___,___	___,___	___,___
To Pt (5)	___,___	___,___	___,___
To Pt (6)	___,___	___,___	___,___
To Pt (7)	___,___	___,___	___,___
To Pt (8)	___,___	___,___	___,___
To Pt (9)	___,___	___,___	___,___
To Pt (10)	___,___	___,___	___,___
To Pt (11)	___,___	___,___	___,___
To Pt (12)	___,___	___,___	___,___
To Pt (13)	___,___	___,___	___,___
To Pt (14)	___,___	___,___	___,___
To Pt (15)	___,___	___,___	___,___
To Pt (16)	___,___	___,___	___,___
To Pt (17)	___,___	___,___	___,___
To Pt (18)	___,___	___,___	___,___
To Pt	Enter	Enter	Enter

Directions for Problems 1–3 through 1–10

Construct one-view drawings from the following figures using the LINE command along with Direct Distance mode, relative coordinates, and/or polar coordinates.

PROBLEM 1-3

PROBLEM 1-4

PROBLEM 1-5

PROBLEM 1-6

PROBLEM 1-7

PROBLEM 1-8

PROBLEM 1-9

PROBLEM 1-10

CHAPTER 2

Drawing Setup and Organization

Chapter 2 will cover a number of drawing setup commands. The user will learn how to assign different units of measure with the UNITS command. The default sheet size can also be increased on the display screen with the LIMITS command. Controlling the grid and snap will be briefly discussed through the Snap and Grid tab located in the Drafting Settings dialog box. The major topic of this chapter is the discussion of layers. All options of the Layer Properties Manager dialog box will be demonstrated, along with the ability to assign color, linetype, and lineweight to layers. The Layer Control box and Properties toolbar will provide easy access to all layers, colors, linetypes, and lineweights used in a drawing. Controlling the scale of linetypes through the LTSCALE command will also be discussed. Advanced layer tools such as Filtering Layers and creating Layer States will be introduced. This chapter concludes with a section on creating template files.

SETTING DRAWING UNITS

The Drawing Units dialog box is available for interactively setting the units of a drawing. Choosing Units from the Format pull-down menu activates the dialog box illustrated in the following image.

Figure 2–1

By default, decimal units are set along with four-decimal-place precision. The following systems of units are available: Architectural, Decimal, Engineering, Fractional, and Scientific (see the left side of the following image). Architectural units are displayed in feet and fractional inches. Engineering units are displayed in feet and decimal inches. Fractional units are displayed in fractional inches. Scientific units are displayed in exponential format.

Methods of measuring angles supported in the Drawing Units dialog box include Decimal Degrees, Degrees/Minutes/Seconds, Grads, Radians, and Surveyor's Units (see the middle of the following image). Accuracy of decimal degree for angles may be set between zero and eight places.

Selecting Direction… in the main Drawing Units dialog box displays the Direction Control dialog box, shown on the right in the following image. This dialog box is used to control the direction of angle zero in addition to changing whether angles are measured in the counterclockwise or clockwise direction. By default, angles are measured in the counterclockwise direction.

Figure 2–2

ENTERING ARCHITECTURAL VALUES FOR DRAWING LINES

The method of entering architectural values in feet and inches is a little different from the method for entering them in decimal places. To designate feet, you must enter the apostrophe symbol (') from the keyboard after the number. For example, "ten feet" would be entered as (10'), as shown in the following image. When feet and inches are necessary, you cannot use the Spacebar to separate the inch value from the foot value. For example, thirteen feet seven inches would be entered as (13'7), as shown in the following image. If you do use the Spacebar after the (13') value, this is interpreted as the ENTER key and your value is accepted as (1'). If you have to enter feet, inches, and fractions of an inch, use the hyphen (-) to separate the inch value from the fractional value. For example, to draw a line seventeen feet eleven and one-quarter inches, you would enter the following value in at the keyboard: (17'11-1/4). See the following image. Placing the inches symbol (") is not required since all numbers entered without the foot symbol are interpreted as inches.

Figure 2–3

 Try It! – Open the drawing file 02_Architectural. Verify that the units setting is in architectural units by activating the Drawing Units dialog box. Using the following image as a guide, follow the prompt sequence below to enter numeric values in feet and inches. The command sequence illustrates the use of polar coordinates. The Direct Distance mode of entry can also be used to construct the shape using architectural values.

Command: **L** *(For LINE)*

Specify first point: **4',2'**
Specify next point or [Undo]: **@10'<0**
Specify next point or [Undo]: **@7'6<90**
Specify next point or [Close/Undo]: **@13'7-1/2<0**
Specify next point or [Close/Undo]: **@4'9-1/2<90**
Specify next point or [Close/Undo]: **@7'<180**
Specify next point or [Close/Undo]: **@8'<90**
Specify next point or [Close/Undo]: **@16'7-1/2<180**
Specify next point or [Close/Undo]: **C** *(To close the shape and exit the LINE command)*

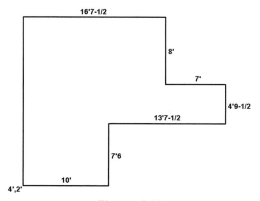

Figure 2–4

SETTING THE LIMITS OF THE DRAWING

By default, the size of the drawing screen in a new drawing file measures 12 units in the X direction and 9 units in the Y direction. This size may be ideal for small objects, but larger drawings require more drawing screen area. Use the LIMITS command for increasing the size of the drawing area. Select this command by picking Drawing Limits from the Format pull-down menu (see the following image); you can also enter this command directly at the command prompt by typing "Limits." Illustrated in the following image is a single-view drawing that fits on a screen size of 24 units in the X direction and 18 units in the Y direction. Follow the next command sequence to change the limits of a drawing.

Figure 2–5

Command: **LIMITS**
ON/OFF/<Lower left corner>: <0.0000,0.0000>: *(Press* ENTER *to accept this value)*
Upper right corner <12.0000,9.0000>: **24,18**

Changing the limits does not change the current viewing area in the display screen. Before continuing, perform a ZOOM-All to change the size of the display screen to reflect the changes in the limits of the drawing. You can find ZOOM-All on the View menu. It is also found on the Zoom toolbar, which you can activate by choosing Toolbars… from the View pull-down menu and placing a check in the box next to Zoom.

Command: **Z** *(For Zoom)*
All/Center/Dynamic/Extents/Left/Previous/Vmax/Window/<Scale(X/XP)>: **A** *(For All)*

CALCULATING THE LIMITS OF THE DRAWING FOR MODEL SPACE

Before you construct any drawing, it is considered good practice to first calculate the limits or sheet size of the drawing. Two items are needed to accomplish this task: the scale of the drawing and the paper size the drawing will be plotted out on. Most design professionals who use CAD systems in industry already have some idea as to the scale of the drawing and whether it will fit on a certain size sheet of paper.

For the purposes of this example, the following steps will illustrate how to determine the drawing limits for a D-size sheet and at a scale of 1/4″ = 1′0″. This scale is commonly used to construct residential house plans. Because the following steps use this scale and sheet, the method works for calculating limits at any scale and on any sheet of paper.

STEP 1

Determine that a D-size drawing sheet measures 36″ × 24″.

STEP 2

Determine the scale factor (essentially a multiplication factor) based on the drawing scale. In computer-aided design, most drawings are created full size. The limits set a full-size drawing sheet used to match the full-size drawing. The scale factor for the scale 1/4″ = 1′0″ is 48 (1′ divided by 1/4). Use this multiplier to increase the size of the drawing space, enabling an operator to produce a full-size drawing on the computer. This is vastly different from manual drawing practices, where an object had to be reduced to 1/48th of its normal size in order to fit on the D-size paper at a scale of 1/4″ = 1′0″.

STEP 3

Using the multiplication factor of 48, increase the size of the D-size sheet of paper:

Paper Size from the Plot dialog box (inches)	36	24
Multiplied by a scale factor of	48	48
New drawing space (inches)	1,728	1,152

Convert the previous values to feet:

1,728/12 = 144′ 1,152/12 = 96′

The limits of the drawing become 144′ wide and 96′ high.

STEP 4

Use the Drawing Units dialog box to set the current units to Architectural (Decimal mode does not accept feet as valid units). Next, use the LIMITS command and enter the lower left corner as 0,0 and the upper right corner as 144′,96′. Once the limits have been set, use the ZOOM command and the All option to display the new drawing space. To view the active drawing area, use the RECTANG command and construct a rectangular border around the drawing using 0,0 as the lower left corner and 144´,96´ as the upper right corner. Construct the drawing inside the rectangular area.

Decimal Scales

SCALE	SCALE FACTOR	ANSI "A" 11"x 8.5"	ANSI "B" 17"x11"	ANSI "C" 22"x17"	ANSI "D" 34"x22"	ANSI "E" 44"x34"
.125=1	8	88,68	136,88	176,136	272,176	352,272
.25=1	4	44,34	68,44	88,68	136,88	176,136
.50=1	2	22,17	34,22	44,34	68,44	88,68
1.00=1	1	11,8.5	17,11	22,17	34,22	44,34
2.00=1	.50	5.5,4.25	8.5,5.5	11,8.5	17,11	22,17

Architectural Scales

SCALE	SCALE FACTOR	ANSI "A" 11"x 8.5"	ANSI "B" 17"x11"	ANSI "C" 22"x17"	ANSI "D" 34"x22"	ANSI "E" 44"x34"
1/8"=1'-0"	96	88',68'	136',88'	76',136'	272',176'	352',272'
1/4"=1'-0"	48	44',34'	68',44'	88',68'	136',88'	176',136'
1/2"=1'-0"	24	22',17'	34',22'	44',34'	68',44'	88',68'
3/4"=1'-0"	16	14.7',11.3'	22.7',14.7'	29.3',22.7'	45.3',29.3'	58.7',45.3'
1"=1'-0"	12	11',8.5'	17',11'	22',17'	34',22'	44',34'
2"=1'-0"	6	5.5',4.25'	8.5',5.5'	11',8.5'	17',11'	22',17'

Metric Scales

SCALE	SCALE FACTOR	ANSI "A" 11"x 8.5"	ANSI "B" 17"x11"	ANSI "C" 22"x17"	ANSI "D" 34"x22"	ANSI "E" 44"x34"
1=1	25.4 mm	279,216	432,279	559,432	864,559	1118,864
1=10	10 cm	110,85	170,110	220,170	340,220	440,340
1=20	20 cm	220,170	340,220	440,340	680,440	880,680
1=50	50 cm	550,425	850,550	1100,850	1700,1100	2200,1700
1=100	100 cm	1100,850	1700,1100	2200,1700	3400,2200	4400,3400

THE QUICK SETUP WIZARD

A number of wizards are available to assist with setting of units and drawing limits. From the Create New Drawing dialog box, click on the Use a Wizard button, as shown in the following image.

Figure 2–6

Two wizards are available that allow you to perform basic setup operations, namely Quick and Advanced. Both wizard screens are illustrated in the following image.

 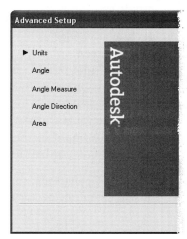

Figure 2–7

The Quick wizard allows you to change the units and drawing area (limits). A more detailed description of these two items is listed in the following table.

Quick Setup Wizard

Category	Description
Units	Select the desired units of measurement. Choices are Decimal, Engineering, Architectural, Fractional, and Scientific.
Area	Drawing area is specified in real-world units.

The Advanced wizard contains five areas that allow you to change the units, angle, angle measure, angle direction, and area (limits) of the drawing. A more detailed description of these five items is listed in the following table.

Advanced Setup Wizard

Category	Description
Units	Select the desired units of measurement. Choices are Decimal, Engineering, Architectural, Fractional, and Scientific.
Angle	Select the measurement style for the precision of angles. Choices are Decimal Degrees, Deg/Min/Sec, Grads, Radians, and Surveyor.
Angle Measure	Select the direction for the angle measurement. Choices are East, North, West, South, and Other.
Angle Direction	Select the orientation for the angle measure. Choices are counterclockwise and clockwise.
Area	Drawing area is specified in real-world units.

USING GRID IN A DRAWING

Use grid to get a relative idea as to the size of objects. Grid is also used to define the size of the display screen originally set by the LIMITS command. The dots that make up the grid will never plot out on paper even if they are visible on the display screen. You can turn grid dots on or off by using the GRID command or by pressing F7, or by single-clicking on the GRID toggle, located in the status bar at the bottom of the display screen. By default, the grid is displayed in 0.50-unit intervals similar to the following image on the left. Illustrated in the following image on the right is a grid that has been set to a value of 0.25, or half its original size.

 Try It! – Open the drawing file 02_Grid. Use the following command sequence and illustrations below for using the GRID command.

Command: **GRID**
Specify grid spacing(X) or [ON/OFF/Snap/Aspect] <0.5000>: **On**

Command: **GRID**
Specify grid spacing(X) or [ON/OFF/Snap/Aspect] <0.5000>: **0.25**

While a grid is a useful aid for construction purposes, it may reduce the overall performance of the computer system. If the grid is set to a small value and it is visible on the display screen, it takes time for the grid to display. If a very small value is used for the grid, a prompt will warn you that the grid value is too small to display on the screen.

Figure 2–8

SETTING A SNAP VALUE

It is possible to have the cursor lock on to or snap to a grid dot, as illustrated in the following image on the left; this is the purpose of the SNAP command. By default, the current snap spacing is 0.50 units. Even though a value is set, the snap must be turned on for the cursor to be positioned on a grid dot. You can accomplish this by using the SNAP command, as shown in the following sequence, by pressing F9, or by single-clicking on SNAP toggle in the status bar at the bottom of the display screen.

Some drawing applications require that the snap be rotated at a specific angular value (see the following image on the right). Changing the snap in this fashion also affects the cursor. Use the following command sequence for rotating the snap.

 Try It! – Open the drawing file 02_Snap. Use the command sequence below and the illustration in the following image for using the SNAP command.

Command: **SN** *(For SNAP)*
Specify snap spacing or [ON/OFF/Aspect/Rotate/Style/Type] <0.5000>: **On**

Command: **SN** *(For SNAP)*
Specify snap spacing or [ON/OFF/Aspect/Rotate/Style/Type] <0.5000>: **R** *(For Rotate)*
Specify base point <0.0000,0.0000>: *(Press* ENTER *to accept this value)*
Specify rotation angle <0>: **30**

One application of rotating snap is for auxiliary views in which an inclined surface needs to be projected in a perpendicular direction. This will be explained and demonstrated in chapter 10.

Figure 2–9

 Tip: To affect both the grid and the snap, set the grid to a value of zero (0). When setting a new snap value, this value is also used for the spacing of the grid.

CONTROLLING SNAP AND GRID

Right-clicking on the SNAP or GRID toggle in the status bar displays the menu as shown in the following image on the left. Clicking on Settings displays the Drafting Settings dialog box, shown in the image on the right. Use this dialog box for making changes to the grid and snap settings. The Snap type & style area controls whether the isometric grid is present or not.

Figure 2–10

GRID BEHAVIOR TOOLS

Additional tools are available under the Snap and Grid tab of the Drafting Settings dialog box to add more control for the display of grid.

Adaptive Grid

When the Adaptive Grid mode is checked and you zoom out of the display, the density of the grid changes. Illustrated on the left in the following image is a display screen where a zoom-out operation was performed. Notice the density, or spacing, of the grid dots when compared to the illustration on the right in the following image. Here the Adaptive Grid control is turned off. When you zoom out, the density, or grid spacing, retains its original values, which are set in the Drafting Settings dialog box. It also means that if you zoom out too much with Adaptive Grid turned off, you will get a message that the grid is too dense to display.

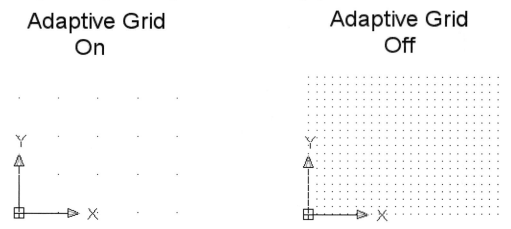

Figure 2–11

Allow Subdivision Below Grid Spacing

The Allow Subdivision Below Grid Spacing grid control mode requires that you change from a 2D to a 3D visual style to see the results. Illustrated on the left is a display that is currently set to a 3D visual style; also, the control that allows subdivision below grid spacing is turned off. When you zoom in to your drawing, the grid spacing remains the same. At some point, if you keep zooming in to your display, the grid will disappear. If you want the grid to show no matter how much you zoom in, place a check in the box to Allow Subdivision Below Grid Spacing. The results are illustrated on the right in the following image. No matter how much you zoom in, the grid changes spacing, allowing it to always be displayed. The only additional requirement is that you change to the 3D visual style.

Figure 2–12

Display Grid Beyond Limits

When the Display Grid Beyond Limits option is checked, the grid will display whenever you zoom beyond the area specified by the LIMITS command.

Follow Dynamic UCS

When the Follow Dynamic UCS option is checked and you create a new user coordinate system to create a 3D model, the grid will change and follow the XY plane.

 Tip: If you change the grid X and Y spacing to 0 and then change the snap X and Y values to the desired spacing value, the grid will conform to the current snap value.

CONTROLLING DYNAMIC INPUT

Right-clicking on DYN, located in the status bar, and picking settings will launch the Dynamic Input tab of the Drafting Settings dialog box, as shown in the following image. Various check boxes are available to turn on or off the Pointer Input (the absolute coordinate display of your cursor position when you are inside a command) and the Dimension Input (the display of distance and angle information for commands that support this type of input). You can also control the appearance of the Dynamic Prompts. This can take the form of changing the background color or even assigning a level of transparency to the display of the dynamic input.

Figure 2–13

Clicking on the Settings button under Pointer Input and Dimension Input will launch the respective dialog boxes, as shown in the following image. Use the settings located in both of these dialog boxes to further control Pointer and Dimension Inputs.

Figure 2–14

THE ALPHABET OF LINES

Engineering drawings communicate information through the use of lines and text, which, if used appropriately, accurately convey a project from design to construction. Before you construct engineering drawings, the quality of the lines that make up the drawing must first be discussed. Some lines of a drawing should be made thick; others need to be made thin. This is to emphasize certain parts of the drawing and it is controlled through a line quality system. Illustrated in the following image is a two-view drawing of an object complete with various lines that will be explained further.

The most important line of a drawing is the Object line, which outlines the basic shape of the object. Because of their importance, Object lines are made thick and continuous so they stand out among the other lines in the drawing. It does not mean that the other lines are considered unimportant; rather, the Object line takes precedence over all other lines.

The Cutting Plane line is another thick line; it is used to determine the placement in the drawing where an imaginary saw will cut into the drawing to expose interior details. It stands out by being drawn as a series of long dashes separated by spaces. Arrowheads determine how the adjacent view will be looked at. This line will be discussed in greater detail in chapter 9, "Creating Section Views."

The Hidden line is a medium-weight line used to identify edges that become invisible in a view. It consists of a series of dashes separated by spaces. Whether an edge is visible or invisible, it still must be shown with a line.

Figure 2–15

The Dimension line is a thin line used to show the numerical distance between two points. The dimension text is placed within the dimension line, and arrowheads are placed at opposite ends of the dimension line.

The Extension line is another thin continuous line, used as a part of the overall dimension. Extension lines show the edge being dimensioned. The vertical extension lines are used to indicate the horizontal dimension distance.

When you use the Cutting Plane line to create an area to cut or slice, the surfaces in the adjacent view are section lined using the Section line, a thin continuous line.

Another important line used to identify the centers of circles is the Center line. It is a thin line consisting of a series of long and short dashes. It is a good practice to dimension to center lines; for this reason center lines, extension lines, and dimensions are made the same line thickness.

The Phantom line consists of a thin line made with a series of two short dashes and one long dash. It is used to simulate the placement or movement of a part or component without actually detailing the component.

The Break line is a thin line with a "zigzag" symbol used to establish where an object is broken to simulate a continuation of the object.

ORGANIZING A DRAWING THROUGH LAYERS

As a means of organizing objects, a series of layers should be devised for every drawing. You can think of layers as a group of transparent sheets that combine to form the completed drawing. The illustration on the left in the following image displays a drawing consisting of object lines, dimension lines, and border. Organizing these three drawing components by layers is illustrated on the right. Only the drawing border occupies a layer called "Border." The object lines occupy a layer called "Object," and the dimension lines are drawn on a layer called "Dimension." At times, it may be necessary to turn off the dimension lines for a clearer view of the object. Creating all dimensions on a specific layer will allow you to turn off the dimensions while viewing all other objects on layers that are still turned on.

Figure 2–16

THE LAYER PROPERTIES MANAGER DIALOG BOX

An efficient way of creating and managing layers is through the Layer Properties Manager dialog box. Choose Layer... from the Format menu, as in the following image, to display the Layer Properties Manager dialog box. Another way of activating this dialog box is by picking the Layers button located in the Layers toolbar, as shown in the following image.

Figure 2–17

Once the dialog box displays, buttons located along the right side of this dialog box are designed for creating new layers, making an existing layer current to draw on, deleting a layer that is unused, or showing a series of details regarding a selected layer (Layer 0 cannot be deleted).

In addition to these buttons, a number of layer states exist that enable you to perform the following operations: turning layers on or off, freezing or thawing layers, locking or unlocking layers, assigning a color, linetype, lineweight, and plot style to a layer or group of layers, and choosing whether to plot a layer or group of layers (see the following image). A brief explanation of each layer state is provided below:

Figure 2–18

On/Off – Makes all objects created on a certain layer visible or invisible on the display screen. The On state is symbolized by a yellow lightbulb. The Off state has a lightbulb icon shaded black.

Freeze – This state is similar to the Off mode; objects frozen will appear invisible on the display screen. Freeze, however, is considered a major productivity tool used to speed up the performance of a drawing. This is accomplished by not calculating any frozen layers during drawing regenerations. A snowflake icon symbolizes this layer state.

Thaw – This state is similar to the On mode; objects on frozen layers will reappear on the display screen when they are thawed. The sun icon symbolizes this layer state.

Lock – This state allows objects on a certain layer to be visible on the display screen while protecting them from accidentally being modified through an editing command. A closed padlock icon symbolizes this layer state.

Unlock – This state unlocks a previously locked layer and is symbolized by an open padlock icon.

Color – This state displays a color that is assigned to a layer and is symbolized by a square color swatch along with the name of the color. By default, the color white is assigned to a layer.

Linetype – This state displays the name of a linetype that is assigned to a layer. By default, the Continuous linetype is assigned to a layer.

Lineweight – This state sets a lineweight to a layer. An image of this lineweight value will be visible in this layer state column.

Plot Style – A plot style allows you to override the color, linetype, and lineweight settings made in the Layer Properties Manager dialog box. Notice how this area is grayed out. When working with a plot style that is color dependent, you cannot change the plot style. Plot styles will be discussed in greater detail later in this book.

Plot – This layer state controls which layers will be plotted. The presence of the printer icon symbolizes a layer that will be plotted. A printer icon with a red circle and diagonal slash signifies a layer that will not be plotted.

CREATING NEW LAYERS

Three buttons are available at the top of the Layer Properties Manager dialog box used for creating and deleting layers and for making a layer current. These layer options can also be activated by pressing the ALT key along with a specific letter, as shown in the following image.

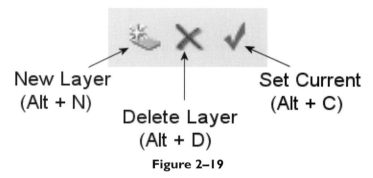

Figure 2–19

Clicking on the New button of the Layer Properties Manager dialog box creates a new layer called Layer1, which displays in the layer list box as shown in the following image. Since this layer name is completely highlighted, you may elect to change its name to something more meaningful, such as a layer called Object or Hidden to hold all object or hidden lines in a drawing.

Figure 2–20

Illustrated in the following image is the result of changing the name Layer1 to Object. You could also have entered "OBJECT" or "object" and these names would appear in the dialog box.

Figure 2–21

You can also be really descriptive with layer names. In the following image, a layer has been created called "Section" (this layer is designed to hold crosshatch lines). You are allowed to add spaces and other characters in the naming of a layer. Because of space limitations, the entire layer name will not display until you move your cursor over the top of the layer name. As a result, the layer name will appear truncated, as in the layer "Object(T...tion lines)" in the following image.

Figure 2–22

If more than one layer needs to be created, it is not necessary to continually click on the New button. Because this operation will create new layers, a more efficient method would be to perform either of the following operations: After creating a new Layer1, change its name to Dimension followed by a comma (,). This will automatically create a new Layer1 (see the following image). Change its name followed by a comma and a new layer is created, and so on. You could also have pressed ENTER twice, which would create another new layer.

Figure 2–23

APPLYING CHANGES MADE TO LAYERS

In the following image of the Layer Properties Manager, notice the appearance of the Apply button near the bottom of the dialog box. This button allows you to update the information contained in the dialog box without actually closing the dialog box and then reopening it to observe the changes. In this figure, the highlighted layers are flagged for deletion. Instead of clicking the OK button and returning to the drawing editor, the Apply button is selected.

Figure 2–24

The results are illustrated in the following image. Clicking the Apply button deleted the layers without you having to exit the dialog box and reenter to view the results.

The Apply button will also update this list when creating new layers, or assigning colors, linetypes, or lineweights.

Figure 2–25

ASSIGNING COLOR TO LAYERS

Once you select a layer from the list box of the Layer Properties Manager dialog box and the color swatch is selected in the same row as the layer name, the Select Color dialog box shown in the following image is displayed. Three tabs allow you to select three different color groupings. The three groupings (Index Color, True Color, and Color Books) are described as follows.

INDEX COLOR TAB

This tab, shown in the following image, allows you to make color settings based on 255 AutoCAD Color Index (ACI) colors. Standard, Gray Shades, and Full Color Palette areas are available for you to choose colors from.

Figure 2–26

TRUE COLOR TAB

Use this tab to make color settings using true colors, also known as 24-bit color. Two color models are available for you to choose from, namely Hue, Saturation, and Luminance (HSL), or Red, Green, and Blue (RGB). Through this tab, you can choose from over sixteen million colors, as shown in the left of the following image.

COLOR BOOKS TAB

Use the Color Books tab to select colors that use third-party color books (such as Pantone) or user-defined color books. You can think of a color book as similar to those available in hardware stores when selecting household interior paints. When you select a color book, the name of the selected color book will be identified in this tab, as shown on the right of the following image.

Note: The Index tab will be used throughout this text. However, you are encouraged to experiment with the True Color and Color Books tabs.

Figure 2–27

ASSIGNING LINETYPES TO LAYERS

Selecting the name Continuous next to the highlighted layer activates the Select Linetype dialog box, as shown in the following image. Use this dialog box to dynamically select preloaded linetypes to be assigned to various layers. By default, the Continuous linetype is loaded for all new drawings.

Figure 2–28

To load other linetypes, click on the Load button of the Select Linetype dialog box; this displays the Load or Reload Linetypes dialog box, as shown in the following image.

Use the scroll bars to view all linetypes contained in the file ACAD.LIN. Notice that, in addition to standard linetypes such as HIDDEN and PHANTOM, a few linetypes are provided that have text automatically embedded in the linetype. As the linetype is drawn, the text is placed depending on how it was originally designed. Notice also three variations of Hidden linetypes; namely HIDDEN, HIDDEN2, and HIDDENX2. The HIDDEN2 represents a linetype where the distances of the dashes and spaces in between each dash are half of the original HIDDEN linetype. HIDDENX2 represents a linetype where the distances of the dashes and spaces in between each dash of the original HIDDEN linetype are doubled. Click on the desired linetypes to load. When finished, click on the OK button.

The loaded linetypes now appear in the Select Linetype dialog box, as in the following image. It must be pointed out that the linetypes in this list are only loaded into the drawing and are not assigned to a particular layer. Clicking on the linetype in this dialog box will assign this linetype to the layer currently highlighted in the Layer Properties Manager dialog box.

Figure 2–29

ASSIGNING LINEWEIGHT TO LAYERS

Selecting the name Default under the Lineweight heading of the Layer Properties Manager dialog box activates the Lineweight dialog box, as shown in the following image. Use this to attach a lineweight to a layer. Lineweights are very important to a drawing file—they give contrast to the drawing. As stated earlier, the object lines should stand out over all other lines in the drawing. A thick lineweight would then be assigned to the object line layer.

Figure 2–30

In the following image on the left, a lineweight of 0.50 mm has been assigned to all object lines and a lineweight of 0.30 has been assigned to hidden lines. However, all lines in this figure appear to consist of the same lineweight. This is due to the lineweight feature being turned off. Notice in this figure the LWT button in the Status bar. Use this button to toggle on or off the display of assigned lineweights. Clicking on the LWT button, as shown in the following image on the right, will turn the lineweight function on. The results are displayed in this illustration.

Figure 2–31

THE LINEWEIGHT SETTINGS DIALOG BOX

You have probably already noticed that when you turn on the lineweights through the LWT button, all lineweights may appear thick. In a complicated drawing, the lineweights could be so thick that it would be difficult to interpret the purpose of the drawing. To give your lineweights a more pleasing appearance, a dialog box is available to control the display of your lineweights. Clicking on Lineweight in the Format pull-down menu, as shown in the following image, will display the Lineweight Settings dialog box. Notice the position of the slider bar in the Adjust Display Scale area of the dialog box. Sliding the bar to the left near the Min setting will reduce the width of all lineweights. Try experimenting with this on any lineweight assigned to a layer. If you think your lineweights appear too thin, slide the bar to the right near the Max setting and observe the results. Continue to adjust your lineweights until they have a pleasing appearance in your drawing.

Figure 2–32

THE LINETYPE MANAGER DIALOG BOX

Choosing Linetype… from the Format pull-down menu, as shown in the following image, displays the Linetype Manager dialog box. This dialog box is designed mainly to preload linetypes.

Clicking on the Load… button activates the Load or Reload Linetypes dialog box illustrated in a previous segment of this chapter. You can select individual linetypes in this dialog box or load numerous linetypes at once by pressing CTRL and clicking on each linetype. Clicking the OK button will load the selected linetypes into the dialog box.

Figure 2–33

The Linetype Manager dialog box, as shown in the following image, also displays a number of extra linetype details. The Global scale factor value has the same effect as the LTSCALE command. Also, setting a different value in the Current object scale box can scale the linetype of an individual object.

Figure 2–34

THE LAYERS TOOLBAR

The Layers toolbar provides an area to better control the layer properties or states. This toolbar is illustrated in the following image. The Layer Control area will now be discussed in greater detail.

Figure 2–35

CONTROL OF LAYER PROPERTIES

Clicking in the long edit box next to the Layer button cascades all layers defined in the drawing in addition to their properties, identified by symbols (see the following image). The presence of the lightbulb signifies that the layer is turned on. Clicking on the lightbulb symbol turns the layer off. The sun symbol signifies that the layer is thawed. Clicking on the sun turns it into a snowflake symbol, signifying that the layer is now frozen. The padlock symbol controls whether a layer is locked or unlocked. By default, all layers are unlocked. Clicking on the padlock changes the symbol to display the image of a locked symbol, signifying that the layer is locked.

Study the following image for a better idea of how the symbols affect the state of certain layers.

Figure 2–36

THE PROPERTIES TOOLBAR

In addition to the Layers toolbar, a Properties toolbar is also available. This toolbar provides three areas to better access options for the control of colors, linetypes, and lineweights. All three areas are illustrated in the following image.

Figure 2–37

 Note: While the Properties toolbar allows you to change color, linetype, and lineweight on the fly, it is considered poor practice to do so. Check to see that each category in the Properties toolbar reads "ByLayer," which means that layers control this category.

MAKING A LAYER CURRENT

Various methods can be employed to make a layer current to draw on. Select a layer in the Layer Properties Manager dialog box and then click on the Current button to make the layer current.

Picking a layer name from the Layer Control box, as shown on the left in the following image, will also make the layer current.

The Make Object's Layer Current button, as shown on the right in the following image, allows you to make a layer the new current layer by just clicking on an object in the drawing. The layer is now made current based on the layer of the selected object.

Figure 2–38

Drawing Setup and Organization

THE STATUS OF LAYERS

An additional layer property called Status is illustrated in the following image. It is used to identify visually which layers contain or hold drawing objects and which are considered empty or unused. If the Status icon is solid blue, objects such as lines, circles, arcs, and so forth reside on this layer. If the Status icon is faded gray in color, the layer is empty of any objects.

Figure 2–39

USING THE LAYER PREVIOUS COMMAND

The Layer Previous button, shown in the following image, is used to undo changes that you have made to layer settings such as color or lineweight. You could even turn a number of layers off and use the Layer Previous command to have the layers turned back on in a single step. This command can also be entered from the keyboard in the following prompt sequence:

Command: **LAYERP**

There are a few exceptions to the use of the Layer Previous command:

- If you rename a layer and change its properties, such as color or lineweight, issuing the Layer Previous command will restore the original properties but will not change back to the original layer name.
- Purged or deleted layers will not be restored by using Layer Previous.
- If you add a new layer, issuing the Layer Previous command will not remove the new layer.

Figure 2–40

RIGHT-CLICK SUPPORT FOR LAYERS

While inside the Layer Properties Manager, right-clicking inside the layer information area displays the shortcut menu in the following image. Use this menu to make the selected layer current, to make a new layer based on the selected layer, to select all layers in the dialog box, or to clear all layers. You can even select all layers except for the current layer. This shortcut menu provides you with easier access to commonly used layer manipulation tools.

Figure 2-41

CONTROLLING THE LINETYPE SCALE

Once linetypes are associated with layers and placed in a drawing, you can control their scale with the LTSCALE command. On the left side of the following image, the default linetype scale value of 1.00 is in effect. This scale value acts as a multiplier for all linetype distances. In other words, if the hidden linetype is designed to have dashes 0.125 units long, a linetype scale value of 1.00 displays the dash of the hidden line at a value of 0.125 units. The LTSCALE command displays the following command sequence:

Command: **LTS** *(For LTSCALE)*
Enter new linetype scale factor <1.0000>: *(Press* ENTER *to accept the default or enter another value)*

 Try It! – Open the drawing file 02_LTScale. Follow the directions, command prompts, and illustrations below for using the LTSCALE command.

In the middle of the following image, a linetype scale value of 0.50 units has been applied to all linetypes. As a result of the 0.50 multiplier, instead of all hidden line dashes measuring 0.125 units, they now measure 0.0625 units.

Command: **LTS** *(For LTSCALE)*
Enter new linetype scale factor <1.0000>: **0.50**

On the right side of the following image, a linetype scale value of 2.00 units has been applied to all linetypes. As a result of the 2.00 multiplier, instead of all hidden line dashes measuring 0.125 units, they now measure 0.25 units. Notice how a large multiplier displays the center lines as what appears to be a continuous linetype.

Command: **LTS** *(For LTSCALE)*
Enter new linetype scale factor <0.5000>: **2.00**

Figure 2–42

Try It! – Open the drawing file 02_Floor Plan illustrated in the following image on the left. This floor plan is designed to plot out at a scale factor of 1/8″ = 1′0″. This creates a scale factor of 96 (found by dividing 1′ by 1/8″). A layer called "Dividers" was created and assigned the Hidden linetype to show all red hidden lines as potential rooms in the plan. However, the hidden lines do not display. For all linetypes to show as hidden, this multiplier should be applied to the drawing through the LTSCALE command.

Since the drawing was constructed in real-world units or full size, the linetypes are converted to these units beginning with the multiplier of 96 in the following image on the right through the LTSCALE command.

Command: **LTS** *(For LTSCALE)*
Enter new linetype scale factor <1.0000>: **96**

Figure 2–43

ADVANCED LAYER TOOLS

LAYER FILTERS

The first button located above the Tree View pane allows you to create a New Property Filter, as shown at left in the following image. Clicking on this button displays the Layer Filter Properties dialog box. You enter information in the Filter definition area (upper half) of the dialog box and observe the results of the filter in the Filter preview area (lower half). In this example, a layer filter has been created in the definition area to identify all layers that have the color BLUE. The results in the preview area show two layers. Notice at the top of the dialog box that this filter has been given a name, namely Color Blue Layers. You can build various named filters as a means of further organizing your layers by function, color, name, and linetype, or even through a combination of these states.

Figure 2–44

LAYER GROUPS

Layer groups allow you to collect a number of layers under a unique name. For instance, you may want to group a number of layers under the name Foundation; or in a mechanical application, you may want to group a number of layers under the name Fasteners. Choose this command by clicking on the second button illustrated on the left in the following image. You could also move your cursor in the Tree View pane and right-click to display the menu shown on the right in the following image. Then click New Group Filter.

Figure 2–45

As shown on the left in the following image, a new layer group called Electrical Plan is created and located in the Tree View pane. To associate layers with this group, select the layers in the List View pane and drag the layers to the layer group. You are not moving these layers; rather, you are grouping them under a unique name.

Figure 2–46

Clicking on the Electrical Plan layer group will display only those layers common to this group in the List View pane of the Layer Properties Manager dialog box, as shown on the left in the following image. Notice also in the Tree View pane the existence of the Floor_Plan and Foundation Plan groups in addition to the following layer filters: All Used Layers, Color_Blue, and Unused Layers.

 Note: Clicking on All in the Tree View pane will list all layers defined in the drawing.

Additional controls allow you to further manipulate layers. With the Electrical Plan group selected, it is possible to invert this filter. This means to select all layers not part of the Electrical Plan group. To perform this operation, place a check in the box next to Invert filter, located in the lower left corner of the Layer Properties Manager dialog box. The result of this inverted list is shown on the right in the following image.

Figure 2–47

LAYER STATES MANAGER

The third button located above the Tree View pane is the Layer States Manager button, which launches the Layer States Manager dialog box, as shown in the following image. In this example, a layer state called All Layers Visible was created based on the layer settings located in the lower half of the dialog box. Rather than manually turning on all layers, clicking on this state will make all layers visible. You can create any number of layer states to better control and manage layers.

Drawing Setup and Organization 121

Figure 2–48

ADDITIONAL LAYER TOOLS

Clicking on Format in the pull-down menu, followed by Layer Tools, displays even more layer commands that can be used to further manage layers. The display of all of these tools is shown on the left in the following image. A special toolbar called Layers II is also available, as shown on the right in the following image.

Figure 2–49

Pull-Down Menu Title	Command	Description
Make Objects Layer Current	LAYMCUR	Makes the layer of a selected object current
Layer Previous	LAYERP	Undoes any changes made to the settings of layers such as color, linetype, or lineweight
Layer Walk	LAYWALK	Activates a dialog box that isolates layers in sequential order
Layer Match	LAYMCH	Changes the layer of selected objects to that of a selected object
Change to Current Layer	LAYCUR	Changes the layer of selected objects to the current layer
Copy Objects to New Layer	COPYTOLAYER	Copies objects to other layers
Layer Isolate	LAYISO	Isolates layers of selected objects by turning all other layers off
Isolate Layer to Current Viewport	LAYVPI	Isolates an object's layer in a viewport
Layer Off	LAYOFF	Turns off the layers of selected objects
Turn All Layers On	LAYON	Turns all layers on
Layer Freeze	LAYFRZ	Freezes the layers of selected objects
Thaw All Layers	LAYTHW	Thaws all layers
Layer Lock	LAYLCK	Locks the layers of selected objects
Layer Unlock	LAYULK	Unlocks the layers of selected objects
Layer Merge	LAYMRG	Merges two layers, and then removes the first layer from the drawing
Layer Delete	LAYDEL	Permanently deletes layers from drawings

MATCH OBJECT'S LAYER (LAYMCH)

The Match Object's Layer command allows you to change the layers of selected objects to match the layer of a selected destination object. The command is LAYMCH; first select the objects to be changed, such as the three chairs shown in the following image. Next select the object on the destination layer, as in the shelf at "D." The three chairs now belong to the same layer as the shelf.

Command: **LAYMCH**

Select objects to be changed:
Select objects: *(Select the three chairs labeled "A," "B," and "C")*
Select objects: *(Press ENTER to continue)*
Select object on destination layer or [Name]: *(Pick the shelf at "D")*
3 objects changed to layer "Furniture".

Figure 2–50

CHANGE TO CURRENT LAYER (LAYCUR)

The Change to Current Layer command is used to change the layer of one or more objects to the current layer. The command is LAYCUR. This command is particularly helpful if objects were constructed on the wrong layer. In the following image, the three chairs labeled "A," "B," and "C" were drawn on the wrong layer. First make the desired layer current; in the following image, the current layer is Furniture. Selecting the three chairs after issuing the LAYCUR command changes the three chairs to the Furniture layer.

Command: **LAYCUR**
Select objects to be changed to the current layer: *(Select the three chairs "A," "B," and "C")*
Select objects: *(Press ENTER to change the three objects to the current layer)*
3 objects changed to layer "Furniture" *(the current layer)*

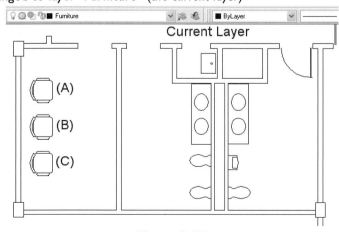

Figure 2–51

ISOLATE OBJECT'S LAYER (LAYISO)

The Isolate Object's Layer command isolates the layer or layers of one or more selected objects by turning all other layers off. The effects of this command are the same as using the Layer Properties Manager dialog box and picking the layer names to turn off. The LAYISO command is yet another tool that manipulates layers in a more efficient manner than through conventional AutoCAD commands. After issuing the command, you are prompted to select the object or objects on the layer or layers to be isolated. On the top in the following image, pick any outside wall and press the ENTER key to execute the command. A message in the command prompt area alerts you that layer Walls is isolated. All objects on the Walls layer are visible, as shown on the bottom in the following image; all other layers are turned off.

Command: **LAYISO**
Current setting: Viewports=Vpfreeze
Select object(s) on the layer(s) to be isolated: *(Pick one of the walls)*
Select objects: *(Press ENTER to isolate the layer based on the object selected)*
Layer "Walls" has been isolated.

Figure 2–52

COPY OBJECTS TO NEW LAYER (COPYTOLAYER)

The Copy Objects to New Layer command copies selected objects to a different layer while leaving the original objects on the original layer. In the following image, three chairs on layer Room 1 need to be copied and have their layer changed to Room 2.

Command: **COPYTOLAYER**

Select objects: *(Select the three chairs illustrated in the following image)*
Select objects: *(Press ENTER to continue)*
Select object on destination layer or [Name] <Name>: *(Press ENTER. When the Copy to Layer dialog box appears, select Room 2)*
3 objects copied and placed on layer "Room 2"
Base point or <return> to finish: *(Pick the endpoint at "A")*
Specify second point of displacement or <use first point as displacement>: *(Pick a point at "B")*

Figure 2–53

The results are illustrated in the following image. The three original objects remain on layer Room 1, while the new set of objects has been copied to a new location and to a new layer (Room 2).

Figure 2–54

LAYER WALK (LAYWALK)

The Layer Walk command displays a dialog box containing all layers in the present drawing. Clicking on a layer in this dialog box will turn off all other layers. If layers are detected as unreferenced (unused), a purge button activates, allowing you to delete the layer.

In the illustration of the office plan shown on the left in the following image, the LAYWALK command has activated the Layer Walk dialog box. Notice in this dialog box that all layers are highlighted, signifying that all layers in the drawing are visible.

Clicking on the layer Walls, as shown on the right in the following image, will turn off all layers except for the Walls layer. You can press the CTRL or SHIFT key while selecting layers in the Layer Walk dialog box to select multiple layers.

Figure 2–55

FREEZE OBJECT'S LAYER (LAYFRZ)

The Freeze Object's Layer command freezes layers by picking objects to control the layers to be frozen. Again, the Layer Properties Manager dialog box and the Layer Control box are normally used to perform all layer freeze operations.

Command: **LAYFRZ**

Select an object on the layer to be frozen or [Settings/Undo]: *(Pick a chair)*
Layer "Chairs" has been frozen.
Select an object on the layer to be frozen or [Settings/Undo]: *(Pick a shelf)*
Layer "Furniture" has been frozen.
Select an object on the layer to be frozen or [Settings/Undo]: *(Pick a sink)*
Layer "Fixtures" has been frozen.
Select an object on the layer to be frozen or [Settings/Undo]: *(Pick a refrigerator)*
Layer "Kitchen" has been frozen.
Select an object on the layer to be frozen or [Settings/Undo]: *(Pick a door)*
Layer "Doors" has been frozen.
Select an object on the layer to be frozen or [Settings/Undo]: *(Pick a computer)*
Layer "Cpu" has been frozen.
Select an object on the layer to be frozen or [Settings/Undo]: *(Press ENTER to complete this command. The results should be similar to the following image.)*

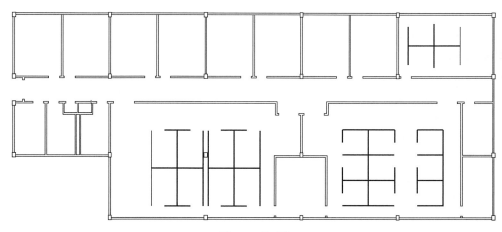

Figure 2–56

TURN OBJECT'S LAYER OFF (LAYOFF)

The Turn Object's Layer Off command is similar to the LAYFRZ command except that, instead of freezing layers, you can turn off a layer or group of layers by selecting an object or group of objects.

Command: **LAYOFF**

Current settings: Viewports=Vpfreeze, Block nesting level=Block
Select an object on the layer to be turned off or [Settings/Undo]: *(Pick a shelf)*
Layer Furniture has been turned off.
Select an object on the layer to be turned off or [Settings/Undo]: *(Pick a partition)*
Layer Partitions has been turned off.
Select an object on the layer to be turned off or [Settings/Undo]: *(Pick a chair)*
Layer Chairs has been turned off.
Select an object on the layer to be turned off or [Settings/Undo]: *(Pick a computer)*
Layer Cpu has been turned off.
Select an object on the layer to be turned off or [Settings/Undo]: *(Press ENTER to exit this command. The results should be similar to the following image.)*

Figure 2–57

LOCK OBJECT'S LAYER (LAYLCK)

The Lock Object's Layer command locks the layer of a selected object. A locked layer is visible on the display screen; however, any object associated with a locked layer is non-selectable. The command to lock a layer is LAYLCK.

Command: **LAYLCK**

Select an object on the layer to be locked: *(Pick an exterior wall in the previous image)*
Layer "Walls" has been locked.

UNLOCK OBJECT'S LAYER (LAYULK)

The Unlock Object's Layer command unlocks the layer of a selected object. Objects on an unlocked layer can now be selected whenever the prompt "Select objects:" appears. The command to unlock a layer is LAYULK.

Command: **LAYULK**

Select an object on the layer to be unlocked: *(Pick an exterior wall in the previous image)*
Layer "Walls" has been unlocked.

MERGING LAYERS INTO ONE (LAYMRG)

The Merging Layers into One command (LAYMRG) merges the contents of one layer with a target layer. The layer containing all chairs (Chairs layer) illustrated in the following image will be merged with the Furniture layer. One of the chairs is first selected as the layer to merge. Then one of the cabinets at "A," as shown in the following image, is selected as the target layer. Before the layers are merged, a warning message appears in the command prompt alerting you that the Chairs layer will be permanently merged into the Furniture layer. After you perform this operation, the Chairs layer will automatically be purged from the drawing.

Command: **LAYMRG**
Select object on layer to merge or [Name]: *(Pick a chair, as shown in the following image)*
Selected layers: "Chairs"
Select object on layer to merge or [Name/Undo] <done>: *(Press ENTER to continue)*
Select object on target layer or [Name]: *(Pick the cabinet at "A" in the following image)*
Selected layers: "Chairs","Furniture".

******* WARNING *******

You are about to merge layer "Chairs" into layer "Furniture".
Do you wish to continue? [Yes/No] <No>: **Y** *(For Yes)*
Deleting layer "Chairs".
1 layer deleted.

Figure 2–58

DELETING ALL OBJECTS ON A LAYER (LAYDEL)

Use the Layer Delete command (LAYDEL) to delete all objects on a specified layer. First select the computer terminal, as shown on the left in the following image. The computer terminal is located on the Cpu layer. Before all objects on this layer are deleted, a message appears asking whether you really want to perform this operation. Answering Yes deletes all objects assigned to the layer in addition to purging the layer from the drawing. Although this tool has many beneficial advantages, you must exercise care in deciding when to use it and when not to use it.

Command: **LAYDEL**
Select object on layer to delete or [Name]: *(Pick the computer at "A" in the following image)*
Selected layers: Cpu
Select object on layer to delete or [Name/Undo]: *(Press ENTER to continue)*

******* WARNING *******

You are about to delete layer "Cpu" from this drawing.
Do you wish to continue? [Yes/No] <No>: **Y** *(For Yes)*
Deleting layer "Cpu"
1 layer deleted.

Figure 2–59

If you are not sure about the layer to delete, you can use the Name option, which will launch the Delete Layers dialog box as shown in the following image. You can then choose the layer by name. The same warning prompts will alert you that the layer name and all of the objects on the layer will be deleted.

Command: **LAYDEL**
Select object on layer to delete or [Name]: **N** *(For Name)*

Figure 2–60

CREATING TEMPLATE FILES

Template files are files that have settings such as units, limits, and layers already created, which enhances the productivity of the user. By default, when you start a new drawing from scratch, AutoCAD uses a template. Two templates are available depending on whether you are drawing in inches or millimeters; they are Acad.Dwt (inches) and Acadiso.Dwt (millimeters). Various settings are made in a template file. For instance, in the Acad.Dwt template file, the drawing units are set to decimal units with four-decimal-place precision. The drawing limits are set to 12,9 units for the upper right corner. Only one layer exists; namely, layer 0. The grid in this template is set to a spacing of 0.50 units.

If you have to create several drawings that have the same limits and units settings and use the same layers, rather than start with the Acad.Dwt file and make changes to the settings at the start of each new drawing file, a better technique would be to create your own template. Templates can be made up of entire drawings or just have a company border already placed in the drawing. Other templates may appear empty of objects such as lines or circles. However, the limits of the drawing may already be set along with grid and snap settings. The following settings are usually assigned to a drawing template:

- Drawing units and precision
- Drawing limits
- Grid and Snap Settings
- Layers, Linetypes, and Lineweights
- Linetype scale factor depending on the final plot scale of the drawing

 Try It! – Create a new drawing file starting from scratch. You will be creating a template designed to draw an architectural floor plan. Make changes to the following settings:

- In the Drawing Units dialog box, change the units type from Decimal to Architectural.
- Set the limits of the drawing to 17′,11′ for the upper right corner and perform a ZOOM All. Change the grid and snap to 3″ for the X and Y values.
- Create the following layer names:
    - Floor Plan
    - Hidden
    - Center
    - Dimension
    - Doors
    - Windows
- Make your own color assignments to these layers.
- Load the Hidden and Center linetypes and assign these to the appropriate layers.
- Assign a lineweight of 0.70mm to the Floor Plan layer.
- Finally, use the LTSCALE command and set the linetype scale to 12.

Now it is time to save the settings in a template format. Follow the usual steps for saving your drawing. However, when the Save Drawing As dialog box appears, first click the arrow on the other side of "Files of type:" in the following image to expose the file types.

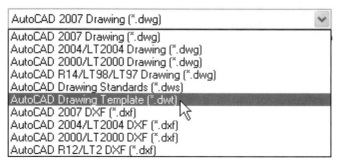

Figure 2–61

Click on the field that states, "AutoCAD Drawing Template (*.dwt)." AutoCAD will automatically take you to the folder, shown in the following image, that holds all template information. It is at this time that you enter the name of the template as AEC_B_1=12. The name signifies an architectural drawing for the B-size drawing sheet at a scale of 1″ = 1′-0″.

Figure 2–62

Before the template file is saved, you have the opportunity to document the purpose of the template. It is always considered good practice to create this documentation, especially if others will be using your template.

TUTORIAL EXERCISE: CREATING LAYERS USING THE LAYER PROPERTIES MANAGER DIALOG BOX

Layer Name	Color	Linetype	Lineweight
Object	White	Continuous	0.60
Hidden	Red	Hidden	0.30
Center	Yellow	Center	Default
Dim	Yellow	Continuous	Default
Section	Blue	Continuous	Default

Purpose

Use the following steps to create layers according to the above specifications.

STEP 1

Activate the Layer Properties Manager dialog box by clicking on the Layer button located on the Object Properties toolbar, as shown in the image on the right.

Figure 2–63

STEP 2

Once the Layer Properties Manager dialog box displays, as shown in the following image, notice on your screen that only one layer is currently listed, namely Layer 0. This happens to be the default layer, the layer that is automatically available to all new drawings.

Figure 2–64

Since it is considered poor practice to construct any objects on Layer 0, new layers will be created not only for object lines but for hidden lines, center lines, dimension objects, and section lines as well. To create these layers, click on the New button and notice that a layer is automatically added to the list of layers. This layer is called Layer1 (see the following image). While this layer is highlighted, enter the first layer name, "Object."

Figure 2–65

Entering a comma after the name of the layer allows more layers to be added to the listing of layers without having to click on the New button. Once you have entered the comma after the layer "Object" and the new layer appears, enter the new name of the layer as "Hidden." Repeat this procedure of using the comma to create multiple layers for "Center," "Dim," and "Section." Press ENTER after typing in "Section." The complete list of all layers created should be similar to the illustration shown in the following image.

Figure 2–66

STEP 3

As all new layers are displayed, the names may be different, but they all have the same color and linetype assignments (see the following image). At this point, the dialog box comes in handy in assigning color and linetypes to layers in a quick and easy manner. First, highlight the desired layer to add color or linetype by picking the layer. A horizontal bar displays, signifying that this is the selected layer. Click on the color swatch identified by the box, as shown in the following image, and follow the next step to assign the color "Red" to the "Hidden" layer name.

Figure 2–67

STEP 4

Clicking on the color swatch in the previous step displays the Select Color dialog box, as shown in the following image. Select the desired color from one of the following areas: Standard Colors, Gray Shades, or Full Color Palette. The standard colors represent colors 1 (Red) through 9 (Gray). On display terminals with a high-resolution graphics card, the full color palette displays different shades of the standard colors. This gives you a greater variety of colors to choose from. For the purposes of this tutorial, the color "Red" will be assigned to the "Hidden" layer. Select the box displaying the color red; a box outlines the color and echoes the color in the bottom portion of the dialog box. Click on the OK button to complete the color assignment; if you select Cancel, the color assignment will be removed, requiring this operation to be used again. Continue with this step by assigning the color Yellow to the Center and Dim layers; assign the color Blue to the Section layer.

Figure 2–68

STEP 5

Once the color has been assigned to a layer, the next step is to assign a linetype, if any, to the layer. The "Hidden" layer requires a linetype called "HIDDEN." Click on the "Continuous" linetype to display the Select Linetype dialog box illustrated in the following image. By default, and to save space in each drawing, AutoCAD initially loads the Continuous linetype. Click on the Load button to load more linetypes.

Figure 2–69

Choose the desired linetype to load from the Load or Reload the Linetype dialog box, as shown in the following image. Scroll through the linetypes until the "Hidden" linetype is found. Click on the OK button to return to the Select Linetype dialog box.

Once you are back in the Select Linetype dialog box, as shown in the following image, notice the "Hidden" linetype listed along with "Continuous." Because this linetype has just been loaded, it still has not been assigned to the "Hidden" layer. Click on the Hidden linetype listed in the Select Linetype dialog box, and click on the OK button. Once the Layer Properties Manager dialog box reappears, notice that the "Hidden" linetype has been assigned to the "Hidden" layer. Repeat this procedure to assign the "Center" linetype to the "Center" layer.

Figure 2–70

STEP 6

Another requirement of the "Hidden" layer is that it use a lineweight of 0.30 mm to have the hidden lines stand out when compared with the object in other layers. Clicking on the highlighted default lineweight displays the Lineweight dialog box in the image on the right. Click on the 0.30 mm lineweight followed by the OK button to assign this lineweight to the "Hidden" line. Use the same procedure to assign a lineweight of 0.60 mm to the Object layer.

Figure 2–71

STEP 7

Once you have completed all color and linetype assignments, the Layer Properties Manager dialog box should appear similar to the following image. Click on the OK button to save all layer assignments and return to the drawing editor.

Figure 2–72

STEP 8

When layers are first created, they are listed in the order they were entered. When the layers are saved and the Layer Properties Manager dialog box is displayed again, all layers are reordered alphabetically (see the following image).

Figure 2–73

STEP 9

Clicking in the Name box at the top of the list of layers reorders the layers. Now, the layers are listed in reverse alphabetical order (see the following image).

This same effect occurs when you click in the On, Freeze, Color, Linetype, Lineweight, and Plot header boxes.

Figure 2–74

The following image displays the results of clicking on the Color header: all colors are reordered, starting with Red (1) and followed by Yellow (2), Blue (5), and White (7).

Figure 2–75

PROBLEMS FOR CHAPTER 2

Create an Architectural template (dwt) that contains the following setup information.

Use the Drawing Units dialog box to set the units of the drawing to Architectural. Keep the remaining default values for all other unit settings.

Use the Layer Properties Manager dialog box to create the layers contained in the table below:

Layer Name	Color	Linetype	Lineweight
Border	Blue (5)	Continuous	1.00
Center	Green (3)	Center	0.25
Dimension	Green (3)	Continuous	0.25
Doors	Red (2)	Continuous	0.50
Electrical	Green (3)	Continuous	0.50
Elevations	White (7)	Continuous	0.70
Foundation	Cyan (4)	Continuous	0.50
Furniture	Red (1)	Continuous	0.50
Hatching	Magenta (6)	Continuous	0.50
Hidden	Red (1)	Hidden	0.50
HVAC	Cyan (4)	Continuous	0.50
Misc	Green (3)	Continuous	0.50
Plumbing	Cyan (4)	Continuous	0.50
Text	Green (3)	Continuous	0.50
Viewports	Gray (9)	Continuous	0.25
Walls	White (7)	Continuous	0.70
Windows	Yellow (2)	Continuous	0.50

When finished making the changes to the drawing units and layer assignments, save this file as a new AutoCAD template called ARCH-Imperial.DWT (imperial units are in feet and inches).

Create a Mechanical template (dwt) that contains the following setup information.

Use the Drawing Units dialog box to set the units of the drawing to Mechanical. Also change the precision to two decimal places. Keep the remaining default values for all other unit settings.

Use the Layer Properties Manager dialog box to create the layers contained in the table below:

Layer Name	Color	Linetype	Lineweight
Border	Blue (5)	Continuous	1.00
Center	Green (3)	Center	0.25
Construction	Gray (8)	Continuous	0.25
Dimension	Green (3)	Continuous	0.25
Hatching	Magenta (6)	Continuous	0.50
Hidden	Red (1)	Hidden	0.50
Misc	Green (3)	Continuous	0.50
Object	White (7)	Continuous	0.70
Phantom	Cyan (4)	Phantom	0.50
Text	Green (3)	Continuous	0.50
Viewports	Gray (9)	Continuous	0.25

When finished making the changes to the drawing units and layer assignments, save this file as a new AutoCAD template called MECH-Imperial.DWT (imperial units are in inches).

Create a Civil template (dwt) that contains the following setup information.

Use the Drawing Units dialog box to set the units of the drawing to Decimal. Also, change the precision to two decimal places. Keep the remaining default values for all other unit settings.

Use the Layer Properties Manager dialog box to create the layers contained in the table below:

Layer Name	Color	Linetype	Lineweight
Border	Blue (5)	Continuous	1.00
Building Outline	White (7)	Continuous	0.70
Center	Green (3)	Center	0.25
Contour Lines – Existing	White (7)	Dashed	0.50
Contour Lines – New	Magenta (6)	Continuous	0.50
Dimension	Green (3)	Continuous	0.25
Drainage	Cyan (4)	Divide	0.50
Fire Line	Red (1)	Continuous	0.50
Gas Line	Red (1)	Gas_Line	0.50
Hatching	Magenta (6)	Continuous	0.50
Hidden	Red (1)	Hidden	0.50
Misc	Green (3)	Continuous	0.50
Parking	Yellow (2)	Continuous	0.50
Property Line	Green (3)	Phantom	0.80
Sewer	White (7)	Continuous	0.50
Text	Green (3)	Continuous	0.50
Viewports	Gray (9)	Continuous	0.25
Water Line	Cyan (4)	Continuous	0.50
Wetlands	Green (3)	Dashdot	0.50

When finished making the changes to the drawing units and layer assignments, save this file as a new AutoCAD template called CIVIL-Imperial.DWT (imperial units are in feet and inches).

PROBLEMS FOR CHAPTER 2

PROBLEM 2–1

Open the existing drawing called 02_Storage.Dwg. Use the Layer Properties Manager along with the Layer Filter Properties dialog box to answer the following layer-related questions:

1. The total number of layers frozen in the database of this drawing is _____.
2. The total number of layers locked in the database of this drawing is _____.
3. The total number of layers turned off in the database of this drawing is _____.
4. The total number of layers assigned the color red and frozen in the database of this drawing is _____.
5. The total number of layers that begin with the letter "L", assigned the color yellow, and turned off in the database of this drawing is _____.

PROBLEM 2–2

Open the existing drawing called 02_Pattern.Dwg. Use the Layer Properties Manager along with the Layer Filter Properties dialog box to answer the following layer-related questions:

1. The total number of layers assigned the color red and turned off in the database of this drawing is _____.
2. The total number of layers assigned the color red and locked in the database of this drawing is _____.
3. The total number of layers assigned the color white and frozen in the database of this drawing is _____.
4. The total number of layers assigned the color yellow and assigned the hidden linetype in the database of this drawing is _____.
5. The total number of layers assigned the color red and assigned the phantom linetype in the database of this drawing is _____.
6. The total number of layers assigned the center linetype in the database of this drawing is _____.

AutoCAD Display and Basic Selection Operations

This chapter introduces you to the various ways of viewing your drawing. A number of options of the ZOOM command will be explained first, followed by a review of panning and understanding how a wheel mouse is used with AutoCAD. User-defined portions of a drawing screen called views will also be introduced in this chapter. You will be introduced to various Object Selection methods such as window, crossing, crossing polygon, window polygon, fence, and previous, to name just a few.

THE ZOOM COMMAND

The ability to magnify details in a drawing or reduce the drawing to see it in its entirety is a function of the ZOOM command. It does not take much for a drawing to become very busy, complicated, or dense when displayed in the drawing editor. Therefore, use the ZOOM command to work on details or view different parts of the drawing. One way to select the options of this command is from the View pull-down menu, as shown in the following image. Choosing Zoom displays the various options of the ZOOM command. These options include zooming in real time, zooming to the previous display, using a window to define a boxed area to zoom to, dynamic zooming, zooming to a user-defined scale factor, zooming based on a center point and a scale factor, or performing routine operations such as zooming in or out, zooming all, or zooming to the extents of the drawing. All these modes will be discussed in the pages that follow.

Another way to choose options of the ZOOM command is through the Standard toolbar area at the top of the display screen. This image shows a series of buttons that perform various options of the ZOOM command. Four main buttons initially appear in the Standard toolbar, namely Realtime PAN, Realtime ZOOM, ZOOM-Window, and ZOOM-Previous. Selecting the ZOOM-Window option displays a series of additional zoom option buttons that cascade down. A third method of performing ZOOM command options, using the Zoom toolbar, is also illustrated in the following image. This toolbar contains the same buttons found in the Standard toolbar; it differs in the fact that the toolbar can be moved to different positions around the display screen.

You can also activate the ZOOM command from the keyboard by entering either ZOOM or the letter Z, which is its command alias.

Figure 3–1

Study the following table for a brief description of each Zoom tool.

Button	Tool	Function
	ZOOM-Window	Zooms to a rectangular area by specifying two diagonal points
	ZOOM-Dynamic	Uses a viewing box to zoom in to a portion of a drawing
	ZOOM-Scale	Zooms the display based on a specific scale factor
	ZOOM-Center	Zooms to an area of the drawing based on a center point and magnification value
	ZOOM-Object	Zooms to the largest possible magnification based on the objects selected
	ZOOM-In	Magnifies the display screen by a scale factor of 2
	ZOOM-Out	De-magnifies the display screen by a scale factor of .5
	ZOOM-All	Zooms to the largest possible magnification based on the current grid setting or drawing extents, whichever is larger
	ZOOM-Extents	Zooms to the largest possible magnification based on all objects in the drawing
	ZOOM-Realtime	Zooms dynamically by moving your cursor up or down
	ZOOM-Previous	Zooms to the previous view

The following image illustrates a complex drawing of a part that consists of all required orthographic views. To work on details of this and other drawings, use the ZOOM command to magnify or reduce the display screen. The following are the options of the ZOOM command (note that nX and nXP refer to the relative scale factors, which will be explained later in the chapter):

Command: **Z** *(For ZOOM)*
Specify corner of window, enter a scale factor (nX or nXP), or
[All/Center/Dynamic/Extents/Previous/Scale/Window/Object] <real time>: *(enter one of the listed options)*

Executing the ZOOM command and picking a blank part of the screen places you in automatic ZOOM-Window mode. Selecting another point zooms in to the specified area. Refer to the following command sequence to use this mode of the ZOOM command on the object illustrated in the following image.

Command: **Z** *(For ZOOM)*
Specify corner of window, enter a scale factor (nX or nXP), or
[All/Center/Dynamic/Extents/Previous/Scale/Window/Object] <real time>: *(Mark a point at "A")*
Other corner: *(Mark a point at "B")*

Figure 3–2

The ZOOM-Window option is automatically invoked once you select a blank part of the screen and then pick a second point. The resulting magnified portion of the screen appears in the following image.

Figure 3–3

ZOOMING IN REAL TIME

A powerful option of the ZOOM command is performing screen magnifications or reductions in real time. This is the default option of the command. Issuing the Realtime option of the ZOOM command displays a magnifying-glass icon with a positive sign and a negative sign above the magnifier icon. Identify a blank part of the drawing editor, press down the pick button of the mouse (the left mouse button), and move in an upward direction to zoom in to the drawing in real time. Identify a blank part of the drawing editor, press down the pick button of the mouse, and move in a downward direction to zoom out of the drawing in real time. Use the following command sequence and image for performing this task.

Command: **Z** *(For ZOOM)*
Specify corner of window, enter a scale factor (nX or nXP), or
[All/Center/Dynamic/Extents/Previous/Scale/Window/Object] <real time>: *(Press* ENTER *to accept Realtime as the default)*

Identify the lower portion of the drawing editor, press and hold down the pick button of the mouse, and move the Realtime cursor up; notice the image zooming in.

Figure 3–4

Once you are in the Realtime mode of the zoom command, press the right mouse button to activate the shortcut menu, as shown in the following image. Use this menu to switch between Realtime ZOOM and Realtime PAN, which gives you the ability to pan across the screen in real time. The ZOOM-Window, Original (Previous), and Extents options are also available in the cursor menu.

Figure 3–5

USING ZOOM-ALL

Another option of the ZOOM command is All. Use this option to zoom to the current limits of the drawing as set by the LIMITS command. In fact, right after the limits of a drawing have been changed, issuing a ZOOM-All updates the drawing file to reflect the latest screen size. To use the ZOOM-All option, refer to the following command sequence.

 Command: **Z** *(For ZOOM)*

Specify corner of window, enter a scale factor (nX or nXP), or
[All/Center/Dynamic/Extents/Previous/Scale/Window/Object] <real time>: **A** *(For All)*

The illustration on the left in the following image shows a zoomed-in portion of a part. Use the ZOOM-All option to zoom to the drawing's current limits, as shown on the right in the following image.

Figure 3–6

USING ZOOM-CENTER

The ZOOM-Center option allows you to specify a new display based on a selected center point, as shown on the left in the following image. A window height controls whether the image on the display screen is magnified or reduced. If a smaller value is specified for the magnification or height, the magnification of the image is increased (you zoom in to the object). If a larger value is specified for the magnification or height, the image gets smaller, or a ZOOM-Out is performed.

 Try It! – Open the drawing file 03_Zoom Center. Follow the illustrations and command sequence below to perform a zoom based on a center point.

 Command: **Z** *(For ZOOM)*

Specify corner of window, enter a scale factor (nX or nXP), or
[All/Center/Dynamic/Extents/Previous/Scale/Window/Object] <real time>: **C** *(For Center)*
Center point: *(Mark a point at the center of circle "A" shown on the left)*
Magnification or Height <7.776>: **2**

(A)

Figure 3–7

 ## USING ZOOM-EXTENTS

The left image of the pump in the following image reflects a ZOOM-All operation. Use this option to display the entire drawing area based on the drawing limits even if the objects that make up the image appear small. Instead of performing a zoom based on the drawing limits, ZOOM-Extents uses the extents of the image on the display screen to perform the zoom. The right image in the following image shows the largest possible image displayed as a result of using the zoom command and the Extents option.

 Try It! – Open the drawing file 03_Zoom Extents. Follow the illustrations and command sequence below to perform a zoom based on the drawing limits (All) and the objects in the drawing (Extents).

 Command: **Z** *(For ZOOM)*

Specify corner of window, enter a scale factor (nX or nXP), or
[All/Center/Dynamic/Extents/Previous/Scale/Window/Object] <real time>: **A** *(For All)*

 Command: **Z** *(For ZOOM)*

Specify corner of window, enter a scale factor (nX or nXP), or
[All/Center/Dynamic/Extents/Previous/Scale/Window/Object] <real time>: **E** *(For Extents)*

Figure 3–8

USING ZOOM-WINDOW

The ZOOM-Window option allows you to specify the area to be magnified by marking two points representing a rectangle, as shown on the left in the following image. The center of the rectangle becomes the center of the new image display; the image inside the rectangle is either enlarged, as shown on the right in the following image, or reduced.

 Try It! – Open the drawing file 03_Zoom Window. Follow the illustrations and command sequence below to perform a zoom based on a window.

 Command: **Z** *(For ZOOM)*
Specify corner of window, enter a scale factor (nX or nXP), or
[All/Center/Dynamic/Extents/Previous/Scale/Window/Object] <real time>: **W** *(For Window)*
First corner: *(Mark a point at "A")*
Other corner: *(Mark a point at "B")*

By default, the Window option of zoom is automatic; in other words, without entering the Window option, the first point you pick identifies the first corner of the window box; the prompt "Other corner:" completes ZOOM-Window, as indicated in the following prompts:

Command: **Z** *(For ZOOM)*
Specify corner of window, enter a scale factor (nX or nXP), or
[All/Center/Dynamic/Extents/Previous/Scale/Window/Object] <real time>: *(Mark a point at "A")*
Other corner: *(Mark a point at "B")*

Figure 3–9

USING ZOOM-PREVIOUS

After magnifying a small area of the display screen, use the Previous option of the zoom command to return to the previous display. The system automatically saves up to ten views when zooming. This means you can begin with an overall display, perform two zooms, and use the ZOOM-Previous command twice to return to the original display, as shown in the following image.

Command: **Z** *(For ZOOM)*
Specify corner of window, enter a scale factor (nX or nXP), or
[All/Center/Dynamic/Extents/Previous/Scale/Window/Object] <real time>: **P** *(For Previous)*

Figure 3–10

USING ZOOM-OBJECT

A zooming operation can also be performed based on an object or group of objects that you select in the drawing. The illustration on the left in the following image displays a facilities plan. The ZOOM-Object option is used to magnify one of the chairs using the following command sequence:

Command: **Z** *(For ZOOM)*
Specify corner of window, enter a scale factor (nX or nXP), or
[All/Center/Dynamic/Extents/Previous/Scale/Window/Object] <real time>: **O** *(For Object)*

Select objects: *(Pick the chair)*

Select objects: *(Press ENTER to perform a ZOOM-Extents based on the chair)*

The results are shown on the right in the following image, where the chair is magnified to fit in the display screen.

Figure 3–11

USING ZOOM-SCALE

In addition to performing zoom operations by picking points on your display screen for such options as zooming to a window or center point, you can also fine-tune your zooming by entering zoom scale factors directly from your keyboard. For example, entering a scale factor of 2 will zoom in to your drawing at a factor of 2 times. Entering a scale factor of 0.50 will zoom out of your drawing.

 Try It! – Open the drawing file 03_Zoom Scale. Use the command sequences and illustration in the following image for performing a zoom based on a scale factor.

If a scale factor of 0.50 is used, the zoom is performed into the drawing at a factor of 0.50, based on the original limits of the drawing. Notice that the image gets smaller.

 Command: **Z** *(For ZOOM)*
Specify corner of window, enter a scale factor (nX or nXP), or
[All/Center/Dynamic/Extents/Previous/Scale/Window/Object] <real time>: **0.50**

If a scale factor of 0.50X is used, the zoom is performed in to the drawing again at a factor of 0.50; however, the zoom is based on the current display screen. The image gets even smaller.

 Command: **Z** *(For ZOOM)*
Specify corner of window, enter a scale factor (nX or nXP), or
[All/Center/Dynamic/Extents/Previous/Scale/Window/Object] <real time>: **0.50X**

Enter a scale factor of 0.90. The zoom is again based on the original limits of the drawing. As a result, the image displays larger.

Command: **Z** *(For ZOOM)*
Specify corner of window, enter a scale factor (nX or nXP), or
[All/Center/Dynamic/Extents/Previous/Scale/Window/Object] <real time>: **0.90**

Figure 3–12

 ## USING ZOOM-IN

Clicking on this button automatically performs a zoom-in operation at a scale factor of 2X; the "X" uses the current screen to perform the zoom-in operation.

 ## USING ZOOM-OUT

Clicking on this button automatically performs a zoom-out operation at a scale factor of 0.5X; the "X" uses the current screen to perform the zoom-out operation.

 ## THE PAN COMMAND

As you perform numerous ZOOM-Window and ZOOM-Previous operations, it becomes apparent that it would be nice to zoom in to a detail of a drawing and simply slide the drawing to a new area without changing the magnification; this is the purpose of the PAN command. In the following image, the ZOOM-Window option is used to construct a rectangle around the Top view to magnify it.

Figure 3–13

The result is shown in the following image. Now the Bottom view needs to be magnified to view certain dimensions. Rather than use ZOOM-Previous and then ZOOM-Window again to magnify the Bottom view, use the PAN command.

 Command: P *(For PAN)*

Press ESC or ENTER to exit, or right-click to display shortcut menu.

Issuing the PAN command displays the Hand symbol. In the following image, pressing the pick button down at "A" and moving the hand symbol to the right at "B" pans the screen and displays a new area of the drawing in the current zoom magnification.

Figure 3–14

In the following image, the Bottom view is now visible after the drawing is panned from the Top view to the Bottom view, with the same display screen magnification. PAN can also be used transparently; that is, while in a current command, you can select the PAN command, which temporarily interrupts the current command, performs the pan, and then restores the current command.

Figure 3–15

SUPPORT FOR WHEEL MOUSE

AutoCAD provides extra support for zooming and panning operations through the mouse. This device consists of the standard Microsoft two-button mouse with the addition of a wheel, as illustrated in the following image. Rolling the wheel forward zooms in to, or magnifies, the drawing. Rolling the wheel backward zooms out of, or reduces, the drawing.

Pressing and holding the wheel down, as shown on the right in the following image, places you in Realtime PAN mode. The familiar hand icon on the display screen identifies this mode.

Figure 3–16

Pressing CTRL and then depressing the wheel places you in Joystick Pan mode (see the following image). This mode is identified by a pan icon similar to that shown in the middle of the following image. This icon denotes all directions in which panning may occur. Moving the mouse in any direction with the wheel depressed displays the icon on the right in the following image, which shows the direction of the pan. The further away from the center dot you move your cursor, the faster the panning takes place.

Figure 3–17

The wheel can also function like a mouse button. Double-clicking on the wheel, as in the following image, performs a ZOOM-Extents and is an extremely productive method of performing this operation. As you can see, the wheel mouse provides for realtime zooming and panning with the additional capability of performing a ZOOM-Extents, all at your fingertips.

Figure 3–18

CREATING NAMED VIEWS

An alternate method of performing numerous zooms is to create a series of views of key parts of a drawing. Then, instead of using the ZOOM command, restore the named view to perform detail work. This named view is saved in the database of the drawing for use in future editing sessions. Selecting Named Views from the View pull-down menu, as shown on the left in the following image, activates the View Manager dialog box, shown on the right. You can activate this same dialog box through the keyboard by entering the following at the command prompt:

Command: **V** *(For VIEW)*

Figure 3–19

The following table describes each view type:

View Type	Description
Current	Displays the current view, view information, and clipping properties.
Model Views	Displays a list of all views created along with general, camera, and clipping properties.
Layout Views	Displays a list of views created in a layout in addition to general and view information (height and width).
Preset Views	Displays a list of orthogonal and isometric views, and lists the general properties for each preset view.

 Try It! – Open the drawing file 03_Views. Follow the next series of steps and illustrations used to create a view called "FRONT."

Clicking the New button in the View Manager dialog box activates the New View dialog box, as shown on the left in the following image. Use this dialog box to guide you in creating a new view. By definition, a view is created from the current display screen. This is the purpose of the Current Display radio button. Many views are created with the Define Window radio button, which will create a view based on the contents of a window that you define. Choosing the Define View Window button returns you to the display screen and prompts you for the first corner and other corner required to create a new view by window.

As illustrated on the right in the following image, a rectangular window is defined around the Front view using points "A" and "B" as the corners.

Figure 3–20

When the window is created, the Define New View dialog box redisplays. Clicking OK saves the view name under the Model Views heading in the View Manager dialog box, as shown in the following image.

Figure 3–21

Try It! – Open the drawing file 03_Views Complete. A series of views have already been created inside this drawing. Activate the View Manager dialog box and experiment with restoring a number of these views.

The following image illustrates numerous views of a drawing already created.

Figure 3–22

Clicking on a defined view name, as shown on the left in the following image, followed by right-clicking the mouse, displays the shortcut menu used to set the view current. You can also create a new view, update layers, edit the boundaries of the view, and delete the view through this shortcut menu. Clicking on Set Current and exiting the View Manager dialog box displays the Front view, as shown on the right in the following image.

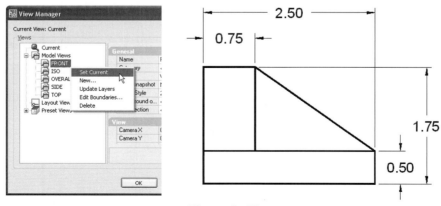

Figure 3–23

CHANGING THE BOUNDARY OF A NAMED VIEW

The View Manager dialog box has an Edit Boundaries button. In the following image, the white area of the drawing illustrates the current view, called Kitchen Elevation. However, this view should also include the cabinet details. Clicking on the view name and then clicking on the Edit Boundaries button will allow you to redefine the view boundary box by picking two new corner points. This is also illustrated in the following image with a rectangle displayed around all kitchen elevations.

Figure 3–24

After you construct this new boundary, the results should be similar to the following image. The new view is displayed in white. Non-view components have the color gray in the background.

Figure 3–25

SAVING LAYER SETTINGS WITH NAMED VIEWS

Another very important function of views is that you can have layer information saved with the view. In the following image, two views are created, namely All Layers Turned On and Only Walls Visible. Selecting the Update Layers button places a "Yes" in the Layer snapshot heading. All current layer settings are saved under this view name. In a way, you can now control layer states through the creation of views.

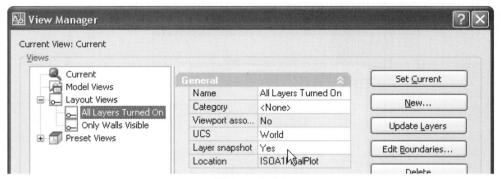

Figure 3–26

The following image shows the results of controlling layers through the View Manager dialog box. When you click on the view name All Layers Turned On, the image on the left displays. When the view name Only Walls Visible is made current, the image on the right displays.

Figure 3–27

CREATING OBJECT SELECTION SETS

Selection sets are used to group a number of objects together for the purpose of modifying. Applications of selection sets are covered in the following pages. Once a selection set has been created, the group of objects may be modified by moving, copying, mirroring, and so on. The operations supported by selection sets will be covered in chapter 4, "Modify Commands." An object manipulation command supports the creation of selection sets if it prompts you to "Select objects."

The commonly used selection set options (how a selection set is made) are briefly described in the following table.

Selection Tool	Selection Key-In	Function
All	All	Selects all objects in the database of the drawing. This mode will even select objects on layers that are turned off. Objects that reside on frozen layers will not be selected.
Crossing	C	You create a rectangular box by dragging your cursor from right to left (transparent green window). This rectangular box will appear as dotted line segments. All objects completely enclosed and touching this rectangular box will be selected.
Crossing Polygon	CP	You create an irregular closed shape. The polygon formed will appear as dotted line segments. All objects completely enclosed and touching this irregular shape will be selected.
Fence	F	You create a series of line segments. These line segments will appear dotted. All objects touching these fence lines will be selected. You cannot close the fence shape; it must remain open.
Last	L	Selects the last object you created.
Previous	P	Selects the object or objects from the most recently created selection set.
Window	W	You create a rectangular box by dragging your cursor from left to right (transparent blue window). This rectangular box will appear as solid lines. All objects completely enclosed in this rectangular box will be selected. Objects touching the edges of the box will be ignored.
Window Polygon	WP	You create an irregular closed shape. The polygon formed will appear as solid line segments. All objects completely enclosed and touching this irregular shape will be selected.

OBJECT PRE-HIGHLIGHT

Whenever you move your cursor on an object, the object pre-highlights. This provides an easy way to select the correct object, especially in busy drawings. In the following image, the cursor is moved over the rectangle that surrounds the 128 text object. This action pre-highlights this rectangle until you move your cursor off this object.

Figure 3–28

SELECTING OBJECTS BY INDIVIDUAL PICKS

When AutoCAD prompts you with "Select objects," a pickbox appears as the cursor is moved across the display screen, as shown on the left in the following image. As you move your cursor over an object, the object will highlight and turn bold to signify that it is selectable, as shown on the right in the following image. An object will remain highlighted once it is selected.

Figure 3–29

 Try It! – Open the drawing file 03_Select. Enter the ERASE command and at the Select objects prompt, pick the arc segment labeled "A" in Figure 3–30. To signify that the object is selected the arc highlights.

Command: **E** *(For ERASE)*
Select objects: *(Pick the object at "A")*
Select objects: *(Press ENTER to execute the ERASE command)*

Figure 3–30

SELECTING OBJECTS BY WINDOW

The individual pick method previously mentioned works fine for small numbers of objects. However, when numerous objects need to be edited, selecting each individual object could prove time-consuming. Instead, you can select all objects that you want to become part of a selection set by using the Window selection mode. This mode requires you to create a rectangular box by picking two diagonal points. In the following image, a selection window has been created with point "A" as the first corner and "B" as the other corner. As an additional aid in selecting, the rectangular box you construct is displayed as a transparent blue box by default. When you use this selection mode, only those objects completely enclosed by the blue window box are selected. Even though the window touches other objects, they are not completely enclosed by the window and therefore are not selected.

Figure 3–31

SELECTING OBJECTS BY CROSSING WINDOW

In the previous example of producing a selection set by a window, the window selected only those objects completely enclosed by it. The following image is an example of selecting objects by a crossing window. The Crossing Window option requires two points to define a rectangle, as does the Window selection option. In the following image, a transparent green rectangle with dashed edges is used to select objects, using "C" and "D" as corners for the rectangle; however, this time the crossing window was used. The highlighted objects illustrate the results. All objects that are touched by or enclosed by the crossing rectangle are selected. Because the transparent green crossing rectangle passes through the three circles without enclosing them, they are still selected by this object selection mode.

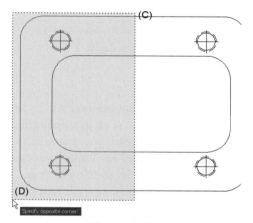

Figure 3–32

SELECTING OBJECTS BY A FENCE

Use this mode to create a selection set by drawing a line or group of line segments called a fence. Any object touched by the fence is selected. The fence does not have to end exactly where it was started. In the following image, all objects touched by the fence are selected, as represented by the dashed lines.

 Try It! – Open the drawing file 03_Fence. Follow the command sequence below and the illustration to select a group of objects using a fence.

 Command: **E** *(For ERASE)*
Select objects: **F** *(For Fence)*
First fence point: *(Pick a first fence point)*
Specify endpoint of line or [Undo]: *(Pick a second fence point)*
Specify endpoint of line or [Undo]: *(Pick a third fence point)*
Specify endpoint of line or [Undo]: *(Press ENTER to exit fence mode)*
Select objects: *(Press ENTER to execute the ERASE command)*

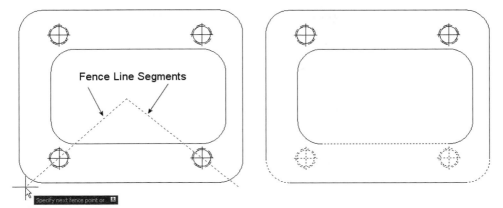

Figure 3–33

REMOVING OBJECTS FROM A SELECTION SET

All the previous examples of creating selection sets have shown you how to create new selection sets. What if you select the wrong object or objects? Instead of canceling out of the command and trying to select the correct objects, you can use the Remove option to remove objects from an existing selection set. As illustrated on the left in the following image, a selection set has been created and made up of all the highlighted objects. However, the outer lines and arcs were mistakenly selected as part of the selection set. The Remove option allows you to remove highlighted objects from a selection set. To activate Remove, press SHIFT and pick the object or objects you want removed; this works only if the Select objects prompt is present. When the highlighted objects are removed from the selection set, as shown on the right in the following image, they regain their original display intensity.

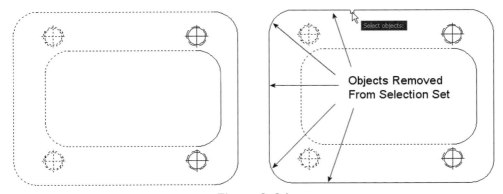

Figure 3–34

SELECTING THE PREVIOUS SELECTION SET

When you create a selection set of objects, this grouping is remembered until another selection set is made. The new selection set replaces the original set of objects. Let's say you moved a group of objects to a new location on the display screen. Now you want to rotate these same objects at a certain angle. Rather than select the same set of objects to rotate, you would pick the Previous option or type P at the Select objects prompt. This selects the previous selection set. The buffer holding the selection set is cleared whenever you use the U command to undo the previous command.

SELECTING OBJECTS BY A CROSSING POLYGON

When you use the Window or Crossing Window mode to create selection sets, two points specify a rectangular box for selecting objects. At times, it is difficult to select objects by the rectangular window or crossing box because in more cases than not, extra objects are selected and have to be removed from the selection set.

 Try It! – Open the drawing file 03_Select CP. The following image shows a mechanical part with a "C"-shaped slot. Rather than use Window or Crossing Window modes, you can pick the Crossing Polygon mode (CPolygon) or type CP at the Select objects prompt. You simply pick points representing a polygon. Any object that touches or is inside the polygon is added to a selection set of objects. As illustrated on the left in the following image, the crossing polygon is constructed using points "1" through "5." A similar but different selection set mode is the Window Polygon (WPolygon). Objects are selected using this mode when they lie completely inside the Window Polygon, which is similar to the regular Window mode.

Figure 3–35

CYCLING THROUGH OBJECTS

At times, the process of selecting objects can become quite tedious. Often, objects lie directly on top of each other. As you select the object to delete, the other object selects instead. To remedy this, press and hold CTRL when prompted to "Select objects." This activates Object Selection Cycling and enables you to scroll through all objects in the vicinity of the pickbox. A message appears in the prompt area when you begin scrolling, alerting you that cycling is on; you can now pick objects until the desired object is highlighted. Pressing ENTER not only accepts the highlighted object but toggles cycling off. In the following image and with selection cycling on, the first pick selects the circle, the second pick selects the line segment, and the third pick selects the rectangle. Keep picking until the desired object highlights.

Figure 3–36

 Try It! – Open the drawing file 03_Cycle and experiment using this feature on the object illustrated in the previous image.

PRE-SELECTING OBJECTS

You can bypass the Select objects prompt by pre-selecting objects at the Command prompt. Notice that when you pick an object, not only does it highlight but a series of blue square boxes also appears. These objects are called grips and will be discussed in greater detail in chapter 7. To cancel or de-select the object, press the ESC key and notice that even the grips disappear. You could also pick a blank part of your screen at the Command prompt. Moving your cursor to the right and picking a point has the same effect as using the Window option of Select objects. Only objects completely inside the window will be selected. If you pick a blank part of your screen at the Command prompt and move your cursor to the left, this has the same effect as using the Crossing option of Select objects in which any items touched by or completely enclosed by the box are selected.

 Tip: Pressing CTRL + A at the Command prompt selects all objects in the entire drawing and displays the blue grip boxes. This will even select objects that are on a layer that has been turned off, but will not select those on a layer that is frozen or locked.

THE QSELECT COMMAND

Yet another way of creating a selection set is by matching the object type and property with objects currently in use in a drawing. This is the purpose of the QSELECT (Quick Select) command. This command can be chosen from the Tools pull-down menu, as shown on the left in the following image, which will launch the dialog box on the right.

Figure 3–37

 Try It! – Open the drawing file 03_Qselect and activate the Quick Select dialog box. This command works only if objects are defined in a drawing; the Quick Select dialog box does not display in a drawing file without any objects drawn. Clicking in the Object type edit box displays all object types currently used in the drawing. This enables you to create a selection set by the object type. For instance, to select all line segments in the drawing file, click on Line in the Object type edit box, as illustrated on the left in the following image.

Clicking the OK button at the bottom of the dialog box returns you to the drawing and highlights all the line segments in the drawing, as shown on the right in the following image.

Figure 3–38

Other controls of Quick Select include the ability to select the object type from the entire drawing or from just a segment of the drawing. You can narrow the selection criteria by adding various properties to the selection mode such as Color, Layer, and Linetype, to name a few. You can also create a reverse selection set. The Quick Select dialog box lives up to its name—it enables you to create a quick selection set.

TUTORIAL EXERCISE: 03_SELECT OBJECTS.DWG

Figure 3–39

Purpose
The purpose of this tutorial is to experiment with the object's selection modes on the drawing shown in the previous image.

System Settings
Keep the current limits settings of (0,0) for the lower left corner and (16,10) for the upper right corner. Keep all remaining system settings.

Layers
Layers have already been created for this tutorial exercise.

Suggested Commands
This tutorial utilizes the ERASE command as a means of learning the basic object selection modes. The following selection modes will be highlighted for this tutorial: Window, Crossing, Window Polygon, Crossing Polygon, Fence, and All. The effects of locking layers and selecting objects will also be demonstrated.

STEP 1

Open the drawing 03_Select Objects.dwg and activate ERASE at the Command prompt. At the Select objects prompt, pick a point on the blank part of your screen at "A." At the next Select objects prompt, pick a point on your screen at "B." Notice that a solid blue rectangular box is formed (see the following image). The presence of this box signifies the Window option of selecting objects.

 Command: **E** (*For ERASE*)

Select objects: (*Pick a point at "A"*)
Specify opposite corner: (*Pick a point at "B"*)

Figure 3–40

The result of selecting objects by a window is illustrated in the following image. Only those objects completely surrounded by the window are highlighted. Pressing ENTER will perform the erase operation. Before continuing on to the next step, issue the U (UNDO) command, which will negate the previous ERASE operation.

Figure 3–41

STEP 2

With the entire object displayed on the screen, activate the ERASE command again. At the Select objects prompt, pick a point on the blank part of your screen at "A." At the Specify opposite corner prompt, pick a point on your screen at "B." Notice that the rectangular box is now dashed and green in appearance (see the following image). The presence of this box signifies the Crossing option of selecting objects.

 Command: **E** (*For ERASE*)

Select objects: (*Pick a point at "A"*)
Specify opposite corner: (*Pick a point at "B"*)

Figure 3–42

The result of selecting objects by a crossing window is illustrated in the following image. Notice that objects touched by the crossing box, as well as those objects completely surrounded by the crossing box, are highlighted. Pressing ENTER will perform the ERASE operation. Before continuing on to the next step, issue the U (UNDO) command, which will negate the previous ERASE operation.

Figure 3–43

STEP 3

Activate the ERASE command again. At the Select objects prompt, pick a point on the blank part of your screen at "A." At the Specify opposite corner prompt, pick a point on your screen at "B" (see the following image). The presence of this blue box signifies the Window option of selecting objects.

 Command: **E** *(For ERASE)*

Select objects: *(Pick a point at "A")*

Specify opposite corner: *(Pick a point at "B")*

Figure 3–44

The result of selecting objects by a window is illustrated in the following image on the left. Notice that a number of circles and rectangles have been selected along with a single dimension. Unfortunately, the rectangles and the dimension were selected by mistake. Rather than cancel the ERASE command and select the objects again, press and hold down the SHIFT key and click on the highlighted rectangles and dimension. Notice how this operation de-selects the objects (see the following image on the right). Pressing ENTER will perform the ERASE operation. Before continuing on to the next step, issue the U (UNDO) command, which will negate the previous ERASE operation.

Figure 3–45

STEP 4

Yet another way to select objects is by using a Window Polygon. This method allows you to construct a closed irregular shape for selecting objects. Again, activate the ERASE command. At the Select objects prompt, enter WP for Window Polygon. At the First polygon point prompt, pick a point on your screen at "A." Continue picking other points until the desired polygon, in blue, is formed, as shown in the following image.

 Note: The entire polygon must form a single closed shape; edges of the polygon cannot cross each other.

 Command: **E** *(For ERASE)*
Select objects: **WP** *(For Window Polygon)*
First polygon point: *(Pick a point at "A")*
Specify endpoint of line or [Undo]: *(Pick at "B")*
Specify endpoint of line or [Undo]: *(Pick at "C")*
Specify endpoint of line or [Undo]: *(Pick at "D")*
Specify endpoint of line or [Undo]: *(Pick at "E")*
Specify endpoint of line or [Undo]: *(Pick at "F")*
Specify endpoint of line or [Undo]: *(Pick at "G")*
Specify endpoint of line or [Undo]: *(Press* ENTER*)*
11 found

Figure 3–46

The result of selecting objects by a window polygon is illustrated in the following image. As with the window selection mode, objects must lie entirely inside the window polygon in order for them to be selected. Pressing ENTER will perform the ERASE operation. Before continuing on to the next step, issue the U (UNDO) command, which will negate the previous ERASE operation.

 Note: A Crossing Polygon (CP) mode is also available to select objects that touch the crossing polygon or are completely surrounded by the polygon. A green polygonal shape displays as it is being created.

Figure 3–47

STEP 5

Objects can also be selected by a fence. This is represented by a crossing line segment. Any object that touches the crossing line is selected. You can construct numerous crossing line segments. You cannot use the Fence mode to surround objects as with the Window or Crossing modes. Activate the ERASE command again. At the Select objects prompt, enter F, for Fence. At the First fence point prompt, pick a point on your screen at "A." Continue constructing crossing line segments until you are satisfied with the objects being selected (see the following image).

 Command: **E** *(For ERASE)*
Select objects: **F** *(For Fence)*
First fence point: *(Pick at "A")*
Specify endpoint of line or [Undo]: *(Pick at "B")*
Specify endpoint of line or [Undo]: *(Pick at "C")*
Specify endpoint of line or [Undo]: *(Pick at "D")*
Specify endpoint of line or [Undo]: *(Pick at "E")*
Specify endpoint of line or [Undo]: *(Press ENTER to create the selection set)*

Figure 3–48

The result of selecting objects by a fence is illustrated in the following image. Notice that objects touched by the fence are highlighted. Pressing ENTER will perform the ERASE operation. Before continuing on to the next step, issue the U (UNDO) command, which will negate the previous ERASE operation.

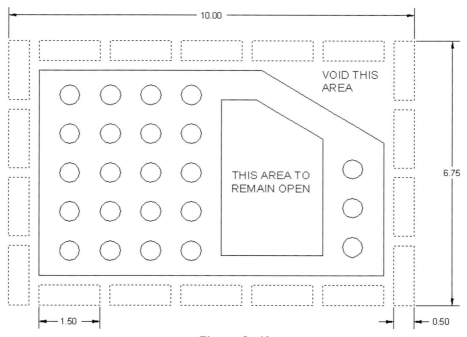

Figure 3–49

STEP 6

When situations require you to select all objects in the entire database of a drawing, the All option is very efficient. Activate the ERASE command again. At the Select objects prompt, enter All. Notice that in the following image all objects are selected by this option.

 Command: **E** *(For ERASE)*
Select objects: **All**

Figure 3–50

Pressing ENTER will perform the ERASE operation. Before continuing on to the next step, issue the U (UNDO) command, which will negate the previous ERASE operation.

 Note: The All option will also select objects even on layers that are turned off. The All option will not select objects on layers that are frozen.

STEP 7

Before performing another ERASE operation on a number of objects, activate the Layer Properties Manager dialog box, select the "Circles" layer, and click the lock icon, as shown in the following image. This operation locks the "Circles" layer.

Figure 3-51

To see what effect this has on selecting objects, activate the ERASE command again. At the Select objects prompt, pick a point on the blank part of your screen at "A." At the Specify opposite corner prompt, pick a point on your screen at "B." You have once again selected a number of objects by a green crossing box (see the following image).

 Command: **E** *(For ERASE)*
Select objects: *(Pick a point at "A")*
Specify opposite corner: *(Pick a point at "B")*
20 were on a locked layer.

Figure 3-52

The result of selecting objects by a crossing window is illustrated in the following image. Notice that even though a group of circles was completely surrounded by the crossing window, they do not highlight because they belong to a locked layer.

Pressing ENTER will perform the ERASE operation. Before continuing on to the next step, issue the U (UNDO) command, which will negate the previous ERASE operation.

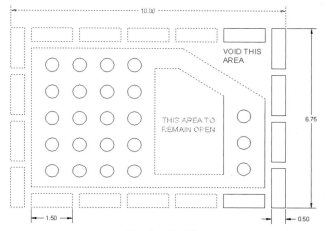

Figure 3–53

STEP 8

This final step illustrates the use of the ZOOM-All and ZOOM-Extents commands and how they differ. First activate the ZOOM command and use the All option (a button is available for performing this operation).

 Command: **Z** *(For ZOOM)*
Specify corner of window, enter a scale factor (nX or nXP), or
[All/Center/Dynamic/Extents/Previous/Scale/Window/Object] <real time>: **A** *(For All)*

The result is illustrated in the following image. Performing a ZOOM-All zooms the display based on the current drawing limits (the limits of this drawing have been established as 0,0 for the lower left corner and 16,10 for the upper right corner).

Figure 3–54

Now activate the ZOOM command and use the Extents option (a button is also available for performing this operation).

Command: **Z** *(For ZOOM)*
Specify corner of window, enter a scale factor (nX or nXP), or [All/Center/Dynamic/Extents/Previous/Scale/Window/Object] <real time>: **E** *(For Extents)*

The result is illustrated in the following image. This zoom is performed to the limits that contain all the objects currently in the drawing.

Figure 3–55

CHAPTER 4

Modifying Your Drawings

The heart of any CAD system is its ability to modify and manipulate existing geometry, and AutoCAD is no exception. Many modify commands relieve the designer of drudgery and mundane tasks, and this allows more productive time for conceptualizing the design. This chapter will break all AutoCAD modify commands down into two separate groupings; the first grouping is called Level I and will cover the MOVE, COPY, SCALE, ROTATE, OFFSET, FILLET, CHAMFER, TRIM EXTEND, and BREAK commands. The second grouping is called Level II and will cover the ARRAY, MIRROR, STRETCH, PEDIT, EXPLODE, LENGTHEN, JOIN, UNDO, and REDO commands. A number of small exercises accompany each command in order to reinforce the importance of its use.

METHODS OF SELECTING MODIFY COMMANDS

As with all commands, you can find the main body of modify commands on the Modify pull-down menu, as shown on the left in the following image, or through the Modify and Modify II toolbars, both shown in the following image. Yet another area to activate modify commands is through the Modify tool palette, as shown on the right in the following image.

Figure 4–1

183

You can also enter all modify commands directly from the keyboard either using their entire name or through command aliasing, as in the following examples.

Enter F for the FILLET command

Enter M for the MOVE command

LEVEL I MODIFY COMMANDS

With all the modify commands available in AutoCAD, the following represent beginning, or Level I, commands, which you will find yourself using numerous times as you make changes to your drawing. These commands are briefly described in the following table:

Button	Tool	Shortcut	Function
	Move	M	Used for moving objects from one location to another
	Copy	Cp or Co	Used for copying objects from one location to another
	Scale	Sc	Used for increasing or reducing the size of objects
	Rotate	Ro	Used for rotating objects to a different angle
	Fillet	F	Used for rounding off the corners of objects at a specified radius
	Chamfer	Cha	Used to connect two objects with an angled line forming a bevel
	Offset	O	Used for copying objects parallel to one another at a specified distance
	Trim	Tr	Used for partially deleting objects based on a cutting edge
	Extend	Ex	Used for extending objects based on a boundary edge
	Break	Br	Creates a gap in an object between two specified points
	Break at Point	Br	Breaks an object into two objects at a specified point without a gap present

⊕ MOVING OBJECTS

The MOVE command repositions an object or group of objects at a new location. Choose this command from one of the following:

    The Modify toolbar

    The pull-down menu (Modify > Move)

    The keyboard (M or MOVE)

    Right-click after selecting object

Once the objects to move are selected, AutoCAD prompts the user to select a base point of displacement (where the object is to move from). Next, AutoCAD prompts the user to select a second point of displacement (where the object is to move to), as shown in the following image.

 Command: **M** *(For MOVE)*
Select objects: *(Select the bed, as shown in the following image)*
Select objects: *(Press ENTER to continue)*
Specify base point or displacement: *(Select the endpoint of the bed at "A")*
Specify second point of displacement or <use first point as displacement>: *(Mark a
   point at "B")*

Figure 4–2

Try It! – Open the drawing file 04_Move. The slot as shown on the left in the following image is incorrectly positioned; it needs to be placed 1.00 unit away from the left edge of the object. You can use the MOVE command in combination with a polar coordinate or Direct Distance mode to perform this operation. Use this illustration and the following command sequence for performing this operation.

Command: **M** *(For MOVE)*
Select objects: *(Select the slot and all centerlines)*
Select objects: *(Press ENTER to continue)*
Specify base point or displacement: *(Select the edge of arc "A")*
Specify second point of displacement or <use first point as displacement>: **@.50<0**
   (or turn Polar on, move your cursor to the right, and type .50 at the command prompt)

As the slot is moved to a new position with the MOVE command, a new horizontal dimension must be placed to reflect the correct distance from the edge of the object to the centerline of the arc, as shown in the following image. The STRETCH command affects a group of objects along with the dimension and will be explained later in this chapter.

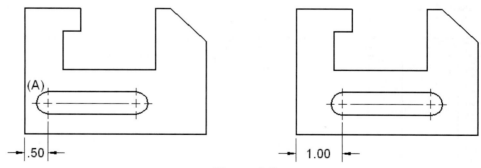

Figure 4–3

PRESS AND DRAG MOVE

If accuracy is not important and you simply need to move an object or group of objects to a new approximate location, you can use a press and drag technique. First, select the objects at the command prompt. Then, press and hold down the left mouse button on one of the highlighted objects (not one of the blue grips), and drag the objects to the new location.

COPYING OBJECTS

The COPY command is used to duplicate an object or group of objects. Choose this command from one of the following:

  The Modify toolbar

  The pull-down menu (Modify > Copy)

  The keyboard (CP or COPY)

  Right-click mouse after selecting object

Once the COPY command is executed, the multiple copy mode is on. To duplicate numerous objects while staying inside the COPY command, simply keep picking new second points of displacement and the objects will copy to these new locations, as shown in the following image. Once the copy is completed, press ENTER or ESC to exit.

Command: **CP** *(For COPY)*
Select objects: *(Select the chair to copy)*
Select objects: *(Press* ENTER *to continue)*
Specify base point or displacement: *(Pick a reference point for the copy operation)*
Specify second point of displacement or <use first point as displacement>: *(Pick a location for the first copy)*
Specify second point of displacement [Exit/Undo] <Exit>: *(Pick a location for the second copy)*
Specify second point of displacement [Exit/Undo] <Exit>: *(Pick a location for the third copy)*
Specify second point of displacement [Exit/Undo] <Exit>: *(Press* ENTER *or* ESC *to exit this command)*

Figure 4–4

Try It! – Open the drawing file 04_Copy Multiple. Follow the command sequence in the previous example to copy the three holes a multiple of times. Use the intersection of "A" as the base point for the copy. Then copy the three holes to the intersections located at "B," "C," "D," "E," "F," "G," "H," and "J," as shown in the following image.

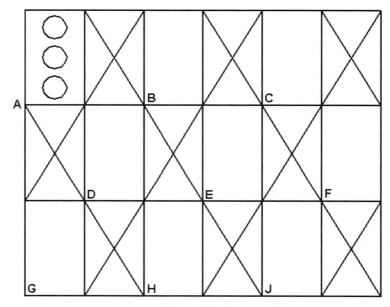

Figure 4–5

PRESS AND DRAG COPY

As with the MOVE command, you can also use the press and drag technique to copy objects to a new approximate location. First, select the objects at the command prompt. Then, press and hold down the right mouse button on one of the highlighted objects (not one of the blue grips), and drag the objects to the new location. When you release the mouse button a menu will display, as shown in the following image. Select the Copy Here item to copy the item or group of items.

Figure 4–6

SCALING OBJECTS

Use the SCALE command to change the overall size of an object. The size may be larger or smaller in relation to the original object or group of objects. The SCALE command requires a base point and scale factor to complete the command. Choose this command from one of the following:

> The Modify toolbar
>
> The pull-down menu (Modify > Scale)
>
> The keyboard (SC or SCALE)
>
> Right-click mouse after selecting object

 Try It! – Open the drawing file 04_Scale1. With a base point at "A" and a scale factor of 0.50, the results of using the SCALE command on a group of objects are shown in the following image.

Command: **SC** *(For SCALE)*
Select objects: **All**
Select objects: *(Press* ENTER *to continue)*
Specify base point: *(Select the endpoint of the line at "A")*

Specify scale factor or [Scale/Reference]: **0.50**

Figure 4–7

Try It! – Open the drawing file 04_Scale2. The example in the following image shows the effects of identifying a new base point in the center of the object.

 Command: **SC** *(For SCALE)*
Select objects: **All**
Select objects: *(Press ENTER to continue)*
Specify base point: *(Pick a point near "A")*
Specify scale factor or [Scale/Reference]: **0.40**

Figure 4–8

 Note: After identifying the base point for the scaling operation, you can use the Copy option to create a scaled copy of the objects you are scaling.

SCALE—REFERENCE

Suppose you are given a drawing that has been scaled down in size. However, no one knows what scale factor was used. You do know what one of the distances should be. In this special case, you can use the Reference option of the SCALE command to identify endpoints of a line segment that act as a reference length. Entering a new length value could increase or decrease the entire object proportionally.

 Try It! – Open the drawing file 04_Scale Reference. Study the following image and the following prompts for performing this operation.

 Command: **SC** *(For SCALE)*
Select objects: *(Pick a point at "A")*
Specify opposite corner: *(Pick a point at "B")*
Select objects: *(Press ENTER to continue)*
Specify base point: *(Select the edge of the circle to identify its center)*
Specify scale factor or [Scale/Reference]: **R** *(For Reference)*
Specify reference length <1>: *(Select the endpoint of the line at "C")*
Specify second point: *(Select the endpoint of the line at "D")*
Specify new length: **2.00**

Because the length of line "CD" was not known, the endpoints were picked after the Reference option was entered. This provided the length of the line to AutoCAD. The final step to perform was to make the line 2.00 units, which increased the size of the object while also keeping its proportions.

Figure 4–9

ROTATING OBJECTS

The ROTATE command changes the orientation of an object or group of objects by identifying a base point and a rotation angle that completes the new orientation. Choose this command from one of the following:

    The Modify toolbar

    The pull-down menu (Modify > Rotate)

    The keyboard (RO or ROTATE)

    Right-click mouse after selecting object

The following image shows an object, complete with crosshatch pattern, that needs to be rotated to a 30° angle using point "A" as the base point.

Try It! – Open the drawing file 04_Rotate. Use the following prompts and image to perform the rotation.

Command: **RO** *(For ROTATE)*
Current positive angle in UCS: ANGDIR=counterclockwise ANGBASE=0
Select objects: **All**
Select objects: *(Press* ENTER *to continue)*
Specify base point: *(Select the endpoint of the line at "A")*
Specify rotation angle or [Scale/Reference]: **30**

Figure 4–10

 Note: After identifying the base point for the rotating operation, you can use the Copy option to create a rotated copy of the objects you are rotating.

ROTATE—REFERENCE

At times it is necessary to rotate an object to a desired angular position. However, this must be accomplished even if the current angle of the object is unknown. To maintain the accuracy of the rotation operation, use the Reference option of the ROTATE command. The following image shows an object that needs to be rotated to the 30°-angle position. Unfortunately, we do not know the angle the object currently lies in. Entering the Reference angle option and identifying two points create a known angle of reference. Entering a new angle of 30° rotates the object to the 30° position from the reference angle.

 Try It! – Open the drawing file 04_Rotate Reference. Use the following prompts and image to accomplish this.

Command: **RO** *(For ROTATE)*
Current positive angle in UCS: ANGDIR=counterclockwise ANGBASE=0
Select objects: *(Select the object in the following image)*
Select objects: *(Press ENTER to continue)*
Specify base point: (Pick either the edge of the circle or two arc segments to locate the center)
Specify rotation angle or [Scale/Reference]: **R** *(For Reference)*
Specify the reference angle <0>: *(Pick either the edge of the circle or two arc segments to locate the center)*
Specify second point: **Mid**
of *(Select the line at "A" to establish the reference angle)*
Specify the new angle: **30**

Figure 4–11

CREATING FILLETS AND ROUNDS

Many objects require highly finished and polished surfaces consisting of extremely sharp corners. Fillets and rounds represent the opposite case, where corners are rounded off, either for ornamental purposes or as required by design. Generally a fillet consists of a rounded edge formed in the corner of an object, as illustrated in the following image. A round is formed at an outside corner. Fillets and rounds are primarily used where objects are cast or made from poured metal. The metal will form more easily around a shape that has rounded corners instead of sharp corners, which usually break away. Some drawings have so many fillets and rounds that a note is used to convey the size of them all, similar to "All Fillets and Rounds .125 Radius."

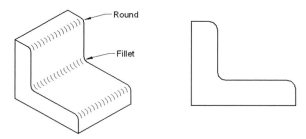

Figure 4–12

AutoCAD provides the FILLET command, which allows you to enter a radius followed by the selection of two lines. The result will be a fillet of the specified radius to the two lines selected. The two lines are also automatically trimmed, leaving the radius drawn from the endpoint of one line to the endpoint of the other line. Choose this command from one of the following:

- The Modify toolbar
- The pull-down menu (Modify > Fillet)
- The keyboard (F or FILLET)

FILLETING BY RADIUS

Illustrated in the following image is an example of setting a radius in the FILLET command for creating rounded-off corners.

 Try It! – Open the drawing file 04_Fillet. Follow the illustration on the left in the following image and command sequence below to place fillets at the three corner locations.

 Command: **F** *(For FILLET)*
Current settings: Mode = TRIM, Radius = 0.0000
Select first object or [Undo/Polyline/Radius/Trim/Multiple]: **R** *(For Radius)*
Specify fillet radius <0.0000>: **0.25**
Select first object or [Undo/Polyline/Radius/Trim/Multiple]: *(Select at "A")*
Select second object or shift-select to apply corner: *(Select at "B")*
Command:
(Press ENTER to re-execute this command)

Repeat this procedure for creating additional fillets using lines "BC" and "CD." When finished, your display should appear similar to the illustration on the right in the following image.

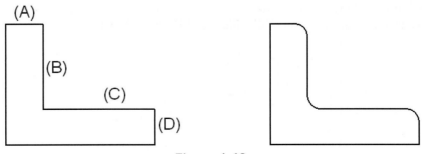

Figure 4–13

FILLET AS A CORNERING TOOL

A very productive feature of the FILLET command is its use as a cornering tool. To accomplish this, set the fillet radius to a value of 0. This produces a corner out of two non-intersecting objects.

 Try It! – Open the drawing file 04_Fillet Corner1. Follow the illustration on the left in the following image and the command sequence below for performing this task.

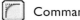 Command: **F** *(For FILLET)*
Current settings: Mode = TRIM, Radius = 0.5000
Select first object or [Undo/Polyline/Radius/Trim/Multiple]: **R** *(For Radius)*
Enter fillet radius <0.5000>: **0**
Select first object or [Undo/Polyline/Radius/Trim/Multiple]: *(Select line "A")*
Select second object or shift-select to apply corner: *(Select line "B")*

Repeat the procedure for the remaining two corners using lines "BC" and "CD." When finished, your display should appear similar to the illustration on the right in the following image.

Figure 4–14

 Tip: Even with a fillet radius set to a positive value, such as .50, you can easily apply a corner to two lines by holding down the SHIFT key when picking the second object. The use of the SHIFT key in this example will temporarily set the fillet radius to zero. Releasing the SHIFT key sets the fillet radius back to the current value, in this case .50.

PERFORMING MULTIPLE FILLETS

Since the filleting of lines is performed numerous times, a multiple option is available that automatically repeats the FILLET command. This option can be found by picking Multiple from the Cursor menu or by typing M for Multiple at the command prompt.

Try It! – Open the drawing file 04_Fillet Multiple. Activate the FILLET command and verify that the radius is set to 0. Use the Multiple option to make the command repeat. Then click on the corners until your object appears similar to the illustration on the right in the following image. If you make a mistake by picking two lines incorrectly, type U to undo this operation and still remain in the FILLET command.

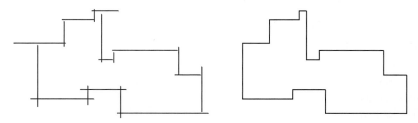

Figure 4–15

FILLETING POLYLINES

In the previous examples of using the FILLET command, you had to pick individual line segments in order to produce one rounded corner. You also had to repeat these picks for additional rounded corners. If the object you are filleting is a polyline, you can have the FILLET command round off all corners of this polyline in a single pick.

 Try It! – Open the drawing file 04_Fillet Pline. Using the FILLET command on a polyline object produces rounded edges at all corners of the polyline in a single operation. Follow the illustration in the following image and the command sequence below for performing this task.

 Command: **F** *(For FILLET)*
Current settings: Mode = TRIM, Radius = 0.0000
Select first object or [Undo/Polyline/Radius/Trim/Multiple]: **R** *(For Radius)*
Enter fillet radius <0.0000>: **0.25**
Select first object or [Undo/Polyline/Radius/Trim/Multiple]: **P** *(For Polyline)*
Select 2D polyline: *(Select the polyline at "A")*

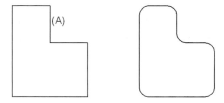

Figure 4–16

NO TRIM FILLETS

The FILLET command can also be used to control whether or not to trim the excess corners after a fillet is placed. The illustration on the left in the following image shows a typical fillet operation where the polyline at "A" is selected. However, instead of the corners of the polyline automatically being trimmed, a new Trim/No trim option allows the lines to remain, as shown on the right in the following image. The following prompts illustrate this operation.

 Try It! – Open the drawing file 04_Fillet No-Trim. Follow the command sequence and illustration in the following image for using the No-Trim option of the FILLET command.

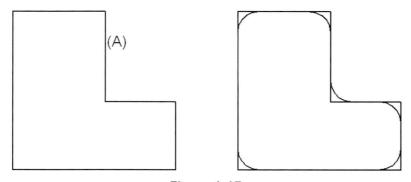

Figure 4–17

Command: **F** *(For FILLET)*
Current settings: Mode = TRIM, Radius = 0.0000
Select first object or [Undo/Polyline/Radius/Trim/Multiple]: **T** *(For Trim)*
Enter Trim mode option [Trim/No trim] <Trim>: **N** *(For No trim)*
Select first object or [Undo/Polyline/Radius/Trim/Multiple]: **R** *(For Radius)*
Specify fillet radius <0.0000>: 0.50
Select first object or [Undo/Polyline/Radius/Trim/Multiple]: **P** *(For Polyline)*
Select 2D polyline: *(Select polyline "A")*
6 lines were filleted

 Note: Reset the Trim setting in the FILLET command back to "Trim" from "No trim." This can be done from the keyboard by entering the command TRIMMODE and setting it to 1 (on).

Command: **TRIMMODE**
Enter new value for TRIMMODE <0>: **1**

FILLETING PARALLEL LINES

Filleting two parallel lines, as shown in the following image, automatically constructs a semicircular arc object connecting both lines at their endpoints. When performing this operation, it does not matter what the radius value is set to.

 Try It! – Open the drawing file 04_Fillet Parallel. Use the illustration on the left in the following image and the command sequence below for performing this task.

Command: **F** *(For FILLET)*
Current settings: Mode = TRIM, Radius = 1.0000
Select first object or [Undo/Polyline/Radius/Trim/Multiple]: *(Select line "A")*
Select second object or shift-select to apply corner: *(Select line "B")*

Continue filleting the remaining parallel lines to complete all slots. Quicken the process by using the Multiple option when prompted to "Select first object."

 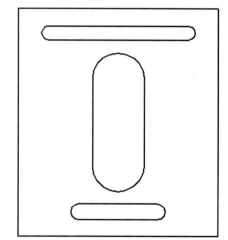

Figure 4–18

CREATING CHAMFERS

Chamfers represent a way to finish a sharp corner of an object. The CHAMFER command produces an inclined surface at an edge of two intersecting line segments. Distances determine how far from the corner the chamfer is made. The following image illustrates two examples of chamfered edges; one edge is created from unequal distances while the other uses equal distances.

Figure 4–19

The CHAMFER command is designed to draw an angle across a sharp corner given two chamfer distances. Choose this command from one of the following:

The Modify toolbar

The pull-down menu (Modify > Chamfer)

The keyboard (CHA or CHAMFER)

CHAMFER BY EQUAL DISTANCES

The most popular chamfer involves a 45° angle, which is illustrated in the following image. You can control this angle by entering two equal distances.

 Try It! – Open the drawing file 04_Chamfer Distances. In the example in the following image, if you specify the same numeric value for both chamfer distances, a 45°-angled chamfer will automatically be formed. As long as both distances are the same, a 45° chamfer will always be drawn. Study the illustration in the following image and the following prompts:

Command: **CHA** *(For CHAMFER)*
(TRIM mode) Current chamfer Dist1 = 0.5000, Dist2 = 0.5000
Select first line or [Undo/Polyline/Distance/Angle/Trim/mEthod/Multiple]: **D** *(For Distance)*
Specify first chamfer distance <0.5000>: **0.15**
Specify second chamfer distance <0.1500>: *(Press* ENTER *to accept the default)*
Select first line or [Undo/Polyline/Distance/Angle/Trim/mEthod/Multiple]: *(Select the line at "A")*
Select second line or shift-select to apply corner: *(Select the line at "B")*

Figure 4–20

Note: When both chamfer distances are set to 0 (zero) and two edges are selected, the effects are identical to setting the fillet radius to 0 (zero); the CHAMFER command is used here as a cornering tool.

CHAMFER BY ANGLE

Another technique of constructing a chamfer is when one distance and the angle are given. When the chamfer is made up of an angle other than 45°, it is commonly referred to as a beveled edge.

 Try It! – Open the drawing file 04_Chamfer Angle. The following image illustrates the use of the CHAMFER command by setting one distance and identifying an angle.

 Command: **CHA** *(For CHAMFER)*
(TRIM mode) Current chamfer Dist1 = 0.5000, Dist2 = 0.5000
Select first line or [Undo/Polyline/Distance/Angle/Trim/mEthod/Multiple]: **A** *(For Angle)*
Specify chamfer length on the first line <0.1500>: **0.15**
Specify chamfer angle from the first line <60>: **60**
Select first line or [Undo/Polyline/Distance/Angle/Trim/mEthod/Multiple]: *(Select the line at "A")*
Select second line or shift-select to apply corner: *(Select the line at "B")*

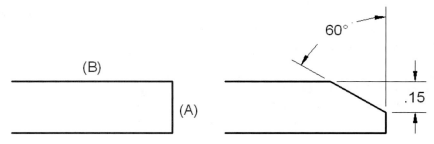

Figure 4–21

CHAMFERING A POLYLINE

When working with polyline objects, you have the opportunity to select only one of the edges of the polyline. All edges of the polyline will be chamfered to the specified distances or angle.

 Try It! – Open the drawing file 04_Chamfer Pline. Because a polyline consists of numerous segments representing a single object, using the CHAMFER command with the Polyline option produces corners throughout the entire polyline, as shown in the following image.

Command: **CHA** *(For CHAMFER)*
(TRIM mode) Current chamfer Dist1 = 0.00, Dist2 = 0.00
Select first line or [Undo/Polyline/Distance/Angle/Trim/mEthod/Multiple]: **D** *(For Distance)*
Enter first chamfer distance <0.00>: **0.50**
Enter second chamfer distance <0.50>: *(Press ENTER to accept the default)*
Select first line or [Undo/Polyline/Distance/Angle/Trim/mEthod/Multiple]: **P** *(For Polyline)*
Select 2D Polyline: (Select the Polyline at "A")

Figure 4–22

Note: A Multiple option of the chamfer command allows you to chamfer edges that share the same chamfer distances without exiting and reentering the command.

CHAMFERING WITHOUT TRIMMING

Special cases come up in the design process in which you may not wish the chamfered edges to be automatically trimmed. Changing the Trim option to No trim in the CHAMFER command and picking two edges will create the chamfered edge but will not trim the lines to the chamfer.

Try It! – Open the drawing file 04_Chamfer No-Trim. The CHAMFER command supports a Trim/No trim option, enabling a chamfer to be placed with lines trimmed or not trimmed, as shown in the following image. Follow the prompt sequence carefully to set the No trim option.

Command: **CHA** *(For CHAMFER)*
(TRIM mode) Current chamfer Dist1 = 0.2000, Dist2 = 0.2000
Select first line or [Undo/Polyline/Distance/Angle/Trim/mEthod/Multiple]: **T** *(For Trim)*
Trim/No trim <Trim>: **N** *(For No trim)*
Undo/Polyline/Distance/Angle/Trim/mEthod/Multiple/<Select first line>: **D** *(For Distance)*
Enter first chamfer distance <0.0000>: **1.00**
Enter second chamfer distance <1.0000>: *(Press ENTER to accept the default)*
Select first line or [Undo/Polyline/Distance/Angle/Trim/mEthod/Multiple]: **P** *(For Polyline)*
Select 2D Polyline: (Select the polyline at "A")

All edges are chamfered without being trimmed.

Note: Reset the Trim setting in the CHAMFER command back to "Trim" from "No trim." This can be done from the keyboard by entering the command TRIMMODE and setting it to 1 (On).

Command: **TRIMMODE**
Enter new value for TRIMMODE <0>: **1**

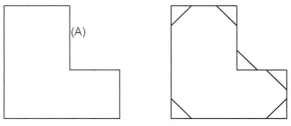

Figure 4–23

CHAMFER PROJECT—BEAM

The following Try it! exercise involves the chamfering of various corners at different distance settings.

 Try It! – Open the drawing file 04_Chamfer Beam. Using the illustration provided in the following image, follow these directions: Apply equal chamfer distances of 0.25 units to corners "AB," "BC," "DE," and "EF." Set new equal chamfer distances to 0.50 units and apply these distances to corners "GH" and "JK." Set a new first chamfer distance to 1.00; set a second chamfer distance to 0.50 units. Apply the first chamfer distance to line "L" and the second chamfer distance to line "H." Complete this object by applying the first chamfer distance to line "M" and the second chamfer distance to line "K."

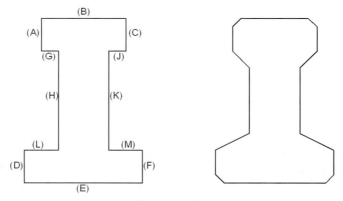

Figure 4–24

OFFSETTING OBJECTS

The OFFSET command is commonly used for creating a copy of one object that is parallel to another. Choose this command from one of the following:

The Modify toolbar

The pull-down menu (Modify > Offset)

The keyboard (O or OFFSET)

OFFSETTING USING A THROUGH POINT

One method of offsetting is to identify a point to offset through, called a through point. Once an object is selected to offset, a through point is identified. The selected object offsets to the point shown in the following image.

 Try It! – Open the drawing file 04_Offset Through. Refer to the following image and command sequence to use this method of the OFFSET command.

 Command: **O** *(For OFFSET)*
Current settings: Erase source=No Layer=Source OFFSETGAPTYPE=0
Specify offset distance or [Through/Erase/Layer] <1.00>: **T** *(For Through)*
Select object to offset or [Exit/Undo] <Exit>: *(Select the line at "A")*
Specify through point or [Exit/Multiple/Undo] <Exit>: **Nod**
of *(Select the point at "B")*
Select object to offset or [Exit/Undo] <Exit>: *(Press ENTER to exit this command)*

Figure 4–25

OFFSETTING BY A DISTANCE

Another method of offsetting is by a specified offset distance, as shown in the following image, where the objects need to be duplicated at a set distance from existing geometry. The COPY command could be used for this operation; a better command would be OFFSET. This allows you to specify a distance and a side for the offset to occur. The result is an object parallel to the original object at a specified distance. All objects in the following image need to be offset 0.50 toward the inside of the original object.

Try It! – Open the drawing file 04_Offset Shape. See the command sequence and following image to perform this operation.

 Command: **O** *(For OFFSET)*
Current settings: Erase source=No Layer=Source OFFSETGAPTYPE=0
Specify offset distance or [Through/Erase/Layer] <1.00>: **0.50**
Select object to offset or [Exit/Undo] <Exit>: *(Select the horizontal line at "A")*
Specify point on side to offset or [Exit/Multiple/Undo] <Exit>: *(Pick a point anywhere on the inside near "B")*

Repeat the preceding procedure for the remaining lines by offsetting them inside the shape, as shown on the left in the following image.

Notice that when all lines were offset, the original lengths of all line segments were maintained. Because all offsetting occurs inside, the segments overlap at their intersection points, as shown in the middle of the following image. In one case, at "A" and "B," the lines did not meet at all. The FILLET command will be used to edit all lines to form a sharp corner. You can accomplish this by setting the fillet radius to 0.

 Command: **F** *(For FILLET)*
Current settings: Mode = TRIM, Radius = 0.5000
Select first object or [Undo/Polyline/Radius/Trim/Multiple]: **R** *(For Radius)*
Enter fillet radius <0.5000>: **0**
Select first object or [Undo/Polyline/Radius/Trim/Multiple]: *(Select line "A")*
Select second object: *(Select line "B")*

Repeat the above procedure for the remaining lines, as shown in the middle in the following image.

Using the OFFSET command along with the FILLET command produces the result shown on the right in the following image. The fillet radius must be set to a value of 0 for this special effect.

Figure 4–26

PERFORMING MULTIPLE OFFSETS

If you know ahead of time that you will be offsetting the same object the same preset distance, you can use the Multiple option of the OFFSET command to work more efficiently.

 Try It! – Open the drawing file 04_Offset Multiple. See the command sequence and following image to perform this operation.

 Command: **O** *(For OFFSET)*
Current settings: Erase source=No   Layer=Source   OFFSETGAPTYPE=0
Specify offset distance or [Through/Erase/Layer] <0.4000>:
Select object to offset or [Exit/Undo] <Exit>: *(Pick vertical line "A")*
Specify point on side to offset or [Exit/Multiple/Undo] <Exit>: M *(For Multiple)*
Specify point on side to offset or [Exit/Undo] <next object>: *(Pick at "B")*
Specify point on side to offset or [Exit/Undo] <next object>: *(Pick at "B")*
Specify point on side to offset or [Exit/Undo] <next object>: *(Pick at "B")*
Specify point on side to offset or [Exit/Undo] <next object>: *(Press ENTER)*

Select object to offset or [Exit/Undo] <Exit>: *(Press ENTER to exit)*

Figure 4–27

OTHER OFFSET OPTIONS

Other options of the OFFSET command include Erase and Layer. When the Erase option is used, the original object you select to offset will be erased after the offset copy is made. When using the Layer option, the object being offset can take on the layer properties of the source object or can be based on the current layer. When using the source, the offset copy will take on the same layer as the source object you pick with offsetting. You could also make a new layer current. Using the Current Layer mode when offsetting will change all offset copies to the current layer. These extra offset modes allow you more flexibility with using this command.

TRIMMING OBJECTS

Use the TRIM command to partially delete an object or a group of objects based on a cutting edge. Choose this command from one of the following:

   The Modify toolbar

   The pull-down menu (Modify > Trim)

   The keyboard (TR or TRIM)

SELECTING INDIVIDUAL CUTTING EDGES

As illustrated on the left in the following image, the four dashed lines are selected as cutting edges. Next, segments of the circles are selected to be trimmed between the cutting edges.

 Try It! – Open the drawing file 04_Trim Basics. Use the following image and the command sequence below to perform this task.

 Command: **TR** *(For TRIM)*
Current settings: Projection=UCS Edge=None
Select cutting edges ...
Select objects or <select all>: *(Select the four dashed lines on the left in the following image)*
Select objects: *(Press ENTER to continue)*
Select object to extend or shift-select to trim or [Fence/Crossing/Project/Edge/eRase/Undo]: *(Select the circle areas at "A" through "D")*
Select object to extend or shift-select to trim or [Fence/Crossing/Project/Edge/eRase/Undo]: *(Press ENTER to exit this command)*

The results of performing trim on this object are illustrated on the right in the following image.

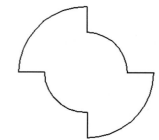

Figure 4–28

SELECTING ALL OBJECTS AS CUTTING EDGES

An alternate method of selecting cutting edges is to press ENTER in response to the prompt "Select objects." This automatically creates cutting edges out of all objects in the drawing. When you use this method, the cutting edges do not highlight. This is a very efficient means of trimming out unnecessary objects. You must, however, examine what you are trimming before using this method.

 Try It! – Open the drawing file 04_Trim All. Enter the TRIM command; press ENTER at the Select Objects prompt to select all objects as cutting edges. In the following image, pick the lines at "A," "B," and "E" and the arc segments at "C" and "D" as the objects to trim.

 Command: **TR** *(For TRIM)*
Current settings: Projection=UCS Edge=None
Select cutting edges ...
Select objects: *(Press ENTER to select all objects as cutting edges)*
Select object to extend or shift-select to trim or [Fence/Crossing/Project/Edge/eRase/Undo]: *(Select segments "A" through "E")*
Select object to extend or shift-select to trim or [Fence/Crossing/Project/Edge/eRase/Undo]: *(Press ENTER to exit this command)*

The results of performing trim on this object are illustrated on the right in the following image.

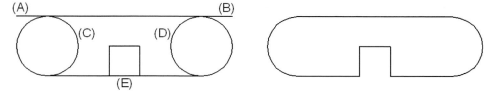

Figure 4–29

TRIMMING BY A CROSSING BOX

When trimming objects, you do not have to pick objects individually. You can erect a crossing box and trim objects out more efficiently.

 Try It! – Open the drawing file 04_Trim Crossing. Yet another application of the TRIM command uses the Crossing option of "Select objects." First, invoke the TRIM command and select the small circle as the cutting edge. Begin the response to the prompt of "Select object to trim" with "Crossing." See the illustration in the middle of the following image. Turn off Running OSNAP before conducting this exercise.

Command: **TR** *(For TRIM)*
Current settings: Projection=UCS Edge=None
Select cutting edges ...
Select objects or <select all>: *(Select the small circle, as shown on the left in the following image)*
Select objects: *(Press ENTER to continue)*
Select object to trim or shift-select to extend or [Fence/Crossing/Project/Edge/eRase/Undo]: **C** *(For Crossing)*
Specify first corner: *(Pick a corner in the middle image)*
Specify opposite corner: *(Pick an opposite corner in the middle image)*
Select object to trim or shift-select to extend or [Fence/Crossing/Project/Edge/eRase/Undo]: *(Press ENTER to exit this command)*

The power of the Crossing option of "Select objects" is shown on the right in the following image. Eliminating the need to select each individual line segment inside the small circle to trim, the Crossing mode trims all objects it touches in relation to the cutting edge.

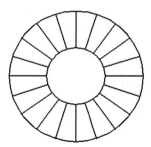

Figure 4–30

EXTENDING A CUTTING EDGE

The TRIM command also allows you to trim to an extended cutting edge. In Extended Cutting Edge mode, an imaginary cutting edge is formed; all objects sliced along this cutting edge will be trimmed if selected individually or by the Fence mode.

 Try It! – Open the drawing file 04_Trim Edge. Study the following image and prompts on this feature of the TRIM command.

Command: **TR** *(For TRIM)*
Current settings: Projection=UCS Edge=None
Select cutting edges ...
Select objects: *(Pick line "A")*
Select objects: *(Press ENTER to continue)*
Select object to extend or shift-select to trim or [Fence/Crossing/Project/Edge/eRase/Undo]: **E** *(For Edge)*
Enter an implied edge extension mode [Extend/No extend] <No extend>: **E** *(For Extend)*
Select object to extend or shift-select to trim or [Fence/Crossing/Project/Edge/eRase/Undo]: *(Pick the line at "B" along with the other segments)*
Select object to extend or shift-select to trim or [Fence/Crossing/Project/Edge/eRase/Undo]: *(Press ENTER to exit this command)*

The Fence mode can also be used to select all line segments at once to trim.

Figure 4–31

 Note: Reset the Edge setting in the TRIM command back to "No Extend" from "Extend." This can be done from the keyboard by entering the command EDGEMODE and setting it to 0 (Zero).

Command: **EDGEMODE**
Enter new value for EDGEMODE <1>: **0**

MORE ABOUT INDIVIDUAL CUTTING EDGES

Care must be taken when it is appropriate to press ENTER and select all objects in your drawing as cutting edges with using the TRIM command. To see this in effect, try the next exercise.

Try It! – Open the drawing file 04_Trim Cut. You need to remove the six vertical lines from the inside of the object. However, if you press ENTER to select all cutting edges, each individual segment would need to be trimmed, which is considered unproductive. Select lines "A" and "B" as cutting edges in the following image and select the inner vertical lines as the objects to trim. This is considered a more efficient way of using this command.

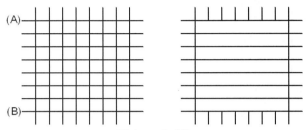

Figure 4–32

TRIMMING EXERCISE—FLOOR PLAN

Use any technique that you have learned to trim the walls of the floor plan in order for the object to appear similar to the illustration in the following image.

 Try It! – Open the drawing file 04_Trim Walls. Using the following image as a guide, use the TRIM command to trim away the extra overshoots and complete the floor plan illustrated in the figure.

Figure 4–33

 Tip: While inside the TRIM command, you can easily toggle to the EXTEND command by holding down the SHIFT key at the following command prompt:

Select object to trim or shift-select to extend or [Fence/Crossing/Project/Edge/eRase/ Undo]:

EXTENDING OBJECTS

The EXTEND command is used to extend objects to a specified boundary edge. Choose this command from one of the following:

    The Modify toolbar

    The pull-down menu (Modify > Extend)

    The keyboard (EX or EXTEND)

SELECTING INDIVIDUAL BOUNDARY EDGES

In the following image, select all dashed objects as the boundary edges. After pressing ENTER to continue with the command, select the lines at "A," "B," "C," and "D" to extend these objects to the boundary edges. If you select the wrong end of an object, use the Undo feature, which is an option of the command, to undo the change and repeat the procedure at the correct end of the object.

 Try It! – Open the drawing file 04_Extend Basics. Use the following illustration and command sequence for accomplishing this task.

 Command: **EX** *(For EXTEND)*

Current settings: Projection=UCS Edge=None
Select boundary edges ...
Select objects: *(Select the objects represented by dashes)*
Select objects: *(Press ENTER to continue)*
Select object to extend or shift-select to trim or [Fence/Crossing/Project/Edge/eRase/ Undo]: *(Select the ends of the lines at "A" through "D")*
Select object to extend or shift-select to trim or [Fence/Crossing/Project/Edge/eRase/ Undo]: *(Press ENTER to exit this command)*

 Tip: An alternate method of selecting boundary edges is to press ENTER in response to the "Select objects" prompt. This automatically creates boundary edges out of all objects in the drawing. When you use this method, however, the boundary edges will not highlight.

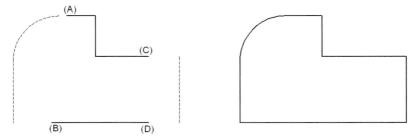

Figure 4–34

EXTENDING MULTIPLE OBJECTS

The EXTEND command can be used for extending multiple objects. After identifying the boundary edges, you can use a number of crossing boxes to identify numerous items to extend. This is a very productive method of using this command.

 Try It! – Open the drawing file 04_Extend Multiple. To extend multiple objects such as the line segments shown in the following image, select the lines at "A" and "B" as the boundary edge and use the Crossing mode to create two crossing boxes, represented by the dashed rectangles. This will extend all line segments to intersect with the boundaries.

Command: **EX** *(For EXTEND)*
Current settings: Projection=UCS Edge=None
Select boundary edges ...
Select objects: *(Select the lines at "A" and "B")*
Select objects: *(Press ENTER to continue)*
Select object to extend or shift-select to trim or [Fence/Crossing/Project/Edge/eRase/Undo]: **C** *(For Crossing)*
Specify first corner: *(Pick a point to start the first rectangle)*
Specify opposite corner: *(Pick a second point for the first rectangle)*
Select object to extend or shift-select to trim or [Fence/Crossing/Project/Edge/eRase/Undo]: *(Press ENTER to exit this command)*

Repeat this procedure to extend the lines on the other side of the concrete block.

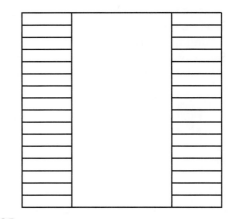

Figure 4–35

EXTENDING A BOUNDARY EDGE

Certain conditions require the boundary edge to be extended where an imaginary edge is projected, enabling objects not in direct sight of the boundary edge to still be extended.

 Try It! – Open the drawing file 04_Extend Edge. Study the following image and prompts on this special case involving the EXTEND command.

 Command: **EX** *(For EXTEND)*
Current settings: Projection=UCS Edge=None
Select boundary edges ...
Select objects: *(Pick line "A")*
Select objects: *(Press ENTER to continue)*
Select object to extend or shift-select to trim or [Fence/Crossing/Project/Edge/eRase/Undo]: **E** *(For Edge)*
Enter an implied edge extension mode [Extend/No extend] <No extend>: **E** *(For Extend the edge)*
Select object to extend or shift-select to trim or [Fence/Crossing/Project/Edge/eRase/Undo]: *(Pick line "B")*

Continue picking the remaining lines to be extended to the extended boundary edge. You could also use the Fence or Crossing mode to select all line segments to extend. After all lines are extended, press ENTER to exit the command.

Figure 4–36

 Note: Reset the Edge setting in the EXTEND command back to "No Extend" from "Extend." This can be done from the keyboard by entering the command EDGEMODE and setting it to 0 (Zero).

Command: **EDGEMODE**
Enter new value for EDGEMODE <1>: **0**

TOGGLING FROM EXTEND TO TRIM

While inside the EXTEND command, you can easily toggle to the TRIM command by holding down the SHIFT key at the following command prompt:

Select object to extend or shift-select to trim or [Fence/Crossing/Project/Edge/eRase/Undo]: *(Pressing* SHIFT *while picking objects activates the* TRIM *command.)*

The next Try It! exercise illustrates this technique.

Try It! – Open the drawing file 04_Extend and Trim. First, activate the EXTEND command and press ENTER, which will select all edges of the object in the following image as boundary edges. When picking the edges to extend, click on the ends of the lines from "A" through "J," as shown in the following image. At this point, do not exit the command. Press and hold down the SHIFT key; this will activate the TRIM command. Now pick all of the ends of the lines until your shape appears like the illustration on the right in the following image. The use of the SHIFT key when trimming or extending provides a quick means of switching between commands.

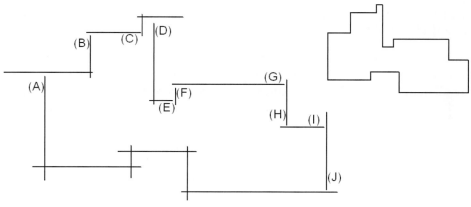

Figure 4–37

EXTEND EXERCISE—PIPING DIAGRAM

Try It! – Open the drawing file 04_Extend Pipe. Enter the EXTEND command and press ENTER when the "Select objects:" prompt appears. This will select all objects as boundary edges. Select the ends of all yellow lines representing pipes as the objects to extend. They will extend to intersect with the adjacent pipe fitting. Your finished drawing should appear similar to the following image.

Figure 4-38

BREAKING OBJECTS

The BREAK command is used to partially delete a segment of an object. Choose this command from one of the following:

    The Modify toolbar

    The pull-down menu (Modify > Break)

    The keyboard (BR or BREAK)

BREAKING AN OBJECT

The following command sequence and image show how the BREAK command is used.

Try It! – Open the drawing file 04_Break Gap. Turn off Running OSNAP prior to conducting this exercise. Use the following prompts and illustrations to break the line segment.

Command: **BR** *(For BREAK)*
Select objects: *(Select the line at "A")*
Specify second break point or [First point]: *(Pick the line at "B")*

Figure 4-39

IDENTIFYING A NEW FIRST BREAK POINT

In the previous example of using the BREAK command, the location where the object was selected became the first break point. You can select the object and then be prompted to pick a new first point. This option resets the command and allows you to select an object to break followed by two different points that identify the break. The following example illustrates this technique.

 Try It! – Open the drawing file 04_Break First. Utilize the First option of the BREAK command along with OSNAP options to select key objects to break. The following command sequence and image demonstrate using the First option of the BREAK command:

Command: **BR** *(For BREAK)*
Select object: *(Select the line)*
Specify second break point or [First point]: **F** *(For First)*
Specify first break point: **Int**
of *(Pick the intersection of the two lines at "A")*
Specify second break point: **End**

of (Pick the endpoint of the line at "B")

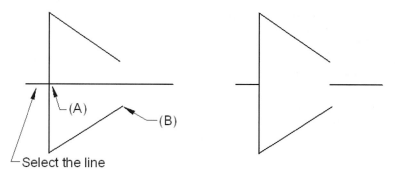

Figure 4–40

BREAKING CIRCLES

Circles can also be broken into arc segments with the BREAK command. There is only one rule to follow when breaking circles; you must pick the two break points in a counterclockwise direction when identifying the endpoints of the segment to be removed.

 Try It! – Open the drawing file 04_Break Circle. Study the following command sequence and image for breaking circles.

Command: **BR** *(For BREAK)*
Select objects: *(Select the circle)*
Specify second break point or [First point]: **F** *(For First)*
Specify first break point: *(Pick at "First Point")*
Specify second break point: *(Pick at "Second Point")*

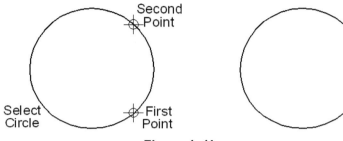

Figure 4–41

BREAK AT POINT

You can also break an object at a selected point. The break is so small that you cannot find it no matter how much you zoom in to the break point. As illustrated on the left in the following image, the line is highlighted to prove that it consists of a continuous object. Clicking on the Break at Point tool in the Modify toolbar activates the following command sequence:

Command: **BR** *(For BREAK)*
Select object: *(Select the line anywhere)*
Specify second break point or [First point]: **F** *(For First point)*
Specify first break point: **Mid**
of *(Pick the midpoint of the line)*
Specify second break point: **@** *(For previous point)*

The results are illustrated on the right in the following image. Here the line is again selected. Notice that only half of the line selects because the line was broken at its midpoint.

Figure 4–42

LEVEL II MODIFY COMMANDS

This second grouping of modify commands is designed to perform more powerful editing operations compared with the Level I modify commands already discussed earlier in this chapter. These commands are briefly described in the following table:

Button	Tool	Key-In	Function
	Array	AR	Creates multiple copies of objects in a rectangular or circular pattern
	Mirror	MI	Creates a mirror image of objects based on an axis of symmetry
	Stretch	S	Used for moving or stretching the shape of an object
	Pedit	PE	Used for editing polylines
	Explode	X	Breaks a compound object such as a polyline, block, or dimension into individual objects
None	Lengthen	LEN	Changes the length of lines and arcs
	Join	JO	Joins collinear objects to form a single unbroken object
	Undo	U	Used for backtracking or reversing the action of the previously used command
	Redo	RD	Reverses the effects of the previously used UNDO command operation

CREATING ARRAYS

If you need to create copies of objects that form rectangular or circular patterns, the ARRAY command is available to help with this task. This is a very powerful command that is dialog-box driven. If performing a rectangular array, you will need to supply the number of rows and columns for the pattern in addition to the spacing between these rows and columns. When performing a circular or polar array, you will need to supply the center point of the array, the number of items to copy, and the angle to fill. The next series of pages documents both methods of performing arrays.

CREATING RECTANGULAR ARRAYS

CREATING RECTANGULAR PATTERNS WITH POSITIVE OFFSET

The Array dialog box allows you to arrange multiple copies of an object or group of objects in a rectangular or polar (circular) pattern. When creating a rectangular array, you are prompted to enter the number of rows and columns for the array. A row is a group of objects that are copied vertically in the positive or negative direction. A column is a group of objects that are copied horizontally, also in the positive or negative direction.

Try It! – Open the drawing file 04_Array Rectangular Positive. Suppose the object illustrated on the left in the following image needs to be copied in a rectangular pattern consisting of three rows and three columns. You also want a row offset of .50 units and a column offset of 1.25 units (measured center to center or from the same location on each object) in order to space the rectangular shapes away from one another. The result is illustrated on the right in the following image.

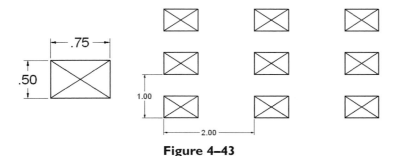

Figure 4–43

Clicking Array in the Modify pull-down menu displays the Array dialog box. When it appears, make the following changes. Be sure the Rectangular Array option is selected at the top left corner of the dialog box. Enter 3 for the number of rows and 3 for the number of columns. For Row Offset, enter a value of 1.00 units. For Column Offset, enter a value of 2.00 units. Click the Select objects button in the upper right corner of the dialog box. This will return you to the drawing. Pick the rectangle and the two diagonal line segments. Pressing ENTER when finished will return you to the Array dialog box. The Array dialog box should now appear similar to the following image.

Figure 4–44

Observe the pattern in the preview image and notice that it has been updated to reflect three rows and three columns. Notice also that the Preview< button in the lower right corner of the dialog box is active. Click this button to preview what the rectangular pattern will look like in your drawing. Clicking the Accept button in the Array alert box in the following image will complete the array operation. If you notice an error in the array results, click the Modify button. This will return you to the Array dialog box and allow you to make changes to any value. Clicking the Cancel button will abort the array operation and return you to the command prompt.

Figure 4–45

Tip: When performing rectangular arrays, a reference point must be established from the object to be arrayed and the distance between the objects determined. Not only must the spacing distance be used, but also the overall size of the object plays a role in determining the spacing distances. With the total height of the original object at 0.50 and a required spacing between rows of 0.50, both object height and spacing result in a distance of 1.00 from one reference point on an object to the next. A center-to-center distance is often specified on engineering drawings, which becomes the row or column spacing required.

CREATING RECTANGULAR PATTERNS WITH NEGATIVE OFFSETS

In the previous array example, the rectangular array illustrates a pattern that runs to the right of and above the original figure. At times these directions change to the left of and below the original object. The only change occurs in the distances between rows and columns, where negative values dictate the direction of the rectangular array, as shown on the right in the following image.

Try It! – Open the drawing file 04_Array Rectangular Negative. Follow the illustration in the image and the command prompt sequence below for performing this operation.

Activate the Array dialog box, as shown on the right in the following image. While in Rectangular Array mode, set the number of rows to 3 and the number of columns to 2. Because the directions of the array will be to the left of and below the original object, both row and column distances will be negative values. When finished, preview the array; if the results are similar to the illustration on the left in the following image, click the OK button to create the rectangular array pattern.

Figure 4–46

Try It! – Open the drawing 04_Array I-Beam and create a rectangular pattern consisting of four rows and five columns. Create a space of 20′ between both rows and columns. Dimensions may be added at a later time. The results are displayed in the following image.

Figure 4–47

Note: The I beam and footing were converted to a block object with the insertion point located at the center of the footing. For this reason, the I beams can be arrayed from their centers.

CREATING POLAR ARRAYS

Polar arrays allow you to create multiple copies of objects in a circular or polar pattern. After selecting the objects to array, you pick a center point for the array in addition to the number of items to copy and the angle to fill.

 Try It! – Open the drawing file 04_Array Polar and activate the Array dialog box. Be sure to select the Polar Array option at the top of the dialog box. Click the Select objects button at the top of the dialog box and pick the circle at "A" and the center line at "B" as the objects to array. For the Center point of the array, click the Select objects button and pick the intersection at "C" (this works only if the Intersection OSNAP mode is checked). Enter 4 for the Total number of items and be sure the Angle to fill is set to 360 (degrees). Also, verify that the box is checked next to Rotate items as copied. The Array dialog box should appear similar to the illustration shown on the right in the following image. Observe the results in the image icon and click the Preview< button to see the results. Click the Accept button to complete the array operation.

Figure 4–48

The results are illustrated on the left in the following image. This image also illustrates the creation of 4 holes that fill an angle of 180° in the clockwise direction. Notice in the Array dialog box illustrated on the right that an angle of −180° is given. This negative value drives the array in the clockwise direction. Remember that positive angles drive polar arrays in the counterclockwise direction.

Figure 4–49

 Try It! – Open the drawing file 04_Array Polar Rotate. Illustrated in the following image are three different results for arraying noncircular objects. The image on the right illustrates a polar array formed by rotating the square object as it is being copied. In the middle image, the square object is not being rotated as it is being copied. This results in the reference point at "B" pulling the arrayed pattern slightly to the right and off the main circular center line. To array rectangular or square objects in a polar pattern without rotating the objects, first convert the square or rectangle to a block with an insertion point located in the center of the square, as shown on the right in the following image (this process will be covered in detail in chapter 16). Now all squares lie an equal distance from their common center.

Figure 4–50

MIRRORING OBJECTS

The MIRROR command is used to create a mirrored copy of an object or group of objects. When performing a mirror operation, you have the option of deleting the original object, which would be the same as flipping the object, or keeping the original object along with the mirror image, which would be the same as flipping and copying. Choose this command from one of the following:

    The Modify toolbar

    The pull-down menu (Modify > Mirror)

    The keyboard (MI or MIRROR)

MIRRORING AND COPYING

The default action of the MIRROR command is to copy and flip the set of objects you are mirroring. After selecting the objects to mirror, you identify the first and second points of a mirror line. You then decide to keep or delete the source objects. The mirror operation is performed in relation to the mirror line. Usually, most mirroring is performed in relation to horizontal or vertical lines. As a result, it is usually recommended to turn on ORTHO mode to force orthogonal mirror lines (horizontal or vertical).

 Try It! – Open the drawing file 04_Mirror Copy. Refer to the following prompts and image for using the MIRROR command:

 Command: **MI** *(For MIRROR)*
Select objects: *(Select a point near "X")*
Specify opposite corner: *(Select a point near "Y")*
Select objects: *(Press* ENTER *to continue)*
Specify first point of mirror line: *(Select the endpoint of the centerline at "A")*
Specify second point of mirror line: *(Select the endpoint of the centerline at "B")*
Delete source objects? [Yes/No] <N>: *(Press* ENTER *for default)*

Because the original object needed to be retained by the mirror operation, the image result is shown on the right in the following image. The MIRROR command works well when symmetry is required.

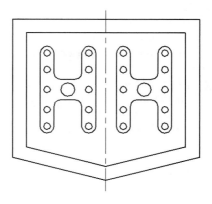

Figure 4–51

MIRRORING BY FLIPPING

The illustration in the following image is a different application of the MIRROR command. It is required to have all items that make up the bathroom plan flip but not copy to the other side. This is a typical process involving "what if" scenarios.

 Try It! – Open the drawing file 04_Mirror Flip. Use the following command prompts to perform this type of mirror operation. The results are displayed on the right in the following image.

Command: **MI** *(For MIRROR)*
Select objects: **All** *(This will select all objects)*
Select objects: *(Press ENTER to continue)*
Specify first point of mirror line: **Mid**
of *(Select the midpoint of the line at "A")*
Specify second point of mirror line: **Per**
to *(Select line "B," which is perpendicular to point "A")*
Delete source objects? [Yes/No] <N>: **Y** *(For Yes)*

Figure 4–52

MIRRORING TEXT

In addition to mirroring other object types, text can also be mirrored, as in the following image of the duplex complex. Rather than copy and move the text into position for the matching duplex half, you can include the text in the mirroring operation.

 Try It! – Open the drawing file 04_Mirror Duplex. Use the MIRROR command and create a mirror image of the Duplex floor plan using line "AB" as the points for the mirror line. Do not delete the source objects. Your finished results should be similar to the following image.

Figure 4–53

MORE INFORMATION ON MIRRORING TEXT

There may be times when you mirror text, but the text is not right-reading. The mirror operation makes the text backwards and difficult to read. This will usually occur with older AutoCAD drawings.

If this happens, a special system variable called MIRRTEXT is available. This variable must be entered at the Command prompt. If this variable is set to a value of 1 (or on), change the value to 0 (Zero, or off).

Command: **MIRRTEXT**
New value for MIRRTEXT <1>: **0**

STRETCHING OBJECTS

Use the STRETCH command to move a portion of a drawing while still preserving the connections to parts of the drawing remaining in place. Choose this command from one of the following:

>The Modify toolbar
>
>The pull-down menu (Modify > Stretch)
>
>The keyboard (S or STRETCH)

THE BASICS OF STRETCHING

To perform this type of operation, you must use the Crossing option of "Select objects." In the following image, a group of objects is selected with the crossing box. Next, a base point is identified at the approximate location of "C." Finally, a second point of displacement is identified with a polar coordinate or by using the Direct Distance mode. Once the objects selected in the crossing box are stretched, the objects not only move to the new location but also mend themselves.

 Try It! – Open the drawing file 04_Stretch. Use the following image and command sequence to perform this task.

 Command: **S** *(For STRETCH)*
Select objects to stretch by crossing-window or crossing-polygon...
Select objects: *(Pick a point at "A")*
Specify opposite corner: *(Pick a point at "B")*
Select objects: *(Press ENTER to continue)*
Specify base point or displacement: *(Select a point at "C")*
Specify second point of displacement: **@.75<180** *(Or use the Direct Distance mode)*

The results of performing this stretching operation are illustrated on the right in the following image.

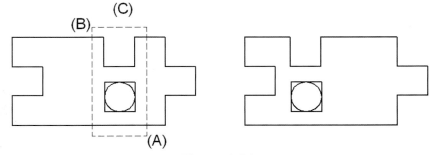

Figure 4–54

HOW STRETCHING AFFECTS DIMENSIONS

Applications of the STRETCH command are illustrated in the following image, in which a number of architectural features need to be positioned at a new location. The key to the success of using the STRETCH command on these features is in the selection of the objects to stretch through the crossing box.

 Try It! – Open the drawing file 04_Stretch Arch. Use the following command sequence and image to stretch the window, wall, and door the designated distances using the Direct Distance mode or by entering a polar coordinate at the keyboard. Since each of the stretch distances is a different value, the STRETCH command must be used three separate times.

Figure 4–55

STRETCHING USING MULTIPLE CROSSING WINDOWS

When identifying items to stretch by crossing box, you are not limited to a single crossing box. You can surround groups of objects with multiple crossing boxes. All items selected in this manner will be affected by the stretching operation.

 Try It! – Open the drawing file 04_Stretch Fence. Enter the STRETCH command and construct three separate crossing boxes to select the top of the fence boards at "A," "B," and "C," as shown in the following image. When prompted for a base point or displacement, pick a point on a blank part of your screen. Stretch these boards up at a distance of 12″ or 1′.

Figure 4–56

The results of performing the stretch operation on the fence are illustrated in the following image. When you have to stretch various groups of objects in a single operation, you can create numerous crossing boxes to better perform this task.

Figure 4–57

EDITING POLYLINES

Editing of polylines can lead to interesting results. A few of these options will be explained in the following pages. Choose this command from one of the following:

> The Modify II toolbar
>
> The pull-down menu (Modify > Object > Polyline)
>
> The keyboard (PE or PEDIT)

CHANGING THE WIDTH OF A POLYLINE

Illustrated on the left in the following image is a polyline of width 0.00. The PEDIT command is used to change the width of the polyline to 0.10 units, as shown on the right in the following image.

 Try It! – Open the drawing file 04_Pedit Width. Refer to the following command sequence to use the PEDIT command with the Width option.

  Command: **PE** *(For PEDIT)*
Select polyline or [Multiple]: *(Select the polyline)*
Enter an option [Open/Join/Width/Edit vertex/Fit/Spline/Decurve/Ltype gen/Undo]: **W**
   (For Width)
Specify new width for all segments: **0.10**
Enter an option [Open/Join/Width/Edit vertex/Fit/Spline/Decurve/Ltype gen/Undo]:
   (Press ENTER *to exit this command)*

Figure 4–58

JOINING OBJECTS INTO A SINGLE POLYLINE

It is very easy to convert regular objects such as lines and arcs into polylines (circles cannot be converted). As you can convert individual objects into polylines, the results are a collection of individual polylines. As long as the polyline endpoints match with one another, these polylines can be easily joined into one single polyline object using the Join option of the PEDIT command.

 Try It! – Open the drawing file 04_Pedit Join. Refer to the following command sequence and image to use this command.

 Command: **PE** *(For PEDIT)*
Select polyline or [Multiple]: *(Select the line at "A")*
Enter an option [Close/Join/Width/Edit vertex/Fit/Spline/Decurve/Ltype gen/Undo]: **J**
  (For Join)
Select objects: *(Pick a point at "B")*
Specify opposite corner: *(Pick a point at "C")*
Select objects: *(Press ENTER to join the lines)*
56 segments added to polyline
Enter an option [Open/Join/Width/Edit vertex/Fit/Spline/Decurve/Ltype gen/Undo]:
  (Press ENTER to exit this command)

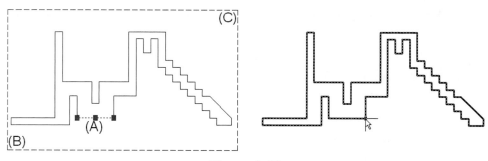

Figure 4–59

CURVE GENERATION

Polylines can be edited to form various curve-fitting shapes. Two curve-fitting modes are available, namely, Splines and Fit Curves.

GENERATING SPLINES

The Spline option produces a smooth-fitting curve based on control points in the form of the vertices of the polyline.

Modifying Your Drawings 231

 Try It! – Open the drawing file 04_Pedit Spline Curve. Refer to the following command sequence and image to use this command.

 Command: **PE** (For PEDIT)
Select polyline or [Multiple]: (Select the polyline frame)
Enter an option [Close/Join/Width/Edit vertex/Fit/Spline/Decurve/Ltype gen/Undo]: **S** (For Spline)
Enter an option [Close/Join/Width/Edit vertex/Fit/Spline/Decurve/Ltype gen/Undo]:
   (Press ENTER to exit this command)

The results of creating a spline curve from a polyline are shown on the right in the following image.

Figure 4–60

GENERATING FIT CURVES

The Fit Curve option passes entirely through the control points, producing a less desirable curve.

 Try It! – Open the drawing file 04_Pedit Fit Curve. Refer to the following command sequence and image to use this command.

 Command: **PE** (For PEDIT)
Select polyline or [Multiple]: (Select the polyline)
Enter an option [Close/Join/Width/Edit vertex/Fit/Spline/Decurve/Ltype gen/Undo]: **F** (For Fit)
Enter an option [Close/Join/Width/Edit vertex/Fit/Spline/Decurve/Ltype gen/Undo]:
   (Press ENTER to exit this command)

The results of creating a fit curve from a polyline are shown on the right in the following image.

Figure 4–61

LINETYPE GENERATION OF POLYLINES

The Linetype Generation option of the PEDIT command controls the pattern of the linetype from polyline vertex to vertex. In the polyline illustrated on the left in the following image, the hidden linetype is generated from the first vertex to the second vertex. An entirely different pattern is formed from the second vertex to the third vertex, and so on. The polyline illustrated on the right in the following image has the linetype generated throughout the entire polyline. In this way, the hidden linetype is smoothed throughout the polyline.

 Try It! – Open the drawing file 04_Pedit Ltype Gen. Refer to the following command sequence and image to use this command.

 Command: **PE** *(For PEDIT)*

Select polyline or [Multiple]: *(Select the polyline)*
Enter an option [Close/Join/Width/Edit vertex/Fit/Spline/Decurve/Ltype gen/Undo]: **Lt** *(For Ltype gen)*
Enter polyline linetype generation option [ON/OFF] <Off>: **On**
Enter an option [Close/Join/Width/Edit vertex/Fit/Spline/Decurve/Ltype gen/Undo]:
   (Press ENTER to exit this command)

Figure 4–62

OFFSETTING POLYLINE OBJECTS

Once a group of objects has been converted to and joined into a single polyline object, the entire polyline can be copied at a parallel distance using the OFFSET command.

 Try It! – Open the drawing file 04_Pedit Offset. Use the OFFSET command to copy the shape in the following image a distance of 0.50 units to the inside.

 Command: **O** *(For OFFSET)*

Specify offset distance or [Through/Erase/Layer] <1.0000>: **0.50**
Select object to offset or [Exit/Undo] <Exit>: *(Select the polyline at "A")*
Specify point on side to offset or [Exit/Multiple/Undo] <Exit>: *(Select a point anywhere near "B")*
Select object to offset or [Exit/Undo] <Exit>: *(Press ENTER to exit this command)*

Because the object was converted to a polyline, all objects are offset at the same time, as shown on the right in the following image.

Figure 4–63

MULTIPLE POLYLINE EDITING

Multiple editing of polylines allows for multiple objects to be converted to polylines. This is accomplished with the PEDIT command and the Multiple option. Illustrated on the left in the following image is a rectangle and four slots, all considered individual objects. When you run the Multiple option of the PEDIT command, not only can you convert all objects at once into individual polylines but you can join the endpoints of common shapes as well. The result of editing multiple polylines and then joining them is illustrated on the right in the following image.

 Try It! – Open the drawing file 04_Pedit Multiple1. Use the prompt sequence below and the following image to illustrate how the MPEDIT command is used.

 Command: **PE** *(For PEDIT)*
Select polyline or [Multiple]: **M** *(For Multiple)*
Select objects: **All**
Select objects: *(Press ENTER to continue)*
Enter an option [Close/Open/Join/Width/Fit/Spline/Decurve/Ltype gen/Undo]: **J** *(For Join)*
Join Type = Extend
Enter fuzz distance or [Jointype] <0.0000>: *(Press ENTER to accept this default value)*
15 segments added to 5 polylines
Enter an option [Close/Open/Join/Width/Fit/Spline/Decurve/Ltype gen/Undo]: *(Press ENTER to exit)*

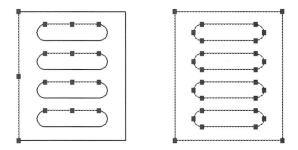

Figure 4–64

As a general rule when joining polylines, you cannot have gaps present or overlapping occurring when performing this operation. This is another feature of using the Multiple option of the PEDIT command. This option works best when joining two objects that have a gap or overlap. After selecting the two objects to join, you will be asked to enter a fuzz factor. This is the distance used by this command to bridge a gap or trim overlapping lines. You could measure the distance between two objects to determine this value. Study the following example for automatically creating corners in objects using a fuzz factor.

 Try It! – Open the drawing file 04_Pedit Multiple2. In this example, one of the larger gaps was measured to be .12 units in length. As a result, a fuzz factor slightly larger than this calculated value is used (.13 units). The completed object is illustrated on the right in the following image. You may have to experiment with various fuzz factors before you arrive at the desired results.

 Command: **PE** *(For PEDIT)*
Select polyline or [Multiple]: **M** *(For Multiple)*
Select objects: **All**
Select objects: *(Press* ENTER *to continue)*
Enter an option [Close/Open/Join/Width/Fit/Spline/Decurve/Ltype gen/Undo]: **J** *(For Join)*
Join Type = Extend
Enter fuzz distance or [Jointype] <0.25>: **.13**
11 segments added to polyline
Enter an option [Close/Open/Join/Width/Fit/Spline/Decurve/Ltype gen/Undo]: *(Press* ENTER *to exit)*

Figure 4–65

EXPLODING OBJECTS

Using the EXPLODE command on a polyline, dimension, or block separates the single object into its individual parts. Choose this command from one of the following:

> The Modify toolbar
>
> The pull-down menu (Modify > Explode)
>
> The keyboard (X or EXPLODE)

Illustrated on the left in the following image is a polyline that is considered one object. Using the EXPLODE command and selecting the polyline breaks the polyline into numerous individual objects, as shown on the right in the following image.

 Command: **X** *(For EXPLODE)*
Select objects: *(Select the polyline)*
Select objects: *(Press ENTER to perform the explode operation)*

Figure 4–66

 Note: Dimensions consist of extension lines, dimension lines, dimension text, and arrowheads, all grouped into a single object. (Dimensions are covered in chapter 11, "Adding Dimensions to Your Drawing.") Using the EXPLODE command on a dimension breaks the dimension down into individual extension lines, dimension lines, arrowheads, and dimension text. This is not advisable, because the dimension value will not be updated if the dimension is stretched along with the object being dimensioned. Also, the ability to manipulate dimensions with a feature called grips is lost. Grips are discussed in chapter 7.

LENGTHENING OBJECTS

The LENGTHEN command is used to change the length of a selected object without disturbing other object qualities such as angles of lines or radii of arcs. Choose this command from one of the following:

The pull-down menu (Modify > Lengthen)

The keyboard (LEN or LENGTHEN)

 Try It! – Open the drawing file 04_Lengthen1. Use the illustration in the following image and the command sequence below for performing this task.

Command: **LEN** *(For LENGTHEN)*
Select an object or [DElta/Percent/Total/DYnamic]: *(Select line "A")*
Current length: 12.3649
Select an object or [DElta/Percent/Total/DYnamic]: **T** *(For Total)*
Specify total length or [Angle] (1.0000)>: **20**
Select an object to change or [Undo]: *(Select the line at "A")*
Select an object to change or [Undo]: *(Press ENTER to exit)*

Figure 4–67

 Tip: After supplying the new total length of any object, be sure to select the desired end to lengthen when you are prompted to select an object to change.

JOINING OBJECTS

For special cases in which individual line segments need to be merged together as a single segment, the JOIN command can be used to accomplish this task. Usually this occurs when gaps occur in line segments and all segments lie in the same line of sight. This condition is sometimes referred to as collinear. Rather than connect the gaps with additional individual line segments, considered very unproductive and poor in practice, use JOIN to connect all segments as one.

Try It! – Open the drawing file 04_Join. Pick one line segment as the source. Then select other line segments to join to the source. Use the following command prompt and image to join various segments using the JOIN command.

Command: **J** *(For Join)*
Select source object: *(Select line "A")*
Select lines to join to source: *(Select line "B")*
Select lines to join to source: *(Select line "C")*
Select lines to join to source: *(Press ENTER to join the segments)*
2 lines joined to source

Continue using the JOIN command on the other line segments. A different JOIN command must be used for each group of line segments that appears broken in the following image.

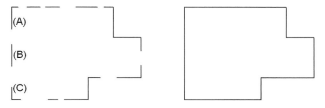

Figure 4–68

UNDOING OPERATIONS

The UNDO command can be used to undo the previous task or command action. Choose this command from one of the following:

   The Standard toolbar

   The pull-down menu (Edit > Undo)

   The keyboard (U or UNDO)

   Select anywhere in the drawing and right-click

For example, if you draw an arc followed by a line followed by a circle, issuing the UNDO command will undo the action caused by the most recent command; in this case, the circle would be removed from the drawing database. This represents one of the easiest ways to remove data or backtrack the drawing process.

Expanding the Undo list found in the Standard toolbar, shown in the following image, allows you to undo several actions at once. From this example, notice that the Line, Trim, Fillet, and Offset actions are highlighted for removal.

Figure 4–69

Note: When grouping actions to be undone, you cannot, in the previous image, for example, highlight Line, skip Trim, and highlight Fillet to be removed. The groupings to undo must be strung together in this dialog box.

REDOING OPERATIONS

You can also reverse the effect of the UNDO command by using REDO immediately after the UNDO operation. Choose this command from one of the following:

    The Standard toolbar

    The pull-down menu (Edit > Redo)

    The keyboard (REDO)

Clicking the REDO command button from the Standard toolbar will negate one UNDO operation. You can click on this button to cancel the effects of numerous UNDO operations.

As with UNDO, you can also REDO several actions at once through the Redo list shown in the following image. This list can be accessed from the Standard toolbar.

Figure 4–70

Note: REDO will only work if you have undone a previous operation. Otherwise, REDO remains inactive.

TUTORIAL EXERCISE: ANGLE.DWG

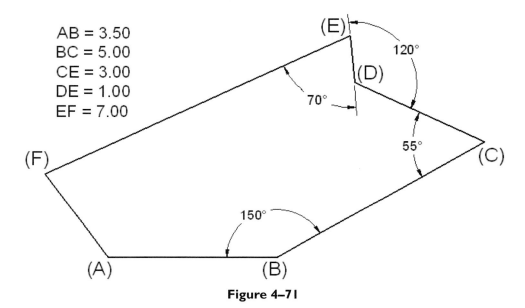

Figure 4-71

Purpose
This tutorial is designed to allow you to construct a one-view drawing of the angle as shown in the previous image using the Array dialog box and LENGTHEN command.

System Settings
Use the Drawing Units dialog box and change the precision of decimal units from 4 to 2 places. Use the current default settings for the limits of this drawing, (0,0) for the lower left corner and (12,9) for the upper right corner. Check to see that the following Object Snap modes are already set: Endpoint, Extension, Intersection, Center.

Layers
Create the following layer with the format:

Name	Color	Linetype
Object	White	Continuous

Suggested Commands
Make the Object layer current. Begin this drawing by constructing line "AB," which is horizontal. Use the Array dialog box to copy and rotate line "AB" at an angle of 150° in the clockwise direction. Once the line is copied and rotated, use the LENGTHEN command and modify the new line to the designated length. Repeat this procedure for lines "CD," "DE," and "EF." Complete the drawing by constructing a line segment from the endpoint at vertex "F" to the endpoint at vertex "A." This object could also have been constructed using a polar angle setting of 5° and a polar setting relative to the previous line.

STEP 1

Draw line "AB" using the Polar coordinate or Direct Distance mode, as shown in the following image. (Line "AB" is considered a horizontal line.)

Figure 4–72

STEP 2

One technique of constructing the adjacent line at 150° from line "AB" is to use the Array dialog box, as shown on the left in the following image, and perform a polar operation. Select line "AB" as the object to array, pick the intersection at "B" as the center of the array, and enter a value of −150° for the angle to fill. Entering a negative angle will copy the line in the clockwise direction. The value of your array center point will be different from the value displayed in the dialog box.

The result is shown on the right in the following image.

Figure 4–73

STEP 3

The array operation allowed line "AB" to be rotated and copied at the correct angle, namely −150°. However, the new line is the same length as line "AB." Use the LENGTHEN command to increase the length of the new line to a distance of 5.00 units. Use the Total option, specify the new total length of 5.00, and select the end of the line at "1" as the object to change, as shown on the left in the following image.

Command: **LEN** *(For LENGTHEN)*
Select an object or [DElta/Percent/Total/DYnamic]: **T** *(For Total)*
Specify total length or [Angle] <1.00)>: **5.00**
Select an object to change or [Undo]: *(Pick the end of the line at "1")*
Select an object to change or [Undo]: *(Press ENTER to exit this command)*

The result is shown on the right in the following image.

Figure 4–74

STEP 4

Use the Array dialog box, shown on the left in the following image, and perform a polar operation. Select line "BC" as the object to array, pick the intersection at "C" as the center of the array, and enter a value of -55° for the angle to fill. Entering a negative angle will copy the line in the clockwise direction.

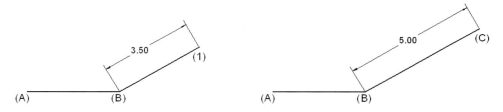

Figure 4–75

STEP 5

Then use the LENGTHEN command to reduce the length of the new line from 5.00 units to 3.00 units. Use the Total option, specify the new total length of 3.00, and select the end of the line at "1" as the object to change, as shown on the left in the following image.

Command: **LEN** *(For LENGTHEN)*
Select an object or [DElta/Percent/Total/DYnamic]: **T** *(For Total)*
Specify total length or [Angle] <5.00)>: **3.00**
Select an object to change or [Undo]: *(Pick the end of the line at "1")*
Select an object to change or [Undo]: *(Press ENTER to exit this command)*

The result is shown on the right in the following image.

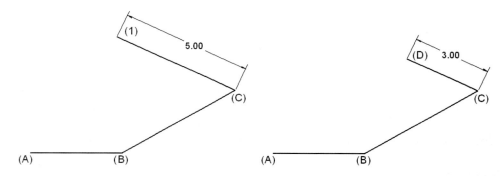

Figure 4–76

STEP 6

Use the Array dialog box and perform another polar operation. Select line "CD" as the object to array, as shown on the left in the following image, pick the intersection at "D" as the center of the array, and enter a value of 120° for the angle to fill.

Entering a positive angle will copy the line in the counterclockwise direction.

Then use the LENGTHEN command to reduce the length of the new line from 3.00 units to 1.00 unit, as shown on the right in the following image.

Command: **LEN** *(For LENGTHEN)*
Select an object or [DElta/Percent/Total/DYnamic]: **T** *(For Total)*
Specify total length or [Angle] <3.00)>: **1.00**
Select an object to change or [Undo]: *(Pick the end of the line at "1")*
Select an object to change or [Undo]: *(Press ENTER to exit this command)*

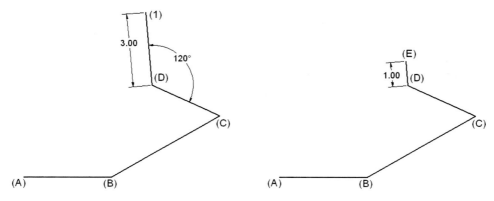

Figure 4–77

STEP 7

Use the Array dialog box and perform the final polar operation. Select line "DE" as the object to array, pick the intersection at "E" as the center of the array, and enter a value of −70° for the angle to fill. Entering a negative angle will copy the line in the clockwise direction.

Then use the LENGTHEN command to increase the length of the new line from 1.00 unit to 7.00 units, as shown on the right in the following image.

Command: **LEN** *(For LENGTHEN)*
Select an object or [DElta/Percent/Total/DYnamic]: **T** *(For Total)*
Specify total length or [Angle] <1.00)>: **7.00**
Select an object to change or [Undo]: *(Pick the end of the line at "1")*
Select an object to change or [Undo]: *(Press* ENTER *to exit this command)*

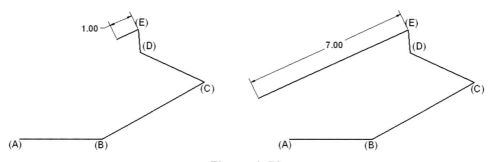

Figure 4–78

STEP 8

Connect endpoints "F" and "A" with a line as shown in the following image.

Command: **L** *(For LINE)*
Specify first point: *(Pick the endpoint of the line at "F")*
Specify next point or [Undo]: *(Pick the endpoint of the line at "A")*
Specify next point or [Undo]: *(Press ENTER to exit this command)*

The completed drawing is illustrated on the right in the following image. You may add dimensions at a later date.

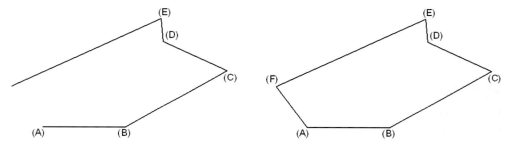

Figure 4–79

TUTORIAL EXERCISE: GASKET.DWG

Figure 4–80

Purpose
This tutorial is designed to allow you to construct a one-view drawing of the gasket using the Array dialog box.

System Settings
Use the current default settings for the units and limits of this drawing, (0,0) for the lower left corner and (12,9) for the upper right corner. Check to see that the following Object Snap modes are already set: Endpoint, Extension, Intersection, Center, and Quadrant.

Layers
Create the following layers with the format:

Name	Color	Linetype
Object	White	Continuous
Center	Yellow	Center

Suggested Commands
Draw the basic shape of the object using the LINE and CIRCLE commands. Lay out a centerline circle; draw one of the gasket tabs at the top of the center circle. Use the Array dialog box to create four copies in the −180° direction and two copies in the 80° direction. Trim out the excess arc segments to form the gasket. Convert the outer profile of the gasket into one continuous polyline object and offset this object .125 units to the outside of the gasket.

STEP 1

Use existing unit and limit settings. Create an Object and a Center layer. Assign the Center linetype to the Center layer.

Make the Object layer current and construct a circle of 2.00 radius with its center at absolute coordinate 4.00,4.00, as shown on the left in the following image.

Make the Center layer current and construct a circle of 2.25 radius using the previous center point, as shown on the right in the following image. Change the linetype scale from 1.00 to 0.50 units using the LTSCALE (LTS) command.

Figure 4–81

STEP 2

Make the Object layer current again and construct a circle of radius .75″ from the quadrant at the top of the centerline circle, as shown on the left in the following image. Also construct a circle of .75″ in diameter from the same quadrant at the top of the centerline circle, as shown on the right in the following image.

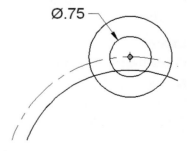

Figure 4–82

STEP 3

Copy the two top circles just created in a polar (circular) pattern using the Array dialog box. Use the center of the 2.00-radius circle as the center point of the array. Change the total number of items to 4 and the angle to fill to −180°, as shown on the left in the following image. The negative angle will drive the array in the clockwise direction, as shown on the right in the following image.

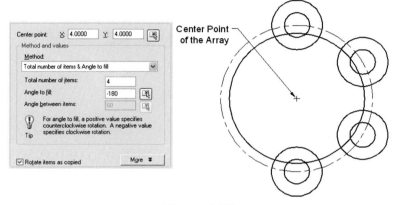

Figure 4–83

STEP 4

Perform another array operation on the top two circles. Again, use the center of the 2.00-radius circle as the center point of the array. Change the total number of items to 2 and the angle to fill to 80°, as shown on the left in the following image. The positive angle will drive the array in the counterclockwise direction, as shown on the right in the following image.

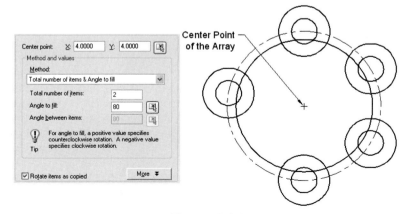

Figure 4–84

STEP 5

Trim out the inside edges of the five circles labeled "A" through "E" using the dashed circle as the cutting edge, as shown on the left in the following image. The results are displayed on the right.

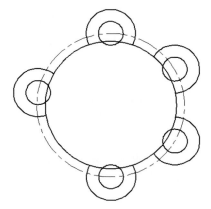

Figure 4–85

STEP 6

Use the TRIM command again. Select the five dashed arc segments as cutting edges and trim away the portions labeled "A" through "E," as shown on the left in the following image. The results are displayed on the right.

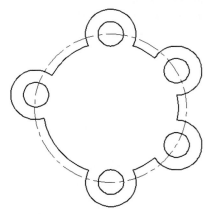

Figure 4–86

STEP 7

Fillet the inside corners of the .75-radius arcs with the 2.00-radius arc using the FILLET command and a radius set to .25˝. These corners are marked by a series of points, shown on the left in the following image. The results are displayed on the right.

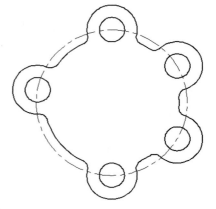

Figure 4–87

STEP 8

Create the large 1.50-radius hole in the center of the gasket using the CIRCLE command. Change the outer perimeter of the gasket into one continuous polyline object using the PEDIT command. Use the Join option of this command to accomplish this task. Finally, create a copy of the outer profile of the gasket a distance of .125˝ using the OFFSET command, as shown on the left in the following image. The completed gasket is displayed on the right.

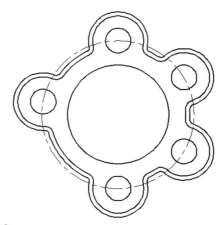

Figure 4–88

TUTORIAL EXERCISE: TILE.DWG

Figure 4–89

Purpose
This tutorial is designed to use the OFFSET and TRIM commands to complete the drawing of the floor tile shown in the previous image.

System Settings
Use the Drawing Units dialog box and change the units of measure from decimal to architectural units. Keep the remaining default settings. Use the LIMITS command and change the limits of the drawing to (0,0) for the lower left corner and (10′,8′) for the upper right corner. Use the ZOOM command and the All option to fit the new drawing limits to the display screen.

Check to see that the following Object Snap modes are currently set: Endpoint, Extension, Intersection, and Center.

Layers
Create the following layer with the format:

Name   Color   Linetype
Object   White   Continuous

Suggested Commands
Make the Object layer current. Use the RECTANGLE command to begin the inside square of the tile. The OFFSET command is used to copy the inner square a distance of 5″ to form the outer square. The ARRAY command is used to copy selected line segments in a rectangular pattern at a specified distance. The TRIM command is then used to form the inside tile patterns.

STEP 1

Verify that the current units are set to architectural and the drawing limits set to 10´, 8´ for the upper right corner. Be sure to perform a ZOOM-All on your screen. Draw the inner 3´-0˝ square using the RECTANGLE command, as shown on the left in the following image. Then offset the square 5˝ to the outside using the OFFSET command, as shown on the right in the following image. Because the square was drawn as a polyline, the entire shape offsets to the outside.

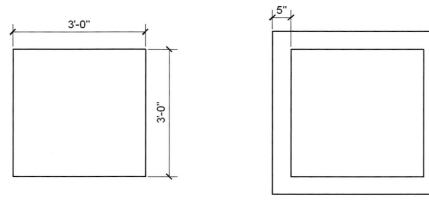

Figure 4–90

STEP 2

Notice that when clicking on the inner square on the left in the following image, the entire object highlights because it consists of a single polyline object. Use the EXPLODE command to break up the inner square into individual line segments. Now when you click on a line that is part of the inner square, only that line highlights, as shown on the right in the following image. This procedure is required in order to perform the next step.

Figure 4–91

STEP 3

You will now begin laying out the individual tiles with a spacing of 3″ between each. The Array dialog box will be used to accomplish this. Activate the Array dialog box, and change the number of rows to 1 and the number of columns to 12. Also change the distance between columns to 3″. Finally, pick the line at "A" as the object to array, as shown on the left in the following image. Your display should appear similar to the illustration on the right in the following image.

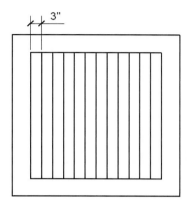

Figure 4–92

STEP 4

The bottom horizontal line needs to be copied multiple times vertically. Activate the Array dialog box; change the number of rows to 12 and the number of columns to 1. Also change the distance between rows to 3″. Finally, pick the line at "A" as the object to array, as shown on the left in the following image. Your display should appear similar to the illustration on the right in the following image.

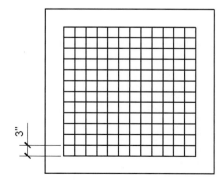

Figure 4–93

STEP 5

The TRIM command will now be used to clean up the inner lines and form the 3″ tiles. When using TRIM, do not press ENTER and select all cutting edges. This would be counterproductive. Instead, select the two vertical dashed lines, as shown on the left in the following image. Then trim away the horizontal segments in zones "A" through "D," as shown on the right in the following image. Remember that if you make a mistake and trim the wrong line, you can enter U in the command line to restore the previous trimmed line and pick the correct line.

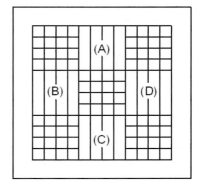

Figure 4–94

STEP 6

Use the TRIM command again to finish cleaning up the object. Select the two horizontal dashed lines as cutting edges, as shown on the left in the following image. Then trim away the vertical segments in zones "A" through "E," as shown on the right in the following image. This completes this exercise on creating the tile.

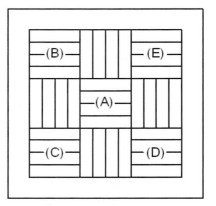

Figure 4–95

PROBLEMS FOR CHAPTER 4

Directions for Problems 4–1 through 4–14

Construct each one-view drawing using the appropriate coordinate mode or Direct Distance mode. Utilize advanced commands such as ARRAY and MIRROR whenever possible.

PROBLEM 4–1

0.125 GASKET THICKNESS

PROBLEM 4–2

WIDTH OF CENTERED RIB IS .25 WIDE

PROBLEM 4–3

PROBLEM 4–4

PROBLEM 4–5

PROBLEM 4–6

PROBLEM 4–7

PROBLEM 4–8

PROBLEM 4–9

PROBLEM 4–10

PROBLEM 4–11

PROBLEM 4–12

PROBLEM 4–13

PROBLEM 4–14

CHAPTER 5

Performing Geometric Constructions

In chapter 1, you were introduced to the LINE, CIRCLE, and PLINE commands. This chapter introduces the remainder of the drawing commands used for object creation. The following commands will be explained in this chapter: ARC, BOUNDARY, CIRCLE-2P, CIRCLE-3P, CIRCLE-TTR, CIRCLE-TTT, DONUT, ELLIPSE, MLINE, MLEDIT, MLSTYLE, POINT, DIVIDE, MEASURE, POLYGON, RAY, RECTANG, REVCLOUD, SPLINE, WIPEOUT, and XLINE.

Various scenarios of the CIRCLE-TTR command will be used to show the power of how AutoCAD can be used to create circles tangent to other object types. When covering the POINT command, the Point Style dialog box will also be discussed as a means of changing the appearance of points in your drawing. Multilines will also be covered, as will the ability to create a multiline style and then edit any intersections of a multiline.

METHODS OF SELECTING OTHER DRAW COMMANDS

You can find the main body of draw commands on the Draw pull-down menu, as shown on the left in the following image, or through the Draw toolbar. Yet another area to activate Draw commands is through the Draw tool palette, as shown on the right in the following image.

Figure 5–1

Refer to the following table for a complete listing of all buttons, tools, shortcuts, and functions of most drawing commands.

Button	Tool	Shortcut	Function
	Arc	A	Constructs an arc object using a number of command options
None	Boundary	BO	Traces a polyline boundary over the top of an existing closed group of objects
	Circle	C	Constructs a circle by radius, diameter, 2 points, 3 points, 2 tangent points and a radius, and 3 tangent points
None	Donut	DO	Creates a filled-in circular polyline object
	Ellipse	EL	Creates an elliptical object
None	Mline	ML	Creates multiline objects that consist of multiple parallel lines
	Point	PO	Creates a point object
None	Divide	DIV	Inserts evenly spaced points or blocks along an object's length or perimeter
None	Measure	ME	Inserts points or blocks along an object's length or perimeter at designated increments
	Polygon	POL	Creates an equilateral polygon shape consisting of various side combinations
None	Ray	RAY	Creates a semi-infinite line used for construction purposes
	Rectang	REC	Creates a rectangular polyline object
	Revcloud	REVCLOUD	Creates a revision cloud consisting of a polyline made up of sequential arc segments
	Spline	SPL	Creates a smooth curve from a sequence of points
None	Wipeout	WIPEOUT	Use to cover or hide objects with a blank area
	Xline	XL	Creates an infinite line used for construction purposes

CONSTRUCTING ARCS

Use the ARC command to construct portions of circular shapes by radius, diameter, arc length, included angle, and direction. Choose this command from one of the following:

> The Draw toolbar
>
> The pull-down menu (Draw > Arc)
>
> The keyboard (A or ARC)

Choosing Arc from the Draw pull-down menu displays the cascading menu shown in the following image. All supported methods of constructing arcs are displayed in the list. By default, the 3 Points Arc mode supports arc constructions in the clockwise as well as the counterclockwise direction. All other arc modes support the ability to construct arcs only in the counterclockwise direction. Three arc examples will be demonstrated in the next series of pages, namely, how to construct an arc by 3 Points, by a Starting point, Center point, and Ending point, and how to Continue arcs.

Figure 5–2

3 POINTS ARC MODE

By default, arcs are drawn with the 3 Points method. The first and third points identify the endpoints of the arc. This arc may be drawn in either the clockwise or counterclockwise direction.

 Try It! – Create a new drawing file starting from scratch. Use the following command sequence and image to construct a 3-point arc.

Command: **A** (For ARC)
Specify start point of arc or [Center]: *(Pick a point at "A")*
Specify second point of arc or [Center/End]: *(Pick a point at "B")*
Specify end point of arc: *(Pick a point at "C")*

Figure 5–3

START, CENTER, END MODE

Use this ARC mode to construct an arc by defining its start point, center point, and endpoint. This arc will always be constructed in a counterclockwise direction.

 Try It! – Open the drawing file 05_Door Swing. Use the following command sequence and image for constructing an arc representing the door swing by start, center, and end points.

 Command: **A** *(For ARC)*
Specify start point of arc or [Center]: *(Pick the endpoint at "A")*
Specify second point of arc or [Center/End]: **C** *(For Center)*
Specify center point of arc: *(Pick the endpoint at "B")*
Specify end point of arc or [Angle/chord Length]: *(Pick a point at "C")*

Figure 5–4

CONTINUE MODE

Use this mode to continue a previously drawn arc. All arcs drawn through Continue mode are automatically constructed tangent to the previous arc. Activate this mode from the Arc menu shown in the following image. The new arc begins at the last endpoint of the previous arc.

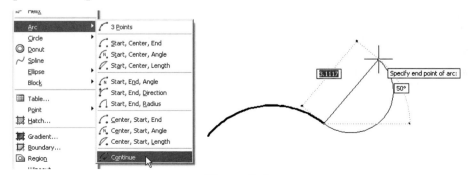

Figure 5–5

CREATING A BOUNDARY

The BOUNDARY command is used to create a polyline boundary around any closed shape. Choose this command from one of the following:

> The pull-down menu (Draw > Boundary)
>
> The keyboard (BO or BOUNDARY)

It has already been demonstrated that the Join option of the PEDIT command is used to join object segments into one continuous polyline. The BOUNDARY command automates this process even more. Start this command by choosing Boundary from the Draw pull-down menu, as shown on the left in the following image. This activates the Boundary Creation dialog box, shown on the right in the following image.

Figure 5–6

Before you use this command, it is considered good practice to create a separate layer to hold the polyline object; this layer could be called "Boundary" or "BP," for Boundary Polyline. Unlike the Join option of the PEDIT command, which converts individual objects to polyline objects, the BOUNDARY command will trace a polyline in the current layer on top of individual objects.

Try It! – Open the drawing file 05_Boundary Extrusion. Activate the Boundary dialog box and click the Pick Points button. Then pick a point inside the object illustrated in the following image. Notice how the entire object is highlighted. To complete the command, press ENTER when prompted to select another internal point, and the polyline will be traced over the top of the existing objects.

Command: **BO** (For BOUNDARY)
(The Boundary Creation dialog box appears. Click the Pick Points button.)
Select internal point: *(Pick a point at "A" in the following image)*
Selecting everything...
Selecting everything visible...
Analyzing the selected data...
Analyzing internal islands...
Select internal point: *(Press ENTER to construct the boundary)*
BOUNDARY created 1 polyline

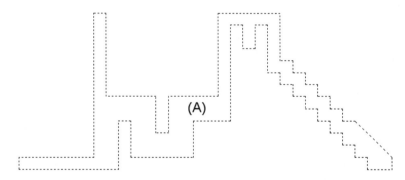

Figure 5–7

Once the boundary polyline is created, the boundary may be relocated to a different position on the screen with the MOVE command. The results are illustrated in the following image. The object on the left consists of the original individual objects. When the object on the left is selected, all objects highlight, signifying that it is made up of a polyline object made through the use of the BOUNDARY command.

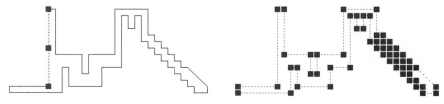

Figure 5–8

When the BOUNDARY command is used on an object consisting of an outline and internal islands similar to the drawing in the following image, a polyline object is also traced over these internal islands.

 Try It! – Open the drawing file 05_Boundary Cover in the following image. Notice that the current layer is Boundary and the color is Magenta. Issue the BOUNDARY command and pick a point inside the middle of the object at "A" without OSNAP turned on. Notice that the polyline will be constructed on the top of all existing objects. Turn the Object layer off to display just the Boundary layer.

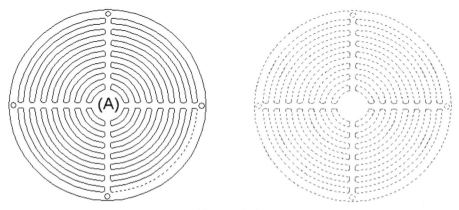

Figure 5–9

In addition to creating a special layer to hold the boundary polyline, another important rule to follow when using the BOUNDARY command is to be sure there are no gaps in the object. In the following image, when the BOUNDARY command encounters the gap at "A," a dialog box informs you that no internal boundary could be found. This is likely because the object is not completely closed. In this case, you must exit the command, close the object, and activate the BOUNDARY command again. It is acceptable, however, to have lines cross at the intersection at "B." While these "overshoots," as they are called, work well when using the Boundary Creation dialog box, this may not be considered a good drawing practice.

Figure 5–10

ADDITIONAL OPTIONS FOR CREATING CIRCLES

Additional options of the CIRCLE command may be selected from the Draw pull-down menu under the Circle heading. The 2 Points, 3 Points, Tan Tan Radius, and Tan Tan Tan modes will be explained in the next series of examples.

3 POINTS CIRCLE MODE

Use the CIRCLE command and the 3 Points mode to construct a circle by three points that you identify. No center point is required when you enter the 3 Points mode. Simply select three points and the circle is drawn.

Try It! – Open the drawing file 05_Circle_3P. Study the following prompts and image for constructing a circle using the 3 Points mode.

Command: **C** *(For CIRCLE)*
Specify center point for circle or [3P/2P/Ttr (tan tan radius)]: **3P**
Specify first point on circle: *(Pick a midpoint at "A")*
Specify second point on circle: *(Pick a midpoint at "B")*
Specify third point on circle: *(Pick a midpoint at "C")*

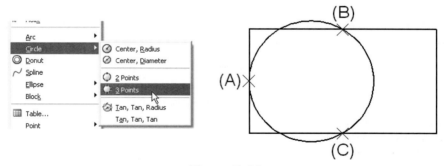

Figure 5–11

2 POINTS CIRCLE MODE

Use the CIRCLE command and the 2 Points mode to construct a circle by selecting two points. These points will form the diameter of the circle. No center point is required when you use the 2 Points mode.

 Try It! – Open the drawing file 05_Circle_2P. Study the following prompts and image for constructing a circle using the 2 Points mode.

Command: **C** *(For CIRCLE)*
Specify center point for circle or [3P/2P/Ttr (tan tan radius)]: **2P**
Specify first end point of circle's diameter: *(Pick a midpoint at "A")*
Specify second end point of circle's diameter: *(Pick a midpoint at "B")*

Figure 5–12

CONSTRUCTING AN ARC TANGENT TO TWO LINES USING CIRCLE-TTR

Illustrated on the left in the following image are two inclined lines. The purpose of this example is to connect an arc tangent to the two lines at a specified radius. The CIRCLE-TTR (Tangent-Tangent-Radius) command will be used here along with the TRIM command to clean up the excess geometry. To assist with this operation, the OSNAP-Tangent mode is automatically activated when you use the TTR option of the CIRCLE command.

First, use the CIRCLE-TTR command to construct an arc tangent to both lines, as shown in the middle in the following image.

 Command: **C** *(For CIRCLE)*
Specify center point for circle or [3P/2P/Ttr (tan tan radius)]: **T** *(For TTR)*
Specify point on object for first tangent of circle: *(Select the line at "A")*
Specify point on object for second tangent of circle: *(Select the line at "B")*
Specify radius of circle: *(Enter a radius value)*

Use the TRIM command to clean up the lines and arc. The completed result is illustrated on the right in the following image. The FILLET command could also have been used for this procedure. Not only will the curve be drawn, but also this command will automatically trim the lines.

Figure 5–13

Try It! – Open the drawing file 05_TTR1. Use the CIRCLE command and the TTR option to construct a circle tangent to lines "A" and "B" in the following image. Use a circle radius of 0.50 units. With the circle constructed, use the TRIM command and select lines "C" and "D" as cutting edges. Trim the circle at "E" and the lines at "C" and "D." Observe the final results of this operation in the following image.

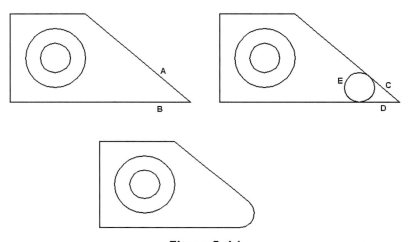

Figure 5–14

CONSTRUCTING AN ARC TANGENT TO A LINE AND ARC USING CIRCLE-TTR

Illustrated on the left in the following image is an arc and an inclined line. The purpose of this example is to connect an additional arc tangent to the original arc and line at a specified radius. The CIRCLE-TTR command will be used here, along with the TRIM command to clean up the excess geometry.

First, use the CIRCLE-TTR command to construct an arc tangent to the arc and inclined line, as shown on the right in the following image.

 Command: **C** *(For CIRCLE)*
Specify center point for circle or [3P/2P/Ttr (tan tan radius)]: **T** *(For TTR)*
Specify point on object for first tangent of circle: *(Select the arc at "A")*
Specify point on object for second tangent of circle: *(Select the line at "B")*
Specify radius of circle: *(Enter a radius value)*

Take note that the radius must be greater than or equal to half the distance between the circle and the line; otherwise, the second circle cannot be constructed. Use the

TRIM command to clean up the arc and line. The completed result is illustrated at the bottom in the following image.

Figure 5–15

 Try It! – Open the drawing file 05_TTR2. Using the following image as a guide, use the CIRCLE command and the TTR option to construct a circle tangent to the line at "A" and arc at "B." Use a circle radius value of 0.50 units. Construct a second circle tangent to the arc at "C" and line at "D" using the default circle radius value of 0.50 units. With the circles constructed, use the TRIM command, press ENTER to select all cutting edges, and trim the lines at "E" and "J" and the arc at "F" and "H," in addition to the circles at "G" and "K." Observe the final results of this operation in the following image.

Figure 5–16

CONSTRUCTING AN ARC TANGENT TO TWO ARCS USING CIRCLE-TTR

Method #1

Illustrated on the left in the following image are two arcs. The purpose of this example is to connect a third arc tangent to the original two at a specified radius. The CIRCLE-TTR command will be used here along with the TRIM command to clean up the excess geometry.

Use the CIRCLE-TTR command to construct an arc tangent to the two original arcs, as shown in the middle in the following image.

 Command: **C** *(For CIRCLE)*
Specify center point for circle or [3P/2P/Ttr (tan tan radius)]: **T** *(For TTR)*
Specify point on object for first tangent of circle: *(Select the arc at "A")*
Specify point on object for second tangent of circle: *(Select the arc at "B")*
Specify radius of circle: *(Enter a radius value)*

Use the TRIM command to clean up the two arcs, using the circle as a cutting edge. The completed result is illustrated on the right in the following image.

Figure 5–17

 Try It! – Open the drawing file 05_TTR3. Use the CIRCLE command and the TTR option to construct a circle tangent to the circles at "A" and "B." Construct a second circle tangent to the circles at "C" and "D" in the following image. Use a circle radius of 4.50 units for both circles. With the circles constructed, use the TRIM command and select both small circles as cutting edges. Trim the circles at "E" and "F." Observe the final results of this operation in the following image.

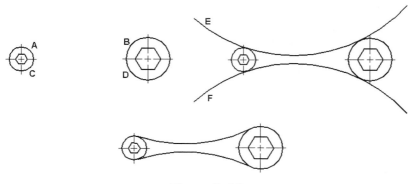

Figure 5–18

CONSTRUCTING AN ARC TANGENT TO TWO ARCS USING CIRCLE-TTR
Method #2

Illustrated on the left in the following image are two arcs. The purpose of this example is to connect an additional arc tangent to and enclosing both arcs at a specified radius. The CIRCLE-TTR command will be used here along with the TRIM command.

First, use the CIRCLE-TTR command to construct an arc tangent to and enclosing both arcs, as shown in the middle in the following image.

 Command: **C** *(For CIRCLE)*
Specify center point for circle or [3P/2P/Ttr (tan tan radius)]: **T** *(For TTR)*
Specify point on object for first tangent of circle: *(Select the arc at "A")*
Specify point on object for second tangent of circle: *(Select the arc at "B")*
Specify radius of circle: *(Enter a radius value)*

Use the TRIM command to clean up all arcs. The completed result is illustrated on the right in the following image.

Figure 5–19

Try It! – Open the drawing file 05_TTR4. Use the CIRCLE command and the TTR option to construct a circle tangent to the circles at "A" and "B." Construct a second circle tangent to the circles at "C" and "D" in the following image. Use a circle radius of 1.50 units for both circles. With the circles constructed, use the TRIM command and trim all circles until you achieve the final results of this operation in the following image.

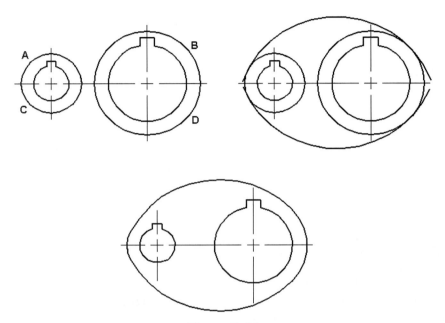

Figure 5–20

CONSTRUCTING AN ARC TANGENT TO TWO ARCS USING CIRCLE-TTR
Method #3

Illustrated on the left in the following image are two arcs. The purpose of this example is to connect an additional arc tangent to one arc and enclosing the other. The CIRCLE-TTR command will be used here along with the TRIM command to clean up unnecessary geometry.

First, use the CIRCLE-TTR command to construct an arc tangent to the two arcs. Study the illustration in the middle in the following image and the following prompts to understand the proper pick points for this operation. It is important to select the tangent points in the approximate location of the tangent to each circle in order to obtain the desired arc tangent orientation.

Command: **C** *(For CIRCLE)*
Specify center point for circle or [3P/2P/Ttr (tan tan radius)]: **T** *(For TTR)*
Specify point on object for first tangent of circle: *(Select the arc at "A")*
Specify point on object for second tangent of circle: *(Select the arc at "B")*
Specify radius of circle: *(Enter a radius value)*

Use the TRIM command to clean up the arcs. The completed result is illustrated on the right in the following image.

Figure 5–21

 Try It! – Open the drawing file 05_TTR5. Use the CIRCLE command and the TTR option to construct a circle tangent to the circles at "A" and "B." Construct a second circle tangent to the circles at "C" and "D" in the following image. Use a circle radius of 3.00 units for both circles. With the circles constructed, use the TRIM command and trim all circles until you achieve the final results of this operation in the following image.

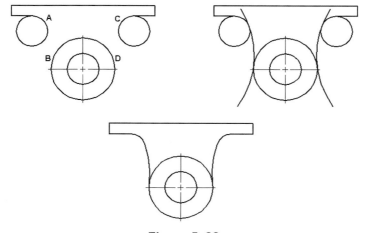

Figure 5–22

CONSTRUCTING A LINE TANGENT TO TWO ARCS OR CIRCLES

Illustrated on the left in the following image are two circles. The purpose of this example is to connect the two circles with two tangent lines. This can be accomplished with the LINE command and the OSNAP-Tangent option.

Use the LINE command to connect two lines tangent to the circles, as shown in the middle in the following image. The following procedure is used for the first line. Use the same procedure for the second.

Command: **L** *(For LINE)*
Specify first point: **Tan**
to *(Select the circle near "A")*
Specify next point or [Undo]: **Tan**
to *(Select the circle near "B")*
Specify next point or [Undo]: *(Press ENTER to exit this command)*

Use the TRIM command to clean up the circles so that the appearance of the object is similar to the illustration on the right in the following image.

Figure 5–23

Try It! – Open the drawing file 05_Tangent Lines. Construct a line tangent to the two circles at "A" and "B." Repeat this procedure for the circles at "C" and "D" in the following image. Use the TRIM command and trim all circles until you achieve the final results of this operation, shown in the following image.

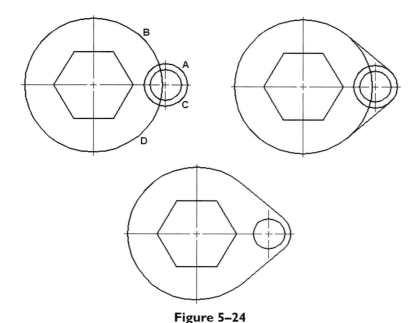

Figure 5–24

CIRCLE—TAN TAN TAN

Yet another mode of the CIRCLE command allows you to construct a circle based on three tangent points. This mode is actually a variation of the 3 Points mode together with using the OSNAP-Tangent mode three times. This mode requires you to select three objects, as in the example of the three line segments shown on the right in the following image.

Command: **C** (For CIRCLE)
Specify center point for circle or [3P/2P/Ttr (tan tan radius)]: **3P**
Specify first point on circle: **Tan**
to (Select the line at "A")
Specify second point on circle: **Tan**
to (Select the line at "B")
Specify third point on circle: **Tan**
to (Select the line at "C")

The result is illustrated on the right in the following image, with the circle being constructed tangent to the edges of all three line segments.

Figure 5–25

 Try It! – Open the drawing file 05_TTT. Use the illustration in the previous image and the previous command sequence for constructing a circle tangent to three lines.

QUADRANT VERSUS TANGENT OSNAP OPTION

Various examples have been given on previous pages concerning drawing lines tangent to two circles, two arcs, or any combination of the two. The object on the left in the following image illustrates the use of the OSNAP-Tangent option when used along with the LINE command.

Command: **L** *(For LINE)*
Specify first point: **Tan**
to *(Select the arc near "A")*
Specify next point or [Undo]: **Tan**
to *(Select the arc near "B")*
Specify next point or [Undo]: *(Press ENTER to exit this command)*

Note that the angle of the line formed by points "A" and "B" is neither horizontal nor vertical. This is a typical example of the capabilities of the OSNAP-Tangent option.

The object illustrated on the right in the following image is a modification of the example on the left with the inclined tangent lines changed to horizontal and vertical tangent lines. This example is to show that two OSNAP options are available for performing tangencies, namely OSNAP-Tangent and OSNAP-Quadrant. However, it is up to you to evaluate under what conditions to use these OSNAP options. In the example on the right, you can use the OSNAP-Tangent or OSNAP-Quadrant option to draw the lines tangent to the arcs. The Quadrant option could be used only because the lines to be drawn are perfectly horizontal or vertical. Usually it is impossible to know this ahead of time, and in this case, the OSNAP-Tangent option should be used whenever possible.

Figure 5–26

CREATING FILLED-IN DOTS

Use the DONUT command to construct a filled-in circle. Choose this command from one of the following:

    The pull-down menu (Draw > Donut)

    The keyboard (DO or DONUT)

This object belongs to the polyline family. The donut illustrated on the left in the following image has an inside diameter of 0.50 units and an outside diameter of 1.00 units. When you place donuts in a drawing, the Multiple option is automatically invoked. This means you can place as many donuts as you like until you choose another command from the menu or issue a Cancel by pressing ESC.

 Try It! – Create a new drawing file starting from scratch. Use the following command sequence for constructing this type of donut.

Command: **DO** (For DONUT)
Specify inside diameter of donut <0.50>: (Press ENTER to accept the default)
Specify outside diameter of donut <1.00>: (Press ENTER to accept the default)
Specify center of donut or <exit>: (Pick a point to place the donut)
Specify center of donut or <exit>: (Pick a point to place another donut or press ENTER to exit this command)

Setting the inside diameter of a donut to a value of zero (0) and an outside diameter to any other value constructs a donut representing a dot, as shown on the right in the following image.

Figure 5–27

 Try It! – Create a new drawing file starting from scratch. Use the following command sequence for constructing this type of donut.

Command: **DO** *(For DONUT)*
Specify inside diameter of donut <0.50>: **0**
Specify outside diameter of donut <1.00>: **0.25**
Specify center of donut or <exit>: *(Pick a point to place the donut)*
Specify center of donut or <exit>: *(Pick a point to place another donut or press* ENTER *to exit this command)*

 Try It! – Open the drawing file 05_Donut. Activate the DONUT command and set the inside diameter to 0 and the outside diameter to .05. Place four donuts at the intersections of the electrical circuit, as shown in the following image.

Figure 5–28

CONSTRUCTING ELLIPTICAL SHAPES

Use the ELLIPSE command to construct a true elliptical shape. Choose this command from one of the following:

    The Draw toolbar

    The pull-down menu (Draw > Ellipse)

    The keyboard (EL or ELLIPSE)

Before studying the three examples for ellipse construction, see the illustration on the right in the following image to view two important parts of any ellipse, namely its major and minor diameters.

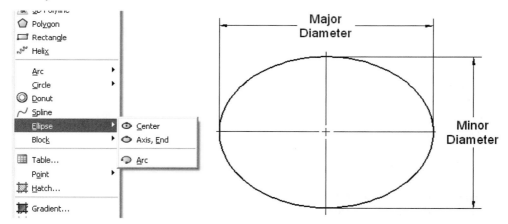

Figure 5–29

You can construct an ellipse by marking two points, which specify one of its axes, as shown in the following image. These first two points also identify the angle with which the ellipse will be drawn. Responding to the prompt "Specify distance to other axis or [Rotation]" with another point identifies half of the other axis. The rubber-banded line is added to assist you in this ellipse construction method.

 Try It! – Create a new drawing file starting from scratch. Use the illustration in the following image and the command sequence to construct an ellipse by locating three points.

Command: **EL** *(For ELLIPSE)*
Specify axis endpoint of ellipse or [Arc/Center]: *(Pick a point at "A")*
Specify other endpoint of axis: *(Pick a point at "B")*
Specify distance to other axis or [Rotation]: *(Pick a point at "C")*

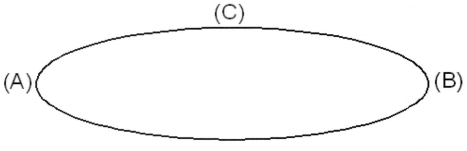

Figure 5–30

You can also construct an ellipse by first identifying its center. You can pick points to identify its axes or use polar coordinates to accurately define the major and minor diameters of the ellipse, as shown on the left in the following image.

 Try It! – Create a new drawing file starting from scratch. Use the illustration on the left in the following image and the following command sequence for constructing an ellipse based on a center. Use a polar coordinate or the Direct Distance mode for locating the two axis endpoints of the ellipse.

Command: **EL** *(For ELLIPSE)*
Specify axis endpoint of ellipse or [Arc/Center]: **C** *(For Center)*
Specify center of ellipse: *(Pick a point at "A")*
Specify endpoint of axis: **@2.50<0** *(To point "B")*
Specify distance to other axis or [Rotation]: **@1.50<90** *(To point "C")*

The last method of constructing an ellipse, as shown on the right in the following image, illustrates constructing an ellipse by way of rotation. Identify the first two points for the first axis. Reply to the prompt "Specify distance to other axis or [Rotation]" with Rotation. The first axis defined is now used as an axis of rotation that rotates the ellipse into a third dimension.

 Try It! – Create a new drawing file starting from scratch. Use the illustration on the right in the following image and the following command sequence for constructing an ellipse based on a rotation around the major axis.

Command: **EL** *(For ELLIPSE)*
Specify axis endpoint of ellipse or [Arc/Center]: *(Pick a point at "A")*
Specify other endpoint of axis: *(Pick a point at "B")*
Specify distance to other axis or [Rotation]: **R** *(For Rotation)*
Specify rotation around major axis: **80**

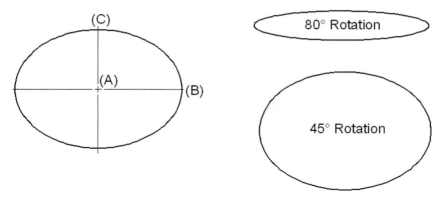

Figure 5–31

CREATING MULTILINE OBJECTS

A multiline consists of a series of parallel line segments. You can group as many as sixteen multilines together to form a single multiline object. Multilines may take on color, may be spaced apart at different increments, and may take on different linetypes. Choose this command from one of the following:

    The pull-down menu (Draw > Multiline)

    The keyboard (ML or MLINE)

In the following image, the floor plan layout was made with the default multiline style of two parallel line segments. The spacing was changed to an offset distance of 4 units. Once the multilines are laid out, as shown on the left in the following image, the Multiline Edit (MLEDIT) command cleans up all corners and intersections, as displayed in the middle in the following image. Gaps for doors and windows can be created by using the TRIM command.

Figure 5–32

THE MLINE COMMAND

Once the element and multiline properties have been set and saved to a multiline style, loading a new current style and clicking the OK button of the Multiline Styles dialog box draws the multiline to the current multiline style configuration. The illustration on the left in the following image is the result of setting the offset distances with the Multiline Style command. This configuration may be used for an architectural application where a block wall 8 units thick and a footing of 16 units thick is needed. Multilines are drawn with the MLINE command and the following command sequence.

Command: **ML** *(For MLINE)*
Current settings: Justification = Top, Scale = 1.00, Style = STANDARD
Specify start point or [Justification/Scale/STyle]: **J** *(For Justification)*
Enter justification type [Top/Zero/Bottom] <top>: **Z** *(For Zero)*
Current settings: Justification = Zero, Scale = 1.00, Style = STANDARD
Specify start point or [Justification/Scale/STyle]: *(Pick a point at "A")*
Specify next point: *(Pick a point at "B")*
Specify next point or [Undo]: *(Pick a point at "C")*
Specify next point or [Close/Undo]: *(Pick a point at "D")*
Specify next point or [Close/Undo]: *(Press ENTER to exit this command)*

The illustration on the right in the following image is an example of how three justification modes can affect the layout of a multiline. Justification modes reference the top of the multiline, the bottom of the multiline, or the zero location of the multiline, which is the multiline's center.

Figure 5–33

 Try It! – Open the drawing file 05_Mline Walls. A multiline style has been created consisting of block walls spaced 8″ apart and foundation walls spaced 16″ apart. Use the MLINE command and the following command sequence to construct the exterior walls in the following image. Use the Direct Distance mode to create this exterior wall.

Command: **ML** *(For MLINE)*
Current settings: Justification = Zero, Scale = 1.00, Style = FDN-WALLS
Specify start point or [Justification/Scale/STyle]: **21´,10´**
Specify next point or [Close/Undo]: **16´** *(To the left)*
Specify next point or [Close/Undo]: **8´** *(Up)*
Specify next point or [Close/Undo]: **26´** *(To the left)*
Specify next point or [Close/Undo]: **28´** *(Down)*
Specify next point or [Close/Undo]: **42´** *(To the right)*
Specify next point or [Close/Undo]: **C** *(To close the shape)*

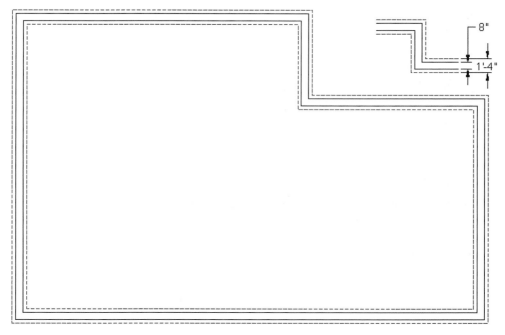

Figure 5–34

CREATING A NEW MULTILINE STYLE

The Multiline Styles dialog box is used to create a new multiline style or to make an existing style current. Choose this command from one of the following:

The pull-down menu (Format > Multiline Style)

The keyboard (MLSTYLE)

Choosing Multiline Style from the pull-down menu, as shown on the left in the following image, will display the Multiline Style dialog box, shown on the right in the following image. The dialog box lists the current multiline style, which by default is named STANDARD. To create a new multiline style, click the New button. When the Create New Multiline Style dialog box appears, add a meaningful name for this new style. Illustrated on the right in the following image is this dialog box with a new name called Walls added.

Figure 5–35

Clicking the Continue button in the previous image displays the New Multiline Style: Walls dialog box, as shown on the left in the following image. This dialog box is divided into two distinct sides; on the left is the Caps area, which controls the following: whether line segments are drawn at each multiline joint, whether the beginning and/or end of the multiline is capped with a line segment, whether the beginning and/or end of the outer parts of a multiline are capped by arcs, whether the beginning and/or end of the inner parts of a multiline are capped by arcs, or whether a user-defined angle is used to cap the start and end of the multiline. With Multiline Fill mode turned on, the entire multiline will be filled in with the selected color.

The area on the right deals with the elements that make up the multiline. This area controls the following: make offset, color, and linetype assignments to the various multiline segments. Clicking the Add button adds a new multiline listing to the Elements box. Offset distances can now be changed along with Color and Linetype assignments to the selected multiline object. If a linetype is not present, a special Linetype button is available at the bottom of the Linetype dialog box. This in turn displays all supported linetypes. Clicking the OK button in the Element Properties dialog box returns you to the Multiline Style dialog box.

Figure 5–36

EDITING MULTILINES

The MLEDIT command is a special feature designed for editing or cleaning up intersections of multiline objects. Choose this command from one of the following:

    The pull-down menu (Modify > Object > Multiline)

    The keyboard (MLEDIT)

Choose Multiline from the Modify pull-down menu area, shown on the left in the following image. This activates the Multiline Edit Tools dialog box, as shown on the right in the following image. Clicking one of the buttons in the dialog box displays the title of the tool in the lower left corner.

Figure 5–37

Various cleanup modes are available, such as creating a closed cross, creating an open cross, or creating a merged cross. Options are available for creating "Tee" sections in multilines. Corner joints may be created along with the adding or deleting of a vertex of a multiline. The following Try It! exercises demonstrate the creation of open intersections and tees or the corner operation. In these figures, the identifying letters "A" and "B" show where to pick the multiline when using the Multiline Edit Tools dialog box.

 Try It! – Open the drawing file 05_Mledit Open Cross. Activate the Multiline Edit Tools dialog box and click the Open Cross icon. Pick the edge at "A" as the first multiline and the edge at "B" as the second multiline, as shown on the left in the following image. The completed cross operation is illustrated on the right in the following image.

Figure 5–38

 Try It! – Open the drawing file 05_Mledit Open Tee. Activate the Multiline Edit Tools dialog box and click the Open Tee icon. Pick the edge at "A" as the first multiline and the edge at "B" as the second multiline, as shown on the left in the following image. The completed tee operation is illustrated on the right in the following image.

Figure 5–39

Try It! – Open the drawing file 05_Mledit Corner Joint. Activate the Multiline Edit Tools dialog box and click the Corner Joint icon. Pick the edge at "A" as the first multiline and the edge at "B" as the second multiline, as shown on the left in the following image. The completed corner operation is illustrated on the right in the following image.

Figure 5–40

 Try It! – Open the drawing file 05_Mledit House, shown on the left in the following image. Activate the Multiline Edit Tools dialog box and click the Corner Joint icon to edit the four corners of the floor plan. Then activate the Multiline Edit Tools dialog box and click the Open Tee icon. Edit the wall intersections using this tool, as shown in the middle in the following image. You may see different results depending on which line you pick first. If this happens, immediately perform an Undo and pick the lines in the correct order. Complete the basic layout of the floor plan by using the TRIM command to trim away any excess lines to form the doorway openings, as shown on the right in the following image.

 Tip: Once the interior and exterior walls of a floor plan are laid out using multilines, these multilines can be exploded into individual line segments. Basic operations such as breaking, trimming, and extending can be used to add windows and doors to the floor plan.

Figure 5–41

CREATING POINT OBJECTS

Use the POINT command to identify the location of a point on a drawing, which may be used for reference purposes. Choose this command from one of the following:

> The Draw toolbar
>
> The pull-down menu (Draw > Point)
>
> The keyboard (PO or POINT)

Choose Point from the Draw pull-down menu, as shown on the left in the following image. Clicking on the Single Point option allows for the creation of one point. If numerous points need to be constructed in the same operation, choose the Multiple Point option. The OSNAP Node or Nearest option is used to snap to points. By default, a point is displayed as a dot on the screen. Illustrated on the right in the following image is an object constructed using points that have been changed in appearance to resemble Xs.

Command: **PO** *(For POINT)*
Current point modes: PDMODE=0 PDSIZE=0.0000
Specify a point: *(Pick the new position of a point)*
Specify a point: *(Either pick another point location or press ESC to exit this command)*

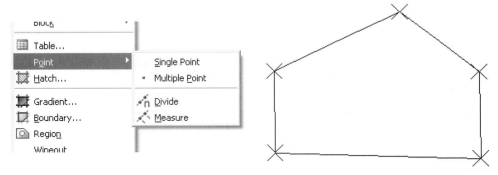

Figure 5–42

SETTING A NEW POINT STYLE

Because the appearance of the default point as a dot may be confused with the existing grid dots already on the screen, a mechanism is available to change the appearance of the point. Choosing Point Style from the Format pull-down menu, as shown on the left in the following image, displays the Point Style dialog box shown on the right. Use this icon menu to set a different point mode and point size.

Tip: Only one point style may be current in a drawing. Once a point is changed to a current style, the next drawing regeneration updates all points to this style.

Figure 5–43

DIVIDING OBJECTS INTO EQUAL SPACES

Illustrated in the following image is an arc constructed inside a rectangular object. The purpose of this example is to divide the arc into an equal number of parts. This was a tedious task with manual drafting methods, but thanks to the DIVIDE command, this operation is much easier to perform. The DIVIDE command instructs you to supply the number of divisions and then performs the division by placing a point along the object to be divided. Choose this command from one of the following:

> The pull-down menu (Draw > Point > Divide)
>
> The keyboard (DIV or DIVIDE)

The Point Style dialog box controls the point size and shape. Be sure the point style appearance is set to produce a visible point. Otherwise, the results of the DIVIDE command will not be obvious.

Try It! – Open the drawing file 05_Divide. Use the DIVIDE command and select the arc as the object to divide, as illustrated on the left in the following image, and enter a value of 6 for the number of segments. The command divides the object by a series of points, as shown in the middle in the following image. Then construct circles with a diameter of .25 using each point as its center, as shown on the right in the following image. The OSNAP-Node mode works with point objects.

Command: **DIV** *(For DIVIDE)*
Select object to divide: *(Select the arc)*
Enter the number of segments or [Block]: **6**

Figure 5–44

For polyline objects, the DIVIDE command will divide the entire polyline object into an equal number of segments. This will occur even if the polyline consists of a series of line and arc segments, as in the following image.

 Try It! – Open the drawing file 05_Divide Pline. Use the DIVIDE command and select the outer polyline as the object to divide, as shown on the left in the following image, and enter a value of 9 for the number of segments. The command divides the entire polyline by a series of points, as shown on the right.

Command: **DIV** *(For DIVIDE)*
Select object to divide: *(Select the outer polyline shape)*
Enter the number of segments or [Block]: **9**

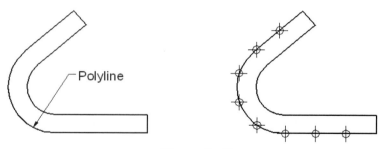

Figure 5–45

MEASURING OBJECTS

The MEASURE command takes an object such as a line or an arc and measures along it depending on the length of the segment. The MEASURE command, similar to the DIVIDE command, places a point on the object at a specified distance given in the MEASURE command. Choose this command from one of the following:

 The pull-down menu (Draw > Point > Measure)

 The keyboard (ME or MEASURE)

It is important to note that as a point is placed along an object during the measuring process, the object is not automatically broken at the location of the point. Rather, the point is commonly used to construct from, along with the OSNAP-Node option. Also, the measuring starts at the endpoint closest to the point you used to select the object.

Choose Point from the Draw pull-down menu, and then choose the MEASURE command.

 Try It! – Open the drawing file 05_Measure. Use the illustration in the following image and the command sequence below for performing this operation.

Command: **ME** *(For MEASURE)*
Select object to measure: *(Select the left end of the diagonal line, as shown on the left in the following image)*
Specify length of segment or [Block]: **1.25**

The results, shown in the following image, illustrate various points placed at 1.25 increments. As with the DIVIDE command, the appearance of the points placed along the line is controlled through the Point Style dialog box.

Figure 5–46

CREATING POLYGONS

The POLYGON command is used to construct a regular polygon. Choose this command from one of the following:

- The Draw toolbar
- The pull-down menu (Draw > Polygon)
- The keyboard (POL or POLYGON)

You create polygons by identifying the number of sides for the polygon, locating a point on the screen as the center of the polygon, specifying whether the polygon is inscribed or circumscribed, and specifying a circle radius for the size of the polygon. Polygons consist of a closed polyline object with width set to zero.

 Try It! – Create a new drawing file starting from scratch. Use the following command sequence to construct the inscribed polygon, as shown on the left in the following image.

 Command: **POL** (For POLYGON)
Enter number of sides <4>: **6**
Specify center of polygon or [Edge]: (Mark a point at "A," as shown on the left in the following image)
Enter an option [Inscribed in circle/Circumscribed about circle] <I>: **I** (For Inscribed)
Specify radius of circle: **1.00**

 Try It! – Create a new drawing file starting from scratch. Use the following command sequence to construct the circumscribed polygon, as shown on the right in the following image.

 Command: **POL** (For POLYGON)
Enter number of sides <4>: **6**
Specify center of polygon or [Edge]: (Pick a point at "A," as shown on the right in the following image)
Enter an option [Inscribed in circle/Circumscribed about circle] <I>: **C** (For Circumscribed)
Specify radius of circle: **1.00**

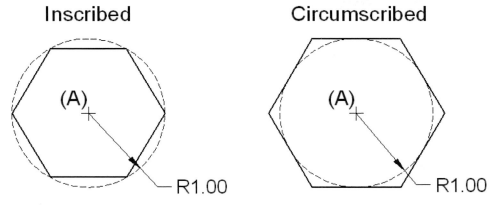

Figure 5–47

Locating the endpoints of one of its edges may specify polygons. The polygon is then drawn in a counterclockwise direction.

 Try It! – Create a new drawing file starting from scratch. Study the following image and the command sequence to construct a polygon by edge.

Command: **POL** (For POLYGON)
Enter number of sides <4>: **5**
Specify center of polygon or [Edge]: **E** (For Edge)
Specify first endpoint of edge: (Mark a point at "A")
Specify second endpoint of edge: (Mark a point at "B")

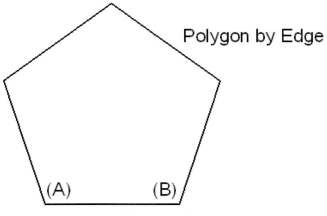

Figure 5–48

CREATING A RAY CONSTRUCTION LINE

A ray is a type of construction line object that begins at a user-defined point and extends to infinity in only one direction. Choose this command from one of the following:

>The pull-down menu (Draw > Ray)
>
>The keyboard (RAY)

In the following image, the quadrants of the circles identify all points where the ray objects begin and are drawn to infinity to the right. You should organize ray objects on specific layers. You should also exercise care in the editing of rays, and take special care not to leave segments of objects in the drawing database as a result of breaking ray objects. Breaking the ray object at "A" in the following image converts one object to an individual line segment; the other object remains a ray. Study the following command sequence for constructing a ray.

Command: **RAY**
Specify start point: *(Pick a point on an object)*
Specify through point: *(Pick an additional point to construct the ray object)*
Specify through point: *(Pick another point to construct the ray object or press* ENTER *to exit this command)*

Figure 5–49

CREATING RECTANGLE OBJECTS

Use the RECTANG command to construct a rectangle by defining two points. Choose this command from one of the following:

>The Draw toolbar
>
>The pull-down menu (Draw > Rectangle)
>
>The keyboard (REC or RECTANG or RECTANGLE)

RECTANGLE BY PICKING TWO DIAGONAL POINTS

As illustrated in the following image, two diagonal points are picked to define the rectangle. The rectangle is drawn as a single polyline object.

Command: **REC** *(For RECTANG)*
Specify first corner point or [Chamfer/Elevation/Fillet/Thickness/Width]: *(Pick a point at "A")*
Specify other corner point or [Area/Dimensions/Rotation]: *(Pick a point at "B")*

Figure 5–50

CHANGING RECTANGLE PROPERTIES

Other options of the RECTANG command enable you to construct a chamfer or fillet at all corners of the rectangle, to assign a width to the rectangle, and to have the rectangle drawn at a specific elevation and at a thickness for 3D purposes. As illustrated in the following image, a rectangle is constructed with a chamfer distance of .20 units; the width of the rectangle is also set at .05 units. A relative coordinate value of 4.00,1.00 will be used to construct the rectangle 4 units in the X direction and 1 unit in the Y direction. The @ symbol resets the previous point at "A" to zero.

 Try It! – Create a new AutoCAD drawing starting from scratch. Use the following command sequence image for constructing a wide rectangle with its corners chamfered.

 Command: **REC** *(For RECTANGLE)*
Specify first corner point or [Chamfer/Elevation/Fillet/Thickness/Width]: **C** *(For Chamfer)*
Specify first chamfer distance for rectangles <0.0000>: **.20**
Specify second chamfer distance for rectangles <0.2000>: *(Press* ENTER *to accept this default value)*
Specify first corner point or [Chamfer/Elevation/Fillet/Thickness/Width]: **W** *(For Width)*
Specify line width for rectangles <0.0000>: **.05**
Specify first corner point or [Chamfer/Elevation/Fillet/Thickness/Width]: *(Pick a point at "A")*
Specify other corner point or [Area/Dimensions/Rotation]: **@4.00,1.00** *(To identify the other corner at "B")*

Figure 5–51

RECTANGLE BY DIMENSIONS

Another method of constructing a rectangle is one in which you supply the length and width dimensions for the rectangle. Before the rectangle is constructed, you indicate a point where the opposite corner of the rectangle will be positioned. There are four possible positions or quadrants the rectangle can be drawn in using this method: upper right, upper left, lower right, and lower left.

RECTANGLE BY AREA

The final method of constructing a rectangle is by area. In this method, you pick a first corner point, as in previous rectangle modes. After entering the Area option, you enter the area of the rectangle based on the current drawing units. You then enter the length or width of the rectangle. If, for instance, you enter the length, the width of the rectangle will automatically be calculated based on the area. In the following command sequence, an area of 250 is entered, along with a length of 20 units. A width of 12.50 units is automatically calculated based on the other two numbers.

 Command: **REC** *(For RECTANGLE)*
Specify first corner point or [Chamfer/Elevation/Fillet/Thickness/Width]: *(Pick a point to start the rectangle)*
Specify other corner point or [Area/Dimensions/Rotation]: **A** *(For Area)*
Enter area of rectangle in current units <100.0000>: **250**
Calculate rectangle dimensions based on [Length/Width] <Length>: **L** *(For Length)*
Enter rectangle length <10.0000>: **20**

 Try It! – Create a new AutoCAD drawing starting from scratch. You can also specify the dimensions (length and width) of the rectangle by using the following command sequence and image.

 Command: **REC** *(For RECTANGLE)*
Specify first corner point or [Chamfer/Elevation/Fillet/Thickness/Width]: *(Pick a point at "A" on the screen)*
Specify other corner point or [Area/Dimensions/Rotation]: **D** *(For Dimensions)*
Specify length for rectangles <0.0000>: **3.00**
Specify width for rectangles <0.0000>: **1.00**
Specify other corner point or [Area/Dimensions/Rotation]: *(Moving your cursor around positions the rectangle in four possible positions. Click the upper right corner of your screen at "B" to anchor the upper right corner of the rectangle)*

Figure 5–52

 Note: Rectangles can also be constructed based on their area and at a user-specified angle.

CREATING A REVISION CLOUD

The REVCLOUD command creates a polyline object consisting of arc segments in a sequence. This feature is commonly used in drawings to identify areas where the drawing is to be changed or revised. Choose this command from one of the following:

    The Draw toolbar

    The pull-down menu (Draw > Revision Cloud)

    The keyboard (REVCLOUD)

In the following image, an area of the house is highlighted with the revision cloud. By default, each arc segment has a minimum and maximum arc length of 0.50 units. In a drawing such as a floor plan, the arc segments would be too small to view. In cases in which you wish to construct a revision cloud and the drawing is large, the arc length can be increased through a little experimentation. Once the arc segment is set, pick a point on your drawing to begin the revision cloud. Then move your cursor slowly in a counterclockwise direction. You will notice the arc segments of the revision cloud being constructed. Continue surrounding the object, while at the same time heading back to the origin of the revision cloud. Once you hover near the original start, the end of the revision cloud snaps to the start and the command ends, leaving the revision cloud as the polyline object. Study the command prompt sequence below and the illustration in the following image for the construction of a revision cloud.

Command: **REVCLOUD**
Minimum arc length: 1/2″ Maximum arc length: 1/2″
Specify start point or [Arc length/Object] <Object>: **A** *(For Arc length)*
Specify minimum length of arc <1/2″>: **24**
Specify maximum length of arc <2′>: *(Press* ENTER*)*
Specify start point or [Object] <Object>: *(Pick a point on your screen to start the revision cloud. Surround the item with the revision cloud by moving your cursor in a counterclockwise direction. When you approach the start point, the revision cloud closes and the command exits.)*
Guide crosshairs along cloud path...
Revision cloud finished.

Figure 5–53

If it is difficult to construct the revision cloud using the previous command prompt sequence, you could create a closed polyline shape around the area you wish to highlight, as shown on the left in the following image. Entering the REVCLOUD command, you will notice the presence of the Object option. Selecting Object and picking the edge of the polyline frame will convert the polyline into a revision cloud, as shown on the right in the following image.

 Command: **REVCLOUD**
Minimum arc length: 2′ Maximum arc length: 2′
Specify start point or [Arc length/Object] <Object>: **O** *(For Object)*
Select object: Reverse direction [Yes/No] <No>: *(Press* ENTER*)*
Revision cloud finished.

 Note: Once a revision cloud is created, you can easily convert it back to its original polyline frame by using the PEDIT command, selecting the edge of the revision cloud, and issuing the Decurve option.

Figure 5–54

CALLIGRAPHY REVISION CLOUDS

The Calligraphy option of creating revision clouds allows you to enhance the use of this command and make your drawings more dramatic. Use the following command sequence and image, which illustrate this feature.

Command: **REVCLOUD**
Minimum arc length: 24.0000   Maximum arc length: 24.0000   Style: Normal
Specify start point or [Arc length/Object/Style] <Object>: **S** *(For Style)*
Select arc style [Normal/Calligraphy] <Normal>: **C** *(For Calligraphy)*
Arc style = Calligraphy
Specify start point or [Arc length/Object/Style] <Object>: *(Pick a starting point)*
Guide crosshairs along cloud path…
Revision cloud finished.

Figure 5–55

CREATING SPLINES

Use the SPLINE command to construct a smooth curve given a sequence of points. Choose this command from one of the following:

The Draw toolbar

The pull-down menu (Draw > Spline)

The keyboard (SPL or SPLINE)

You have the option of changing the accuracy of the curve given a tolerance range. The basic command sequence follows, which constructs the spline segment shown on the left in the following image.

 Command: **SPL** *(For SPLINE)*

Specify first point or [Object]: *(Pick a first point)*
Specify next point: *(Pick another point)*
Specify next point or [Close/Fit tolerance] <start tangent>: *(Pick another point)*
Specify next point or [Close/Fit tolerance] <start tangent>: *(Pick another point)*
Specify next point or [Close/Fit tolerance] <start tangent>: *(Press* ENTER *to continue)*
Specify start tangent: *(Press* ENTER *to accept)*
Specify end tangent: *(Press* ENTER *to accept the end tangent position, which exits the command and places the spline)*

The spline may be closed to display a continuous segment, as shown on the right in the following image. Entering a different tangent point at the end of the command changes the shape of the curve connecting the beginning and end of the spline.

 Command: **SPL** *(For SPLINE)*

Specify first point or [Object]: *(Pick a first point)*
Specify next point: *(Pick another point)*
Specify next point or [Close/Fit tolerance] <start tangent>: *(Pick another point)*
Specify next point or [Close/Fit tolerance] <start tangent>: *(Pick another point)*
Specify next point or [Close/Fit tolerance] <start tangent>: **C** *(To Close)*
Specify tangent: *(Press* ENTER *to exit the command and place the spline)*

Figure 5–56

Try It! – Open the drawing file 05_ Spline Rasp Handle, as shown on the left in the following image. Turn OSNAP on and set Running OSNAP to Node. This will allow you to snap to points when drawing splines or other objects. Construct a spline by connecting all points between "A" and "B." Construct an arc with its center at "C," the start point at "D," and the ending point at "B." Connect points "D" and "E" with a line segment. Construct another spline by connecting all points between "E" and "F." Construct another arc with its center at "G," the start point at "H" and the ending point at "F." Connect points "H" and "A" with a line segment.

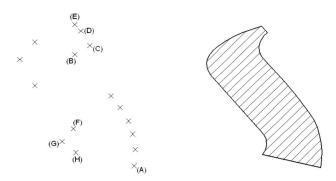

Figure 5–57

MASKING TECHNIQUES WITH THE WIPEOUT COMMAND

To mask or hide objects in a drawing without deleting them or turning off layers, the WIPEOUT command could be used. Choose this command from one of the following:

The pull-down menu (Draw > Wipeout)

The keyboard (WIPEOUT)

This command reads the current drawing background color and creates a mask over anything defined by a frame. In the illustration on the left in the following image, a series of text objects needs to be masked over. The middle image shows a four-sided frame that was created over the text using the following command prompt sequence:

Command: **WIPEOUT**
Specify first point or [Frames/Polyline] <Polyline>: *(Pick a first point)*
Specify next point: *(Pick a second point)*
Specify next point or [Undo]: *(Pick a third point)*
Specify next point or [Close/Undo]: *(Pick a fourth point)*
Specify next point or [Close/Undo]: *(Press* ENTER *to exit the command and create the wipeout)*

As the text seems to disappear, the wipeout frame is still visible. A visible frame is important if you would like to unmask or delete the wipeout. If you want to hide all wipeout frames, use the following command sequence:

Command: **WIPEOUT**
Specify first point or [Frames/Polyline] <Polyline>: **F** *(For Frames)*
Enter mode [ON/OFF] <ON>: **Off**

Now all wipeout frames in the current drawing will be turned off, as shown in the illustration on the right in the following image.

 Note: You could also create a predefined polyline object and then convert it into a wipeout using the Polyline option of the WIPEOUT command.

Figure 5–58

CREATING CONSTRUCTION LINES WITH THE XLINE COMMAND

Xlines are construction lines drawn from a user-defined point. Choose this command from one of the following:

- The Draw toolbar
- The pull-down menu (Draw > Construction Line)
- The keyboard (XL or XLINE)

You are not prompted for any length information because the Xline extends an unlimited length, beginning at the user-defined point and going off to infinity in opposite directions from the point. Xlines can be drawn horizontal, vertical, and angular. You can bisect an angle using an Xline or offset the Xline at a specific distance. Xlines are particularly useful in constructing orthogonal drawings in establishing the surfaces between views. As illustrated on the left in the following image, the circular view represents the Front view of a flange. To begin the creation of the Side views, lines are usually projected from key features on the adjacent view. In the case of the Front view, the key features are the top of the plate in addition to the other circular features. In this case, the Xlines were drawn with the Horizontal mode from the Quadrant of all circles. The following prompts outline the XLINE command sequence:

 Command: **XL** *(For XLINE)*
Specify a point or [Hor/Ver/Ang/Bisect/Offset]: **H** *(For Horizontal)*
Specify through point: *(Pick a point on the display screen to place the first Xline)*
Specify through point: *(Pick a point on the display screen to place the second Xline)*

Since the Xlines continue to be drawn in both directions, care must be taken to manage these objects. Construction management techniques of Xlines could take the form of placing all Xlines on a specific layer to be turned off or frozen when not needed. When editing Xlines (especially with the BREAK command), you need to take special care to remove all excess objects that will remain on the drawing screen. As illustrated on the right in the following image, breaking the Xline converts the object to a ray object. Use the ERASE command to remove any excess Xlines.

Figure 5–59

 Try It! – Open the drawing file 05_Xline. Another application of Xlines is illustrated on the left in the following image. Three horizontal and vertical Xlines are constructed. You must corner the Xlines to create the object illustrated on the right in the following image. Use the FILLET command and set the radius to zero. Pick the lines at "A" and "B" to create the first corner. Repeat this sequence on the remaining lines to form the other corners.

 Note: The Multiple option of the FILLET command has no effect on XLINE objects.

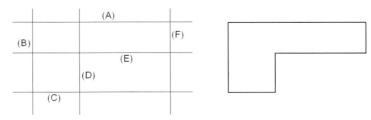

Figure 5–60

OGEE OR REVERSE CURVE CONSTRUCTION

An ogee curve connects two parallel lines with a smooth, flowing curve that reverses itself in a symmetrical form.

 Try It! – Open the drawing file 05_Ogee. To begin constructing an ogee curve to line segments "AB" and "CD," a line was drawn from "B" to "C," which connects both parallel line segments, as shown in the following image.

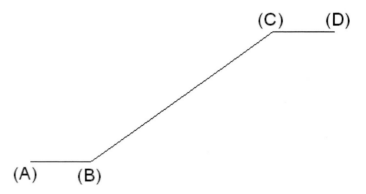

Figure 5–61

Use the DIVIDE command to divide line segment "BC" into four equal parts. Be sure to set a new point mode by picking a new point from the Point Style dialog box. Construct vertical lines from "B" and "C." Complete this step by constructing line segment "XY," which is perpendicular to line "BC," as shown on the left in the following image. Do not worry about where line "XY" is located at this time.

Move line "XY" to the location identified by the point, as shown on the right in the following image. Complete this step by copying line "XY" to the location identified by point "Z," also shown on the right.

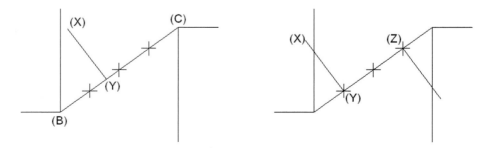

Figure 5–62

Construct two circles with centers located at points "X" and "Y," as shown in the following image. Use the OSNAP-Intersection mode to accurately locate the centers. If an intersection is not found from the previous step, use the EXTEND command to find the intersection and continue with this step. The radii of both circles are equivalent to distances "XB" and "YC."

Use the TRIM command to trim away any excess arc segments to form the ogee curve, as shown on the right in the following image.

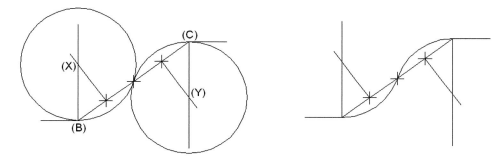

Figure 5–63

This forms the frame of the ogee for the construction of objects such as the wrench illustrated in the following image.

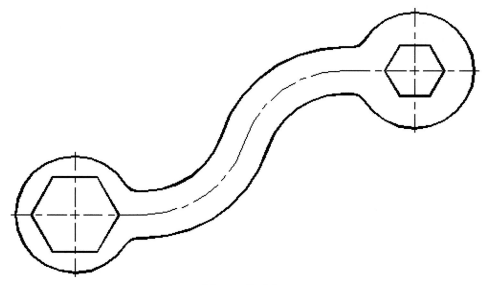

Figure 5–64

TUTORIAL EXERCISE: GEAR-ARM.DWG

Figure 5–65

Purpose
This tutorial is designed to use geometric commands to construct a one-view drawing of the gear-arm in metric format, as illustrated in the previous image.

System Settings
Use the Drawing Units dialog box to change the number of decimal places past the zero from four to two. Keep the remaining default unit values. Using the LIMITS command, keep (0,0) for the lower left corner and change the upper right corner from (12,9) to (265.00,200.00). Perform a ZOOM-All after changing the drawing limits. Since a layer called "Center" must be created to display centerlines, use the LTSCALE command and change the default value of 1.00 to 15.00. This will make the long and short dashes of the centerlines appear on the display screen. Check to see that the following Object Snap modes are already set: Endpoint, Extension, Intersection, Center.

Layers
Create the following layers with the format:

Name	Color	Linetype
Center	Yellow	Center
Construction	Gray	Center
Dimension	Yellow	Continuous
Object	White	Continuous

Suggested Commands
Begin a new drawing called Gear-arm. The object consists of a combination of circles and arcs along with tangent lines and arcs. Before beginning, be sure to set Object as the new current layer. Construction lines will be used as a layout tool to mark the centers of key circles and arcs. The CIRCLE-TTR command will be used for constructing tangent arcs to existing geometry. Use the ARC command to construct a series of arcs for the left side of the gear-arm. The TRIM command will be used to trim circles, lines, and arcs to form the basic shape.

STEP 1

Begin constructing the gear arm by creating construction geometry that will be used to create all circles. First make the Construction layer current and use the LINE command to lay out the long horizontal line and middle vertical line. Then use the OFFSET command to copy the vertical line at the specified distances. Also, create a line 70 units long at an 80° angle, as shown in the following image.

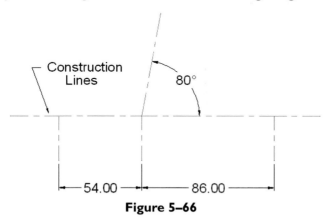

Figure 5–66

STEP 2

Continue using the Construction layer to create an arc that will be used to locate the slot detail. This arc is to be constructed using the Center, Start, Angle mode of the arc command, as shown on the left in the following image. Pick "A" as the center of the arc, "B" as the start of the arc, as shown on the right in the following image, and for angle, enter a value of −90°. The negative degree value is needed to construct the arc in the clockwise direction.

Create another construction object; this time use the OFFSET command to offset the long horizontal construction 19.05 units down. The line shown on the right in the following image was shortened on each end using the BREAK command.

Figure 5–67

STEP 3

One final item of construction geometry needs to be created, namely, the circle at a diameter of 76.30. Construct this circle from the intersection of the construction lines at "A," as shown on the left in the following image. Convert the circle to an arc segment by using the BREAK command and picking the first break point at "B" and the second break point at "C." This will convert the circle of diameter 76.30 to an arc of radius 38.15, as shown on the right in the following image.

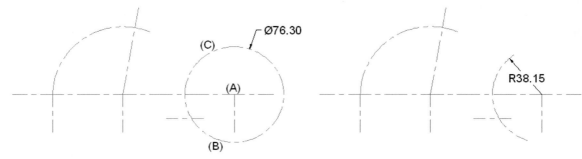

Figure 5–68

STEP 4

Using the following image as a guide, create all circles using the intersection of the construction line geometry as the centers for the circles. All circles are given with diameter dimensions.

Figure 5–69

STEP 5

Use the Arc Center, Start, End command to create the arc shown on the left in the following image. Pick "A" as the center of the arc, "B" as the start, and "C" as the end of the arc. You must use a combination of Object Snaps in order for the arcs to be drawn correctly. Following these steps will construct the outer arc of the gear arm. Use the same Arc Center, Start, End command to create the inside arc using a different set of points.

When finished creating the outer arc segments, create the two inner arc segments, as shown on the right in the following image, using the same Arc Center, Start, End command. Continue to use Object Snap to pick valid intersections in order to create the most accurate set of arcs.

Figure 5–70

STEP 6

Use the CIRCLE-TTR command to construct a circle tangent to two existing circles. Pick the two dashed circles as the objects to be made tangent, as shown on the left in the following image, and enter a radius value of 51.

After the circle is created, enter the TRIM command, pick the two dashed circles as cutting edges, and pick the upper edge of the large 51-radius circle. The results of this operation are illustrated on the right in the following image.

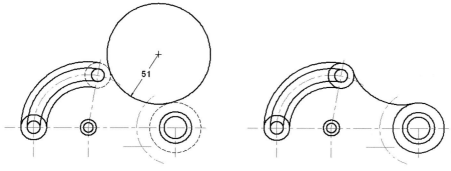

Figure 5–71

STEP 7

Enter the TRIM command again, pick the two dashed arcs as cutting edges, and pick the two inside edges of the 25.4-diameter circles, as shown on the left in the following image. The results of this operation are illustrated on the right in the following image.

Figure 5–72

STEP 8

Do the same procedure to clean up the geometry that makes up the inner slot. Enter the TRIM command, pick the two dashed arcs as cutting edges, and pick the two inside edges of the 12.7-diameter circles, as shown on the left in the following image. The results of this operation are illustrated on the right in the following image.

Figure 5–73

STEP 9

Construct a new circle of diameter 25.50 from the intersection of the construction line and arc, as shown on the left in the following image. Then draw a line segment that is tangent with the bottom of the arc and top of the circle just constructed, as shown on the right in the following image.

Figure 5–74

STEP 10

Use the TRIM command to clean up all unnecessary segments of geometry. Select all dashed objects as cutting edges, as shown on the left in the following image. Trim objects until your display appears similar to the illustration on the right in the following image.

Figure 5–75

STEP 11

Erase all unnecessary construction geometry, as shown on the left in the following image. Notice that the arc and 70° angle line both remain and will be used later to define the centers of the gear arm.

Next create centerlines to mark the centers of all circles. Switch to the Center layer. A system variable that controls centerlines and marks called DIMCEN must first be entered from the keyboard and changed to a value of −.05. The negative value will construct the small dash and long segment, all representing the centerline. Entering a value of .05 would only mark the center of circles by constructing two short intersecting line segments. Then use the DIMCENTER (or DCE for short) command and pick the edge of a circle or arc segment to place the centerline. The results should appear similar to the illustration on the right in the following image.

Figure 5–76

STEP 12

The completed gear arm drawing is illustrated in the following image.

Figure 5–77

STEP 13

The finished object may be dimensioned as an optional step, as shown in the following image.

Figure 5–78

TUTORIAL EXERCISE: PATTERN1.DWG

Figure 5–79

Purpose
This tutorial is designed to use various Draw commands to construct a one-view drawing of Pattern1, as shown in the previous image. Refer to the following special system settings and suggested command sequences.

System Settings
Use the Drawing Units dialog box to change the number of decimal places past the zero from four to two. Keep the remaining default unit values. Using the LIMITS command, keep (0,0) for the lower left corner and change the upper right corner from (12,9) to (21.00,16.00). Perform a ZOOM-All after changing the drawing limits. Check to see that the following Object Snap modes are already set: Endpoint, Extension, Intersection, and Center.

Setting OSNAP Mode would also be helpful with completing this exercise.

Layers
Create the following layers with the format:

Name	Color	Linetype
Object	White	Continuous
Center	Yellow	Center
Dimension	Yellow	Continuous

Suggested Commands:
Begin constructing Pattern1 by first creating four circles by relative coordinate mode. Then use the CIRCLE-TTR command/option to draw tangent arcs to the circles already drawn. Use the TRIM command to clean up and partially delete circles to obtain the outline of the pattern. Then add the 2.00-diameter holes followed by the center markers, using the DIMCENTER (DCE) command.

STEP 1

Check that the current layer is set to Object. See the following image for the dimensions and table for locating the centers of all circles. Use the CIRCLE command to create the circle at "A," whose center point is located at coordinate 9.50, 4.50 and which has a diameter of 4.00 units. Locate circle "B" at relative coordinate @-2.25,3.00 and assign a diameter of 5.00 units. The relative coordinate is based on the previous circle. Continue by drawing circle "C" using the center point at relative coordinate @5.25,3.50 and assign a diameter of 3.50 units. Construct the last circle, "D," with center point @-.75,-2.50 and a diameter of 3.00 units. Notice that when using relative coordinate mode, you do not need to draw construction geometry and then erase it later.

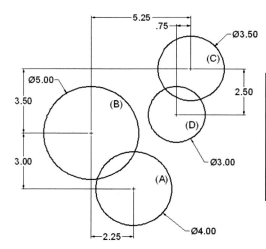

CIRCLE	Center Point Coordinate
A	9.50,4.50
B	@-2.25,3.00
C	@5.25,3.50
D	@-.75,-2.50

Figure 5–80

STEP 2

Use the CIRCLE-TTR command/option to construct a 4.00-radius circle tangent to the two dashed circles, as shown on the left in the following image. Then use the TRIM command to trim away part of the circle so your display appears similar to the illustration on the right in the following image.

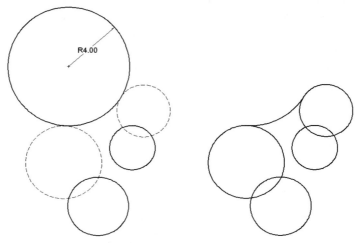

Figure 5–81

STEP 3

Use the CIRCLE-TTR command/option to construct another 4.00-radius circle tangent to the two dashed circles, as shown on the left in the following image. Then use the TRIM command to trim away part of the circle so your display appears similar to the illustration on the right in the following image.

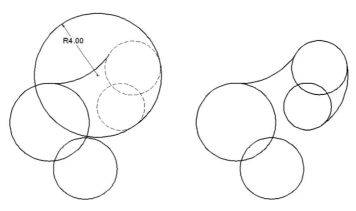

Figure 5–82

STEP 4

Use the CIRCLE-TTR command/option to construct a 6.00-radius circle tangent to the two dashed circles, as shown on the left in the following image. Then use the TRIM command to trim away part of the circle so your display appears similar to the illustration on the right in the following image.

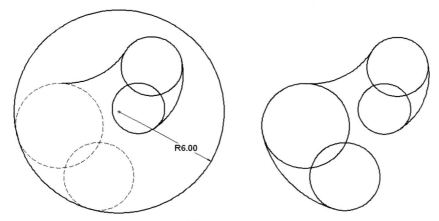

Figure 5–83

STEP 5

Use the CIRCLE-TTR command/option to construct a .75-radius circle tangent to the two dashed circles, as shown on the left in the following image. Then use the TRIM command to trim away part of the circle so your display appears similar to the illustration on the right in the following image.

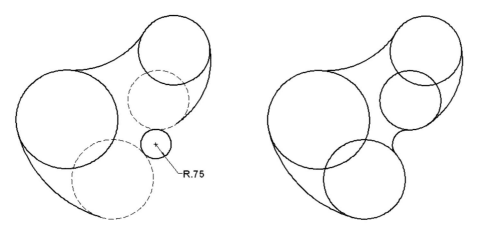

Figure 5–84

STEP 6

Use the TRIM command, select all dashed arcs, shown on the left in the following image, as cutting edges, and trim away the circular segments to form the outline of the Pattern1 drawing. When finished, your display should appear similar to the illustration on the right in the following image.

 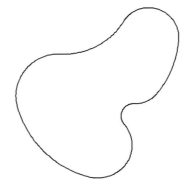

Figure 5–85

STEP 7

Use the CIRCLE command to construct a total of three circles of 2.00 diameter. Use the centers of arcs "A," "B," and "C," as shown on the left in the following image, as the centers for the circles. Switch to the Center layer, change the DIMCEN system variable to a value of 0.07, and use the DIMCENTER command to place center marks identifying the centers of all circles and arcs, as shown on the right in the following image.

 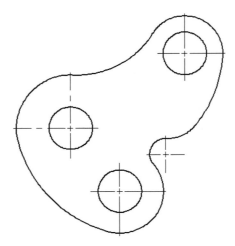

Figure 5–86

PROBLEMS FOR CHAPTER 5

Directions for Problems 5–1 through 5–15:

Construct these geometric construction figures using existing AutoCAD commands.

PROBLEM 5–1

PROBLEM 5–2

PROBLEM 5–3

PROBLEM 5–4

PROBLEM 5-5

PROBLEM 5-6

PROBLEM 5-7

PROBLEM 5-8

PROBLEM 5–9

PROBLEM 5–10

PROBLEM 5-11

PROBLEM 5-12

PROBLEM 5-13

PROBLEM 5-14

CHAPTER 6

Working with Text, Fields, and Tables

AutoCAD provides a robust set of text commands that allow you to place different text objects, edit those text objects, and even globally change the text height without affecting the justification. The heart of placing text in a drawing is through the MTEXT command. The chapter continues with a discussion of the uses of the DTEXT command, which stands for Dynamic Text or Single Line Text. You will also be given information on how to edit text created with the MTEXT and DTEXT commands. The SCALETEXT and JUSTIFYTEXT commands will be explained in detail, as will the ability to create custom text styles and assign different text fonts to these styles. Intelligent text in the form of fields will also be discussed in this chapter, as will the creation of tables.

AUTOCAD TEXT COMMANDS

All text commands can be located easily by launching the Text toolbar, illustrated in the following image.

Figure 6–1

Study the information in the following table for a brief description of each text command.

Button	Tool	Shortcut	Function
A	Multiline Text	MT	Creates paragraphs of text as a single object
AI	Single Line Text	DT	Creates one or more lines of text; each line of text is considered a separate object
A	Edit Text	ED	Used for editing multiline and single line text in addition to dimension and attribute text
abc	Find and Replace	FIND	Used for finding, replacing, selecting, or zooming in to a particular text object
A	Text Style	ST	Launches the Text Style dialog box used for creating different text styles
A	SCALETEXT	None	Used for scaling selected text objects without affecting their locations
A	JUSTIFYTEXT	None	Used for justifying selected text objects without affecting their locations
	SPACETRANS	None	Converts text heights between Model Space and Paper Space (Layout mode)

Text can be added to your AutoCAD drawing file through two methods: Multiline Text (the MTEXT command) and Single Line Text (the DTEXT command). In addition to being able to select these commands from the Text toolbar, you can also find them by choosing the Text cascading menu from the Draw pull-down menu, as shown on the left in the following image. Text modification commands can be found under the Modify pull-down menu, as shown on the right in the following image.

Figure 6–2

ADDING MULTILINE TEXT

The MTEXT command allows for the placement of text in multiple lines. Entering the MTEXT command from the command prompt displays the following prompts:

A Command: **MT** *(For MTEXT)*
Current text style:"GENERAL NOTES" Text height: 0.2000
Specify first corner: *(Pick a point at "A" to identify one corner of the mtext box)*
Specify opposite corner or [Height/Justify/Line spacing/Rotation/Style/Width]: *(Pick another corner forming a box).*

Picking a first corner displays a user-defined box with an arrow at the bottom, as shown in the following image. This box will define the area where the multiline text will be placed. If the text cannot fit on one line, it will wrap to the next line automatically.

Figure 6–3

After you click on a second point marking the other corner of the insertion box, the Text Formatting toolbar will appear, along with the Multiline Text Editor. The parts of these items are shown in the following image. One of the biggest advantages of this tool is the transparency associated with the Multiline Text Editor. As you begin entering text, you can still see your drawing in the background.

As you begin to type, your text will appear in the box shown in the following image. As the multiline text is entered, it will automatically wrap to the next line depending on the size of the initial bounding box. The current text style and font are displayed, in addition to the text height. You can make the text bold, italicized, or underlined by using the B, I, and U buttons present in the toolbar. Notice in the following image other important areas of the Text Formatting toolbar. When you have finished entering the text, click on the OK button to dismiss the Text Formatting toolbar and text editor and place the text in the drawing.

Figure 6–4

Note: Picking a blank part of your screen after you are finished entering text in the previous example will also exit the Text Formatting toolbar.

Another method of displaying additional options of the Multiline Text Editor is to right-click inside the text box. This will display the menu shown in the following image.

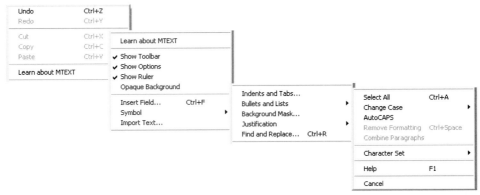

Figure 6–5

FORMATTING FRACTIONAL TEXT

Fractional text can be reformatted through the Multiline Text Editor. Entering a space after the fraction value in the following image displays the AutoStack Properties dialog box, also displayed in the following image. Use this dialog box to enable the autostacking of fractions. You can also remove the leading space in between the whole number and the fraction. In the following image, the fraction on the left was converted to the fraction on the right using the AutoStack Properties dialog box.

Figure 6–6

TEXT JUSTIFICATION MODES

The following image illustrates a sample text item and the various locations by which the text can be justified. By default, when you place text, it is left justified. These justification modes are designed to work in combination with one another. For example, one common justification is to have the text middle centered. Options in text commands allow you to pick Middle Center from a menu or type MC in at the keyboard. The same type of combination is available for the other justification modes.

Figure 6–7

JUSTIFYING MULTILINE TEXT

Right-clicking in the Multiline Text Editor and picking Justification displays the information illustrated in the following image. Various justification modes are available to change the appearance of the text in your drawing. To change the justification of existing text, first highlight the text for the changes to take place.

Figure 6–8

INDENTING TEXT

Multiline text can be indented in order to align text objects to form a list or table. A ruler, illustrated in the following image, consisting of short and longer tick marks displays the current paragraph settings.

The tabs and indents that you set before you start to enter text apply to the whole multiline text object. If you want to set tabs and indents to individual paragraphs, click in a paragraph or select multiple paragraphs to apply the indentations.

Two sliders are available on the left side of the ruler to show the amount of indentation applied to the various multiline text parts. In the following image, the top slider is used to indent the first line of a paragraph. A tooltip is available to remind you of this function.

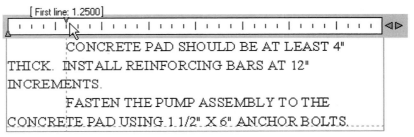

Figure 6–9

In the following image, a bottom slider is also available. This slider controls the amount of indentation applied to the other lines of the paragraph. You can see the results of using this slider in the following image.

Figure 6–10

Note: If you set tabs and indents prior to starting to enter multiline text, the tabs are applied to the whole multiline object.

You can also set indentations and tabs through the Indents and Tabs dialog box, illustrated on the right in the following image. Right-clicking in the ruler displays a menu. Pick on Indents and Tabs, as shown on the left in the following image, to activate the dialog box.

Figure 6–11

FORMATTING WITH TABS

You can further control your text by adding tabs. The long tick marks on the ruler show the default tab stops. When you click the ruler to set your own tabs, the ruler displays a small, L-shaped marker at each custom tab stop, as shown in the following image. When you anchor your cursor in front of a sentence and press the TAB key on your keyboard, the text aligns to the nearest tab marker. To remove a custom tab stop, drag the marker off the ruler.

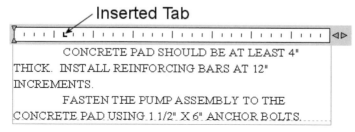

Figure 6–12

BULLETING AND NUMBERING TEXT

Multiline text can also be formatted in a numbered, bulleted, or alphabetic character list. The three types of lists are illustrated in the following image.

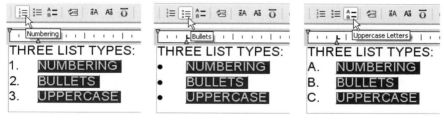

Figure 6–13

As lists are created, times occur when you need to change the format for new lists generated in the same multiline text command. In the following image on the left, an alphanumeric list was interrupted to generate a numbered list. The menu shown on the right can assist with this operation. When going from the lettered to the numbered list, the Restart mode was utilized. Notice also in the illustration on the left that a paragraph was added with more numbers continuing beneath it. This demonstrates the ability to continue a list.

Figure 6–14

CHANGING THE MTEXT WIDTH

When you first construct a rectangle that defines the boundary of the mtext object, you are not locked into this boundary. Right-clicking inside the tab marker area, as illustrated in the following image, activates the Set Mtext Width dialog box. The current value in the Width field is the current width of the rectangle used to construct the mtext object. Enter a larger or smaller value to expand or contract the mtext object.

Figure 6–15

Note: Two other methods are available to change text box width. Clicking on the multiline text object as it appears in your drawing will cause blue boxes called grips to appear in the four corners of the text box. Picking one of these grips and moving to the left or right will increase or decrease the size of the text box. Grips will be covered in greater detail in chapter 7. Also, located at the right edge of the tab marker area are left and right arrows. Pressing on either of these arrows and dragging your cursor will increase or decrease the width of the multiline text box.

IMPORTING TEXT INTO YOUR DRAWING

If you have existing files created in a word processor, you can import these files directly into AutoCAD and have them converted into mtext objects. As illustrated in the following image, right-click in the Multiline Text Editor box to display the menu; then pick Import Text. An open file dialog box will display, allowing you to pick the desired file to import. The Import Text feature of the Multiline Text Editor supports two file types; those ending in a TXT extension and those ending in the RTF (rich text format) extension. Text with the TXT extension will conform to the current AutoCAD text style regarding font type. Text with the RTF extension will remember formatting and font setting from the word-processing document that it was originally created in.

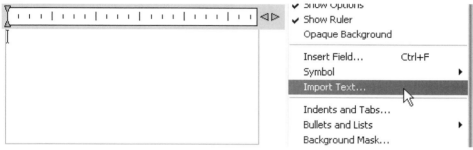

Figure 6–16

The finished result of this text-importing operation is illustrated in the following image. The text on the left, imported with the TXT extension, conformed to the present text style, which has a font of Arial assigned to it. The text on the right was imported with the RTF extension. Even with the text style being set to Arial, the RTF-imported text displays in the Times New Roman font. This is the font in which it was originally created in the word-processing document.

TXT Extension	RTF Extension
The Import Text feature of the Multiline Text supports two file types; those ending in a TXT extension and the other in the RTF (Rich Text Format) extension.	The Import Text feature of the Multiline Text supports two file types; those ending in a TXT extension and the other in the RTF (Rich Text Format) extension.

Figure 6–17

BACKGROUND MASKING

In busy, complicated drawings, you want text to be legible even though it lies on the top of other objects, as shown on the left in the following image. You can apply a background mask by first selecting the text objects.

With the text objects highlighted, activate the Properties Window, as shown on the right in the following image. Under the Text heading, find the Background listing and change the value to Yes. This will apply a background to the selected text objects. The affected text objects are identified in the following image by the gray color.

Figure 6–18

MULTILINE TEXT SYMBOLS

While in the Text Formatting toolbar, right-clicking in the text window and selecting Symbol, as shown in the following image, will display various symbols that can be incorporated into your drawing.

Figure 6–19

The table illustrated in the following image shows all available multiline text symbols along with their meanings.

\multicolumn{4}{c}{MTEXT SYMBOLS}			
SYMBOL	DESCRIPTION	SYMBOL	DESCRIPTION
°	Degrees	≡	Identity
±	Plus/Minus	⌒	Initial Length
⌀	Diameter	M	Monument Line
≈	Almost Equal	≠	Not Equal
∠	Angle	Ω	Ohm
℄	Boundary Line	Ω	Omega
℄	Center Line	ℙ	Property Line
Δ	Delta	$H_2 0$	Subscript 2
φ	Electrical Phase	4^2	Squared
℄	Flow Line	4^3	Cubed

Figure 6–20

CREATING SINGLE LINE TEXT

Single line text is another method of adding text to your drawing. The actual command used to perform this operation is DTEXT, which stands for Dynamic Text mode, and which allows you to place a single line of text in a drawing and view the text as you type it. Choose the Draw menu and then choose Single Line Text, as shown in the following image.

Figure 6–21

When using the DTEXT command, you are prompted to specify a start point, height, and rotation angle. You are then prompted to enter the actual text. As you do this, each letter displays on the screen. When you are finished with one line of text, pressing ENTER drops the Insert bar to the next line, where you can enter more text. Again pressing ENTER drops the Insert bar down to yet another line of text, as shown on the left in the following image. Pressing ENTER at the "Enter text" prompt exits the DTEXT command and permanently adds the text to the database of the drawing, as shown on the right in the following image.

Command: **DT** *(For DTEXT)*
Current text style: "Standard" Text height: 0.2000
Specify start point of text or [Justify/Style]: *(Pick a point at "A")*
Specify height <0.2000>: **0.50**
Specify rotation angle of text <0>: *(Press ENTER to accept this default value)*
Enter text: **ENGINEERING** *(After this text is entered, press ENTER to drop to the next line of text)*
Enter text: **GRAPHICS** *(After this text is entered, press ENTER to drop to the next line of text)*
Enter text: *(Either add more text or press ENTER to exit this command and place the text)*

Figure 6–22

ADDITIONAL DYNAMIC TEXT APPLICATIONS

When you place a line of text using the DTEXT command and press ENTER, the Insert bar drops down to the next line of text. This sequence continues until you press ENTER at the "Enter text" prompt, which exits the command and places the text permanently in the drawing. You can also control the placement of the Insert bar by clicking a new location in response to the "Enter text" prompt. In the following image, various labels need to be placed in the pulley assembly. The first label, "BASE PLATE," is placed with the DTEXT command. Without pressing ENTER, immediately pick a new location at "B" and place the text "SUPPORT." Continue this process with the other labels. The Insert bar at "E" denotes the last label that needs to be placed. When you perform this operation, pressing ENTER one last time at the "Specify rotation angle of text <0>:" prompt places the text and exits the command. Follow the command sequence below for a better idea of how to perform this operation.

Working with Text, Fields, and Tables 341

 Try It! – Open the drawing file 06_Pulley Text. Use the following image and command sequence to perform this operation.

Command: DT *(For DTEXT)*
Current text style: "Standard" Text height: 0.2500
Specify start point of text or [Justify/Style]: *(Pick a point at "A")*
Specify height <0.2500>: *(Press ENTER to accept this default value)*
Specify rotation angle of text <0>: *(Press ENTER to accept this default value)*
Enter text: BASE PLATE *(After this text is entered, pick approximately at "B")*
Enter text: SUPPORT *(After this text is entered, pick approximately at "C")*
Enter text: SHAFT *(After this text is entered, pick approximately at "D")*
Enter text: PULLEY *(After this text is entered, pick approximately at "E")*
Enter text: *(Enter BUSHING for this part and then press ENTER)*
Specify rotation angle of text <0>: *(Press ENTER to exit this command)*

Figure 6–23

EDITING TEXT

Text constructed with the MTEXT and DTEXT commands is easily modified with the DDEDIT command. Start this command by choosing Edit…, which is found in the Text category of the Modify pull-down menu, as shown on the left in the following image. Selecting the multiline text object displays the Multiline Text Editor, as shown on the right in the following image. This is the same dialog box as that used to initially create text through the MTEXT command. Use this dialog box to change the text height, font, color, and justification.

 Note: A quicker way to edit multiline text is to double-click on the text object. This will launch the Multiline Text Editor box and Text Formatting toolbar.

 Try It! – Open the drawing file 06_Edit Text1. Find the Edit text command in the Modify pull-down menu or use the following command sequence to display the Multiline Text Editor.

 Command: **ED** *(For DDEDIT)*
Select an annotation object or [Undo]: *(Select the mtext object and the Multiline Text Editor dialog box will appear. Perform any text editing task and click the OK button)*
Select an annotation object or [Undo]: *(Pick another mtext object to edit or press* ENTER *to exit this command)*

Figure 6–24

 Try It! – Open the drawing file 06_Edit Text2. Activate the Multiline Text Editor dialog box by entering the ED command and picking any text object. In this dialog box, the text is currently drawn in the RomanD font and needs to be changed to Arial. To accomplish this, first highlight all the text. Next, change to the desired font, as shown in the following image.

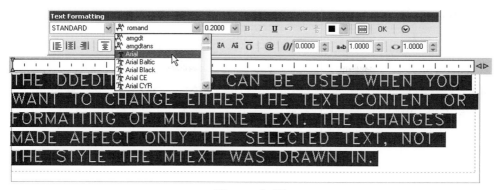

Figure 6–25

Since all text was highlighted, changing the font to Arial updates all text, as shown in the following image. Clicking on the OK button dismisses the dialog box and changes the text font in the drawing.

Figure 6–26

When editing mtext objects, you can selectively edit only certain words that are part of a multiline text string. To perform this task, highlight the text to change and apply the new formatting style, such as font, underscore, or height, to name a few.

 Try It! – Open the drawing file 06_Edit Text3. Activate the Multiline Text Editor dialog box by entering the ED command and picking any text object. In the following image, the text "PLACING" needs to be underscored; the text "AutoCAD" needs to be changed to a Swiss font; the text "MTEXT" needs to be increased to a text height of 0.30 units. When performing this type of operation, you need to highlight only the text object you want to change. Clicking on the OK button dismisses the dialog box and changes the individual text objects in the drawing.

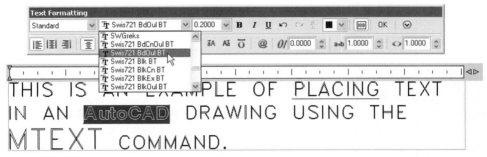

Figure 6–27

Double-clicking on a text object created with the DTEXT command, as shown on the left in the following image, displays an edit box, as shown on the right in the following image. Use this to change text in the edit box provided. Font, justification, and text height are not supported in this edit box.

Figure 6–28

 ## GLOBALLY MODIFYING THE HEIGHT OF TEXT

The height of a text object can be easily changed through a command found in the Modify pull-down menu. Click on Object followed by Text and then Height. This command allows you to pick a text object and change its text height. The important feature of this command is that scaling has no effect on the justification of the text. This means that after scaling the text, you should not have to move each individual text item to a new location.

 Try It! – Open the drawing file 06_Scale Text, illustrated in the following image. All offices in this facilities plan have been assigned room numbers. One of the room numbers (ROOM 114) has a text height of 12″, while all other room numbers are 8″ in height. You could edit each individual room number until all match the height of ROOM 114; or you could use the SCALETEXT command by using the following command sequence:

Command: **SCALETEXT**
Select objects: *(Pick all 16 mtext objects except for ROOM 114)*
Select objects: *(Press* ENTER *to continue)*
Enter a base point option for scaling
[Existing/Left/Center/Middle/Right/TL/TC/TR/ML/MC/MR/BL/BC/BR] <Existing>: *(Press* ENTER *to accept this value)*
Specify new height or [Match object/Scale factor] <8.0000>: **M** *(For Match object)*
Select a text object with the desired height: *(Pick the text identified by ROOM 114)*
Height=12.0000

Figure 6–29

The results are displayed in the following image. All the text was properly scaled without affecting the justification locations.

Figure 6–30

 Note: The Quick Select dialog box could be used to select all mtext objects.

GLOBALLY MODIFYING THE JUSTIFICATION OF TEXT

Yet another text editing mode can be found in the Modify pull-down menu. Click on Object followed by Text and then Justify. This command will allow you to pick a text item and change its current justification. If you pick a multiline text object, all text in this mtext object will have its justification changed. In the case of regular text placed with the DTEXT command, you would have to select each individual line of text for it to be justified.

 Try It! – Open the drawing file 06_Justify Text, illustrated in the following image. All the text justification points in this facilities drawing need to be changed from left justified to top center justified. Use the following command prompt and illustration in the following image to accomplish this task.

Figure 6–31

Command: **JUSTIFYTEXT**

Select objects: *(Select all 17 text objects)*
Select objects: *(Press* ENTER *to continue)*
Enter a justification option
[Left/Align/Fit/Center/Middle/Right/TL/TC/TR/ML/MC/MR/BL/BC/BR] <Left>: **TC** *(For Top Center)*

The results are illustrated in the following image. All text objects have been globally changed from left justified to top center justified.

Figure 6–32

SPELL-CHECKING TEXT

Issuing the SPELL command and selecting the multiline text object displays the Check Spelling dialog box, shown in the following image, with the following components: the Current dictionary and the Current word identified by the spell-checker as being misspelled. The Suggestions area displays all possible alternatives to the word identified as being misspelled. The Ignore button allows you to skip the current word; this would be applicable especially in the case of acronyms such as CAD and GDT. Clicking on Ignore All skips all remaining words that match the current word. The Change button replaces the word in the Current word box with a word in the Suggestions box. The Add button adds the current word to the current dictionary. The Lookup button checks the spelling of a selected word found in the Suggestions box. The Change Dictionaries button allows you to change to a different dictionary containing other types of words. The Context area displays a phrase showing how the current word is being used in a sentence. Use this area to check proper sentence structure. In the following image, the word "DIRECION" was identified as being misspelled. Notice that the correct word, "DIRECTION," is listed as a suggested correction. Clicking on the Change button replaces the misspelled word and continues with the spell-checking operation until completed.

PLACE THE YOKE
PERPENDICULAR
TO THE DIRECION
OF THE SUSPECTED
WELD DEFECT.

Figure 6–33

After undergoing spell-checking, the mtext object is displayed in the following image.

PLACE THE YOKE
PERPENDICULAR
TO THE DIRECTION
OF THE SUSPECTED
WELD DEFECT.

Figure 6–34

 Try It! – Open the drawing 06_Spell Check1 and perform a spell-check operation on this mtext object.

Before	After
THIS IS AN EXAMPL OF PLACING TEXT IN AN AUTOCAD DRAWNG USING THE MTEXT COMMAND.	THIS IS AN EXAMPLE OF PLACING TEXT IN AN AUTOCAD DRAWING USING THE MTEXT COMMAND.

Figure 6–35

CREATING DIFFERENT TEXT STYLES

A text style is a collection of settings that are applied to text placed with the DTEXT or MTEXT command. These settings could include presetting the text height and font, in addition to providing special effects, such as an oblique angle for inclined text. Choose Text Style... from the Format pull-down menu or from the Text Style toolbar, shown on the left in the following image, to activate the Text Style dialog box, shown on the right.

Use the Text Style dialog box, as shown on the right in the following image, to create new text styles. As numerous styles are created, this dialog box can also be used to make an existing style current in the drawing. By default, when you first begin a drawing, the current Style Name is STANDARD. It is considered good practice to create your own text style and not rely on STANDARD. Once a new style is created, a font name is matched with the style. Clicking in the edit box for Font Name displays a list of all text fonts supported by the operating system. These fonts have different extensions, such as SHX and TTF. TTF or TrueType Fonts are especially helpful in AutoCAD because these fonts display in the drawing in their true form. If the font is bold and filled-in, the font in the drawing file displays as bold and filled-in.

When a Font Name is selected, it displays in the Preview area located in the lower right corner of the dialog box. The Effects area allows you to display the text upside down, backwards, or vertically. Other effects include a width factor, explained later, and the oblique angle for text displayed at a slant.

Figure 6–36

Clicking on the New... button of the Text Style dialog box displays the New Text Style dialog box, shown in the following image. A new style is created called General Notes. Clicking on the OK button returns you to the Text Style dialog box. Clicking in the Font Name edit box displays all supported fonts. Clicking on Arial assigns the font to the style name General Notes. Clicking on the Apply button saves the font to the database of the current drawing file.

Figure 6–37

THE TEXT STYLE CONTROL BOX

A special Text Style toolbar is available next to the Standard toolbar on your display screen. Use the Text Style Control box illustrated in the following image to make an existing text style current. This control box allows you to change from one text style to another easily.

 Note: You can also highlight a text object and use the Text Style Control box to change the selected text to a different text style. This action is similar to changing an object from one layer to another through the Layer Control box.

Figure 6–38

FIELDS

Fields are a type of intelligent text that you can add to your drawings. In the following image, fields are identified by a distinctive gray background, which is not present on regular text. Fields are intended for items that tend to change during the life cycle of a drawing, such as the designer's name, checker, creation date of the drawing, and drawing number, as shown in the following image.

Figure 6–39

The FIELD command is selected from the Insert pull-down menu, as shown on the left in the following image. This activates the Field dialog box, which consists of Field categories, Field names, Author, and Format information. Rather than cycle through all field names, you can pick a field category, which will limit the amount of items in the Field name area.

Figure 6–40

By default, all fields are available under the Field names area. You can specify a certain Field category to narrow down the list of field names. Illustrated on the left in the following image is a list of all field categories; the Date & Time category is selected. This activates specific field names only related to this category, as shown on the right in the following image. With the Field name CreateDate highlighted, notice all the examples that pertain to the creation date of a drawing. There is also a Hints area that gives examples of various month, day, and year notations.

Figure 6–41

FIELDS AND MULTILINE TEXT

While placing multiline text, you can create a field through the Text Formatting toolbar, as shown in the following image. This allows you to place numerous fields at one time. Right-clicking in the text box and then picking Insert Field from the menu will also activate the Field dialog box.

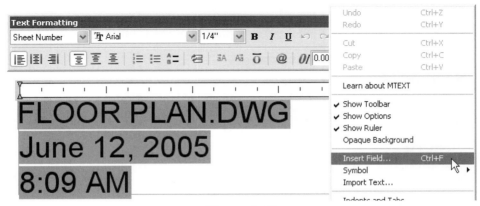

Figure 6–42

The results of creating a field are illustrated on the left in the following image. In this example, the following field categories were used: Filename and Date. The time entry is a subset of the Date field.

Whenever field information changes, such as the date and time, you may need to update the field in order to view the latest information. Updating fields is accomplished by clicking on Update Fields, which is located under the Tools pull-down menu area, as shown in the middle in the following image. When picking this command, you are prompted to select the field. In the following image on the right, the date automatically changed. Whenever you open a drawing that contains fields, the fields will automatically update to reflect the most updated information.

Figure 6–43

FIELDS AND DRAWING PROPERTIES

When creating some fields, information is displayed as a series of four dashed lines. This is to signify that the information contained in the field name either has not been entered or has not been updated. In the following image, the Author, Subject, and Title fields need to be completed before the correct information displays.

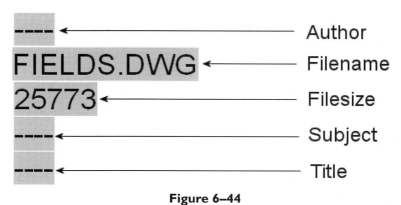

Figure 6–44

In the previous illustration of the Author, Subject, and Title field names, activating the Drawing Properties dialog box, as shown on the left in the following image, will allow you to fill in this information. When returning to the drawing, click on the Tools pull-down menu, select Update Fields, and pick the field that needs updating. The information listed in the Drawing Properties dialog box will be transferred to the specific fields, as shown on the right in the following image.

Figure 6–45

CREATING TABLES

As a means of further organizing your work, especially text, you can create cells organized in rectangular patterns that consist of rows and columns. The TABLE command is used to accomplish this and can be selected using one of the following methods, as shown in the following image.

> From the Draw pull-down menu.
>
> From the Draw toolbar.
>
> By typing TABLE in at the Command prompt.

Using one of the above methods will activate the Insert Table dialog box, as shown in the following image. It is here that you specify the number of rows and columns that make up the table, in addition to the column width and row height.

Figure 6–46

Once you decide on the number of rows and columns for the table, clicking the OK button returns you to the drawing editor. Here is where you locate the position of the table in your drawing. This is a similar operation to inserting a block. After the table is positioned, the table appears in the drawing editor and the Text Formatting toolbar displays above the table, as shown in the following image. In this example, the text "WINDOW SCHEDULE" was added as the title of the table. Notice also that the text height automatically adjusts to the current table style. This will be discussed later in this chapter.

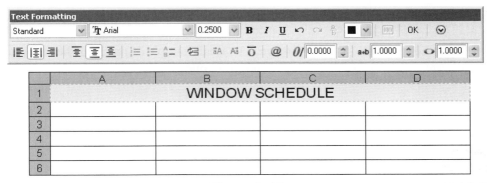

Figure 6–47

After you have entered text into the title cell, pressing the TAB key will move you from one cell to another, as shown in the following image. In this example, cell headers are added to identify the various categories of the table.

Figure 6–48

 Note: Pressing the TAB key will move you from one cell to another; however, you cannot reverse this direction through the TAB key. Instead, press SHIFT + TAB to reverse the direction in moving from one cell to another inside a table.

MOVING A TABLE

One method of moving a table is to use the MOVE command. The table will reposition itself to a new location since it is considered an AutoCAD object. Another method of moving a table is to first select the table, which will display various grips. Pick the grip in the upper left corner and move your cursor. Notice that the table will move to where your cursor is positioned.

MERGING CELLS IN A TABLE

There are times when you need to merge the rows or columns in a table. To accomplish this, first pick inside a cell, then hold down the SHIFT key as you pick the cells you want to merge with it. Then right-click to display the menu and pick Merge Cells followed by All, as shown in the following image. The final example of merging the cells that contain the "FLOOR" text is also shown in the following image.

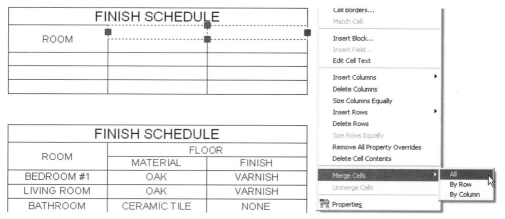

Figure 6–49

TABLES AND GRIPS

The modification of table cells, table rows, and even the overall size of the table is greatly enhanced through the use of grips. You can achieve different results depending on the table grip that is selected. For instance, first select the table to display all grips. Then, select the grip in the upper left corner of the table and move your cursor. This action will move the table to a new location. Selecting the grip in the upper right corner of the table and then dragging your cursor left or right will increase or decrease the spacing of all table columns. Selecting the grip in the lower left corner of the table will increase the spacing between rows. Study the following image to see how other grips affect the sizing of a table. Grips will be covered in greater detail in chapter 7.

Figure 6–50

 Note: When using grips to stretch rows or columns in a table, you may want to turn running OSNAP off. Otherwise, if OSNAP Endpoint is set, your table may stretch back to its original location, giving the appearance that the operation did not work.

TABLES AND FIELDS

Fields can also be inserted into table cells. Begin this operation by first double-clicking inside the cell to activate the multiline text box. Place your cursor in the cell where you want the field to occur. Then right-click to display the menu and click on Insert Field, as shown in the following image.

Figure 6–51

TABLES AND MICROSOFT EXCEL

Microsoft Excel spreadsheets can easily be imported into AutoCAD drawings as tables. The illustration on the left in the following image represents information organized in an Excel spreadsheet. With the spreadsheet open, copy the appropriate data to the Windows Clipboard. Switch to AutoCAD and while in the drawing editor, click on the Edit pull-down menu and pick Paste Special, as shown in the middle of the following image. This will display the Paste Special dialog box, as shown on the right in the following image. Click on AutoCAD Entities to merge the spreadsheet into an AutoCAD drawing as a table.

Figure 6–52

The initial results of importing an Excel spreadsheet into an AutoCAD drawing are illustrated on the left in the following image. The Excel spreadsheet was converted into an AutoCAD drawing object. However, a few columns need to be lengthened and the rows shortened in order to organize the data in a single line. Also, headings need to be added to the top of the table by inserting a row above the top line of data. The completed table is illustrated on the right in the following image.

Figure 6–53

CREATING TABLE STYLES

Table styles can be used to organize different properties that make up a table. In the following image on the left, click Table Style in the Format pull-down menu to display the Table Style dialog box, shown on the right. Clicking on the New button will display the Create New Table Style dialog box, in which you enter the name of the table style. The Table Style dialog box can also be displayed by picking Table Style from the Styles toolbar, as shown in the following image.

Figure 6–54

Clicking on Continue allows you to make various changes to a table based on the type of cell. For example, in the following image three tabs can be used to make changes to the text that occupies a cell in a table based on whether the text is cell data, the column heads of a cell, or even the title cell of the table. These text settings can be used to emphasize the information in the cells. For example, you would want the title of the table to stand out with a larger text height than the column headings. You might also want the column headings to stand out over the data in the cells below. Also, as in the following image, you can change the direction of the table. By default, a table is created in the downward direction. You may want to change this direction to upward.

Figure 6–55

The following image illustrates the information required whenever you change the column heads and title of a table.

Figure 6–56

TUTORIAL EXERCISE: 06_PUMP_ASSEMBLY.DWG

Figure 6–57

Purpose

This tutorial is designed to create numerous text styles and add different types of text objects to the title block illustrated in the previous image.

System Settings

Since this drawing is already provided on the CD, open an existing drawing file called 06_Pump_Assembly. Follow the steps in this tutorial for creating a number of text styles and then placing text in the title block area. Check to see that the following Object Snap modes are already set: Endpoint, Extension, Intersection, Center.

Layers

Various layers have already been created for this drawing. Since this tutorial covers the topic of text, the current layer is Text.

Suggested Commands:

Open the drawing called 06_Pump_Assembly. Create four text styles called Company Info, Title Block Text, Disclaimer, and General Notes. The text style Company Info will be applied to one part of the title block. Title Block Text will be applied to a majority of the title block, where questions such as Drawn By, Date, and Scale are asked. Also, the Drawn By, Date, and Drawing Number will be supplied by creating fields. A disclaimer will be imported into the title block in the Disclaimer text style. Finally, a series of notes will be imported in the General Notes text style. Once these general notes are imported, a spell-check operation will be performed on the notes.

STEP 1

Opening the drawing file 06_Pump_Assembly displays an image similar to the following image. The purpose of this tutorial is to fill in the title block area with a series of text, multiline text, and field objects. Also, a series of general notes will be placed to the right of the pump assembly. Before you place the text, four text styles will be created to assist with the text creation. The four text styles are also identified in the following image. The text style Company Info will be used to place the name of the company and drawing title. Information in the form of scale, date, and who performed the drawing will be handled by the text style Title Block Text. A disclaimer will be imported from an existing .TXT file available on the CD; this will be accomplished in the Disclaimer text style. A listing of six general notes is also available on the CD. It is in .RTF format, originally created in Microsoft Word. It will be imported into the drawing and placed in the General Notes text style.

Since you will be locating various justification points for the placement of text, turn OSNAP off by single-clicking on OSNAP in the Status bar at the bottom of the display screen.

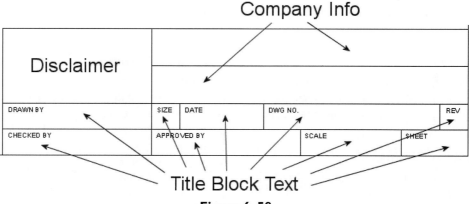

Figure 6–58

STEP 2

Before creating the first text style, first see what styles are already defined in the drawing by choosing Text Style... from the Format pull-down menu, as shown on the left in the following image. This displays the Text Style dialog box, as shown on the right in the following image. Click in the Style Name edit box to display the current text styles defined in the drawing. Every AutoCAD drawing contains the STANDARD text style. This is created by default and cannot be deleted. Also, the Title Block Headings text style was already created. This text style was used to create the headings in each title block box, for example, "Scale" and "Date."

Figure 6–59

STEP 3

Create the first text style by clicking on the New... button, as shown in the following image. This will activate the New Text Style dialog box, also shown in the following image. For Style Name, enter Company Info. When finished, click the OK button. This will take you back to the Text Style dialog box, as shown on the right in the following image. In the Font area, change the name of the font to Arial. Notice the font appearing in the Preview area. When finished, click the Apply button to complete the text style creation process.

Figure 6–60

STEP 4

Using the same procedure outlined in Step 3, create the following new text styles: Disclaimer, General Notes, and Title Block Text. Assign the Arial font to all of these style names. Keep all other remaining default text settings. When finished, your display should appear similar to the following image.

Figure 6–61

STEP 5

Add multiline text under the Company Info text style. Change the text height for both entries to .25 units. Center justify the text. Add the text PUMP ASSEMBLY inside the text formatting box. Click the OK button to place the text. Repeat this procedure for placing the text THE K-GROUP in the next title block space. When finished, move the text entries to a better location inside the title block space.

Command: **MT** *(For MTEXT)*
MTEXT Current text style: "Company Info" Text height: 0.20
Specify first corner: *(Pick a point at "A")*
Specify opposite corner or [Height/Justify/Line spacing/Rotation/Style/Width]: **H** *(For Height)*
Specify height <0.20>: **.25**
Specify opposite corner or [Height/Justify/Line spacing/Rotation/Style/Width]: *(Pick a point at "B")*

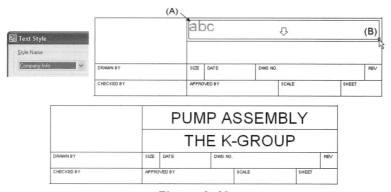

Figure 6–62

STEP 6

Change the current text style to Title Block Text, as shown in the following image. Begin placing fields into the title block. Click on Insert Field under the Insert pull-down menu, as shown in the following image. When the Field dialog box appears, pick Author under the Field name and click the OK button to dismiss the dialog box. Place this in the space identified as Drawn By, as shown on the right in the following image. Use a height of .10 for all of the next series of entries. Notice the appearance of four dashes inside a gray rectangle. This will be filled in when you enter your name in the appropriate area of the Properties dialog box later on in this exercise. Use the following command sequence to assist in the placing of this first field.

Command: **FIELD**
MTEXT Current text style: "Title Block Text" Text height: 0.20
Specify start point or [Height/Justify]: **H** *(For Height)*
Specify height <0.20>: **.10**
Specify start point or [Height/Justify]: *(Pick a point inside of area identified as Drawn By)*

Figure 6–63

STEP 7

Place a new field for the Date category. Activate the Field dialog box, as shown in the following image, and pick the Date Field name. Also select the format of the date from the list of examples located in this dialog box. Place this field under the DATE heading, as shown in the following image. Notice that the date automatically calculated based on the current date setting on your computer.

Command: **FIELD**
MTEXT Current text style: "Title Block Text" Text height: 0.10
Specify start point or [Height/Justify]: *(Pick a point inside of the area identified as Date)*

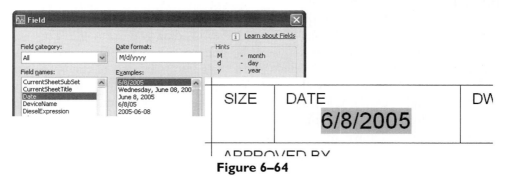

Figure 6–64

STEP 8

Place a new field for the Filename category. Activate the Field dialog box, as shown in the following image, and pick the Filename Field name. Pick Uppercase for the Format. Also click on the radio button next to Filename only. This will limit the number of letters in the drawing name. Locate this field under the DWG NO. heading name, as shown in the following image.

Command: **FIELD**
MTEXT Current text style: "Title Block Text" Text height: 0.10
Specify start point or [Height/Justify]: *(Pick a point inside of the area identified as DWG NO.)*

Figure 6–65

STEP 9

While still in the Title Block Text style, use the Single Line text command and add the following entries to the title block using the following prompts and image as guides.

Command: **DT** *(For DTEXT)*
Current text style: "Title Block Text" Text height: 0.10
Specify start point of text or [Justify/Style]: *(Pick a point approximately at "A")*
Specify rotation angle of text <0d0'>: *(Press ENTER to accept this default value)*
Enter text: **ADRIAN CULPEPPER** *(After this text is entered, pick approximately at "A")*
Enter text: *(Press ENTER to exit this command)*

Continue using Single Line text to fill in the following areas of the title block:

Text	Title Block Area
B	SIZE (at "B")
JOHNNY MOSS	APPROVED BY (at "C")
1"=1'-0"	SCALE (at "D")
0	REV (at "E")
1 OF 1	SHEET (at "F")

Figure 6–66

STEP 10

Open the Text Style dialog box and click on the Disclaimer text style, as shown in the following image. Activate the MTEXT command, pick a first corner at "A," change the text height to .08, and pick a second corner at "B." While in the Text Formatting dialog box, right-click and pick Import Text from the menu, as shown in the following image.

[A] Command: **MT** *(For MTEXT)*
MTEXT Current text style: "Disclaimer" Text height: 0.20
Specify first corner: *(Pick a point at "A")*
Specify opposite corner or [Height/Justify/Line spacing/Rotation/Style/Width]: **H** *(For Height)*
Specify height <0.20>: **.08**
Specify opposite corner or [Height/Justify/Line spacing/Rotation/Style/Width]: *(Pick a point at "B")*

Figure 6–67

When the Select File dialog box appears, click on the location of your CD and pick the file 06_Disclaimer.txt, as shown on the left in the following image. Since this information was already created in an application outside AutoCAD, it will be imported into the Multiline Text Editor dialog box. Click the OK button to place the text in the title block, as shown on the right in the following image.

Figure 6–68

STEP 11

Before placing the last text object, zoom to the extents of the drawing. Then click on the General Notes text style located in the Text Style Control box, as shown on the left in the following image. Use the MTEXT command to create a rectangle to the right of the Pump Assembly, as shown on the right in the following image. This will be used to hold a series of general notes in multiline text format.

Command: **MT** *(For MTEXT)*
Current text style: "General Notes" Text height: 0.12
Specify first corner: *(Pick a point at "A")*
Specify opposite corner or [Height/Justify/Line spacing/Rotation/Style/Width]: *(Pick a point at "B")*

Figure 6–69

When the Multiline Text Editor dialog box appears, click on the Import Text... button, as shown on the left in the following image. Then click on the location of your CD and select the file 06_GENERAL_NOTES.rtf, as shown on the right in the following image. This RTF file (rich text file) was created outside AutoCAD in Microsoft Word.

Figure 6–70

STEP 12

The results of this operation are displayed on the left in the following image. However, the text appears to be displayed in a different font. This is due to the fact that when files are imported in RTF format, the original format of the text is kept. This means that the Times New Roman font was used even though the current text style uses the Arial font.

While still inside the Text Formatting toolbar, highlight all the general notes and change the font to Arial, as shown on the right in the following image.

Figure 6–71

One more item needs to be taken care of before leaving the Text Formatting toolbar. Highlight all the text directly under the heading of GENERAL NOTES, as shown in the following image. Click on the Numbering button. This will create a numbered list based on all highlighted lines of text, as shown in the following image. When finished, click the OK button to return to the drawing and place the general notes.

Figure 6–72

STEP 13

Use the SPELL command and click on the multiline text object that holds the six general notes. When the Check Spelling dialog box appears in the following image, make the corrections to the words CONCRETE, MOTOR, and PROPERLY.

Figure 6–73

STEP 14

In a previous step, you created a field for the author of the drawing. However, this field displayed as empty. You will now complete the title block by filling in the information and completing the Author field. Click on Drawing Properties, found under the File pull-down menu, as shown on the left in the following image. When the 06_Pump Assembly.dwg Properties dialog box appears, enter your name in the Author text box, as shown in the middle in the following image.

When you return to your drawing, the Author field located in the title block is still not reflecting your name. Click Update Fields, found under the Tools pull-down menu area, to place your name in the title block.

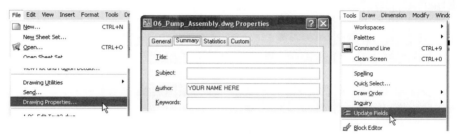

Figure 6–74

STEP 15

The completed title block and general notes area are displayed in the following image.

Figure 6–75

TUTORIAL EXERCISE: 06_TABLE.DWG

DOOR SCHEDULE

MARK	WIDTH	HEIGHT	THICKNESS	MATERIAL	NOTES
1	9'-0"	8'-0"	1 3/4"	VINYL	~~
2	9'-0"	8'-0"	1 3/4"	VINYL	~~
3	9'-0"	8'-0"	1 3/4"	VINYL	~~
4	3'-0"	6-8"	1 3/8"	HCW	~~
5	2'-4"	6-8"	1 3/8"	HCW	~~
6	1'-6"	6-8"	1 3/8"	HCW	~~
7	5'-0"	6-8"	1 3/4"	METAL CLAD	FRENCH DOOR
8	5'-0"	6-8"	1 3/4"	METAL CLAD	FRENCH DOOR
9	2-8"	6-8"	1 3/4"	METAL	SCREENED DOOR

Figure 6-76

Purpose

This tutorial is designed to create a new table style and then a new table object, as shown in the previous image.

System Settings

Since this drawing is already provided on the CD, open an existing drawing file called 06_Table. Two text styles are already created for this exercise; namely Architectural and Title.

Layers

A Table layer is already created for this drawing.

Suggested Commands

Open the drawing called 06_Table. Create a new table style called Door Schedule. Changes will be made to the Data, Heading, and Data fields through the Table Style dialog box. Next, create a new table consisting of six columns and eleven rows. Information related to the door schedule will be entered in at the Title, Heading, and Data fields.

STEP 1

Open the existing drawing file 06_Table. Activate the Text Style dialog box (STYLE command) and notice the two text styles already created, as shown on the left in the following image, namely, Title and Architectural. The Title text style uses the Arial font while Architectural uses the CityBlueprint font.

Next, activate the Table Style dialog box and create a new table style called Door Schedule, as shown on the right in the following image.

Figure 6–77

STEP 2

Clicking on the Continue button located in the previous image will take you to the main New Table Style dialog box. Use the information in the following image to change the properties under the Data, Column Heads, and Title tabs.

Figure 6–78

STEP 3

Click the OK button to return to the Table Style dialog box. Select the Door Schedule style from the list and click the Set Current button to make it the current table style, as shown in the following image (you could also double-click on any table style name from the list to make it current). When finished, close the Table Style dialog box.

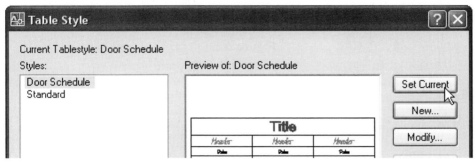

Figure 6–79

STEP 4

Activate the TABLE command and while in the Insert Table dialog box, create a table consisting of six columns and eleven rows using the new Door Schedule table style. Click the OK button to place the table in a convenient location on your screen.

Figure 6–80

STEP 5

When the Text Formatting toolbar appears, fill out the title of the table as DOOR SCHEDULE, as shown in the following image.

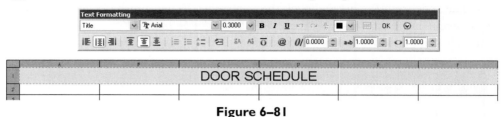

Figure 6–81

STEP 6

Next, double-click inside the first column on the next row, and when the Text Formatting toolbar appears, label this cell MARK. Then press the TAB key to advance to the next column header and label this WIDTH. Repeat this procedure for the next series of column headers, namely HEIGHT, THICKNESS, MATERIAL, and NOTES, as shown in the following image.

DOOR SCHEDULE					
MARK	WIDTH	HEIGHT	THICKNESS	MATERIAL	NOTES

Figure 6–82

STEP 7

As the table appears to be stretched too long, click on the table and pick the grip in the upper right corner. Then stretch the table to the left, which will shorten the length of all rows, as shown in the following image.

DOOR SCHEDULE					
MARK	WIDTH	HEIGHT	THICKNESS	MATERIAL	NOTES

Figure 6–83

STEP 8

Complete the table by filling in all rows and columns that deal with the data, as shown in the following image.

DOOR SCHEDULE

MARK	WIDTH	HEIGHT	THICKNESS	MATERIAL	NOTES
1	9'-0"	8'-0"	1 3/4"	VINYL	~~
2	9'-0"	8'-0"	1 3/4"	VINYL	~~
3	9'-0"	8'-0"	1 3/4"	VINYL	~~
4	3'-0"	6-8"	1 3/8"	HCW	~~
5	2'-4"	6-8"	1 3/8"	HCW	~~
6	1'-6"	6-8"	1 3/8"	HCW	~~
7	5-0"	6-8"	1 3/4"	METAL CLAD	FRENCH DOOR
8	5'-0"	6-8"	1 3/4"	METAL CLAD	FRENCH DOOR
9	2-8"	6-8"	1 3/4"	METAL	SCREENED DOOR

Figure 6–84

Note: To copy similar data from one cell to another, first click once inside the cell, then right-click in the cell and select Copy from the menu. Move your cursor to another cell and click once inside it, right-click in the cell, and select Paste from the menu.

TUTORIAL EXERCISE: 06_TABLE SUMMATIONS.DWG

Figure 6–85

Purpose
This tutorial is designed to perform mathematical calculations and summations on a table.

System Settings
No special system settings need to be made for this drawing.

Layers
Layers are already created for this drawing.

Suggested Commands
Open the drawing file 06_Table Summations as shown in the previous image. Notice how the drawing opens up in a layout called Floor Plan, as shown in the following image. In addition to the floor plan being displayed, a table is present. Basic mathematical calculations such as addition, subtraction, multiplication, and division can be made on information in a table. Order of operations can also be used on the basic math calculations. You can also perform sum, average, and counting operations. You will insert formulas into this table that will calculate the area of each room. You will then perform a summation on all room areas.

STEP 1

Zoom in to the table. You will first assign a formula that will calculate the area of the Hallway. Right click on the cell that will hold the area of the Hallway. When the menu displays, as shown in the following image, click on Insert Formula followed by Equation.

Figure 6–86

STEP 2

The format of the table will turn into a spreadsheet, with letters identifying columns and numbers identifying rows, as shown in the following image. In the highlighted cell, you will need to multiply the Hallway length (38) by the width (8). You should not enter these numbers directly into the equation cell. Rather, enter the column and row number of each value. For example, the value 38 is identified by B3 and the value 8 is identified by C3. Enter these two cell identifiers separated by the multiplication symbol (*). Your display should appear similar to the following image.

Figure 6–87

STEP 3

Exiting the Text Formatting dialog box will create a field in the cell the calculation was being performed in, as shown in the following image. This field is easily recognized by the typical gray text background. Placing information as a field means if the information in the length and width cells changes, the field information will automatically recalculate itself.

ROOM AREA CHART			
ROOM NUMBER	LENGTH	WIDTH	ROOM AREA (SQ FT)
HALLWAY	38	8	304
ROOM 1	22	15	
ROOM 2	15	15	

Figure 6–88

STEP 4

Follow steps 1 and 2 to calculate the area of Room 1. Be careful to use the correct cell identifying letters and numbers for this area calculation, as shown in the following image.

	A	B	C	D
1		ROOM AREA CHART		
2	ROOM NUMBER	LENGTH	WIDTH	ROOM AREA (SQ FT)
3	HALLWAY	38	8	304
4	ROOM 1	22	15	=B4*C4
5	ROOM 2	15	15	

Figure 6–89

STEP 5

Continue inserting equations to calculate the area of Rooms 2, 3, and 4, as shown in the following image.

ROOM AREA CHART			
ROOM NUMBER	LENGTH	WIDTH	ROOM AREA (SQ FT)
HALLWAY	38	8	304
ROOM 1	22	15	330
ROOM 2	15	15	225
ROOM 3	17	25	425
ROOM 4	20	25	500
		TOTAL AREA	

Figure 6–90

STEP 6

With the individual room areas calculated, you will now perform a summation that will add all current area fields. First, right click in the cell the summation will be performed in. When the menu appears, click Insert Formula followed by Sum, as shown in the following image.

Figure 6–91

STEP 7

After selecting Sum from the menu in the previous step, you will be prompted for the following:

- Select first corner of table cell range:
    - Answer this prompt by picking a point inside the upper left corner of the cell that displays an area of 304. You will be prompted a second time:
- Select second corner of table cell range:
    - Answer this prompt by picking a point inside of the lower right corner of the cell that displays an area of 500.
    - This action identifies the cells on which to perform the summations.

ROOM AREA CHART			
ROOM NUMBER	LENGTH	WIDTH	ROOM AREA (SQ FT)
HALLWAY	38	8	304
ROOM 1	22	15	330
ROOM 2	15	15	225
ROOM 3	17	25	425
ROOM 4	20	25	500
		TOTAL AREA	

Figure 6–92

STEP 8

After you select the second corner of the table cell range in the previous step, the equation is automatically created to add the values of cells D3 through D7, as shown in the following image.

Figure 6–93

STEP 9

The final results are displayed in the following image. This completes this tutorial exercise.

ROOM AREA CHART			
ROOM NUMBER	LENGTH	WIDTH	ROOM AREA (SQ FT)
HALLWAY	38	8	304
ROOM 1	22	15	330
ROOM 2	15	15	225
ROOM 3	17	25	425
ROOM 4	20	25	500
		TOTAL AREA	1784

Figure 6–94

CHAPTER 7

Object Grips and Changing the Properties of Objects

This chapter begins with a discussion of what grips are and how they are used to edit portions of your drawing. Various Try It! exercises are available to practice using grips. This chapter continues by examining a number of methods used to modify objects. These methods are different from the editing commands learned in chapter 4. Sometimes you will want to change the properties of objects such as layer, color, and even linetype. This is easily accomplished through the Properties Palette. The Match Properties command will also be introduced in this chapter. This powerful command allows you to select a source object and have the properties of the source transferred to other objects that you pick.

USING OBJECT GRIPS

An alternate method of editing is to use object grips. A grip is a small box appearing at key object locations such as the endpoints and midpoints of lines and arcs or the center and quadrants of circles. Once grips are selected, the object may be stretched, moved, rotated, scaled, or mirrored. Grips are at times referred to as visual Object Snaps because your cursor automatically snaps to all grips displayed along an object.

 Try It! – Open the drawing file called 07_Grip Objects. While in the Command: prompt, click on each object type displayed in the following image to activate the grips. Examine the grip locations on each object.

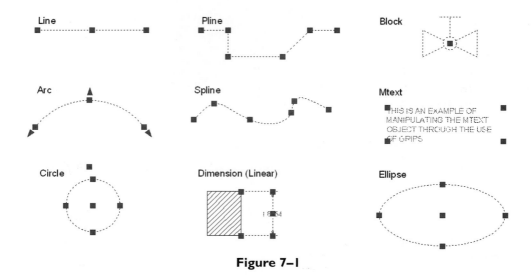

Figure 7–1

Changing the settings, color, and size of grips is accomplished under the Selection tab of the Options dialog box. Right-clicking on a blank part of your screen and choosing Options… from the menu, as shown on the left in the following image, will display the main Options dialog box. Click on the Selection tab to display the grip settings, as shown on the right in the following image.

Figure 7–2

The following grip settings are explained as follows:

Grip size—Use the Grip Size area to move a slider bar to the left to make the grip smaller or to the right to make the grip larger.

Unselected grip color—Controls the color for an unselected grip. It is considered good practice to change the Unselected grip color to a light color if you are using a black screen background.

Selected grip color—Controls the color for a grip that is selected.

Hover grip color—Controls the color of an unselected grip when your cursor pauses over it.

Enable grips—By default, grips are enabled; a check in the Enable grips box means that grips will display when you select an object.

Enable grips within blocks—Normally a single grip is placed at the insertion point when a block is selected. Check the Enable grips within blocks box if you want grips to be displayed along with all individual objects that make up the block.

Enable grip tips—Some custom objects support grip tips. This controls the display of these tips when your cursor hovers over the grip on a custom object that supports grip tips.

Object selection limit for display of grips—By default, this value is set to 100. This means that when you select less than 100 objects, grips will automatically appear on the highlighted objects. When you select more than 100 objects, the display of grips is suppressed.

OBJECT GRIP MODES

The following image shows the main types of grips. When an object is first selected, the object highlights and the square grips are displayed in the default color of blue. In this example of a line, the grips appear at the endpoints and midpoint. The entire object is subject to the many grip edit commands. As you pause your cursor over a grip, the grip will turn green. You will also see the angle and distance of this line, as shown in the middle of the following image, provided that dynamic input is turned on. When one of the grips is selected, it turns red by default, as shown on the right side of the following image. As you move your cursor, dynamic input appears in the form of direction and angle edit boxes used to assist in making changes to this line.

If dynamic input is turned off, distance and angle feedback do not appear. Once a grip is selected, the following prompts appear in the command prompt area:

** STRETCH **
Specify stretch point or [Base point/Copy/Undo/eXit]: *(Press the* SPACEBAR*)*
** MOVE **
Specify move point or [Base point/Copy/Undo/eXit]: *(Press the* SPACEBAR*)*
** ROTATE **
Specify rotation angle or [Base point/Copy/Undo/Reference/eXit]: *(Press the* SPACEBAR*)*
** SCALE **
Specify scale factor or [Base point/Copy/Undo/Reference/eXit]: *(Press the* SPACEBAR*)*
** MIRROR **
Specify second point or [Base point/Copy/Undo/eXit]: *(Press the* SPACEBAR *to begin STRETCH mode again or enter X to exit grip mode)*

Figure 7–3

 Note: When you hover your cursor over an unselected grip, the grip turns green underneath your cursor to signify that it is selectable.

To move from one edit command mode to another, press the Spacebar. Once an editing operation is completed, pressing ESC removes the highlight and removes the grips from the object. The following image shows various examples of each editing mode.

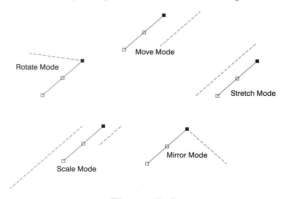

Figure 7–4

In the case of using grips with arcs, directional arrows display when grips appear. Use these arrows to guide you through the various directions in which the grips can be stretched, as shown in the following image.

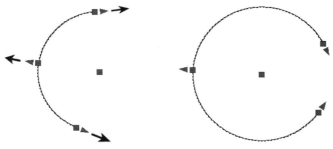

Figure 7–5

ACTIVATING THE GRIP SHORTCUT MENU

In the following image, a horizontal line has been selected and grips appear. The rightmost endpoint grip has been selected. Rather than use the Spacebar to scroll through the various grip modes, click the right mouse button. Notice that a shortcut menu on grips appears. This provides an easier way of navigating from one grip mode to another.

 Try It! – Open the drawing file 07_Grip Shortcut. Click on any object, pick on a grip, and then press the right mouse button to activate the grip shortcut menu.

Figure 7–6

USING THE GRIP—STRETCH MODE

The STRETCH mode of grips operates similarly to the normal STRETCH command. Use STRETCH mode to move an object or group of objects and have the results mend themselves similar to the following image. The line segments "A" and "B" are both too long by two units. To decrease these line segments by two units, use the STRETCH mode by selecting lines "A," "B," and "C" at the command prompt. Next, while holding down SHIFT, select the grips "D," "E," and "F." This will select multiple grips and turn the grips red and ready for the stretch operation. Release SHIFT and pick the grip at "E" again. The STRETCH mode appears in the command prompt area. Since the last selected grip is considered the base point, entering a polar coordinate value of @2.00<180 will stretch the three highlighted grip objects to the left a distance of two units. The Direct Distance mode could also be used with grips. To remove the object highlight and grips, press ESC at the command prompt.

 Try It! – Open the drawing file 07_Grip Stretch. Use the illustration in the following image and the prompt sequence below to perform this task.

Command: *(Select the three dashed lines shown in the following image. Then, while holding down SHIFT, select the grips at "D," "E," and "F." Release the SHIFT key and pick the grip at "E" again)*
STRETCH
Specify stretch point or [Base point/Copy/Undo/eXit]: **@2<180** (or turn on Ortho mode, move your cursor to the right, and type 2)
Command: *(Press ESC to remove the object highlight and grips)*

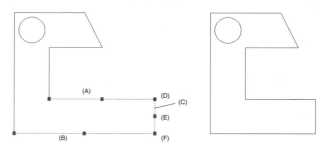

Figure 7–7

USING THE GRIP—SCALE MODE

Using the SCALE mode of object grips allows an object to be uniformly scaled in both the X and Y directions. This means that a circle, such as the one shown in the following image, cannot be converted to an ellipse through different X and Y values. As the grip is selected, any cursor movement will drag the scale factor until a point is marked where the object will be scaled to that factor. The following image and the following prompt illustrate the use of an absolute value to perform the scaling operation of half the circle's normal size.

 Try It! – Open the drawing file 07_Grip Scale. Use the illustration in the following image and the prompt sequence below to perform this task.

Command: *(Select the circle to display grips, and then select the grip at the center of the circle. Press the* SPACEBAR *until the SCALE mode appears at the bottom of the prompt line or press the right mouse button to activate the grip cursor menu to choose Scale)*

SCALE
Specify scale factor or [Base point/Copy/Undo/Reference/eXit]: **0.50**
Command: *(Press* ESC *to remove the object highlight and grips)*

Figure 7–8

USING THE GRIP—MOVE/COPY MODE

The Multiple Copy option of the MOVE mode is demonstrated with the circle shown on the left in the following image. The circle is copied with polar coordinates at distances 2.50 and 5.00 in the 270° direction. The Direct Distance mode could also be used to accomplish this task. This Multiple Copy option is actually disguised under the command options of object grips.

First, select the circle to display the grips at the circle center and quadrants, and then select the center grip. Use the Spacebar to scroll past the STRETCH mode to the MOVE mode. Issue a Copy within the MOVE mode to be placed in Multiple MOVE mode. See the following image.

 Try It! – Open the drawing file 07_Grip Copy. Use the illustration in the middle of the following image and the prompt sequence below to perform this task.

Command: *(Select the circle to activate the grips at the center and quadrants; select the grip at the center of the circle. Then press the* SPACEBAR *until the MOVE mode appears at the bottom of the prompt line or press the right mouse button to activate the grip cursor menu to choose Move)*

MOVE
Specify move point or [Base point/Copy/Undo/eXit]: **C** *(For Copy)*
MOVE (multiple)
Specify move point or [Base point/Copy/Undo/eXit]: **@2.50<270**
MOVE (multiple)
Specify move point or [Base point/Copy/Undo/eXit]: **@5.00<270**
MOVE (multiple)
Specify move point or [Base point/Copy/Undo/eXit]: **X** *(For exit)*
Command: *(Press* ESC *to remove the object highlight and grips)*

  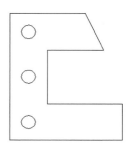

Figure 7–9

USING THE GRIP—MIRROR MODE

Use the grip-MIRROR mode to flip an object along a mirror line similar to the one used in the regular MIRROR command. Follow the prompts for the MIRROR option if an object needs to be mirrored but the original does not need to be saved. This performs the mirror operation but does not produce a copy of the original. If the original object needs to be saved during the mirror operation, use the Copy option of MIRROR mode. This places you in multiple MIRROR mode. Locate a base point and a second point to perform the mirror operation. See the following image.

 Try It! – Open the drawing file 07_Grip Mirror. Use the illustration in the following image and the prompt sequence below to perform this task.

Command: *(Select the circle in the following image to enable grips, and then select the grip at the center of the circle. Press the* SPACEBAR *until the MIRROR mode appears at the bottom of the prompt line or press the right mouse button to activate the grip cursor menu to choose Mirror)*

MIRROR
Specify second point or [Base point/Copy/Undo/eXit]: **C** *(For Copy)*
MIRROR (multiple)
Specify second point or [Base point/Copy/Undo/eXit]: **B** *(For Base Point)*
Base point: **Mid**
of *(Pick the midpoint at "A")*
MIRROR (multiple)
Specify second point or [Base point/Copy/Undo/eXit]: **@1<90**
MIRROR (multiple)
Specify second point or [Base point/Copy/Undo/eXit]: **X** *(For Exit)*
Command: *(Press ESC to remove the object highlight and grips)*

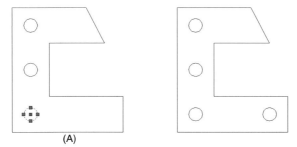

Figure 7–10

USING THE GRIP—ROTATE MODE

Numerous grips may be selected with window or crossing boxes. At the command prompt, pick a blank part of the screen; this should place you in Window/Crossing selection mode. Picking up or below and to the right of the previous point places you in Window selection mode; picking up or below and to the left of the previous point places you in Crossing selection mode. This method is used on all objects shown in the following image. Selecting the lower left grip and using the Spacebar to advance to the ROTATE option allows all objects to be rotated at a defined angle in relation to the previously selected grip.

 Try It! – Open the drawing file 07_Grip Rotate. Use the illustration in the following image and the prompt sequence below to perform this task.

Command: *(Pick near "X," then near "Y" to create a window selection set and enable all grips in all objects. Select the grip at the lower left corner of the object. Then press the* SPACEBAR *until the ROTATE mode appears at the bottom of the prompt line or click the right mouse button to activate the grip cursor menu to choose Rotate)*

ROTATE
Specify rotation angle or [Base point/Copy/Undo/Reference/eXit]: **30**
Command: *(Press* ESC *to remove the object highlight and grips)*

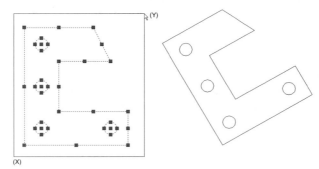

Figure 7–11

USING THE GRIP—MULTIPLE ROTATE MODE

With object grips, you may use ROTATE mode to rotate and copy an object. The following image illustrates a line that needs to be rotated and copied at a 40° angle. With a positive angle, the direction of the rotation will be counterclockwise.

Selecting the line on the left in the following image enables grips located at the endpoints and midpoint of the line. Select the grip at "B" in the middle of the following image. This grip also locates the vertex of the required angle. Distance and angle feedback will appear if dynamic input is turned on. Press the Spacebar until the ROTATE mode is reached. Enter Multiple ROTATE mode by entering C for Copy when you are prompted in the following command sequence. Finally, enter a rotation angle of 40 to produce a copy of the original line segment at a 40° angle in the counterclockwise direction, as shown on the right in the following image.

 Try It! – Open the drawing file 07_Grip Multiple Rotate. Use the illustration in the following image and the prompt sequence below to perform this task.

Command: *(Select line segment "A"; then select the grip at "B." Press the* SPACEBAR *until the ROTATE mode appears at the bottom of the prompt line or click the right mouse button to activate the grip cursor menu to choose Rotate)*
ROTATE
Specify rotation angle or [Base point/Copy/Undo/Reference/eXit]: **C** *(For Copy)*
ROTATE (multiple)
Specify rotation angle or [Base point/Copy/Undo/Reference/eXit]: **40**
Specify rotation angle or [Base point/Copy/Undo/Reference/eXit]: **X** *(For Exit)*
Command: *(Press* ESC *to remove the object highlight and grips)*

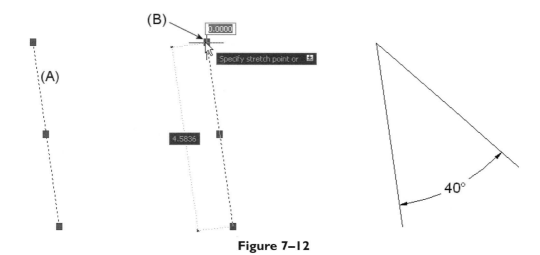

Figure 7-12

USING GRIP OFFSET SNAP FOR ROTATIONS

All Multiple Copy modes within grips may be operated in a snap location mode while you hold down the CTRL key. Here's how it works. On the left in the following image, the vertical centerline and circle are selected. The objects highlight and the grips appear. A multiple copy of the selected objects needs to be made at an angle of 45°. The grip ROTATE option is used in Multiple Copy mode.

 Try It! – Open the drawing file 07_Grip Offset Snap Rotate. Use the illustration in the following image and the prompt sequence below to perform this task.

Command: *(Select centerline segment "A" and circle "B"; and then select the grip near the center of circle "C." Press the* SPACEBAR *until the ROTATE mode appears at the bottom of the prompt line)*
ROTATE
Specify rotation angle or [Base point/Copy/Undo/Reference/eXit]: **C** *(For Copy)*
ROTATE (multiple)
Specify rotation angle or [Base point/Copy/Undo/Reference/eXit]: **B** *(For Base Point)*
Base point: **Cen**
of *(Select the circle at "C" to snap to the center of the circle)*
ROTATE (multiple)
Specify rotation angle or [Base point/Copy/Undo/Reference/eXit]: **45**

Rather than enter another angle to rotate and copy the same objects, hold down the CTRL key, which places you in offset snap location mode. Moving the cursor snaps the selected objects to the angle just centered, namely 45°, as shown in the middle of the following image.

ROTATE (multiple)
Specify rotation angle or [Base point/Copy/Undo/Reference/eXit]: *(Hold down the* CTRL *key and move the circle and centerline until it snaps to the next 45° position shown in the middle of the following image)*
ROTATE (multiple)
Specify rotation angle or [Base point/Copy/Undo/Reference/eXit]: **X** *(For Exit)*
Command: *(Press* ESC *to remove the object highlight and grips)*

The Rotate-Copy-Snap Location mode could allow you to create the circles illustrated on the right side of the following image without the aid of the ARRAY command. Since all angle values are 45°, continue holding down CTRL to snap to the next 45° location, and mark a point to place the next group of selected objects.

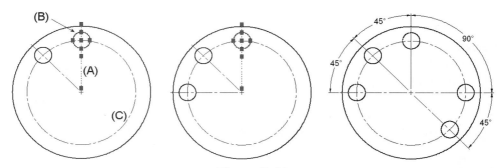

Figure 7–13

USING GRIP OFFSET SNAP FOR MOVING

As with the previous example of using Offset Snap Locations for ROTATE mode, these same snap locations apply to MOVE mode. The illustration on the left in the following image shows two circles along with a common centerline. The circles and centerline are first selected, which highlights these three objects and activates the grips. The intent is to move and copy the selected objects at 2-unit increments.

 Try It! – Open the drawing file 07_Grip Offset Snap Move. Use the illustration in the following image and the prompt sequence below to perform this task.

Command: *(Select the two circles and centerline to activate the grips; select the grip at the midpoint of the centerline. Then press the* SPACEBAR *until the MOVE mode appears at the bottom of the prompt line or click the right mouse button to activate the grip cursor menu to choose Move)*

MOVE
Specify move point or [Base point/Copy/Undo/eXit]: **C** *(For Copy)*
MOVE (multiple)
Specify move point or [Base point/Copy/Undo/eXit]: **@2.00<270**

In the middle of the following image, instead of remembering the previous distance and entering it to create another copy of the circles and centerline, hold down SHIFT and move the cursor down to see the selected objects snap to the distance already specified.

MOVE (multiple)
Specify move point or [Base point/Copy/Undo/eXit]: *(Hold down the* CTRL *key and move the cursor down to have the selected objects snap to another 2.00-unit distance)*
MOVE (multiple)
Specify move point or [Base point/Copy/Undo/eXit]: **X** *(To exit)*
Command: *(Press* ESC *to remove the object highlight and grips)*

The illustration on the right side of the following image shows the completed hole layout, the result of using the Offset Snap Location method of object grips.

Figure 7–14

GRIP PROJECT—LUG.DWG

 Try It! – Open the drawing file 07_Lug. Use the illustration in the following image and the prompt sequence below to perform this task.

Using the illustration in the following image as a guide, make the following changes to the lug:

1. Use the Grip-Stretch mode to join the endpoint of the inclined line at (A) with the endpoint of the horizontal line also at (A).

2. Use the Grip-Stretch mode to join the endpoint of the inclined line at (B) with the endpoint of the horizontal line also at (B).

3. Use the Grip-Stretch mode, pick the 1.50 diameter dimension, make the grip located on the dimension text hot, and position the dimension text in the lower right corner of the object.

4. Use the Grip-Scale mode, pick the 1.50 diameter dimension and the circle, make the grip located at the center of the circle hot, and use a scale factor of .50 to reduce the size of the circle by half. Notice that the dimension will recalculate to reflect this change.

5. Use the Grip-Mirror mode, and pick the circle, 1.50 diameter dimension, two vertical lines, both fillets, and the horizontal line at (E). Pick one of the grips to make it hot. Use the midpoint of the long horizontal line of the object as the new base point. Turn Ortho on and move your cursor down to establish the mirror line.

6. Use the Grip-Stretch mode to increase the size of the .50 throat at (F) to a new distance of 1.25. Pick both fillets and the three line segments attached to the fillets. Hold down the SHIFT key and select the grips at the endpoints of both fillets and the middle of the horizontal line. Release the SHIFT key and pick the grip at the midpoint of the horizontal line again. Move your cursor up (either Ortho or Polar mode needs to be turned on) and type a value of .75.

7. The completed exercise is illustrated on the right in the following image.

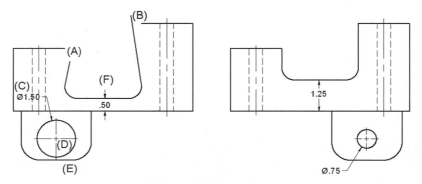

Figure 7–15

MODIFYING THE PROPERTIES OF OBJECTS

At times, objects are drawn on the wrong layer, color, or even in the wrong linetype. The lengths of line segments are incorrect, or the radius values of circles and arcs are incorrect. Eliminating the need to erase these objects and reconstruct them to their correct specifications, a series of tools are available to modify the properties of these objects. Illustrated in the following image are three line segments. One of the line segments has been pre-selected, as shown by the highlighted appearance and the presence of grips.

Clicking on the Properties button on the Standard toolbar in the following image displays the Properties Palette. This palette displays information about the object already selected; in this case the information is about the line segment, which is identified at the top of the palette.

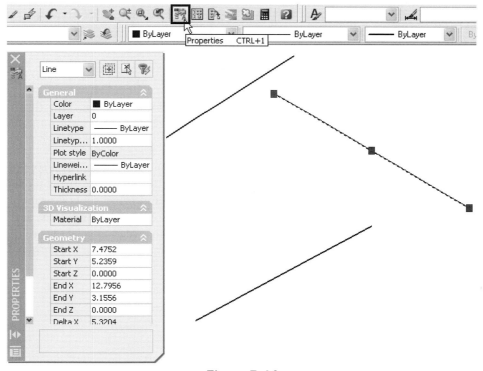

Figure 7–16

When changing a selected object to a different layer, click in the layer field on the left in the following image and the current layer displays (in this example, Layer Object). Clicking the down arrow displays all layers defined in the drawing. Clicking on one of these layers changes the selected object to a different layer.

A number of options that control the Properties Palette are illustrated on the right in the following image. You can elect to move, size, or close the Properties Palette from this menu. You can allow or prevent the Properties Palette from docking. Auto-hiding displays the Properties Palette when your cursor lies anywhere inside the window. When your cursor moves off of the Properties Palette, the window collapses to display only the blue side strip. Checking Description controls the display of a description area located at the bottom of the palette.

Figure 7–17

The same three line segments are illustrated in the following image. This time, all three segments have been pre-selected, as shown by their highlighted appearance and the presence of grips.

Clicking on the Properties button on the Standard toolbar displays the Properties Palette, as shown in the following image. At the top of the dialog box, the three lines are identified. You can change the color, layer, linetype, and other general properties of all three lines. However, you are unable to enter the Start and End X, Y, and Z values, and there is no length or angle information. Whenever you select more than one of the same object, you can change the general properties of the object but not individual values that deal with the object's geometry.

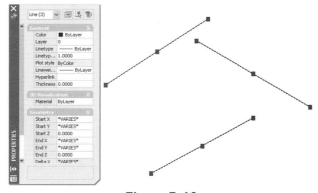

Figure 7–18

On the left side of the following image, an arc, circle, and line are pre-selected as identified by the appearance of grips. Activating the Properties Palette shown on the right side of the following image displays a number of object types at the top of the dialog box. You can click which object or group of objects to modify. With "All (3)" highlighted, you can change the general properties, such as layer and linetype, but not any geometry settings.

Figure 7–19

 Try It! – Open the drawing file 07_Clutch Properties. What if you need to increase the radius of the circle to 1.25 units? Click on the inner circle and then select the Properties button. Click on the Circle object type at the top of the Properties Palette. Since only one object was selected, the full complement of general and geometry settings is present for you to modify. Click in the Radius field and change the current value to 1.25 units. Pressing ENTER automatically updates the other geometry settings in addition to the actual object in the drawing, as shown in the following image. When finished, dismiss this dialog box to return to the drawing.

Figure 7–20

You can also open the Properties Palette by choosing Properties from the Modify or Tools pull-down menus, as shown in the following image. This displays the Properties Palette shown on the right in the following image. Notice that "No selection" is listed at the top of the palette, meaning that no objects have been selected to modify. Even though nothing is selected, you can still change the current color or linetype, and even make a layer current.

Figure 7–21

USING THE QUICK SELECT DIALOG BOX

At various times in a drawing, you need to build a selection set and modify the properties of these selected objects. However, there can be numerous objects in a drawing and if you attempt to select objects through the conventional methods such as window or crossing, you may accidentally ignore a few objects. A tool is available in which you enter certain parameters and AutoCAD selects the objects based on these parameters; this tool is called Quick Select. Clicking on Quick Select from the Tools pull-down menu in the following image displays the Quick Select dialog box, also shown in the following image.

To have AutoCAD create a selection set by object type and property, first choose the appropriate object type in the dialog box. Notice in the following example that you can choose from Block Reference or Line. Once you determine the object type, the property area of this dialog box will change depending on the object. In the following example, study the two rows of properties on the right in the following image. Notice that the properties for the Block Reference object type greatly differ from those for the Line object type. This, however, is what makes Quick Select so powerful. Not only can you create a selection set based on object type, but you can also narrow your search down based on layer and color for most objects or even by block name.

By default, whenever you apply the object type and property information, this information becomes the basis for the new selection set. You could also create the inverse of this information; this will occur when you click on the radio button next to Exclude from new selection set entry. If this button is picked and you build a selection set based on a certain object and property, all other objects will be selected except this object and property. As you can see, the Quick Select dialog box is a very powerful tool that actually works for you and saves you from having to build tedious selection sets.

Figure 7–22

 Try It! – Open the drawing file 07_Change Text Height. In the following image, the room numbers in the rectangular boxes are currently set to a height of 18″. All text items need to be changed to a new height of 12″. Rather than individually change each text item, the Quick Select dialog box will be used to assist with this operation.

Figure 7–23

First, activate the Properties Palette and click on the Quick Select button, as shown on the left in the following image. This represents another method of activating Quick Select. Once inside the Quick Select dialog box, click in the Object type window and select "Text," as shown on the right side of the following image.

Figure 7–24

Clicking the OK button at the bottom of the Quick Select dialog box will return you to your drawing. Notice that all text is highlighted and a text height of 18.0000 is listed in the Properties Palette. Change this value to 12, as shown in the following image.

Figure 7–25

When finished changing the text height in the Properties Palette, press the ENTER key followed by the ESC key and notice that all text in the following image has been changed to the new height of 12˝.

Figure 7–26

MODIFYING THE PROPERTIES OF INDIVIDUAL OBJECTS

This segment of the chapter begins a study of the Properties Palette and all the properties displayed on the particular object selected. You can modify the following General properties of all objects through this dialog box: Color, Layer, Linetype, Linetype scale, Plot style, Lineweight, Hyperlink, and Thickness. The remainder of this segment will concentrate on the individual geometric properties of an arc object that can be modified through the Properties Palette, which include:

- Name of the object being listed (Arc) at the top of the dialog box
- Starting and Ending X, Y, and Z values of the arc (grayed out)
- X, Y, and Z values of the center of the arc
- Radius of the arc
- Start and End angles of the arc
- Total angle (grayed out)
- Total arc length (grayed out)
- Length of the arc (grayed out)
- Normal X, Y, and Z values (grayed out)

Try It! – Open the drawing file 07_Arc Properties. Pick the arc object and then click the Properties button. The Properties Palette for an arc provides you with the following information, which is displayed in the following image. Values in bold can be modified; items grayed out cannot.

Figure 7–27

 Note: The following drawing files are provided on CD. Open each one individually, select the object, click on the Properties button, and observe the results displayed in the Properties Palette.

07_Block Properties	07_Mtext Properties
07_Circle Properties	07_Polyline Properties
07_Dim Properties	07_Points Properties
07_Ellipse Properties	07_Ray Properties
07_Hatch Properties	07_Spline Properties
07_Line Properties	07_Text Properties
07_Mline Properties	07_Xline Properties

USING THE PICKADD FEATURE OF THE PROPERTIES PALETTE

Another very interesting feature is available to you inside the Properties Palette; it is the PICKADD button located in the following image. To see how it operates, follow the next exercise.

Try It! – Open the drawing file 07_PickAdd. A number of object types ranging from lines to text to multiline text and polylines with dimensions are displayed. Activate the Properties Palette and pick on one object. Notice that it highlights, the grips appear, and information about the object is displayed in the Properties Palette. Suppose, however, that you want to list information about another object. You must first press ESC to deselect the original object. Now select a different object and this information is displayed in the Properties Palette. This time, select the PICKADD button on the left in the following image. Notice that the button changes in appearance. Instead of the "plus" sign, a "1" is displayed on the right in the following image. Click on any object and notice the information displayed in the Properties Palette. Without pressing ESC, click on another object. The original object deselects and the new object highlights with its information displayed in the Properties Palette. Very simply, the PICKADD button eliminates the need for the ESC key when displaying information about individual objects.

Figure 7–28

Note: Changing the PICKADD button will affect all drawings. Since this change is global, click the PICKADD button to set it back to the "plus" sign before continuing on.

PERFORMING MATHEMATICAL CALCULATIONS

A mathematical calculator is available to assist in making a variety of calculations while still remaining in an AutoCAD drawing. The calculator is activated by clicking on QuickCalc, found under the Tools pull-down menu, as shown on the left in the following image. This launches the calculator, as shown on the right in the following image. The calculator resembles the Properties Palette and can be manipulated in the same fashion. When you perform calculations throughout the design cycle, these calculations are stored in a History area. Calculations in this area can be retrieved for later use. As you enter numbers and mathematical functions such as addition and subtraction from the supplied keypads, they are displayed in the Input area of the calculator. Four tabbed areas control the methods by which calculations are made. They are Number Pad, Scientific, Units Conversion, and Variables.

Figure 7–29

Each of the four tabbed areas of the calculator are illustrated in the following image. A brief description of each follows:

Numeric Pad: This area of the calculator is used to perform the most basic of mathematical functions. Clicking on a number or function will place this information in the Input area of the calculator.

Scientific: Use this tabbed area to enter trigonometric and other advanced functions.

Units Conversion: This area is used to convert units from one system to another. You enter the units type, the unit type to convert from, and the units to convert to. Entering a numeric value to convert will display the converted value.

Variables: This area contains a series of shortcut commands that can be used for performing specialized operations.

Figure 7–30

The following table illustrates the variable names and their functions:

Variable	Function
Phi	Golden Ratio (1.61803399)
dee	Finds the distance between two endpoints
illi	Finds the intersection of two lines defined by four endpoints
mee	Finds the midpoint between two endpoints
nee	Finds the unit vector in the XY plane normal (perpendicular) to two endpoints
rad	Obtains the radius from a circle, arc, or polyline arc segment
vee	Finds a vector from two endpoints
Vee1	Finds a unit vector from two endpoints

When you are using the Properties Palette and you need to change the value of one of the Geometry listings, a calculator icon is present. Clicking on this icon will launch the QuickCalc dialog box, as shown on the right in the following image. Entering a mathematical formula in the Input area and clicking the Apply button will send the calculated value back to the Properties Palette and change the highlighted object.

Figure 7–31

USING THE LAYER CONTROL BOX TO MODIFY OBJECT PROPERTIES

If all you need to do is to change an object or a group of objects from one layer to another, the Layer Control box can easily perform this operation.

 Try It! – Open the drawing file 07_Change Layer. In the following image, select the arc and two line segments. Notice that the current layer is 0 in the Layer Control box. These objects need to be on the OBJECT layer.

Click in the Layer Control box to display all layers defined in the drawing. Then click on the desired layer for all highlighted objects (in this case, the OBJECT layer in the following image).

Figure 7–32

Notice that in the following image, with the objects still highlighted, the layer listed is OBJECT. This is one of the quickest and most productive ways of changing an object from one layer to another.

Figure 7–33

Try It! – Open the drawing file 07_Mosaic. In the following image, this drawing consists of 5 different types of objects all drawn on Layer 0: circles, squares drawn as closed polylines, lines, text (the letter X), and text (the letter Y). The Quick Select dialog box will be used to select these object types individually.

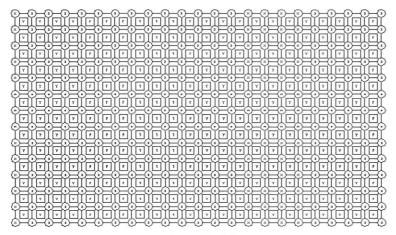

Figure 7–34

Once you have selected them, change the objects to the correct layers, which are also supplied with this drawing and identified on the left in the following image.

First, activate the Quick Select dialog box from the Tools pull-down menu, shown in the middle in the following image. In the Object type box, select Circle, as you see on the right in the following image, and click the OK button.

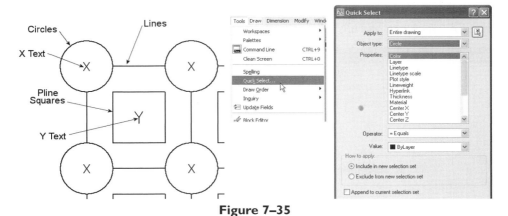

Figure 7–35

With all circles selected, click the Layer Control box and pick the Circles layer as shown in the following image. All circles in the mosaic pattern should turn red. Press ESC to remove the object highlight and grips.

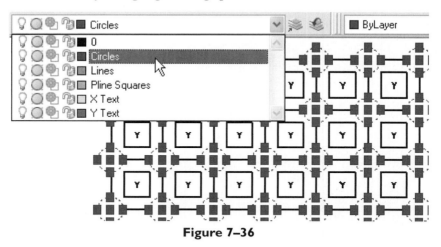

Figure 7–36

Next, you need to select all the text with the letter "X" and change this text to the layer called X Text. Activate the Quick Select dialog box and make the following changes: Change the Object type to Text, set the Properties to Contents, and enter "X" as the value. Your display should be similar to the dialog box on the left in the following image. Click the OK button to dismiss the Quick Select dialog box and select all text with the letter "X".

Change all letters to the X Text layer in the Layer Control box, as shown on the right in the following image. Press ESC to remove the object highlight and grips. Follow the same procedures for changing all line segments to the Lines layer, all polylines to the Pline Squares layer, and all "Y" text to the Y Text layer.

Figure 7–37

Try It! – Open the drawing file 07_Qselect Duplex. Use the illustration in the following image and the information below to change the items to their proper layers.

Change all lines to the Walls layer.
Change all text to the Room Labels layer.
Change the blocks "Door" and "Louver" to the Door layer
Change the block "Window" to the Window layer
Change the blocks "Countertop," "Range," "Sink," and "Refrigerator" to the Countertop layer

Figure 7–38

 Tip: For the blocks, see the illustration in the following image for supplying the correct information in the Quick Select dialog box; change the Object type to Block Reference. Change the Properties to Name. Click the down arrow next to the Value heading and select the desired block from the list provided. Then, after clicking the OK button and all desired blocks are selected, change the highlighted blocks to the correct layer. Follow this procedure for selecting the Window, Countertop, Range, Sink, Door, Louver, and Refrigerator blocks.

Figure 7–39

DOUBLE-CLICK EDIT ON ANY OBJECT

Double-clicking on any object provides you with a quick way of launching the Properties Palette or other related dialog boxes depending on the object type. For instance, double-clicking on a line segment launches the Properties Palette, which displays information about the line. Double-clicking on a text object will display the Edit Text dialog box, allowing you to enter or delete words. Whenever the Properties Palette is launched and you want to modify a different object, first press ESC to deselect the current object. You may also want to dismiss the Properties Palette.

 Try It! – Open the drawing file 07_Double Click Edit. In the following image, double-click on the magenta center line and the Properties Palette launches with information about the line. Press ESC and dismiss the Properties Palette. Double-click on the word "BLOCK" to launch the Edit Text function. Press ESC to exit the edit function when finished. Double-click on the sentence "THE OBJECT SHOWN ABOVE IS A WINDOW SYMBOL" to launch the Multiline Text Editor dialog box. Press ESC to dismiss the dialog box when finished. Double-click on the hatch pattern and the Hatch Edit dialog box will launch. Pressing ESC dismisses this dialog box when finished. Continue by double-clicking on the circle, dimension, and rectangle and observe the type of dialog box launched through this method. Press ESC to dismiss the dialog boxes when finished with each operation.

Figure 7–40

MATCHING THE PROPERTIES OF OBJECTS

At times objects are drawn on the wrong layers or the wrong color scheme is applied to a group of objects. Text objects are sometimes drawn with an incorrect text style. You have just seen how the Properties Palette and the Layer Control box provide quick ways to fix such problems. Yet another tool is available for changing the properties of objects—the MATCHPROP command. Choose this command from the Standard toolbar, as shown on the left in the following image, or from the Modify pull-down menu, as shown on the right in the following image.

Figure 7–41

 Try It! – Open the drawing file 07_Matchprop Flange. Use the illustration in the following image and the command sequence below for performing this task.

When you start the command, a source object is required. This source object transfers all its current properties to other objects designated as "Destination Objects." As shown on the left in the following image, the flange requires the object lines located at "B," "C," "D," and "E" to be converted to hidden lines. Using the MATCHPROP command, select the existing hidden line "A" as the source object. Notice the appearance of the Match Properties icon. Select lines "B" through "E" as the destination objects using this icon.

 Command: **MA** *(For MATCHPROP)*
Select source object: *(Select the hidden line at "A")*
Current active settings: Color Layer Ltype Ltscale Lineweight Thickness
PlotStyle Text Dim Hatch Polyline Viewport Table
Select destination object(s) or [Settings]: *(Select line "B")*
Select destination object(s) or [Settings]: *(Select line "C")*
Select destination object(s) or [Settings]: *(Select line "D")*
Select destination object(s) or [Settings]: *(Select line "E")*
Select destination object(s) or [Settings]: *(Press ENTER to exit this command)*

The results appear in the flange illustrated on the right in the following image, where the continuous object lines were converted to hidden lines. Not only did the linetype property get transferred, but the color, layer, lineweight, and linetype scale information did as well.

Figure 7–42

To get a better idea of what object properties are affected by the MATCHPROP command, re-enter the command, pick a source object, and instead of picking a destination object immediately, enter S for Settings. This will display the Property Settings dialog box in the following image.

 Command: **MA** *(For MATCHPROP)*
Select source object: *(Select the hidden line at "A" in the previous image)*
Current active settings: Color Layer Ltype Ltscale Lineweight Thickness PlotStyle Dim Text Hatch Polyline Viewport Table Material Shadow display
Select destination object(s) or [Settings]: **S** *(For Settings; this displays the Property Settings dialog box in the following image)*

Any box with a check displayed in it will transfer that property from the source object to all destination objects. If you need to transfer only the layer information and not the color and linetype properties of the source object, remove the checks from the Color and Linetype properties before you select the destination objects. This will prevent these properties from being transferred to any destination objects.

Figure 7–43

MATCHING DIMENSION PROPERTIES

The MATCHPROP command can control special properties of dimensions, text, hatch patterns, polylines, viewports, and tables. The Dimension property will be featured next (see the following image).

Figure 7–44

 Try It! – Open the drawing file 07_Matchprop Dim. The following image shows two blocks: the block assigned a dimension value of 46.6084 was dimensioned with the METRIC dimension style with the Arial font applied. The block assigned a dimension value of 2.3872 was dimensioned with the STANDARD dimension style with the TXT font applied. Both blocks need to be dimensioned with the METRIC dimension style. Issue the MATCHPROP command and select the 46.6084 dimension as the source object and then select the 2.3872 dimension as the destination object.

 Command: **MA** *(For MATCHPROP)*
Select source object: *(Select the dimension at "A" in the following image)*
Current active settings: Color Layer Ltype Ltscale Lineweight Thickness PlotStyle Dim Text Hatch Polyline Viewport Table Material Shadow display
Select destination object(s) or [Settings]: *(Select the dimension at "B")*
Select destination object(s) or [Settings]: *(Press ENTER to exit this command)*

The results are shown on the right in the following image, with the METRIC dimension style applied to the STANDARD dimension style through the use of the MATCHPROP command. Because the text font was associated with the dimension style, it also changed in the destination object.

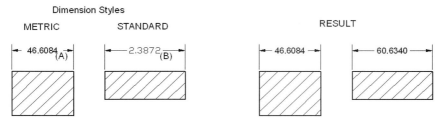

Figure 7–45

MATCHING TEXT PROPERTIES

The following is an example of how the MATCHPROP command affects a text object with the Text property shown in the following image.

Figure 7–46

 Try It! – Open the drawing file 07_Matchprop Text. The following image shows two text items displayed in different fonts. The text "Coarse Knurl" at "A" was constructed with a text style called Arial. The text "Medium Knurl" at "B" was constructed with the default text style called STANDARD. Use the following command sequence to match the STANDARD text style with the Arial text style using the MATCHPROP command.

 Command: **MA** *(For MATCHPROP)*
Select source object: *(Select the text at "A")*
Current active settings: Color Layer Ltype Ltscale Lineweight Thickness PlotStyle Dim Text Hatch Polyline Viewport Table Material Shadow display
Select destination object(s) or [Settings]: *(Select the text at "B")*
Select destination object(s) or [Settings]: *(Press ENTER to exit this command)*

The result is shown on the right in the following image. Both text items now share the same text style. Notice that the text string stays intact when text properties are matched. Only the text style of the source object is applied to the destination object.

Figure 7–47

MATCHING HATCH PROPERTIES

A source hatch object can also be matched to a destination pattern with the MATCHPROP command and the Hatch property (see the following image).

Figure 7-48

 Try It! — Open the drawing file 07_Matchprop Hatch. In the following image, the crosshatch patterns at "B" and "C" are at the wrong angle and scale. They should reflect the pattern at "A" because it is the same part. Use the MATCHPROP command, select the hatch pattern at "A" as the source object, and select the patterns at "B" and "C" as the destination objects.

Command: **MA** *(For MATCHPROP)*
Select source object: *(Select the hatch pattern at "A")*
Current active settings: Color Layer Ltype Ltscale Lineweight Thickness PlotStyle Dim Text Hatch Polyline Viewport Table Material Shadow display
Select destination object(s) or [Settings]: *(Select the hatch pattern at "B")*
Select destination object(s) or [Settings]: *(Press* ENTER *to exit this command)*

The results appear on the right in the following image, where the source hatch pattern property was applied to all destination hatch patterns.

Figure 7-49

MATCHING POLYLINE PROPERTIES

A source polyline object can also be matched to a destination pattern with the MATCH-PROP command and the Polyline property (see the following image).

Figure 7–50

 Try It! – Open the drawing file 07_Matchprop Pline. In the following image, the polyline at "B" is at the wrong width. It should match the width of the polyline at "A." Using the MATCHPROP command, select the polyline at "A" as the source object, and select the polyline at "B" as the destination object.

 Command: **MA** *(For MATCHPROP)*
Select source object: *(Select the polyline at "A")*
Current active settings: Color Layer Ltype Ltscale Lineweight Thickness PlotStyle Dim Text Hatch Polyline Viewport Table Material Shadow display
Select destination object(s) or [Settings]: *(Select the polyline at "B")*
Select destination object(s) or [Settings]: *(Press ENTER to exit this command)*

The results appear on the right in the following image, where the source polyline property (width, in this example) was applied to the destination polyline.

Figure 7–51

MATCHING VIEWPORT PROPERTIES

A source viewport object can also be matched to a destination viewport with the MATCHPROP command and the Viewport property (see the following image). Even though

Object Grips and Changing the Properties of Objects 421

the topic of viewports will not be covered until chapter 14, this concept is introduced here. Viewports are used to arrange images of a drawing in layout mode before they are plotted out. Matching the properties of a viewport can transfer the scale of one viewport to another. The following Try It! exercise illustrates this.

Figure 7–52

 Try It! – Open the drawing file 07_Matchprop Viewport. In the following image, the viewport image at "A" is at the correct scale. The viewport at "B" should have the same scale as that of viewport "A." Using the MATCHPROP command, select the viewport at "A" as the source object, and select the viewport at "B" as the destination object.

 Command: **MA** *(For MATCHPROP)*
Select source object: *(Select the edge of the viewport at "A")*
Current active settings: Color Layer Ltype Ltscale Lineweight Thickness PlotStyle Dim Text Hatch Polyline Viewport Table Material Shadow display
Select destination object(s) or [Settings]: *(Select the viewport at "B")*
Select destination object(s) or [Settings]: *(Press* ENTER *to exit this command)*

Figure 7–53

The results appear in the following image, where the source viewport property (scale of the image) was applied to the destination viewport. Notice also that the layer of the viewport "B" was changed from 0 to Vports.

Figure 7–54

MATCH PROPERTIES AND TABLES

A source table object can be matched to a destination table using the Table property, as shown in the following image. When one table is matched with another, the table style of the source table updates the table style of the destination table.

Figure 7–55

Try It! – Open the drawing file 07_Matchprop Table. In the following image, one table has the title strip originating at the top and the other table title originates at the bottom. Using the MATCHPROP command, select the table on the left as the source object, and select the table on the right as the destination object.

Command: **MA** *(For MATCHPROP)*
Select source object: *(Select the table on the left in the following image)*
Current active settings: Color Layer Ltype Ltscale Lineweight Thickness PlotStyle Dim Text Hatch Polyline Viewport Table Material Shadow display
Select destination object(s) or [Settings]: *(Select the table on the right in the following image)*
Select destination object(s) or [Settings]: *(Press ENTER to exit this command)*

Figure 7–56

In the following image, notice how the table on the right changes to the downward direction arrangement as controlled by the table style.

Figure 7–57

 Note: You may have to move the table into position.

TUTORIAL EXERCISE: 07_MODIFY-EX.DWG

Figure 7–58

Purpose
This tutorial exercise is designed to change the properties of existing objects displayed in the above image.

System Settings
Since this drawing is provided on the CD, open an existing drawing file called "07_Modify-Ex." Follow the steps in this tutorial for changing various objects to the correct layer, text style, and dimension style.

Layers
Layers have already been created in this drawing.

Suggested Commands
Begin this tutorial by using the Properties Palette to change the isometric object to a different layer. Continue by changing the text height and layer of the view identifiers (FRONT, TOP, SIDE, ISOMETRIC). The MATCHPROP command will be used to transfer the properties from one dimension to another, one text style to another, and one hatch pattern to another. The Layer Control box will be used to change the layer of various objects located in the Front and Top views.

STEP 1

Loading this drawing displays the objects in a page layout called "Orthographic Views." Page layout is where the drawing will be plotted out in the future. This name is present next to the Model tab in the bottom portion of the drawing screen. Since a majority of changes will be made in Model mode, click on the Model tab near the bottom of the display screen. Your drawing will appear similar to the following image.

Figure 7–59

STEP 2

While in the Model environment, select all lines that make up the isometric view in the following image. You can accomplish this by using the Window mode from the Command prompt. If you accidentally select the word ISOMETRIC, de-select this word. Activate the Properties Palette and click in the Layer field. This will display the current layer the objects are drawn on (DIM) in the following image. Click the down arrow to display the other layers and pick the OBJECT layer. This will change all selected objects that make up the isometric view to the OBJECT layer. When finished, dismiss the Properties Palette. Press ESC to remove the object highlight and the grips from the drawing.

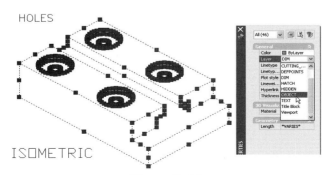

Figure 7–60

STEP 3

Select the view titles (FRONT, TOP, SIDE, ISOMETRIC). These text items need to be changed to a height of 0.15 and the TEXT layer. With all four text objects highlighted, activate the Properties Palette, click in the Layer field, and click the down arrow to change the selected objects to the TEXT layer in the following image.

Figure 7–61

With the text still highlighted, change the text height in the Properties Palette from 0.20 to a new value of 0.15. Pressing ENTER after making this change automatically updates all selected text objects to this new height in the following image. Dismiss this dialog box when finished. Press ESC to remove the object highlight and the grips from the drawing.

Figure 7–62

STEP 4

When you examine the view titles, TOP and SIDE are in one text style while FRONT and ISOMETRIC are in another. To remain consistent in the design process, you should make sure all text identifying the view titles has the same text style as TOP. Activate the MATCHPROP command. Click the text object TOP as the source object. When the Match Property icon appears in the following image, select ISOMETRIC and FRONT as the destination objects. All properties associated with the TOP text object will be transferred to ISOMETRIC and FRONT, including the text style.

 Command: **MA** *(for MATCHPROP)*
Select source object: *(Select the text object "TOP," which should highlight in the following image)*
Current active settings: Color Layer Ltype Ltscale Lineweight Thickness PlotStyle Dim Text Hatch Polyline Viewport Table Material Shadow display
Select destination object(s) or [Settings]: *(Select the text object "ISOMETRIC")*
Select destination object(s) or [Settings]: *(Select the text object "FRONT")*
Select destination object(s) or [Settings]: *(Press ENTER to exit this command)*

Figure 7–63

STEP 5

Notice the dimensions in this drawing. Two dimensions stand out above the rest (the 1.5000 vertical dimension in the Top view and the 2.5000 horizontal dimension in the Side view). Again, to remain consistent in the design process, you should make sure all dimensions have the same appearance (dimension text height, number of decimal places, broken inside instead of placed above the dimension line). Activate the MATCHPROP command again by clicking on the button in the Standard toolbar. Click the 4.00 horizontal dimension in the Top view as the source object. When the Match Property icon appears, as in the following image, select the 1.5000 and 2.5000 vertical dimensions as the destination objects. All dimension properties associated with the 4.00 dimension will be transferred to the 1.5000 and 2.5000 dimensions.

 Command: **MA** *(for MATCHPROP)*
Select source object: *(Select the 4.00 dimension, which should highlight)*
Current active settings: Color Layer Ltype Ltscale Lineweight Thickness
PlotStyle Dim Text Hatch Polyline Viewport Table Material Shadow display
Select destination object(s) or [Settings]: *(Select the 2.5000 dimension at "A")*
Select destination object(s) or [Settings]: *(Select the 1.5000 dimension at "B")*
Select destination object(s) or [Settings]: *(Press ENTER to exit this command)*

Figure 7–64

STEP 6

In the Front view, an area is crosshatched. However, both sets of crosshatching lines need to be drawn in the same direction, rather than opposing each other. Activate the MATCHPROP command again by clicking on the button in the Standard toolbar. Click the left hatch pattern as the source object. When the Match Property icon appears, as in the following image, select the right hatch pattern as the destination object. All hatch properties associated with the left hatch pattern will be transferred to the right hatch pattern.

Command: **MA** (for MATCHPROP)
Select source object: *(Select the left hatch pattern, which should highlight)*
Current active settings: Color Layer Ltype Ltscale Lineweight Thickness PlotStyle Dim Text Hatch Polyline Viewport Table Material Shadow display
Select destination object(s) or [Settings]: *(Pick the right hatch pattern)*
Select destination object(s) or [Settings]: *(Press ENTER to exit this command.)*

Figure 7–65

STEP 7

Your display should appear similar to the following image. All view titles (FRONT, TOP, SIDE, ISOMETRIC) share the same text style and are at the same height. All dimensions share the same parameters and text orientation. Both crosshatch patterns are drawn in the same direction.

Figure 7–66

STEP 8

A different method will now be used to change the specific layer properties of objects. The two highlighted lines in the Top view in the following image were accidentally drawn on Layer OBJECT and need to be transferred to the HIDDEN layer. With the lines highlighted, click in the Layer Control box to display all layers. Click on the HIDDEN layer to change the highlighted lines to the HIDDEN layer. Press ESC to remove the object highlight and the grips from the drawing.

Note: You may need to REGEN the drawing in order to get the dashes to appear in the hidden lines.

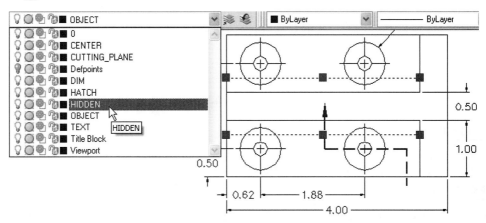

Figure 7–67

STEP 9

The two highlighted lines in the Front view in the following image were accidentally drawn on the TEXT Layer and need to be transferred to the CENTER layer. With the lines highlighted, click in the Layer Control box to display all layers. Click on the CENTER layer to change the highlighted lines to the CENTER layer. Press ESC to remove the object highlight and the grips from the drawing.

Figure 7–68

STEP 10

Your display should appear similar to the following image. Notice the hidden lines in the Top view and the centerlines in the Front view.

Figure 7–69

STEP 11

Click on the Orthographic Views Tab. Select the rectangular viewport and change this object's layer to "Viewport" in the following image.

Figure 7–70

STEP 12

Turn off the Viewport layer. Your display should appear similar to the following image. This completes this tutorial exercise.

Figure 7–71

Shape Description/ Multiview Projection

Before any object is made in production, some type of drawing needs to be created. This is not just any drawing, but rather an engineering drawing consisting of overall object sizes with various views of the object organized on the computer screen. This chapter introduces the topic of shape description or how many views are really needed to describe an object. The art of multiview projection includes methods of constructing one-view, two-view, and three-view drawings using AutoCAD commands.

SHAPE DESCRIPTION

Begin constructing an engineering drawing by first analyzing the object being drawn. To accomplish this, describe the object by views or by how an observer looks at the object. The illustration on the left in the following image shows a simple wedge; this object can be viewed at almost any angle to get a better idea of its basic shape. However, to standardize how all objects are to be viewed, and to limit the confusion usually associated with complex multiview drawings, some standard method of determining how and where to view the object must be exercised.

Even though the simple wedge is easy to understand because it is currently being displayed in picture or isometric form, it would be difficult to produce this object because it is unclear what the sizes of the front and top faces are. Illustrated on the right in the following image are six primary ways or directions in which to view an object. The Front view begins the shape description and is followed by the Top view and Right Side view. Continuing on, the Left Side view, Back view, and Bottom view complete the primary ways to view an object.

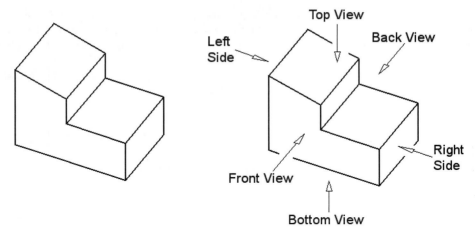

Figure 8–1

Now that the primary ways of viewing an object have been established, the views need to be organized to promote clarity and have the views reference one another. Imagine the simple wedge positioned in a transparent glass box similar to the illustration on the left in the following image. With the entire object at the center of the box, the sides of the box represent the ways to view the object. Images of the simple wedge are projected onto the glass surfaces of the box.

With the views projected onto the sides of the glass box, we must now prepare the views to be placed on a two-dimensional (2D) drawing screen. For this to be accomplished, the glass box, which is hinged, is unfolded as shown on the right in the following image. All folds occur from the Front view, which remains stationary.

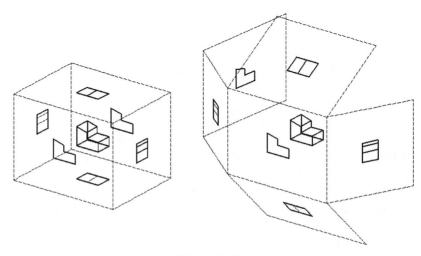

Figure 8–2

The following image shows the views in their proper alignment to one another. However, this illustration is still in a pictorial view. These views need to be placed flat before continuing.

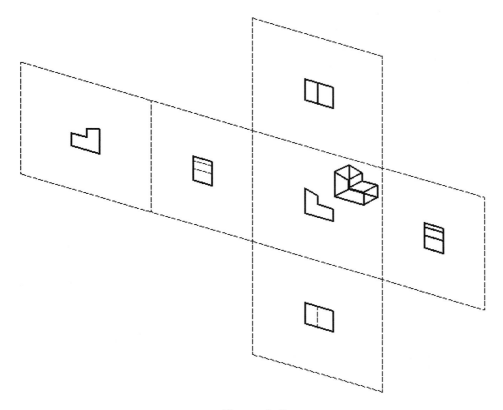

Figure 8–3

As the glass box is completely unfolded and laid flat, the result is illustrated on the left in the following image. The Front view becomes the main view, with other views placed in relation to it. Above the Front view is the Top view. To the right of the Front view is the Right Side view. To the left of the Front view is the Left Side view, followed by the Back view. Underneath the Front view is the Bottom view. This becomes the standard method of laying out the views needed to describe an object. But are all the views necessary? Upon closer inspection we find that, except for being a mirror image, the Front and Back views are identical. The Top and Bottom views appear similar, as do the Right and Left Side views. One very important rule to follow in multiview objects is to select only those views that accurately describe the object and discard the remaining views.

The complete multiview drawing of the simple wedge is illustrated on the right in the following image. Only the Front, Top, and Right Side views are needed to describe this object. When laying out views, remember that the Front view is usually the most important—it holds the basic shape of the object being described. Directly above the Front view is the Top view, and to the right of the Front view is the Right Side view. All three views are separated by a space of varying size. This space is commonly called a dimension space because it is a good area in which to place dimensions describing the size of the object. The space also acts as a separator between views; without it, the views would touch one another, making them difficult to read and interpret. The minimum distance of this space is usually 1.00 unit.

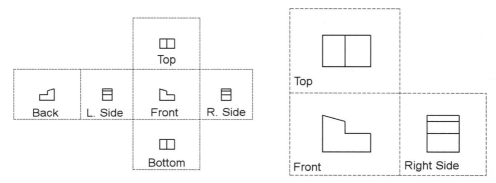

Figure 8–4

RELATIONSHIPS BETWEEN VIEWS

Some very interesting and important relationships are set up when the three views are placed in the configuration illustrated on the left in the following image. Notice that because the Top view is directly above the Front view, both share the same width dimension. The Front and Right Side views share the same height. The relationship of depth on the Top and Right Side views can be explained with the construction of a 45°-projection line at "A" and the projection of the lines over and down or vice versa to get the depth. Another principle illustrated by this example is that of projecting lines up, over, across, and down to create views. Editing commands such as ERASE and TRIM are then used to clean up unnecessary lines.

With the three views identified as shown on the right in the following image, the only other step—and it is an important one—is to annotate the drawing, or add dimensions to the views. With dimensions, the object drawn in multiview projection can now be produced. Even though this has been a simple example, the methods of multiview projection work even for the most difficult and complex of objects.

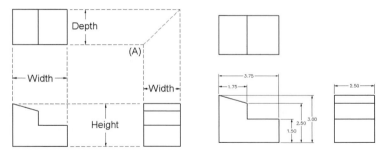

Figure 8–5

REVIEW OF LINETYPES AND CONVENTIONS

At the heart of any engineering drawing is the ability to assign different types of lines to convey meaning to the drawing. When plotted, all lines of a drawing are dark; border and title block lines are the thickest. Object lines outline visible features of a drawing and are made thick and dark, but not as thick as a border line. Features that are behind a surface but important to the description of the object are identified by a hidden line. This line is a series of dashes 0.12 units in length with a spacing of 0.06 units. Centerlines identify the centers of circular features, such as holes, or show that a hidden feature in one view is circular in another. The centerline consists of a series of long and short dashes. The short dash measures approximately 0.12 units, whereas the long dash may vary from 0.75 to 1.50 units. A gap of 0.06 units separates dashes. Study the examples of these lines, shown on the left in the following image.

The object shown on the right in the following image is a good illustration of the use of phantom lines in a drawing. Phantom lines are especially useful where motion is applied. The arm on the right is shown by standard object lines consisting of the continuous linetype. To show that the arm rotates about a center pivot, the identical arm is duplicated through the ARRAY command, and all lines of this new element are converted to phantom lines through the Properties dialog box, MATCHPROP command, or Layer Control box. Notice that the smaller segments are not shown as phantom lines due to their short lengths.

Figure 8–6

Use of linetypes in a drawing is crucial to the interpretation of the views and the final design before the object is actually made. Sometimes the linetype appears too long; in other cases, the linetype does not appear at all, even though using the LIST command on the object will show the proper layer and linetype. The LTSCALE or Linetype Scale command is used to manipulate the size of all linetypes loaded into a drawing. By default, all linetypes are assigned a scale factor of 1.00. This means that the actual dashes and/or spaces of the linetype are multiplied by this factor. The view illustrated on the left in the following image shows linetypes that use the default value of 1.00 from the LTSCALE command.

If a linetype dash appears too long, use the LTSCALE command and set a new value to less than 1.00. If a linetype appears too short, use the LTSCALE command and set a new value to greater than 1.00. The view illustrated in the middle of the following image shows the effects of the LTSCALE command set to a new value of 0.75. Notice that this view has one more series of dashes than the same object illustrated previously. The 0.75 multiplier affects all dashes and spaces defined in the linetype.

The view on the right in the following image shows how using the LTSCALE command with a new value of 0.50 shortens the linetype dashes and spaces even more. Now even the center marks identifying the circles have been changed to centerlines. When you use the LTSCALE command, the new value, whether larger or smaller than 1.00, affects all linetypes visible on the display screen.

Figure 8–7

ONE-VIEW DRAWINGS

An important rule to remember concerning multiview drawings is to draw only enough views to accurately describe the object. In the drawing of the gasket in the following image, Front and Side views are shown. However, the Side view is so narrow that it is difficult to interpret the hidden lines drawn inside. A better approach would be to leave out the Side view and construct a one-view drawing consisting of just the Front view.

Figure 8–8

Begin the one-view drawing of the gasket by first laying out centerlines marking the centers of all circles and arcs, as shown on the left in the following image. A layer containing centerlines could be used to show all lines as centerlines.

Use the CIRCLE command to lay out all circles representing the bolt holes of the gasket, as shown on the right in the following image. A layer containing continuous object lines could be used for these circles. You could use the OFFSET command to form the large rectangle on the inside of the gasket. If lines of the rectangle extend past each other, use the FILLET command set to a value of 0. Selecting two lines of the rectangle will form a corner. Repeat this procedure for any other lines that do not form exact corners.

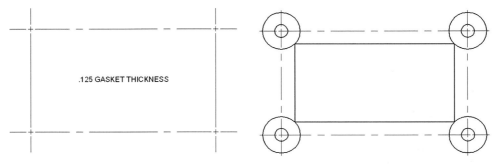

Figure 8–9

Use the TRIM command to begin forming the outside arcs of the gasket, as shown on the left in the following image.

Use the FILLET command set to the desired radius to form a smooth transition from the arcs to the outer rectangle, as shown on the right in the following image.

Figure 8–10

TWO-VIEW DRAWINGS

Before attempting any drawing, determine how many views need to be drawn. A minimum number of views are needed to describe an object. Drawing extra views is not only time-consuming, but it may result in two identical views with mistakes in each view. You must interpret which is the correct set of views. The illustration in the following image is a three-view multiview drawing of a coupler. The circles and circular hidden circle identify the Front view. Except for their rotation angles, the Top and Right Side views are identical. In this example, or for other symmetrical objects, only two views are needed to accurately describe the object being drawn. The Top view has been deleted to leave the Front and Right Side views. The Side view could have easily been deleted in favor of leaving the Front and Top views. This decision is up to the designer, depending on sheet size and which views are best suited for the particular application.

Figure 8–11

Try It! – Open the drawing file 08_R-Guide. Use the following images and descriptions for constructing a two-view drawing of this object.

To illustrate how AutoCAD is used as the vehicle for creating a two-view engineering drawing, study the two-view drawing along with the pictorial drawing illustrated in the following image to get an idea of how the drawing will appear.

Figure 8–12

Begin the two-view drawing by using the LINE command to lay out the Front and Top views. You can find the width of the Top view by projecting lines up from the front because both views share the same width, as shown on the left in the following image. Provide a space of 1.50 units between views to act as a separator and allow for dimensions at a later time. With the two views laid out, use the TRIM command to trim unnecessary lines in order for your drawing to appear similar to the illustration on the right in the following image.

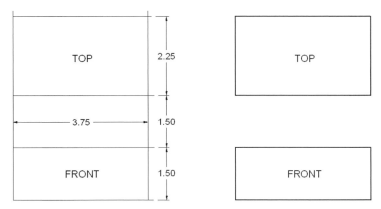

Figure 8–13

Next, add visible details to the views, such as arcs, filleted corners, and angles, as shown on the left in the following image. Use various editing commands such as TRIM, EXTEND, and OFFSET to clean up unnecessary geometry.

From the Front view, project corners up to the Top view, as shown on the right in the following image. These corners will form visible edges in the Top view.

Figure 8–14

Use the same projection technique to project features from the Top view to the Front view, as shown on the left in the following image. Then use the TRIM command to delete any geometry that appears in the 1.50 dimension space. The views now must conform to engineering standards by showing which lines are visible and which are invisible, as shown on the right in the following image. Use the Layer Control box to change the line in the Top view from the Object layer to the Hidden layer. In the same manner, the slot visible in the Top view is hidden in the Front view. Again change the line in the Front view to the Hidden layer. Since the slot in the Top view represents a circular feature, use the DIMCENTER command to place a center marker at the center of the semicircle. To show in the Front view that the hidden line represents a circular feature, add one centerline consisting of one short dash and two short dashes. Be sure this line is drawn on the Center layer. If the slot in the Top view were square instead of circular, centerlines would not be necessary.

Figure 8–15

THREE-VIEW DRAWINGS

If two views are not enough to describe an object, draw three views. This consists of Front, Top, and Right Side views.

 Try It! – Open the drawing file 08_Guide Block. Use the following images and descriptions for constructing a three-view drawing of this object.

A three-view drawing of the guide block, as illustrated in orthographic and pictorial formats in the following image, will be the focus of this segment. Notice the broken section exposing the Spotface operation above a drill hole.

Figure 8–16

Begin this drawing by laying out all views using overall dimensions of width, depth, and height, as shown in the following image. The LINE and OFFSET commands are popular commands used to accomplish this. Provide a space between views to accommodate dimensions at a later time.

Figure 8–17

Next, draw features in the views where they are visible, as shown in the following image. Since the Spotface holes appear above, draw these in the Top view. The notch appears in the Front view; draw it there. A slot is visible in the Right Side view and is drawn there.

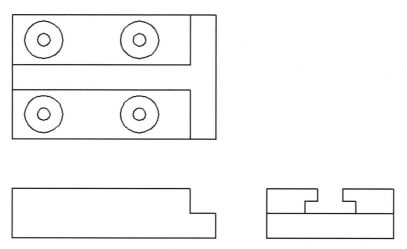

Figure 8–18

As in two-view drawings, all features are projected down from the Top to the Front view. To project depth measurements from the Top to the Right Side view, construct the 45° line as shown in the following image.

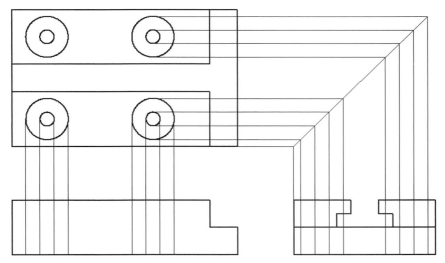

Figure 8–19

Use the 45° line to project the slot from the Right Side view to the Top view, as shown in the following image. Project the height of the slot from the Right Side view to the Front view.

Figure 8–20

Change the continuous lines to the Hidden layer where features appear invisible, such as the holes in the Front and Right Side views as shown in the following image. Change the remaining lines to the Hidden layer. Erase any construction lines, including the 45°-projection line.

Figure 8–21

Begin adding centerlines to label circular features, as shown in the following image. Either construct these center lines on the Center layer or change them later on to the Center layer. The DIMCENTER (type DCE at the Command prompt) command is used where the circles are visible. Where features are hidden but represent circular features, the single centerline consisting of one short dash and two long dashes is used.

Figure 8–22

RUNOUTS

Where flat surfaces become tangent to cylinders, there must be some method of accurately representing this using fillets. In the object illustrated on the left in the following image, the Front view shows two cylinders connected to each other by a tangent slab. The Top view is complete; the Front view has all the geometry necessary to describe the object, with the exception of the exact intersection of the slab with the cylinder.

The illustration on the right in the following image displays the correct method for finding intersections or runouts—areas where surfaces intersect others and blend in, disappear, or simply run out. A point of intersection is found at "A" in the Top view with the intersecting slab and the cylinder. This actually forms a 90° angle with the line projected from the center of the cylinder and the angle made by the slab. A line is projected from "A" in the Top view to intersect with the slab found in the Front view.

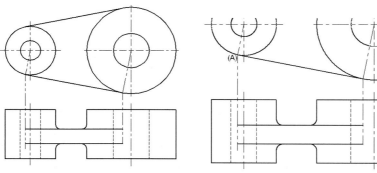

Figure 8–23

Illustrated on the left in the following image, fillets are copied to represent the slab and cylinder intersections. This forms the runout.

The resulting two-view drawing, complete with runouts, is illustrated on the right in the following image.

Figure 8–24

TUTORIAL EXERCISE: ORTHOGRAPHIC BLOCK.DWG

Figure 8–25

Purpose
Using an isometric grid, construct the Front, Top, and Right Side views of the Orthographic Block, as shown in the previous image.

System Settings
Begin a new drawing called Orthographic Block .dwg. Use the Drawing Units dialog box to change the precision from four to two decimal places. Keep the remaining default unit values.

Using the LIMITS command, keep (0,0) for the lower left corner and change the upper right corner from (12,9) to (15.50,9.50).

It is very important to check and see that the following Object Snap modes are already set: Endpoint, Extension, Intersection, and Center.

Layers
Create the following layers with the format:

Name	Color	Linetype
Object	Green	Continuous
Hidden	Red	Hidden
Center	Yellow	Center

Suggested Commands
Begin this tutorial by laying out the three primary views using the LINE and OFFSET commands. Use a grid spacing of 0.25 units for all distance calculations.

Use the TRIM command to clean up any excess line segments. As an alternative projection method, use temporary tracking points in combination with Object Snap options to add features in other views.

STEP 1

Begin constructing the Orthographic Block by laying out the Front, Top, and Right Side views using only the overall dimensions. Do not be concerned about details such as holes or slots; these will be added to the views at a later step. The length of the object shown on the left in the following image is 8 grid units (2 inches). The height of the object is 5 grid units (1.25 inches); the depth of the object is 4 units (1 inch). The distance between views is 5 grid units (1.25 inches).

STEP 2

Once the overall dimensions have been used to lay out the Front, Top, and Right Side views, begin adding visible details to the views. The "L" shape is added to the Front view; the hole and corner fillets are added to the Top view; the rectangular slot is added to the Right Side view, as shown on the right in the following image. Refer to the isometric view of this object at the beginning of this tutorial for the dimensions of the "L" shape, the hole, corner fillets, and the rectangular slot.

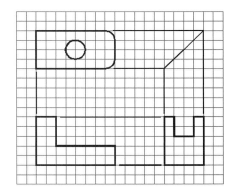

Figure 8–26

STEP 3

Begin projecting the visible edges from hole and slot features onto other views. Slot information is added to the Top view and height information is projected onto the Right Side view from the Front view. At this point, only add visible information to other views where required, as shown on the left in the following image.

STEP 4

Now project all hidden features to the other views. The hole projection is hidden in the Front view along with the slot visible in the Right Side view. The hole is also hidden in the Right Side view, as shown on the right in the following image. Notice how the 45° angle is used to project the hole from the Top view to the Right Side view.

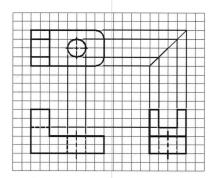

Figure 8–27

STEP 5

The completed multiview drawing solution is illustrated on the left in the following image. Dimensions are added to document the exact size of the object, as shown on the right in the following image. Proper placement of dimensions will be discussed in chapter 11.

Figure 8–28

TUTORIAL EXERCISE: 08_SHIFTER.DWG

Figure 8–29

Purpose
This tutorial is designed to allow the user to construct a three-view drawing of the 08_Shifter as shown in the previous image.

System Settings
Use the settings from the previous tutorial and check to see that the following Object Snap modes are already set: Endpoint, Extension, Intersection, Center.

Layers
Create two additional layers:

Name	Color	Linetype
Dimension	Yellow	Continuous
Projection	Cyan	Continuous

Suggested Commands
The primary commands used during this tutorial are OFFSET and TRIM. The OFFSET command is used for laying out all views before the TRIM command is used to clean up excess lines. Since different linetypes represent certain features of a drawing, the Layer Control box is used to convert to the desired linetype needed as set in the Layer Properties Manager dialog box. Once all visible details are identified in the primary views, project the visible features to the other views using the LINE command. A 45° inclined line is constructed to project lines from the Top view to the Right Side view and vice versa.

STEP 1

Open the drawing file 08_Shifter. Make the Object layer current. Then use the LINE and OFFSET commands and create the three views displayed in the following image. The dimensions in this view represent the overall width, height, and depth of the Shifter. Space the views a distance of 1.50 units away from each other.

Figure 8–30

STEP 2

Continue using the OFFSET command to add the various features to the Front, Top, and Right Side views using the dimensions provided in the following image. Use the TRIM command to clean up all intersections in order for your display to appear similar to the following image.

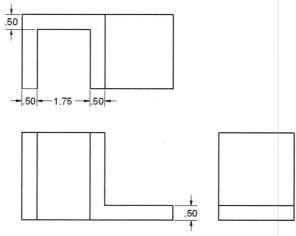

Figure 8–31

STEP 3

Add both circles to their respective views, as shown in the following image. One method of finding the circle centers is to use the OFFSET command. Another method would be to use the OSNAP-From mode and the dimensions displayed in the following image.

Figure 8–32

STEP 4

Project lines from the quadrants of the circles from the Top and Side views into the Front view, as shown in the following image. These projected lines in the Front view represent the hidden edges of the circles.

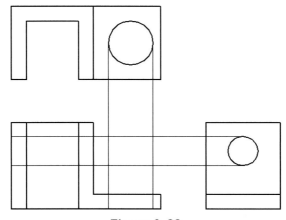

Figure 8–33

STEP 5

Use the following image to guide you along the trimming of unnecessary projected lines representing the circle edges in the Front view. Then change the highlighted lines in the following image to the Hidden layer.

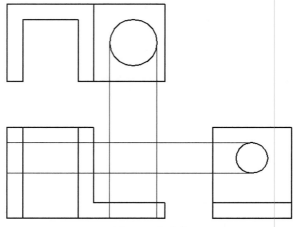

Figure 8–34

STEP 6

With the Center layer current, add center marks to both circles using the Dimension Center (DCE) command by touching the edge of each circle to place the center mark, as shown in the following image.

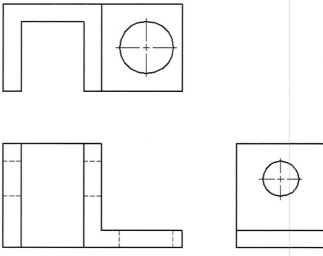

Figure 8–35

STEP 7

Project the endpoints of the center marks into the Front view, as shown in the following image.

Figure 8–36

STEP 8

Use the BREAK command to form the center lines in the Front view. Then change these lines to the Center layer, as shown in the following image.

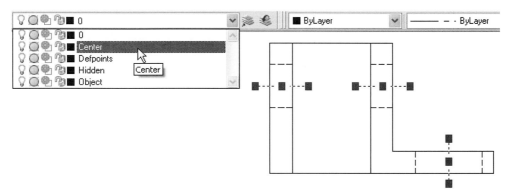

Figure 8–37

STEP 9

Use the FILLET command to create a corner between the Top and Side views, as shown in the following image. Then create a 45° line from the corner at "A" at a distance of 4 units. This distance is an approximation and will be used to project edges to the Top and Side views.

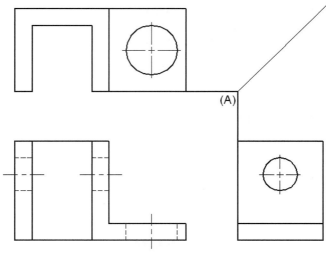

Figure 8–38

STEP 10

Project lines from the edges of the circles in the Top view to intersect with the 45° angle. Then continue constructing lines into the Side view, as shown in the following image.

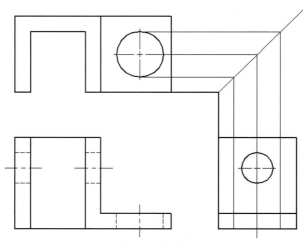

Figure 8–39

STEP 11

Use the TRIM and BREAK commands to clean up lines in the Side view, as shown in the following image. Then change these lines to the Hidden and Center layers.

Figure 8–40

STEP 12

Project lines from the edges of the circles in the Side view to intersect with the 45° angle. Then continue constructing lines into the Top view, as shown in the following image.

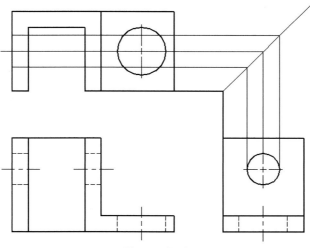

Figure 8–41

STEP 13

Use the TRIM and BREAK commands to clean up lines in the Top view, as shown in the following image. Then change these lines to the Hidden and Center layers.

Figure 8–42

STEP 14

Erase the 45° angle line, trim the corner connecting the Top and Side views, and extend the hidden line in the Side view to the top of the line at "A". This line represents a hidden edge. The completed orthographic drawing of the Shifter is shown in the following image.

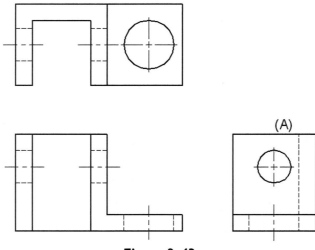

Figure 8–43

PROBLEMS FOR CHAPTER 8

Directions for Problems 8–1 and 8–2:

Open each drawing file individually (Pr08-01.dwg and Pr08-02.dwg). Create two new layers called Missing Visible Lines and Missing Hidden Lines. Assign the color red to both of these layers. Assign the Hidden linetype to the Missing Hidden Lines layer. Using projection techniques, find the missing lines for each object and draw in the correct solution on the proper layer.

PROBLEM 8–1

PROBLEM 8–2

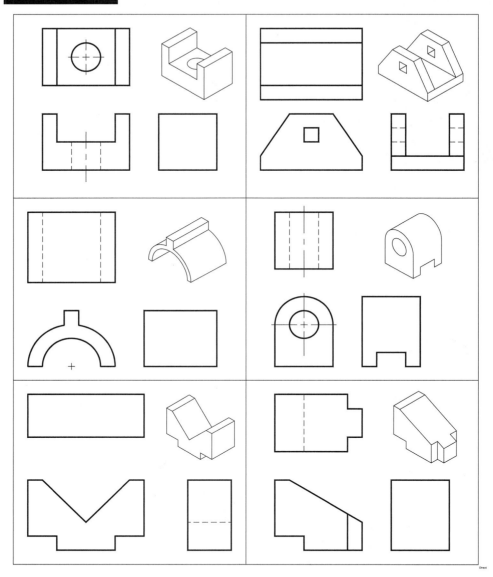

Directions for Problems 8–3 through 8–20:

Construct a multiview drawing by sketching the Front, Top, and Right Side views. Use a grid of 0.25 units to assist in the construction of the sketches. These sketching problems can also be constructed directly in AutoCAD using typical commands and projection techniques.

Shape Description/Multiview Projection 461

PROBLEM 8–3

PROBLEM 8–6

PROBLEM 8–4

PROBLEM 8–7

PROBLEM 8–5

PROBLEM 8–8

PROBLEM 8-9

PROBLEM 8-12

PROBLEM 8-10

PROBLEM 8-13

PROBLEM 8-11

PROBLEM 8-14

PROBLEM 8–15

PROBLEM 8–18

PROBLEM 8–16

PROBLEM 8–19

PROBLEM 8–17

PROBLEM 8–20

Directions for Problems 8–21 through 8–35:

Construct a multiview drawing of each object. Be sure to construct only those views that accurately describe the object.

PROBLEM 8–21

PROBLEM 8–22

PROBLEM 8–23

PROBLEM 8–24

PROBLEM 8–25

PROBLEM 8–26

PROBLEM 8–27

PROBLEM 8–28

PROBLEM 8-29

PROBLEM 8-30

PROBLEM 8–31

PROBLEM 8–32

PROBLEM 8-33

PROBLEM 8-34

PROBLEM 8-35

CHAPTER 9

Creating Section Views

This chapter will cover Section Views, which are created by slicing an object along a cutting plane in order to view its interior details. The basics of section views will be covered first. Full, half, assembly, aligned, and offset sections will be shown. Other special section topics include sectioning ribs, creating broken sections, revolved sections, removed sections, and isometric sections. After covering the basics of sections, the chapter will shift to a discussion on how AutoCAD crosshatches objects through the Hatch and Gradient dialog box (BHATCH command). You will be able to select from a collection of many hatch patterns including gradient patterns for special effects when adding hatch patterns to your drawings. Hatch scaling and angle considerations will also be covered along with associative crosshatching. Once hatching exists in a drawing, it can be easily modified through the Hatch Edit dialog box. Automatic hatch scaling in a drawing layout (Paper Space) will also be covered.

SECTION VIEW BASICS

Principles of orthographic projections remain the key method for the production of engineering drawings. As these drawings get more complicated in nature, the job of the designer becomes more challenging in the interpretation of views, especially where hidden features are involved. The concept of slicing a view to expose these interior features is the purpose of performing a section. On the left in the following image is a pictorial representation of a typical flange consisting of eight bolt holes and counterbore hole in the center.

The drawings on the right in the following image show a typical solution to a multiview problem complete with Front and Side views. The Front view displaying the eight bolt holes is obvious to interpret; however, the numerous hidden lines in the Side view make the drawing difficult to understand, and this is considered a relatively simple drawing. To relieve the confusion associated with a drawing too difficult to understand because of numerous hidden lines, a section drawing is made for the part.

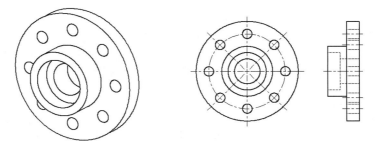

Figure 9–1

To understand section views better, see the illustration on the left in the following image. Creating a section view slices an object in such a way as to expose what used to be hidden features and convert them to visible features. This slicing or cutting operation can be compared to that of using a glass plate or cutting plane to perform the section. In the object shown on the left in the following image, the glass plate cuts the object in half. It is the responsibility of the designer or CAD operator to convert one half of the object to a section and to discard the other half. Surfaces that come in contact with the glass plane are crosshatched to show where material was cut through.

A completed section view drawing is shown on the right in the following image. Two new types of lines are also shown: namely, a cutting plane line and section lines. The cutting plane line performs the cutting operation on the Front view. In the Side view, section lines show the surfaces that were cut. The counterbore edge is shown because this corner would be visible in the section view. Notice that holes are not section lined because the cutting plane passes across the center of the hole. Notice also that hidden lines are not displayed in the Side view. It is poor practice to merge hidden lines into a section view, although there are always exceptions. The arrows of the cutting plane line tell the designer to view the section in the direction of the arrows and discard the other half.

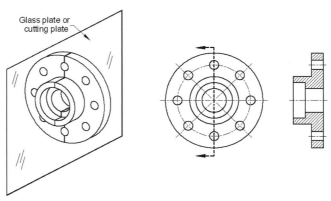

Figure 9–2

The cutting plane line consists of a very thick line with a series of dashes approximately 0.25″ in length, as shown on the left in the following image. A polyline may be used to create this line of 0.05″ thickness. The arrows point in the direction of sight used to create the section, with the other half generally discarded. Assign this line one of the dashed linetypes; the hidden linetype is reserved for detailing invisible features in views. The section line, in contrast with the cutting plane line, is a very thin line, as shown on the right in the following image. This line identifies the surfaces being cut by the cutting plane line. The section line is usually drawn at an angle and at a specified spacing.

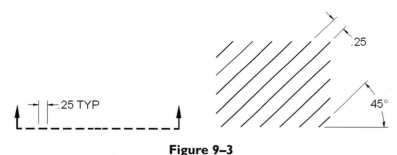

Figure 9–3

A wide variety of hatch patterns is supplied with the software. One of these patterns, ANSI31, is displayed on the left in the following image. This is one of the more popular patterns, with lines spaced 0.125″ apart and at a 45° angle.

The object on the right in the following image illustrates proper section lining techniques. Much of the pain of spacing the section lines apart from each other and at angles has been eased considerably with the use of the computer as a tool. However, the designer must still practice proper section lining techniques at all times for clarity of the section.

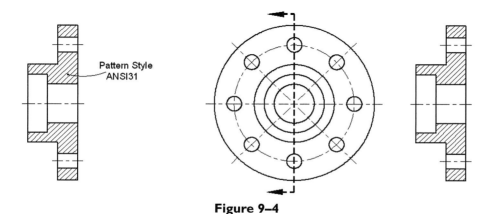

Figure 9–4

The following image illustrates common errors involved when applying section lines to a cutaway view. At "A," the section lines run in the correct direction and at the same angle; however, the hidden lines have not been converted to object lines. This will confuse the more experienced designer because the presence of hidden lines in the section means more complicated invisible features.

At "B" is another error, encountered when sections are created. Again, the section lines are properly placed; however, all surfaces representing holes have been removed, which displays the object as a series of sectioned blocks unconnected, implying four separate parts.

At "C," the object appears to be properly section lined; however, upon closer inspection, notice that the angle of the crosshatch lines in the upper half differs from the angle of the same lines in the lower left half. This suggests two different parts when, in actuality, it is the same part.

At "D," all section lines run in the correct direction. The problem is that the lines run through areas that were not sliced by the cutting plane line. These areas in "D" represent drill and counterbore holes and are typically left unsectioned.

We use standard orthographic projection to show features of the object in areas where we are not cutting through material.

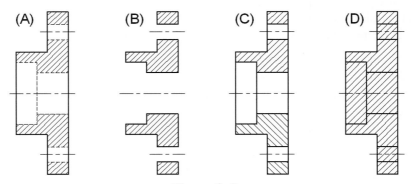

Figure 9–5

These have been identified as the most commonly made errors when an object is crosshatched. Remember just a few rules to follow:

- Section lines are present only on surfaces that are cut by the cutting plane line.
- Section lines are drawn in one direction when the same part is crosshatched.
- Hidden lines are usually omitted when a section view is created.
- Areas such as holes are not sectioned because the cutting line only passes across this feature.

FULL SECTIONS

When the cutting plane line passes through the entire object, a full section is formed. As illustrated on the left in the following image, a full section would be the same as taking an object and cutting it completely down the middle. Depending on the needs of the designer, one half is kept while the other half is discarded. The half that is kept is shown with section lines.

The multiview solution to this problem is shown on the right in the following image. The Front view is drawn with lines projected across to form the Side view. To show that the Side view is in section, a cutting plane line is added to the Front view. This line performs the physical cut. You have the option of keeping either half of the object. This is the purpose of adding arrowheads to the cutting plane line. The arrowheads define the direction of sight in which you view the object to form the section. You must then interpret what surfaces are being cut by the cutting plane line in order to properly add crosshatching lines to the section that is located in the Right Side view. Hidden lines are not necessary once a section has been made.

Figure 9–6

Numerous examples illustrate a cutting plane line with the direction of sight off to the left. This does not mean that a cutting plane line cannot have a direction of sight going to the right, as shown in the following image. In this example, the section is formed from the Left Side view if the circular features are located in the Front view.

Figure 9–7

HALF SECTIONS

When symmetrical-shaped objects are involved, sometimes it is unnecessary to form a full section by cutting straight through the object. Instead, the cutting plane line passes only halfway through the object, which makes the pictorial drawing as shown on the left in the following image a half section. The rules for half sections are the same as for full sections; namely, once a direction of sight is established, part of the object is kept, and part is discarded.

The views are laid out, as shown on the right in the following image, in the usual multiview format. To prepare the object as a half section, the cutting plane line passes halfway through the Front view before being drawn off to the right. Notice that there is only one direction-of-sight arrow and a centerline is used to depict where the sectioned portion of the object ends in the Right Side view. The Right Side view is converted to a half section by the crosshatching of the upper half of the Side view while hidden lines remain in the lower half.

Figure 9–8

Depending on office practices, some designers prefer to omit hidden lines entirely from the Side view, similar to the drawing on the left in the following image. In this way, the lower half is drawn showing only what is visible.

The illustration on the right in the following image shows another way of drawing the cutting plane line to conform to the Right Side view drawn in section. Hidden lines have been removed from the lower half; only those lines visible are displayed. This is consistent with the practice of omitting hidden lines from section views.

Figure 9–9

ASSEMBLY SECTIONS

It would be unfair to give designers the impression that section views are used only for displaying internal features of individual parts. Another advantage of using section views is that they permit the designer to create numerous objects, assemble them, and then slice the assembly to expose internal details and relationships of all parts. This type of section is an assembly section similar to the illustration on the left in the following image. For all individual parts, notice the section lines running in the same direction. This follows one of the basic rules of section views: keep section lines at the same angle for each individual part.

The illustration on the right in the following image shows the difference between assembly sections and individual parts that have been sectioned. For parts in an assembly that contact each other, it is good practice to alternate the directions of the section lines to make the assembly much clearer and to distinguish the parts from each other. You can accomplish this by changing the angle of the hatch pattern or even the scale of the pattern.

Figure 9–10

To identify parts in an assembly, an identifying part number along with a circle and arrowhead line are used, as shown on the left in the following image. The line is very similar to a leader line used to call out notes for specific parts on a drawing. The addition of the circle highlights the part number. Sometimes this type of callout is referred to as a "bubble."

In the enlarged assembly shown on the right in the following image, the large area in the middle is actually a shaft used to support a pulley system. With the cutting plane passing through the assembly including the shaft, refrain from crosshatching features such as shafts, screws, pins, or other types of fasteners. The overall appearance of the assembly is actually enhanced when these items are not crosshatched.

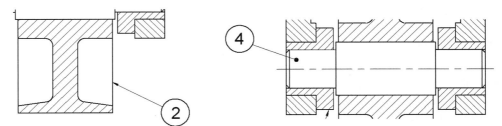

Figure 9–11

ALIGNED SECTIONS

Aligned sections take into consideration the angular position of details or features of a drawing. Instead of the cutting plane line being drawn vertically through the object, as shown on the left in the following image, the cutting plane is angled or aligned with the same angle as the elements it is slicing. Aligned sections are also made to produce better clarity of a drawing. As illustrated on the left in the following image, with the cutting plane forming a full section of the object, it is difficult to obtain the true size of the angled elements. In the Side view, they appear foreshortened or not to scale. Hidden lines were added as an attempt to better clarify the view.

Instead of the cutting plane line being drawn all the way through the object, the line is bent at the center of the object before being drawn through one of the angled legs, as shown on the right in the following image. The direction of sight arrows on the cutting plane line not only determines the direction in which the view will be sectioned, but also shows another direction for rotating the angled elements so they line up with the upper elements. This rotation is usually not more than 90°. As lines are projected across to form the Side view, the section appears as if it were a full section. This is only because the features were rotated and projected in section for greater clarity of the drawing. It is interesting to note that the aligned section is drawn higher than the Front view from which it is drawn. Standard orthographic projection is usually not applicable to aligned sections.

Figure 9–12

OFFSET SECTIONS

Offset sections take their name from the offsetting of the cutting plane line to pass through certain features in a view, as shown in the following image. If the cutting plane line passes straight through any part of the object, some features would be exposed while others would remain hidden. By offsetting the cutting plane line, the designer controls its direction and which features of a part it passes through. The view to section follows the basic section rules. Notice that the changes in direction of the cutting plane as it passes through the object are shown in the Top view but not in the Front view.

Figure 9–13

SECTIONING RIBS

Parts made out of cast iron with webs or ribs used for reinforcement do not follow the basic rules of sections. As illustrated on the left in the following image, the Front view has the cutting plane line passing through the entire view; the Side view at "A" is crosshatched according to basics of section views. However, it is difficult to read the thickness of the base because the crosshatching includes the base along with the web. A more efficient method is to leave off crosshatching webs, as in "B." In this example, eliminating the crosshatching of the web exposes other important details such as thickness of bases and walls.

The object shown on the right in the following image is another example of performing a full section on an area consisting of webbed or ribbed features. If you do not crosshatch the webbed areas, more information is available, such as the thickness of the base and wall areas around the cylindrical hole. While this may not be considered true projection, it is considered good practice.

Figure 9–14

BROKEN SECTIONS

At times, only a partial section of an area needs to be created. For this reason, a broken section might be used. The object illustrated on the left in the following image at "A" shows a combination of sectioned areas and conventional areas outlined by the hidden lines. When converting an area to a broken section, you create a break line, crosshatch one area, and leave the other area in a conventional drawing. Break lines may take the form of short freehanded line segments, as shown at "B" in the following image, or a series of long lines separated by break symbols, as shown at "C" in the following image. You can use the AutoCAD LINE command with Ortho off to produce the desired effect, as in "A" and "B."

Figure 9–15

REVOLVED SECTIONS

Section views may be constructed as part of a view with revolved sections. As illustrated on the left in the following image, the elliptical shape is constructed while it is revolved into position and then crosshatched. The crosshatched shape represents the cross section shape of the arm.

Illustrated on the right in the following image is another example of a revolved section where a cross section of the C-clamp was cut away and revolved to display its shape.

Figure 9–16

REMOVED SECTIONS

Removed sections are very similar to revolved sections except that instead of the section being drawn somewhere inside the object, as is the case with a revolved section, the section is placed elsewhere or removed to a new location in the drawing. The cutting plane line is present, with the arrows showing the direction of sight. Identifying letters are placed on the cutting plane and underneath the section to keep track of the removed sections, especially when there are a number of them on the same drawing sheet. See the illustration on the left in the following image.

Another way of displaying removed sections is to use centerlines as a substitute for the cutting plane line. As illustrated on the right in the following image, the centerlines determine the three shapes of the chisel and display the basic shapes. Identification numbers are not required in this particular example.

Figure 9–17

ISOMETRIC SECTIONS

Section views may be incorporated into pictorial drawings that appear in the following image. The illustration on the left is an example of a full isometric section with the cutting plane passing through the entire object. In keeping with basic section rules, only those surfaces sliced by the cutting plane line are crosshatched. Isometric sections make it easy to view cut edges compared to holes or slots. The illustration on the right is an example of an isometric drawing converted to a half section.

Figure 9–18

ARCHITECTURAL SECTIONS

Mechanical representations of machine parts are not the only type of drawings where section views are used. Architectural drawings rely on sections to show the building materials and construction methods for the construction of foundation plans, roof details, or wall sections, as shown in the following image. Here, numerous types of crosshatching symbols are used to call out the different types of building materials, such as brick veneer at "A," insulation at "B," finished flooring at "C," floor joists at "D," concrete block at "E," earth at "F," and poured concrete at "G." Section hatch patterns that are provided with AutoCAD may be used to crosshatch most building components.

Figure 9–19

THE BHATCH COMMAND

With the basics of sections having been covered, it is time now to see how AutoCAD is used to generate section lines. This is accomplished by using the Boundary Hatch, or BHATCH, command. Choosing Hatch… from the Draw pull-down menu or picking the Hatch button activates the main Hatch and Gradient dialog box, as shown in the following image. Use the Hatch tab to pick points identifying the area or areas to be crosshatched, select objects using one or more of the popular selection set modes, select hatch patterns supported in the software, and change the scale and angle of the hatch pattern. Any areas grayed out in the following image are considered inactive. Only when additional parameters have been satisfied will these buttons activate for use.

Figure 9-20

 Try It! – Open the drawing file 09_Hatch Basics. The object shown on the left in the following image will be used to demonstrate the boundary crosshatching method. The object needs to have areas "A," "B," and "C" crosshatched.

Areas "A," "B," and "C" first need to be identified as the boundaries to be hatched. The BHATCH command is a very efficient method used for automatically highlighting areas to be crosshatched, as shown on the right in the following image.

Figure 9-21

The BHATCH command opens the Hatch tab of the Hatch and Gradient dialog box. Click on the Add: Pick points button found in the upper right corner of this dialog box. The illustration on the left in the following image shows three areas with internal points identified by the Xs. The dashed areas identify the highlighted areas to be crosshatched.

In the final step of the BHATCH command, the crosshatch pattern is applied to the areas, and the highlighted outlines deselect, leaving the crosshatch patterns. See the illustration on the right in the following image.

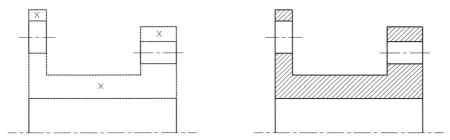

Figure 9–22

AVAILABLE HATCH PATTERNS

The Hatch tab holds numerous hatch patterns in a drop-down list box, shown in the following image. Clicking on a pattern in this listing makes the pattern current.

Figure 9–23

Selecting the pattern (…) button or ellipses next to the pattern name, as shown on the left in the following image, activates the Hatch Pattern Palette dialog box. Here, a series of tabs are present that display a number of crosshatching patterns already created and ready for use, as shown on the right in the following image. Select a particular pattern by clicking on the pattern itself. If the wrong pattern was selected, simply choose the correct one or move to another tab to view other patterns for use.

Figure 9–24

Once a hatch pattern is selected, the name is displayed in the Hatch and Gradient dialog box, as shown in the following image. If the scaling and angle settings look favorable, click on the Add: Pick points button. The command prompt requires you to select an internal point or points.

Figure 9–25

Continue working with the drawing file 09_Hatch Basics. Click in the three areas marked by the X, as shown on the left in the following image. Notice that these areas highlight. If these areas are correct, press ENTER to return to the Hatch and Gradient dialog box.

When you are back in the Hatch and Gradient dialog box, you now have the option of first previewing the hatch to see if all settings and the appearance of the pattern are desirable. Click on the Preview button in the lower left corner of the dialog box to accomplish this. The results appear on the right in the following image. At this point in the process, the pattern is still not permanently placed on the object. Pressing the ENTER key to exit Preview mode returns you to the main Hatch and Gradient dialog box. If the hatch pattern is correct in appearance, click the OK button to place the pattern with the drawing.

Figure 9–26

SOLID FILL HATCH PATTERNS

Activating the Hatch and Gradient dialog box and clicking on the pattern (…) button or ellipses followed by the Other Predefined tab displays the dialog box as shown on the left in the following image. Additional patterns are available and include brick, roof, and other construction-related materials. One powerful pattern is called SOLID. This pattern is used to fill in an enclosed area with a solid color based on the current layer. The illustration on the right in the following image shows a typical application of filling in the walls of a floor plan with a solid color.

Figure 9–27

Try It! – Open the drawing file 09_Hatch Solid1. The object shown on the left in the following image illustrates a thin wall that needs to be filled in. The BHATCH command used with the SOLID hatch pattern can perform this task very efficiently. When you pick an internal point, the entire closed area will be filled with a solid pattern and placed on the current layer with the same color and other properties of the current layer.

The solid hatch pattern shown on the right in the following image was placed with one internal point pick.

Figure 9–28

 Try It! – Open the drawing file 09_Hatch Solid2. Applying the SOLID hatch pattern to the object, as shown on the left in the following image, can easily be accomplished with the BHATCH command. This object also is a thin wall example, with the addition of various curved edges and fillets.

The solid hatch pattern also works for curved outlines. The illustration on the right in the following image displays the solid pattern, completely filling in the outline where all corners consist of some type of curve generated by the FILLET command.

Figure 9–29

GRADIENT PATTERNS

A gradient hatch pattern is a solid hatch fill that makes a smooth transition from a lighter shade to a darker shade. Predefined patterns such as linear, spherical, and radial sweep are available to provide different effects. As with the vector hatch patterns that have always been supplied with AutoCAD, the angle of the gradient patterns can also be controlled. Gradient patterns can also be associative; the pattern will update if the boundary is changed. To edit a gradient pattern, simply double-click on it to launch the Hatch Edit dialog box. Gradient patterns can be selected and applied by picking the Gradient tab of the Hatch and Gradient dialog box, as shown in the following image. A few of the controls for this tab of the dialog box are explained as follows:

- One Color—When this option is selected, a color swatch with a Browse button and a Shade and Tint slider appears. The One Color option designates a fill that uses a smooth transition between darker shades and lighter tints of a single color.

- Two Color—When this option is selected, a color swatch and Browse button display for colors 1 and 2. This option allows the fill to transition smoothly between two colors.

- Color Swatch—This is the default color displayed as set by the current color in the drawing. This option specifies the color to be used for the gradient fill. The presence of the three dots, called ellipses (...), next to the color swatch signifies the Browse button. Use this to display the Select Color dialog box, similar to the dialog used for assigning colors to layers. Use this dialog box to select color based on the AutoCAD Index Color, True Color, or Color Book.

- Shade and Tint Slider—This option designates the amount of tint and shade applied to the gradient fill of one color. Tint is defined as the selected color mixed with white. Shade is defined as the selected color mixed with black.

- Centered Option—Use this option for creating special effects with gradient patterns. When this option is checked, the gradient fill will appear symmetrical. If the option is not selected, AutoCAD shifts the gradient fill up and to the left. This position creates the illusion that a light source is located to the left of the object.

- Angle—Set an angle that affects the gradient pattern fill. This angle setting is independent of the angle used for regular hatch patterns.

- Gradient Patterns—Nine gradient patterns are available for you to apply. These include linear sweep, spherical, and parabolic.

Figure 9–30

The example of the house elevation in the following image illustrates parabolic linear sweep being applied above and below the house.

Figure 9–31

The garage detail in the following image illustrates one of the parabolic gradient patterns applied to the windows.

Figure 9–32

HATCH PATTERN SYMBOL MEANINGS

Listed below are a number of hatch patterns and their purposes. For example, if you are constructing a mechanical assembly in which you want to distinguish plastic material from steel, use the patterns ANSI34 and ANSI32, respectively. Patterns that begin with AR- have architectural applications. Refer to the following list for the purposes of other materials and their associated hatch patterns.

ANSI31	ANSI Iron, Brick, Stone masonry
ANSI32	ANSI Steel
ANSI33	ANSI Bronze, Brass, Copper
ANSI34	ANSI Plastic, Rubber
ANSI35	ANSI Fire brick, Refractory material
ANSI36	ANSI Marble, Slate, Glass
ANSI37	ANSI Lead, Zinc, Magnesium, Sound/Heat/Elec Insulation
ANSI38	ANSI Aluminum
AR-B816	8x16 Block elevation stretcher bond
AR-B816C	8x16 Block elevation stretcher bond with mortar joints
AR-B88	8x8 Block elevation stretcher bond
AR-BRELM	Standard brick elevation English bond with mortar joints
AR-BRSTD	Standard brick elevation stretcher bond
AR-CONC	Random dot and stone pattern
AR-HBONE	Standard brick herringbone pattern @ 45 degrees
AR-PARQ1	2x12 Parquet flooring: pattern of 12x12

HATCHING ISLANDS

 Try It! – Open the drawing file 09_Hatch Islands. When charged with the task of crosshatching islands, first activate the Hatch and Gradient dialog box and select a pattern. Select the Add: Pick points button in the upper right corner of the dialog box. Next pick a point at "A" on the left in the following image, which will define not only the outer perimeter of the object but the inner shapes as well. This result is due to Flood mode in the Island Detection area being activated in the Advanced tab of the Hatch and Gradient dialog box.

Selecting the Preview button will display the hatch pattern shown on the right in the following image. If changes need to be made, such as a change in the hatch scale or angle, the preview allows these changes to be made. After making changes, be sure to preview the pattern once again to check whether the results are desirable. Choosing OK places the pattern and exits to the command prompt.

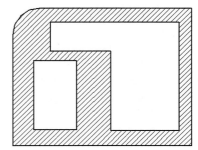

Figure 9–33

HATCH PATTERN SCALING

Hatching patterns are predefined in size and angle. When you use the BHATCH command, the pattern used is assigned a scale value of 1.00, which will draw the pattern exactly the way it was originally created.

 Try It! – Open the drawing file 09_Hatch Scale. Activate the Hatch and Gradient dialog box, accept all default values, and pick internal points to hatch the object as shown on the left in the following image.

Entering a different scale value for the pattern in the Hatch and Gradient dialog box will either increase or decrease the spacing between crosshatch lines. The illustration on the right in the following image is an example of the ANSI31 pattern with a new scale value of 0.50.

Figure 9–34

As you can decrease the scale of a pattern to hatch small areas, so also can you scale the pattern up for large areas. The illustration in the following image has a hatch scale of 2.00, which doubles all distances between hatch lines.

Use care when hatching large areas. In this image, the total length of the object measures 190.50 millimeters. If the hatch scale of 1.00 were used, the pattern would take on a filled appearance. This results in numerous lines being generated, which will increase the size of the drawing file. A value of 25.4, the number of millimeters in an inch, is used to scale hatch lines for metric drawings.

 Try It! – Open the drawing file 09_Hatch Scale mm. Activate the Hatch and Gradient dialog box and change the pattern scale to 25.4 units. Select the Add: Pick points button to pick internal points to crosshatch the object illustrated in the following image.

Figure 9–35

HATCH PATTERN ANGLE MANIPULATION

As with the scale of the hatch pattern, depending on the effect, you can control the angle for the hatch pattern within the area being hatched. By default, the BHATCH command displays a 0° angle for all patterns.

 Try It! – Open the drawing file 09_Hatch Angle. Activate the Hatch and Gradient dialog box and hatch the object as shown on the left in the following image, keeping all default values. The angle for ANSI31 is drawn at 45°—the angle in which the pattern was originally created.

Experiment with the angle setting of the Hatch and Gradient dialog box by entering any angle different from the default value of "0" to rotate the hatch pattern by that value. This means that if a pattern were originally designed at a 45° angle, like ANSI31, entering a new angle for the pattern would begin rotating the pattern starting at the 45° position.

Entering an angle other than the default rotates the pattern from the original angle to a new angle. In the illustration on the right in the following image, an angle of 90° has been applied to the ANSI31 pattern in the Hatch and Gradient dialog box. Providing different angles for patterns is useful when you create section assemblies in which different parts are in contact with each other because it makes the different parts easy to see.

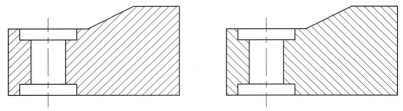

Figure 9–36

TRIMMING HATCH AND FILL PATTERNS

Hatch patterns can also be trimmed when you add objects to a drawing. In the illustration on the left in the following image, the rectangle is constructed and hatched with the ANSI31 pattern. In the image on the right, two circles have been added to the ends of the rectangle. You now want to remove the hatch pattern from both halves of the circles.

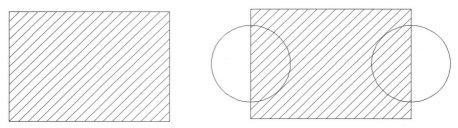

Figure 9–37

To accomplish this, enter the TRIM command and select the edges of both circles as cutting edges, as shown on the left in the following image. Then pick the hatch patterns inside both circles. The results are illustrated on the right, with the hatch pattern trimmed out of both circles.

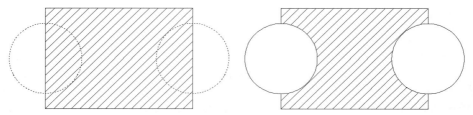

Figure 9–38

MODIFYING ASSOCIATIVE HATCHES

Try It! – Open the drawing file 09_Hatch Assoc. The following is a review of associative hatching. In the following image, the plate needs to be hatched in the ANSI31 pattern at a scale of one unit and an angle of zero; the two slots and three holes are to be considered islands. Enter the BHATCH command, make the necessary changes in the Hatch and Gradient dialog box, including making sure Associative is picked in the Options part of the dialog box, click on the Add: Pick points button, and mark an internal point somewhere inside the object, such as at "A," as shown in the following image.

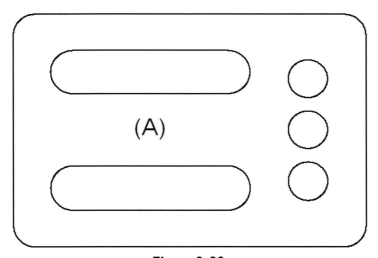

Figure 9–39

When all objects inside and including the outline highlight, press ENTER to return to the Hatch and Gradient dialog box. It is very important to realize that all objects highlighted are tied to the associative crosshatch object. Each shape works directly with the hatch pattern to ensure that the outline of the object is being read by the hatch pattern and that the hatching is performed outside the outline. You have the option of first previewing the hatch pattern or applying the hatch pattern. In either case, the results should appear similar to the object on the left in the following image.

Associative hatch objects may be edited, and the results of this editing will have an immediate impact on the appearance of the hatch pattern. For example, the two outer holes need to be increased in size by a factor of 1.5; also, the middle hole needs to be repositioned to the other side of the object, as shown on the right in the following image. Not only does using the MOVE command allow you to reposition the hole, but when the move is completed, the hatch pattern mends itself automatically to the moved circle. In the same manner, using the SCALE command to increase the size of the two outer circles by a value of 1.5 units makes the circles larger and automatically mends the hatch pattern.

Figure 9–40

On the left in the following image, the length of the slots needs to be shortened. Also, hole "A" needs to be deleted. Use the STRETCH command to shorten the slots. Use the crossing box at "B" to select the slots and stretch them one unit to the right. Use the ERASE command to delete hole "A."

The result of the editing operations is shown on the right in the following image. In this figure, associative hatching has been enhanced for members to be erased while the hatch pattern still maintains its associativity. The inner hole was deleted with the ERASE command. After the ERASE command was executed, the hatch pattern updated itself and hatched through the area once occupied by the hole.

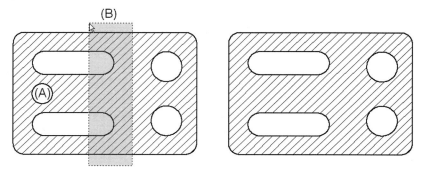

Figure 9–41

Once a hatch pattern is placed in the drawing, it will not update itself to any new additions in the form of closed shapes in the drawing. On the left in the following image, a rectangle was added at the center of the object. Notice, however, that the hatch pattern cuts directly through the rectangle. The hatch pattern does not have the intelligence to recognize the new boundary.

On the other hand, the TRIM command could be used to select the rectangle as the cutting edge and select the hatch pattern inside the rectangle to remove it, as shown on the right in the following image. After this trimming operation is performed, the rectangle is now recognized as a valid member of the associative hatch.

Figure 9–42

EDITING HATCH PATTERNS

 Try It! – Open the drawing file 09_Hatchedit. In the illustration on the left in the following image, the pattern needs to be increased to a new scale factor of 3 units and the angle of the pattern needs to be rotated 90°. The current scale value of the pattern is 1 unit and the angle is 0°. Issuing the HATCHEDIT command prompts you to select the hatch pattern to edit. Clicking on the hatch pattern anywhere inside the object, as shown on the left in the following image, displays the Hatch Edit dialog box, as shown on the right in the following image.

With the dialog box displayed, click in the Scale edit box to change the scale from the current value of 1 unit to the new value of 3 units. Next, click in the Angle edit box and change the angle of the hatch pattern from the current value of 0° to the new value of 90°. This will increase the spacing between hatch lines and rotate the pattern 90° in the counterclockwise direction.

Figure 9–43

Clicking the OK button in the Hatch Edit dialog box returns you to the drawing editor and updates the hatch pattern to these changes, as shown in the following image. Be aware of the following rules for the HATCHEDIT command to function: it only works on an associative hatch pattern. If the pattern loses associativity through the use of the MIRROR command or if the hatch pattern is exploded, the HATCHEDIT command will cease to function.

 Note: Double-clicking on any associative hatch pattern will automatically launch the Hatch Edit dialog box. You could also pre-select the hatch pattern, right-click, and click on Hatch edit... to launch the Hatch Edit dialog box.

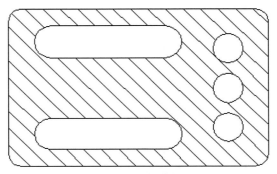

Figure 9–44

ADVANCED HATCHING TECHNIQUES

Advanced features are available inside the Hatch and Gradient dialog box. To view these features, click on the arrow located in the lower right corner of the dialog box, as shown in the following image. This will expand the Hatch and Gradient dialog box to include the advanced features, which include the following areas: Islands, Boundary Retention, Boundary Set, Gap Tolerance, and Inherit Options.

Figure 9–45

The illustration on the left in the following image shows three boundary styles and how they affect the crosshatching of Islands. The Normal boundary style is the default hatching method in which, when you select all objects within a window, the hatching begins with the outermost boundary, skips the next inside boundary, hatches the next innermost boundary, and so on. The Outer boundary style hatches only the outermost boundary of the object. The Ignore style ignores the default hatching methods of alternating hatching and hatches the entire object.

You have the option of retaining the hatch boundary by clicking in the check box for Retain boundaries, as shown on the right in the following image.

Figure 9–46

GAP TOLERANCE

Your success in hatching an object relies heavily on making sure the object being hatched is completely closed. This requirement can be relaxed with the addition of a Gap tolerance edit box, as shown on the left in the following image. In this example, a gap tolerance of .20 has been entered. However, the object in the middle has a gap of .40. Since the object's gap is larger than the tolerance value, an alert dialog box appears, as shown on the right in the following image. You must set the gap tolerance larger than the gap in the object or completely close the shape.

Figure 9–47

The illustration on the left in the following image shows a gap of .19. With a gap tolerance setting of .20, another alert dialog box warns you that the shape is not closed but if the gap setting is larger than the gap in the object, the object will be hatched. The results are shown on the right in the following image.

Figure 9–48

PRECISION HATCH PATTERN PLACEMENT

At times, you want to control where the hatch pattern begins inside a shape. In the following image, a brick pattern was applied without any special insertion base points. In this image, the pattern seems to be laid out based on the center of the rectangle. As the brick pattern reaches the edges of the rectangle, the pattern just ends. This occurs because the hatch pattern uses a current origin point to be constructed from. This control can be found in the main Hatch and Gradient dialog box under the Hatch origin area, as shown on the left in the following image.

Figure 9–49

To have more control over the origin of the hatch pattern, you can click on the Specific origin radio button, as shown on the left in the following image, and pick the button Click to set new origin. You can then select the new origin of the hatch pattern back in your drawing. The results are shown on the right in the following image, with the hatch pattern origin located at the lower left corner of the rectangle.

 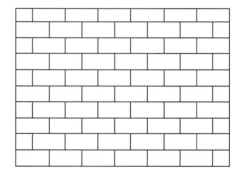

Figure 9–50

INHERIT HATCHING PROPERTIES

 Try It! – Open the drawing file 09_Hatch Inherit. The illustration on the left in the following image consists of a simple assembly drawing. At least three different hatch patterns are displayed to define the different parts of the assembly. Unfortunately, a segment of one of the parts was not hatched, and it is unclear what pattern, scale, and angle were used to place the pattern.

Whenever you are faced with this problem, click on the Inherit Properties button in the main Hatch and Gradient dialog box, as shown in the following image. Clicking on the pattern at "A" sets the pattern, scale, and angle to match the selected pattern.

To complete the hatch operation, click an internal point in the empty area. Right-click and select ENTER from the shortcut menu. The hatch pattern is placed in this area, and it matches that of the other patterns to identify the common parts, as shown on the right in the following image.

Figure 9–51

EXCLUDE HATCH OBJECTS FROM OBJECT SNAPS

At times, it is easy to mistakenly snap to the endpoint of a hatch pattern instead of the endpoint of the perimeter of an object. This usually happens when adding linear dimensions to a drawing, as shown on the left in the following image. To prevent this from happening, activate the Options dialog box, select the Drafting tab, and place a check in the box next to Ignore hatch objects, as shown on the right in the following image. Object Snap modes will now ignore endpoints and intersections of hatch patterns.

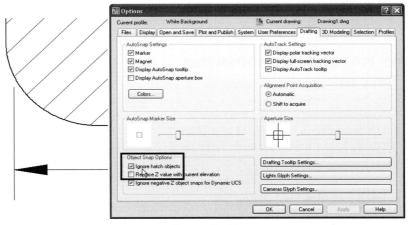

Figure 9–52

TUTORIAL EXERCISE: 09_COUPLER.DWG

Figure 9–53

Purpose
This tutorial is designed to use the MIRROR and BHATCH commands to convert 09_Coupler to a half section, as shown in the previous image.

System Settings
Since this drawing is provided on CD, edit an existing drawing called 09_Coupler. Follow the steps in this tutorial for converting the upper half of the object to a half section. All Units, Limits, Grid, and Snap values have been previously set.

Layers
The following layers are already created:

Name	Color	Linetype
Object	White	Continuous
Center	Yellow	Center
Hidden	Red	Hidden
Hatch2	Magenta	Continuous
Cutting Plane Line	Yellow	Dashed
Dimension	Yellow	Continuous

Suggested Commands
Convert one-half of the object to a section by erasing unnecessary hidden lines. Use the Layer Control box and change the remaining hidden lines to the Object layer. Issue the BHATCH command and use the ANSI31 hatch pattern to hatch the upper half of the 09_Coupler on the Hatch layer.

STEP 1

Use the MIRROR command to copy and flip the upper half of the Side view and form the lower half. When in Object Selection mode, use the Remove option to deselect the main centerline, hole centerlines, and hole. If these objects are included in the mirror operation, a duplicate copy of these objects will be created. See the following image.

Command: **MI** *(For MIRROR)*
Select objects: *(Pick a point at "A")*
Specify opposite corner: *(Pick a point at "B")*
Select objects: *(Hold down the SHIFT key and pick centerlines "C" and "D" to remove them from the selection set)*
Select objects: *(With the SHIFT key held down, pick a point at "E")*
Specify opposite corner: *(With the SHIFT key held down, pick a point at "F")*
Select objects: *(Press ENTER to continue)*
Specify first point of mirror line: *(Select the endpoint of the centerline near "C")*
Specify second point of mirror line: *(Select the endpoint of the centerline near "D")*
Delete source objects? [Yes/No] <N>: *(Press ENTER to perform the mirror operation)*

Figure 9–54

STEP 2

Begin converting the upper half of the Side view to a half section by using the ERASE command to remove any unnecessary hidden lines and centerlines from the view. See the following image.

 Command: **E** *(For ERASE)*

Select objects: *(Carefully select the hidden lines labeled "A," "B," "C," and "D")*
Select objects: *(Select the centerline labeled "E")*
Select objects: *(Press ENTER to execute the ERASE command)*

Figure 9–55

STEP 3

Since the remaining hidden lines actually represent object lines when shown in a full section, use the Layer Control box in the following image to convert all highlighted hidden lines from the Hidden layer to the Object layer.

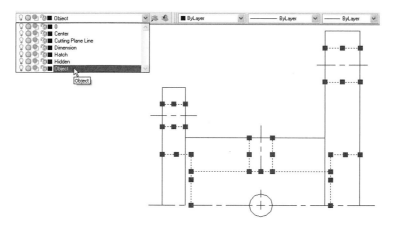

Figure 9–56

STEP 4

Remove unnecessary line segments from the upper half of the converted section using the TRIM command. Use the horizontal line at "A" as the cutting edge, and select the two vertical segments at "B" and "C" as the objects to trim. See the following image.

 Command: **TR** *(For TRIM)*
Current settings: Projection=UCS Edge=None
Select cutting edges ...
Select objects or <select all>: *(Select the horizontal line at "A")*
Select objects: *(Press ENTER to continue)*
Select object to trim or shift-select to extend or
    [Fence/Crossing/Project/Edge/eRase/Undo]: *(Select the vertical line at "B")*
Select object to trim or shift-select to extend or
    [Fence/Crossing/Project/Edge/eRase/Undo]: *(Select the vertical line at "C")*
Select object to trim or shift-select to extend or
    [Fence/Crossing/Project/Edge/eRase/Undo]: *(Press ENTER to exit this command)*

Figure 9–57

STEP 5

Make the Hatch layer the current layer. Then, use the BHATCH command to display the Hatch and Gradient dialog box shown on the left in the following image. Use the pattern "ANSI31" and keep all default settings. Click on the Add: Pick points button. When the drawing reappears, click inside areas "A," "B," "C," and "D," as shown on the right in the following image. When finished selecting these internal points, press ENTER to return to the Hatch and Gradient dialog box.

 Command: **BH** *(For BHATCH)*
(The Hatch and Gradient dialog box appears. Make changes to match the following image. When finished, click on the Add: Pick points < button.)
Select internal point: *(Pick a point at "A")*
Selecting everything...
Selecting everything visible...
Analyzing the selected data...
Analyzing internal islands...

Select internal point: *(Pick a point at "B")*

Analyzing internal islands...
Select internal point: *(Pick a point at "C")*
Analyzing internal islands...
Select internal point: *(Pick a point at "D")*
Analyzing internal islands...
Select internal point: *(Press ENTER to exit this area and return to the Boundary Hatch and Gradient dialog box)*

While in the Hatch and Gradient dialog box, click on the Preview button to view the results. If the hatch results are acceptable, right-click to place the hatch pattern.

Figure 9–58

STEP 6

The complete hatched view is shown in the following image.

Figure 9–59

TUTORIAL EXERCISE: 09_ELEVATION.DWG

Figure 9–60

Purpose
This tutorial is designed to use the Inherit Properties button of the Hatch and Gradient dialog box on 09_Elevation. Gradient hatch patterns will also be applied to this drawing for presentation purposes.

System Settings
Since this drawing is provided on CD, edit an existing drawing called 09_Elevation. Follow the steps in this tutorial for creating an elevation of the residence.

Layers
The following layers are already created:

Name	Color	Linetype
Object	White	Continuous
Center	Yellow	Center
Hidden	Red	Hidden
Hatch2	Magenta	Continuous
Cutting Plane Line	Yellow	Dashed
Dimension	Yellow	Continuous

Suggested Commands
Open up the existing drawing file 09_Elevation and notice the appearance of existing roof and brick patterns. Use the Inherit Properties feature of the Hatch and Gradient dialog box to transfer the roof and brick patterns to the irregular-shaped areas of this drawing. Next, a gradient pattern will be added to the upper part of the drawing for the purpose of creating a background. Another gradient pattern will be added to the lower portion of the drawing. Finally, the last gradient pattern will be added to the garage door windows. In this way, the gradient patterns will be used to enhance the elevation of the house.

STEP 1

Open the drawing 09_Elevation. Notice the appearance of a roof and brick hatch patterns in the following image. The roof and brick patterns need to be applied to the other irregular areas of this house elevation. First, make the Roof Hatch layer current through the Layer Control box.

Figure 9–61

STEP 2

Activate the Hatch and Gradient dialog box and click on the Inherit Properties button, as shown on the left in the following image. This button will allow you to click on an existing hatch pattern and have the pattern name, scale, and angle transfer to the proper fields in the dialog box. In other words, you will not have to figure out the name, scale, or angle of the pattern. The only requirement is that a hatch pattern already exists in the drawing. When you return to the drawing, shown on the right in the following image, notice the appearance of a glyph that is similar to the Match Properties icon. Click on the existing roof pattern in the figure.

Figure 9–62

STEP 3

After you pick the existing hatch pattern in the previous image, your cursor changes appearance again. Now pick internal points inside every irregular shape in the following image. Notice that each one highlights. When you're finished, press ENTER to return to the Hatch and Gradient dialog box. Click the Preview button to examine the results. If the hatch pattern appearance is correct, right-click to place the roof hatch patterns.

Figure 9–63

STEP 4

Next, make the Exterior Brick layer current using the Layer Control box. Then activate the Hatch and Gradient dialog box, click on the Inherit Properties button, pick the existing brick pattern, and click inside every irregular shape in the following image until each one highlights. Preview the hatch pattern and right-click if the results are correct.

Figure 9–64

STEP 5

To give the appearance that the hatch patterns are floating on the roof and wall, use the Layer Control box to turn off the two layers that control the boundaries of these patterns, namely, Roof Boundaries and Wall Boundaries. Your display should appear similar to the following image.

Figure 9–65

STEP 6

A number of gradient hatch patterns will now be applied to the outer portions of the elevation. First perform a ZOOM-EXTENTS. Then change the current layer to Gradient Background using the Layer Control box, as shown in the following image.

Command: **Z** *(For ZOOM)*
Specify corner of window, enter a scale factor (nX or nXP), or
[All/Center/Dynamic/Extents/Previous/Scale/Window/Object] <real time>:
 E *(For Extents)*

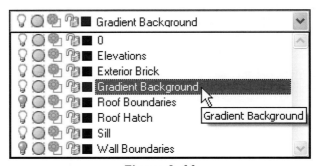

Figure 9–66

STEP 7

You will now apply a gradient hatch pattern to the upper portion of the elevation plan. Activate the Hatch and Gradient dialog box, click on the Gradient tab, and pick the pattern as shown on the left in the following image. Then click on the Add: Pick points button and click an internal point in the figure of the elevation. Once the boundary highlights, press ENTER, preview the hatch, and right-click to place the hatch pattern.

Figure 9–67

STEP 8

Apply another gradient hatch pattern to the lower portion of the elevation plan. Activate the Hatch and Gradient dialog box, click on the Gradient tab, and pick the pattern as shown on the left in the following image. Then click on the Add: Pick points button and click an internal point in the figure of the elevation. Once the boundary highlights, press ENTER, preview the hatch, and right-click to place the hatch pattern.

Figure 9–68

STEP 9

Apply gradient patterns to each of the garage windows, as shown in the following image, by using the Boundary Hatch command for each of the four windows.

Figure 9–69

STEP 10

If the boundary lines that define the elevation disappear, send the gradient patterns back using the Draw Order feature, as shown in the following image.

Figure 9–70

PROBLEMS FOR CHAPTER 9

PROBLEM 9–1

Center a three-view drawing and make the Front view a full section.

PROBLEM 9–2

Center two views within the work area and make one view a full section. Use correct drafting practices for the ribs.

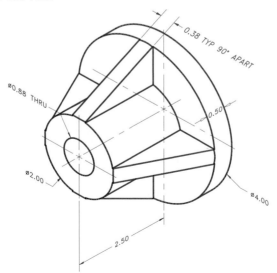

PROBLEM 9–3

Center two views within the work area and make one view a half section.

PROBLEM 9–4

Center two views within the work area and make one view a half section.

PROBLEM 9-5

Center three views within the work area and make one view an offset section.

PROBLEM 9-6

Center three views within the work area and make one view an offset section.

PROBLEM 9-7

Center three views within the work area and make one view an offset section.

PROBLEM 9-8

Center two views within the work area and make one view a half section.

PROBLEM 9-9

Center the required views within the work area and make one view a broken-out section to illustrate the complicated interior area.

PROBLEM 9-10

Create the required views within the work area and add a full removed section called A-A.

PROBLEM 9–11

Center the required views within the work area and add a full removed section called A-A.

Directions for Problems 9–12 through 9–14

Center the required views within the work area. Leave a 1″ or 25-mm space between the views. Make one view a full section to fully illustrate the object. Consult your instructor if you need to add dimensions.

PROBLEM 9–12

PROBLEM 9–13

PROBLEM 9–14

CHAPTER 10

Auxiliary Views

During the discussion of multiview drawings, we discovered that you need to draw enough views of an object to accurately describe it. In most cases, this requires a Front, Top, and Right Side view. Sometimes additional views are required, such as Left Side, Bottom, and Back views, to show features not visible in the three primary views. Other special views, such as sections, are created to expose interior details for better clarity. Sometimes all these views are still not enough to describe the object, especially when features are located on an inclined surface. To produce a view perpendicular to this inclined surface, an auxiliary view is drawn. This chapter will describe where auxiliary views are used and how they are projected from one view to another. A tutorial exercise is presented to show the steps in the construction of an auxiliary view.

AUXILIARY VIEW BASICS

The illustration on the left in the following image presents interesting results if constructed as a multiview drawing or orthographic projection.

The illustration on the right in the following image should be quite familiar; it represents the standard glass box with objects located in the center. Again, the purpose of this box is to prove how orthographic views are organized and laid out. This illustration is no different. First the primary views, Front, Top, and Right Side, are projected from the object to intersect perpendicularly with the glass plane. Under normal circumstances, this procedure would satisfy most multiview drawing cases. Remember, only those views necessary to describe the object are drawn. However, upon closer inspection, we notice that the object in the Front view consists of an angle forming an inclined surface.

Figure 10–1

When you lay out the Front, Top, and Right Side views, a problem occurs. The Front view shows the basic shape of the object, the angle of the inclined surface, as shown in the following image. The Top view shows the true size and shape of the surface formed by the circle and arc. The Right Side view shows the true thickness of the hole from information found in the Top view. However, there does not exist a true size and shape of the features found in the inclined surface at "A." We see the hexagonal hole going through the object in the Top and Right Side views. These views, however, show the detail not actual size; instead, the views are foreshortened. For that matter, the entire inclined surface is foreshortened in all views. Dimensioning the size and location of the hexagonal hole would be challenging with these three views. This is one case in which the Front, Top, and Right Side views are not enough to describe the object. An additional view, or auxiliary view, is used to display the true shape of surfaces along an incline.

Figure 10–2

To demonstrate the creation of an auxiliary view, let's create another glass box, this time with an inclined plane on one side. This plane is always parallel to the inclined surface of the object. Instead of just the Front, Top, and Right Side views being projected, the geometry describing the features along the inclined surface is projected to the auxiliary plane, as shown on the left in the following image.

As in multiview projection, the edges of the glass box are unfolded, with the edges of the Front view plane as the pivot, as shown on the right in the following image.

Figure 10–3

All planes are extended perpendicular to the Front view, where the rotation stops. The result is the organization of the multiview drawing, complete with an auxiliary viewing plane, as shown on the left in the following image.

The illustration on the right in the following image represents the final layout, complete with auxiliary view. This figure shows the auxiliary being formed as a result of the inclined surface present in the Front view. An auxiliary view may be created in relation to any inclined surface located in any view. Also, the figure displays circles and arcs in the Top view that appear as ellipses in the auxiliary view.

Figure 10–4

It is usually not required that elliptical shapes be drawn in one view where the feature is shown at true size and shape in another. The resulting view minus these elliptical features is called a partial view, used extensively in auxiliary views. An example of the Top view converted to a partial view is shown on the left in the following image.

A few rules to follow when constructing auxiliary views are displayed pictorially on the right in the following image. First, the auxiliary plane is always constructed parallel to the inclined surface. Once this is established, visible as well as hidden features are projected from the incline to the auxiliary view. These projection lines are always drawn perpendicular to the inclined surface and perpendicular to the auxiliary view.

Figure 10–5

CONSTRUCTING AN AUXILIARY VIEW

Try It! – Open the drawing file 10_Aux Basics. The following image is a basic multiview drawing consisting of Front, Top, and Right Side views. The inclined surface in the Front view is displayed in the Top and Right Side views; however, the surface appears foreshortened in both adjacent views. An auxiliary view of the incline needs to be made to show its true size and shape. Follow the next series of images that illustrate one suggested method for projecting to find auxiliary views.

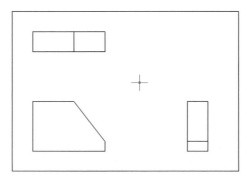

Figure 10–6

To assist with the projection process, the cursor needs to be rotated to a position perpendicular to the inclined surface. This is accomplished through the SNAP command and the Rotate option. The rotation angle is established by picking points "A" and "B," as shown on the left in the following image.

Command: **SN** *(For SNAP)*
Specify snap spacing or [ON/OFF/Aspect/Rotate/Style/Type] <0.5000>: **R** *(For Rotate)*
Specify base point <0.0000,0.0000>: *(Select the endpoint of the line at "A")*
Specify rotation angle <0>: *(Select the endpoint of the line at "B")*

The illustration on the right in the following image shows the result of rotating the cursor. This operation has no effect on the already-existing views; however, the cursor is now rotated in relation to the incline located in the Front view. Lines are easily drawn perpendicular to the incline with ORTHO mode on. This will draw lines in relation to the current cursor.

While the cursor is rotated, it also snaps from one predefined increment to another. To prevent the cursor from bouncing around from one snap point to another, turn SNAP off by clicking in the status bar located at the bottom of the display screen. This action will keep the cursor in the rotated position.

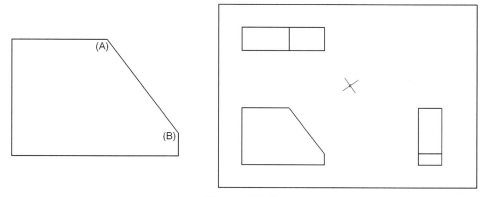

Figure 10–7

Make the Objects layer the current layer. Use the OFFSET command to construct a reference line at a specified distance from the incline in the Front view, as shown on the left in the following image. This reference line becomes the start for the auxiliary view.

Use the LENGTHEN command to extend the two endpoints of the previous line, as shown on the right in the following image. The exact distances are not critical; however, the line should be long enough to accept projector lines from the Front view.

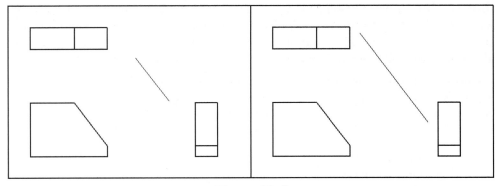

Figure 10–8

Use the OFFSET command to copy the auxiliary reference line to a location that indicates the thickness of the object, as shown on the left in the following image. This distance may be retrieved from the depth of the Top or Right Side views because they both show the depth measurement of the object.

Use the LINE command and connect each intersection on the Front view perpendicularly to the outer line on the auxiliary view, as shown on the right in the following image. Draw the lines starting with the OSNAP-Intersect mode and ending with the OSNAP-Perpend mode.

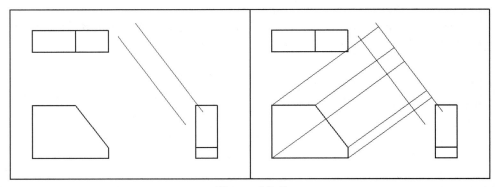

Figure 10–9

Before editing the auxiliary view, analyze the drawing to see whether any corners in the Front view are to be represented as hidden lines in the auxiliary view, as shown on the left in the following image. It turns out that the lower left corner of the Front view is hidden in the auxiliary view. Use the Layer Control box to convert this projection line to a layer containing hidden lines.

Use the TRIM command to partially delete all projection lines and corners of the auxiliary view. Another method would be to use the FILLET command set to a radius of 0. Selecting two lines in a corner automatically trims the excess lines away, as shown on the right in the following image.

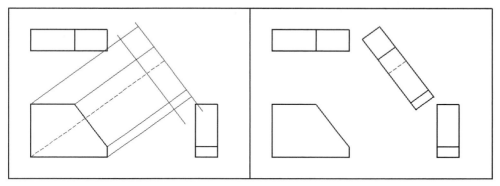

Figure 10–10

The result is a multiview drawing complete with auxiliary view displaying the true size and shape of the inclined surface, as shown on the left in the following image. For dimensioning purposes, use the SNAP command and set the cursor appearance back to normal, as shown on the right in the following image.

Command: **SN** *(For SNAP)*
Specify snap spacing or [ON/OFF/Aspect/Rotate/Style/Type] <0.5000>: **R** *(For Rotate)*
Specify base point <Current default>: **0,0**
Specify rotation angle <Current default>: **0**

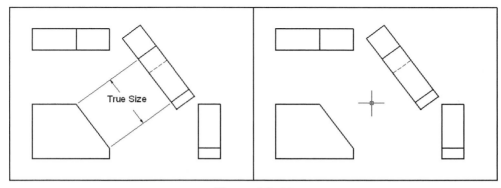

Figure 10–11

TUTORIAL EXERCISE: 10_BRACKET.DWG

Figure 10–12

Purpose
This tutorial is designed to allow you to construct an auxiliary view of the inclined surface for the bracket shown in the previous image.

System Settings
Since this drawing is provided on CD, edit an existing drawing called 10_Bracket. Follow the steps in this tutorial for the creation of an auxiliary view. All Units, Limits, Grid, and Snap values have been previously set.

Layers
The following layers have already been created with the following format:

Name	Color	Linetype
CEN	Yellow	Center
DIM	Yellow	Continuous
HID	Red	Hidden
OBJ	Cyan	Continuous

Suggested Commands
Begin this tutorial by using the OFFSET command to construct a series of lines parallel to the inclined surface containing the auxiliary view. Next construct lines perpendicular to the inclined surface. Use the CIRCLE command to begin laying out features that lie in the auxiliary view. Use ARRAY to copy the circle in a rectangular pattern. Add centerlines using the DIMCENTER command. Insert a predefined view called Top. A three-view drawing consisting of Front, Top, and auxiliary views is completed.

STEP 1

Before you begin, understand that an auxiliary view will be taken from a point of view illustrated on the left in the following image. This direction of sight is always perpendicular to the inclined surface. This perpendicular direction ensures that the auxiliary view of the inclined surface will be of true size and shape. Begin this tutorial by checking to see that Ortho mode is turned off for the next few steps by clicking on the ORTHO button located in the status bar of the display screen. Also, make the OBJ layer current. Use the VIEW command (or V for short) to activate the View Manager dialog box, as shown on the left in the following image, and set a previously saved view called FRONT current.

Figure 10–13

STEP 2

Use the SNAP command to rotate the grid perpendicular to the inclined surface. For the base point, identify the endpoint of the line at "A," as shown on the right in the following image. For the rotation angle, mark a point at the endpoint of the line at "B." The snap angle should change along with the appearance of the standard cursor. Next, activate the View Manager dialog box, as shown on the left in the following image, and set the view called OVERALL current.

Command: **SN** *(For SNAP)*
Specify snap spacing or [ON/OFF/Aspect/Rotate/Style/Type] <0.50>: **R** *(For Rotate)*
Specify base point <0.00,0.00>: *(Select the endpoint at "A")*
Specify rotation angle <0>: *(Select the endpoint at "B")*

Figure 10–14

STEP 3

To prevent the cursor from bouncing around from one snap point to another, turn SNAP mode off by clicking the SNAP button located in the status bar. Begin the construction of the auxiliary view by using the OFFSET command to copy a line parallel to the inclined line located in the Front view, as shown on the left in the following image. Use an offset distance of 8.50, pick line "A" as the object to offset, and pick near "B" as the side to perform the offset. Then use the OFFSET command again to offset line "B" to the side at "C" at a distance of 6.00 units.

The previous two lines formed using the OFFSET command define the depth of the auxiliary view. To determine the width of the auxiliary view, project two lines from the endpoints of the Front view at "D" and "F," as shown on the right in the following image. To better assist with this operation, turn ORTHO mode on. The endpoints of both lines are illustrated on the right at "E" and "G." The location of the endpoints of the lines is not critical. This will be fixed in the next step as these lines are cornered.

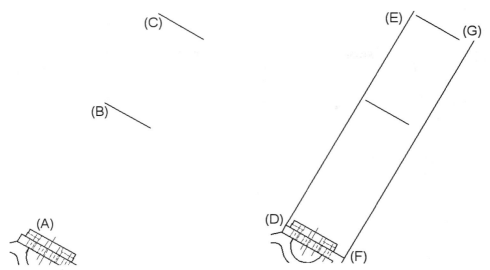

Figure 10–15

STEP 4

Use the ZOOM-Window option to magnify the display of the auxiliary view similar to the illustration on the left in the following image. Then use the Multiple option of the FILLET command to create four corners using the lines labeled "A" through "D." The fillet radius should be set to 0 in order to form the corners. The results are displayed on the right in the following image.

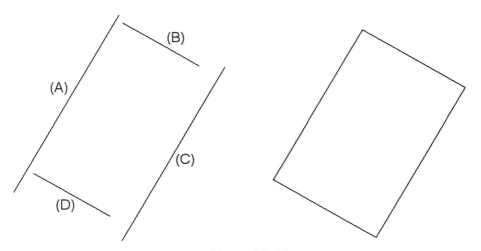

Figure 10–16

STEP 5

When you are finished with the filleting operation, activate the View Manager dialog box, as shown on the left in the following image, and click on the New button. When the New View dialog box appears, as shown on the right in the following image, verify that the view will be defined based on the Current display; a radio button should be active for this mode. Then save the display to a new name, AUX.

Figure 10–17

STEP 6

Use the ZOOM-Previous option or other ZOOM mode to demagnify the screen back to the original display. Then draw a line from the endpoint of the centerline in the Front view at "A" to a point past the auxiliary view at "B," as shown on the left in the following image. Check to see that ORTHO mode is turned on. This is one of the construction lines that will be used for finding the centers of the circular features located in the auxiliary view.

Then use the OFFSET command to create the other construction line, as shown on the right in the following image. Use an offset distance of 3.00, pick line "C" as the object to offset, and pick a point near "D."

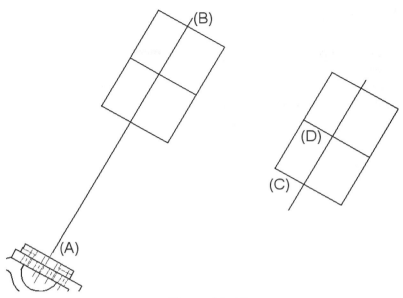

Figure 10–18

STEP 7

Activate the View Manager dialog box and set the view AUX current, as shown on the left in the following image. Then draw two circles of diameters 3.00 and 1.50 from the center at "A," as shown on the right in the following image, using the CIRCLE command. For the center of the second circle, you can use the @ option to pick up the previous point that was the center of the first circle.

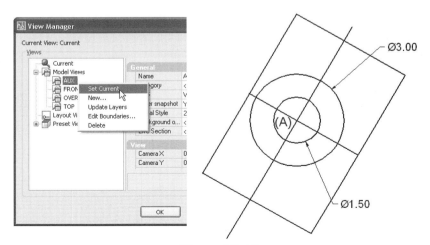

Figure 10–19

STEP 8

Use the OFFSET command to offset the centerline the distance of .25 units, as shown on the left in the following image. Perform this operation on both sides of the centerline. Both offset lines form the width of the .50 slot. Then use the TRIM command to trim away portions of the offset lines, using the two circles as cutting edges. When finished, your display should resemble the illustration on the right in the following image.

 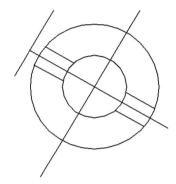

Figure 10–20

STEP 9

Use the ERASE command to delete the two lines at "A" and "B," as shown on the left in the following image. Standard centerlines will be placed here later, marking the center of both circles.

Two more construction lines need to be made. These lines will identify the center of the small .37-diameter circle. Use the OFFSET command to create offsets for the lines illustrated on the right in the following image. Offset distances of .75 and 1.00 are used to perform this task. Also be careful to offset the correct line at the specified distance.

 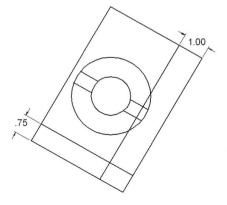

Figure 10–21

STEP 10

Draw a circle of .37 units in diameter from the intersection of the two lines created in the last OFFSET command, as shown on the left in the following image. Then use the ERASE command to delete the two lines labeled "A" and "B." A standard center marker will be placed later at the center of this circle.

For the next phase of this step, make the CEN layer current through the Layer Control Box. The DIMCENTER command (or DCE for short) will be used to create the center mark for the .37-diameter circle. Type DCE at the command prompt and pick the edge of the circle at "C," as shown in the middle of the following image, to place this center mark.

Next, use the ROTATE command to rotate the center marker parallel to the edges of the auxiliary view. Turn ORTHO mode on, activate the ROTATE command, and pick the center lines. For the base point of the rotation, pick a point at the center of the circle at "D" using OSNAP Center or Intersection mode. For the rotation angle, pick a point near "E." With ORTHO on, the center lines should rotate parallel to the edge of the auxiliary, as shown on the right in the following image.

Figure 10–22

STEP 11

Since the remaining seven holes form a rectangular pattern, use the Array dialog box, as shown on the left in the following image, and perform a rectangular array based on the .37-diameter circle and center marks. The number of rows is two and the number of columns is four. The distance between rows is −4.50 units and between columns is 0.75 units. Set the angle of the array to 150 degrees. Also be sure to select the circle and center lines as the objects to array. Performing a preview will result in the array displayed on the right in the following image.

Figure 10–23

STEP 12

Use the FILLET command set to a radius of .75 to place a radius along the two corners of the auxiliary, as shown on the left in the following image.

Place a center mark in the center of the two large circles using the DIMCEN (or DCE for short) command. When prompted to select the arc or circle, pick the edge of the large circle. The results are displayed in the middle of the following image.

Then rotate the center mark using the ROTATE command. ORTHO mode should be turned on to assist in performing this operation. The results are displayed on the right in the following image.

Figure 10–24

STEP 13

Activate the View Manager dialog box, as shown in the following image, and set the view named OVERALL current. This should return your display to view all views of the Bracket.

Figure 10–25

STEP 14

Complete the drawing of the bracket by activating the Insert dialog box as shown in the following image and inserting an existing block called TOP into the drawing. This block represents the complete Top view of the drawing. Use an insertion point of 0,0 for placing this view in the drawing. Your display should appear similar to the following image.

One final step is to return the snap to its original orthographic form using the Rotate option of the SNAP command. Use a base point of 0,0 and a rotation angle of 0 degrees.

Command: **SN** *(For SNAP)*
Specify snap spacing or [ON/OFF/Aspect/Rotate/Style/Type] <0.50>: **R** *(For Rotate)*
Specify base point <17.33,5.85>: **0,0**
Specify rotation angle <330>: **0**

Figure 10–26

PROBLEMS FOR CHAPTER 10

Directions for Problems 10–1 through 10–10

Draw the required views to fully illustrate each object. Be sure to include an auxiliary view.

PROBLEM 10–1

PROBLEM 10–2

PROBLEM 10–3

PROBLEM 10–4

PROBLEM 10-5

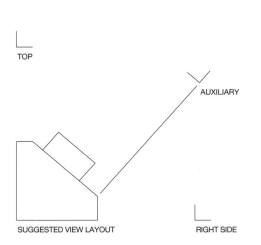

TOP

AUXILIARY

SUGGESTED VIEW LAYOUT

RIGHT SIDE

PROBLEM 10-6

SUGGESTED VIEW LAYOUT

PROBLEM 10–7

SUGGESTED VIEW LAYOUT

PROBLEM 10–8

PROBLEM 10-9

PROBLEM 10-10

Directions for Problems 10–11 through 10–13

Create the missing Auxiliary view through projection methods.

PROBLEM 10–11

PROBLEM 10–12

PROBLEM 10-13

CHAPTER 11

Adding Dimensions to Your Drawing

Once orthographic views have been laid out, a design is not ready for the production line until dimensions describing the width, height, and depth of the object are added to the drawing. However, these numbers must be added in a certain organized fashion; otherwise, the drawing becomes difficult to read. That may lead to confusion and the possible production of a part that is incorrect according to the original design. This chapter begins with an introduction to the proper placement of dimensions in a drawing. The chapter continues with a discussion of the commands used in AutoCAD for creating dimensions. These commands include Linear, Aligned, Ordinate, Radius, Diameter, Angular, Quick, Baseline, and Continue dimensions. A section in this chapter deals with the placement of geometric dimensioning and tolerancing symbols. A short segment on the effects grips have on dimensions is included in this chapter.

DIMENSIONING BASICS

Before discussing the components of a dimension, remember that object lines (at "A") continue to be the thickest lines of a drawing, with the exception of border or title blocks (see the following image). To promote contrasting lines, dimensions become visible, yet thin, lines. The heart of a dimension is the dimension line (at "B"), which is easily identified by the location of arrow terminators at both ends (at "D"). In mechanical cases, the dimension line is broken in the middle, which provides an excellent location for the dimension text (at "E"). For architectural applications, dimension text is usually placed above an unbroken dimension line. The extremities of the dimension lines are limited by the placement of lines that act as stops for the arrow terminators. These lines, called extension lines (at "C"), begin close to the object without touching the object. Extension lines will be discussed in greater detail in the pages that follow. For placing diameter and radius dimensions, a leader line consisting of an inclined line with a short horizontal shoulder is used (at "F"). Other applications of leader lines are for adding notes to drawings.

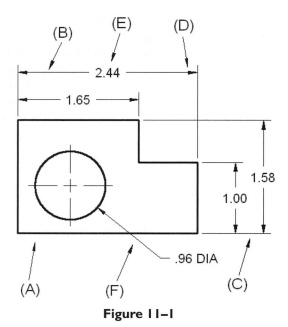

Figure 11–1

When you place dimensions in a drawing, provide a spacing of at least 0.38 units between the first dimension line and the object being dimensioned (at "A" in the following image). If you're placing stacked or baseline dimensions, provide a minimum spacing of at least 0.25 units between the first and second dimension line (at "B") or any other dimension lines placed thereafter. This will prevent dimensions from being placed too close to each other.

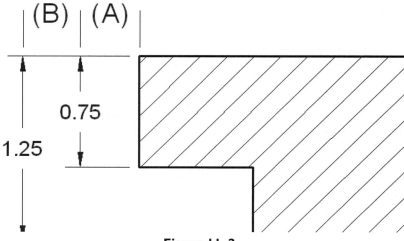

Figure 11–2

It is recommended that extensions never touch the object being dimensioned and that they begin between approximately 0.03 and 0.06 units away from the object (at "A" in the following image). As dimension lines are added, extension lines should extend no further than 0.12 units beyond the arrow or any other terminator (at "B"). The height of dimension text is usually 0.125 units (at "C"). This value also applies to notes placed on objects with leader lines. Certain standards may require a taller lettering height. Become familiar with your office's standard practices, which may deviate from these recommended values.

Figure 11–3

PLACEMENT OF DIMENSIONS

When placing multiple dimensions on one side of an object, place the shorter dimension closest to the object, followed by the next larger dimension, as shown on the left in the following image. When you are placing multiple horizontal and vertical dimensions involving extension lines that cross other extension lines, do not place gaps in the extension lines at their intersection points.

It is acceptable for extension lines to cross each other, but it is considered unacceptable practice for extension lines to cross dimension lines, as in the example shown on the right in the following image. The shorter dimension is placed closest to the object, followed by the next larger dimension.

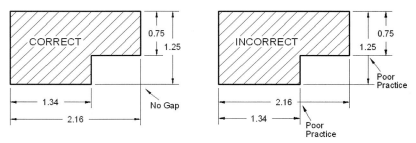

Figure 11–4

Dimensions should be added to the outside areas of the view being dimensioned, as shown on the left in the following image. It is considered poor practice to dimension inside an object when there is sufficient room to place dimensions on the outside; however, there may be exceptions to this rule. It is also considered poor practice to cross dimension lines, because this may render the drawing confusing and possibly result in the inaccurate interpretation of the drawing. It is also considered poor practice to superimpose extension lines over object lines. See the illustration on the right in the following image.

Figure 11–5

PLACING EXTENSION LINES

Because two extension lines may intersect without providing a gap, so also may extension lines and object lines intersect with each other without the need for a gap between them. The practice is the same when centerlines extend beyond the object without gapping, as shown on the left in the following image.

On the right in the following image, the gaps in the extension lines may appear acceptable; however, in a very complex drawing, gaps in extension lines would render a drawing confusing. Draw extension lines as continuous lines without providing breaks in the lines.

Figure 11–6

In the same manner, when centerlines are used as extension lines for dimension purposes, no gap is provided at the intersection of the centerline and the object, as shown on the left in the following image. As with extension lines, the centerline should extend no further than 0.125 units past the arrow terminator.

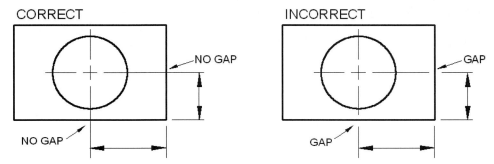

Figure 11–7

GROUPING DIMENSIONS

To promote ease of reading and interpretation, group dimensions whenever possible, as shown on the left in the following image. In addition to making the drawing and dimensions easier to read, this promotes good organizational skills and techniques. As in previous examples, always place the shorter dimensions closest to the drawing, followed by any larger or overall dimensions.

Avoid placing dimensions to an object line substituting for an extension line, as shown on the right in the following image. The drawing is more difficult to follow, with the dimensions placed at different levels instead of being grouped. In some cases, however, this practice of dimensioning is unavoidable.

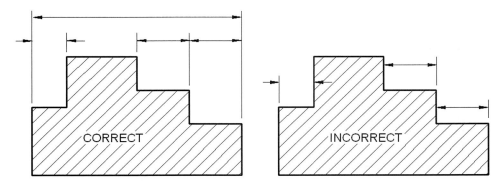

Figure 11–8

DIMENSIONING TO VISIBLE FEATURES

On the left in the following image, the object is dimensioned correctly. However, the problem is that dimensions have been added to hidden features. Although there are always exceptions, try to avoid dimensioning to any hidden surfaces or features.

The object shown on the right in the following image has been converted to a full section. Surfaces that were previously hidden are now exposed. This example illustrates a better way to dimension details that were previously hidden.

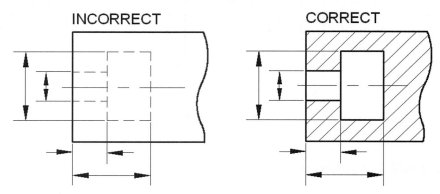

Figure 11–9

DIMENSIONING TO CENTERLINES

Centerlines are used to identify the center of circular features, as in "A" in the following image. You can also use centerlines to indicate an axis of symmetry, as in "B." Here, the centerline, consisting of a short dash flanked by two long dashes, signifies that the feature is circular in shape and form. Centerlines may take the place of extension lines when dimensions are placed in drawings.

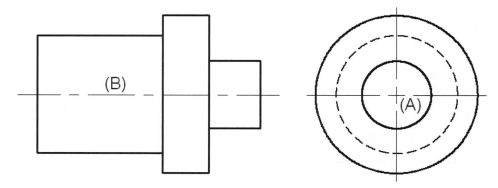

Figure 11–10

The following image on the left represents the top view of a U-shaped object with two holes placed at the base of the U. It also represents the correct way of utilizing centerlines as extension lines when dimensioning to holes. What makes this example correct is the rule of always dimensioning to visible features. The example in this image uses centerlines to dimension to holes that appear as circles.

The following image on the right represents the front view of the U-shaped object. The hidden lines display the circular holes passing through the object along with centerlines. Centerlines are being used as extension lines for dimensioning purposes; however, it is considered poor practice to dimension to hidden features or surfaces. Always attempt to dimension to a view where the features are visible before dimensioning to hidden areas.

Figure 11-11

CONTINUE AND BASELINE DIMENSIONS

Additional dimensioning techniques are illustrated in the following image. In the illustration on the left, the dimensions are arranged alongside each other. This technique is commonly referred to as continue or chain dimensions. In the illustration on the right, a series of dimensions can be placed by stacking method. Also, all dimensions begin from the left edge of the part. This technique is called baseline dimensioning.

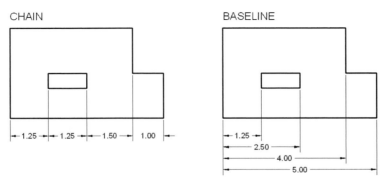

Figure 11-12

DIMENSIONING CYLINDERS

When dimensioning the diameters of cylinders, it is preferred to add these dimensions to the noncircular view, as shown on the right in the following image. Notice also that these dimensions are placed between views for clarity. While the illustration on the left is dimensionally correct, it is difficult to interpret the diameter dimensions placed in the Top view. Also, radius dimensions should never be used to dimension full diameters, since measurement tools such as micrometers and calipers are designed to check diameters.

Figure 11–13

DIMENSIONING SYSTEMS

A typical example of placing dimensions using AutoCAD is illustrated on the left in the following image. Here, all text items are aligned horizontally. This applies to all vertical, aligned, angular, and diameter dimensions. When all dimension text can be read from one direction, this is the Unidirectional Dimensioning System. By default, all AutoCAD settings are set to dimension in the Unidirectional System.

The same object is also illustrated on the right in the following image. Notice that all horizontal dimensions have the text positioned in the horizontal direction, as in the unidirectional system. However, vertical and aligned dimension text is rotated or aligned with the direction being dimensioned. This is the most notable feature of the Aligned Dimensioning System. Text along vertical dimensions is rotated in such a way that the drawing must be read from the right.

Figure 11–14

ORDINATE DIMENSIONS

A different type of dimensioning system is illustrated in the following image. Notice how all hole locations are designated by two numbers. Actually, these sets of numbers are coordinates. One of the numbers measures a distance in the X direction, while the other number measures a distance in the Y direction. No arrows or dimension lines are used. This type of dimensioning technique is referred to as Ordinate dimensioning. All hole location coordinates reference a known 0,0 location in the drawing. As shown in the following image, this type of dimensioning is well suited for locating hole positions.

Figure 11–15

ARCHITECTURAL DIMENSIONS

The following image illustrates another type of dimensioning technique. Architectural dimensions usually have their text drawn directly above the dimension line. Instead of an arrow, architectural dimensions are terminated with a tick mark. Architectural dimension text for vertical dimensions is read from the left side; this makes all vertical architectural dimensions aligned or parallel to the dimension line.

Figure 11–16

REPETITIVE DIMENSIONS

Throughout this chapter, numerous methods of dimensioning are discussed, supported by many examples. Just as it was important in multiview projection to draw only those views that accurately describe the object, so also is it important to dimension these views. In this manner, the actual production of the part may start. However, take care when placing dimensions. It takes planning to better place a dimension. The problem with the illustration on the left in the following image is that, even though the views are correct and the dimensions call out the overall sizes of the object, there are too many cases in which dimensions are duplicated. Once a feature has been dimensioned, such as the overall width of 3.50 units shown between the Front and Top view, this dimension does not need to be repeated in the Top view. This is the purpose of understanding the relationship between views and what dimensions the views have in common with each other. Adding unnecessary dimensions also makes the drawing very busy and cluttered in addition to being very confusing to read. Also if a size changes in production and a dimension needs to change on the drawings, more instances of the dimension must be found and corrected. Compare the illustration on the left with that on the right in the following image, which shows only those dimensions needed to describe the size of the object.

Figure 11-17

METHODS OF CHOOSING DIMENSION COMMANDS

The Dimension toolbar, illustrated in the following image, makes the selection of dimensioning commands easier.

The Dimension pull-down menu, also illustrated in the following image, is an additional area used for selecting the more commonly used dimension commands. Another way of activating dimension commands is through the keyboard. These commands tend to get long; the DIMLINEAR command is one example. To spare you the effort of entering the entire command, all dimension commands have been abbreviated to three letters. For example, DLI is the alias for the DIMLINEAR command.

Figure 11-18

Study the following table for a brief description of each tool.

Button	Tool	Shortcut	Function
	Linear	DLI	Creates a horizontal, vertical, or rotated dimension
	Aligned	DAL	Creates a linear dimension parallel to an object
	Arc Length	DAR	Dimensions the total length of an arc
	Ordinate	DOR	Creates an ordinate dimension based on the current position of the User Coordinate System (UCS)
	Radius	DRA	Creates a radius dimension
	Jogged	DJO	Creates a jogged dimension
	Diameter	DIA	Creates a diameter dimension
	Angular	DAN	Creates an angular dimension
	Quick Dimension	QDIM	Creates a quick dimension
	Continue	DCO	Creates a continued dimension
	Baseline	DBA	Creates a baseline dimension
	Quick Leader	LEA	Creates a leader line
	Tolerance	TOL	Activates the Tolerance dialog box
	Center Mark	DCE	Places a center mark inside a circle or an arc
	Edit	DED	Used to edit a dimension
	Text Edit	—	Used to edit the text of a dimension
	Update	—	Used for updating existing dimensions to changes in the current dimension style
	Dimstyle Dialog	D	Activates the Dimension Style Manager dialog box

BASIC DIMENSION COMMANDS

LINEAR DIMENSIONS

The Linear Dimensioning mode generates either a horizontal or vertical dimension depending on the location of the dimension. The following prompts illustrate the generation of a horizontal dimension with the DIMLINEAR command. Notice that identifying the dimension line location at "C" in the following image automatically generates a horizontal dimension.

 Try It! – Open the drawing file 11_Dim Linear1. Verify that OSNAP is on and set to Endpoint. Use the following command sequence and image for performing this dimensioning task.

Command: **DLI** (For DIMLINEAR)
Specify first extension line origin or <select object>: (Select the endpoint of the line at "A")
Specify second extension line origin: (Select the other endpoint of the line at "B")
Specify dimension line location or [Mtext/Text/Angle/Horizontal/Vertical/Rotated]: (Pick a point near "C" to locate the dimension)
Dimension text = 2.00

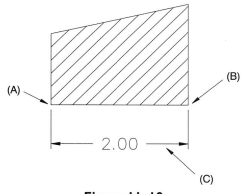

Figure 11–19

The linear dimensioning command is also used to generate vertical dimensions. The following prompts illustrate the generation of a vertical dimension with the DIMLINEAR command. Notice that identifying the dimension line location at "C" in the following image automatically generates a vertical dimension.

Try It! – Open the drawing file 11_Dim Linear2. Verify that OSNAP is on and set to Endpoint. Use the following command sequence and image for performing this dimensioning task.

Command: **DLI** *(For DIMLINEAR)*
Specify first extension line origin or <select object>: *(Select the endpoint of the line at "A")*
Specify second extension line origin: *(Select the endpoint of the line at "B")*
Specify dimension line location or [Mtext/Text/Angle/Horizontal/Vertical/Rotated]:
   (Pick a point near "C" to locate the dimension)
Dimension text = 2.10

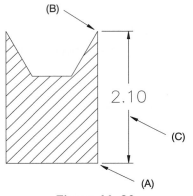

Figure 11–20

Rather than select two separate endpoints to dimension to, certain situations allow you to press ENTER and select the object (in this case, the line). This selects the two endpoints and prompts you for the dimension location. The completed dimension is illustrated in the following image.

 Try It! – Open the drawing file 11_Dim Linear3. Use the following command sequence and image for performing this dimensioning task.

Command: **DLI** *(For DIMLINEAR)*
Specify first extension line origin or <select object>: *(Press ENTER to select an object)*
Select object to dimension: *(Select the line at "A")*
Specify dimension line location or [Mtext/Text/Angle/Horizontal/Vertical/Rotated]:
   (Pick a point near "B" to locate the dimension)
Dimension text = 5.5523

Figure 11-21

ALIGNED DIMENSIONS

The Aligned Dimensioning mode generates a dimension line parallel to the distance specified by the location of two extension line origins, as shown in the following image.

 Try It! – Open the drawing file 11_Dim Aligned. Verify that OSNAP is on and set to Endpoint. The following prompts and image illustrate the creation of an aligned dimension with the DIMALIGNED command.

Command: **DAL** *(For DIMALIGNED)*
Specify first extension line origin or <select object>: *(Select the endpoint of the line at "A")*
Specify second extension line origin: *(Select the endpoint of the line at "B")*
Specify dimension line location or [Mtext/Text/Angle]: *(Pick a point at "C" to locate the dimension)*
Dimension text = 5.6525

Figure 11-22

ROTATED LINEAR DIMENSIONS

This Linear Dimensioning mode can also create a dimension line rotated at a specific angle, as shown in the following image. The following prompts illustrate the generation of a rotated dimension with the DIMLINEAR command and a known angle of 45°. If you do not know the angle, you could easily establish the angle by clicking the endpoints at "A" and "D" in the following image when prompted to "Specify angle of dimension line <0>."

 Try It! – Open the drawing file 11_Dim Rotated. Verify that OSNAP is on and set to Endpoint. Use the following command sequence and image for creating a rotated dimension.

 Command: **DLI** (For DIMLINEAR)

Specify first extension line origin or <select object>: (Select the endpoint of the line at "A")
Specify second extension line origin: (Select the endpoint of the line at "B")
Specify dimension line location or [Mtext/Text/Angle/Horizontal/Vertical/Rotated]: **R** (For Rotated)
Specify angle of dimension line <0>: **45**
Specify dimension line location or [Mtext/Text/Angle/Horizontal/Vertical/Rotated]: (Pick a point at "C" to locate the dimension)
Dimension text = 2.8284

Figure 11–23

CONTINUE DIMENSIONS

The power of grouping dimensions for ease of reading has already been explained. The following image shows yet another feature of dimensioning in AutoCAD: the practice of using continue dimensions. With one dimension already placed with the DIMLINEAR command, you issue the DIMCONTINUE command, which prompts you for the second extension line location. Picking the second extension line location strings the dimensions next to each other, or continues the dimension.

 Try It! – Open the drawing file 11_Dim Continue. Verify that Running OSNAP is on and set to Endpoint. Use the following command sequence and image for creating continue dimensions.

Command: **DLI** *(For DIMLINEAR)*
Specify first extension line origin or <select object>: *(Select the endpoint of the line at "A")*
Specify second extension line origin: *(Select the endpoint of the line at "B")*
Specify dimension line location or [Mtext/Text/Angle/Horizontal/Vertical/Rotated]: *(Locate the 1.75 horizontal dimension)*
Dimension text = 1.75

Command: **DCO** *(For DIMCONTINUE)*
Specify a second extension line origin or [Undo/Select] <Select>: *(Select the endpoint of the line at "C")*
Dimension text = 1.25
Specify a second extension line origin or [Undo/Select] <Select>: *(Select the endpoint of the line at "D")*
Dimension text = 1.50
Specify a second extension line origin or [Undo/Select] <Select>: *(Select the endpoint of the line at "E")*
Dimension text = 1.00
Specify a second extension line origin or [Undo/Select] <Select>: *(Press ENTER when finished)*
Select continued dimension: *(Press ENTER to exit this command)*

Figure 11–24

BASELINE DIMENSIONS

Yet another aid in grouping dimensions is the DIMBASELINE command. Continue dimensions place dimensions next to each other; baseline dimensions establish a base or starting point for the first dimension, as shown in the following image. Any dimensions that follow in the DIMBASELINE command are calculated from the common base point already established. This is a very popular mode to use when one end of an object acts as a reference edge. When you place dimensions using the DIMBASELINE command, a baseline spacing setting of 0.38 units controls the spacing of the dimensions from each other. The DIMBASELINE command is initiated after the DIMLINEAR command.

Try It! – Open the drawing file 11_Dim Baseline. Verify that Running OSNAP is on and set to Endpoint. Use the following command sequence and image for creating baseline dimensions.

 Command: **DLI** *(For DIMLINEAR)*
Specify first extension line origin or <select object>: *(Select the endpoint of the line at "A")*
Specify second extension line origin: *(Select the endpoint of the line at "B")*
Specify dimension line location or [Mtext/Text/Angle/Horizontal/Vertical/Rotated]: *(Locate the 1.75 horizontal dimension)*
Dimension text = 1.75

 Command: **DBA** *(For DIMBASELINE)*
Specify a second extension line origin or [Undo/Select] <Select>: *(Select the endpoint of the line at "C")*
Dimension text = 3.00
Specify a second extension line origin or [Undo/Select] <Select>: *(Select the endpoint of the line at "D")*
Dimension text = 4.50
Specify a second extension line origin or [Undo/Select] <Select>: *(Select the endpoint of the line at "E")*
Dimension text = 5.50
Specify a second extension line origin or [Undo/Select] <Select>: *(Press ENTER when finished)*
Select base dimension: *(Press ENTER to exit this command)*

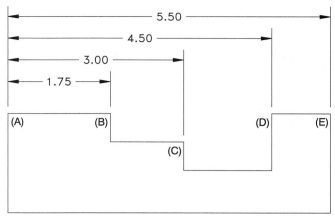

Figure 11–25

THE QDIM COMMAND

A more efficient means of placing a series of continued or baseline dimensions is the QDIM, or Quick Dimension, command. You identify a number of valid corners representing intersections or endpoints of an object and all dimensions are placed.

CONTINUOUS MODE

In the following image, a crossing box is used to identify all corners to dimension to.

Figure 11–26

When you have finished identifying the crossing box and press the ENTER key, a preview of the dimensioning mode appears, as in the following image. During this preview mode, you can right-click and have a shortcut menu appear. This allows you to select other dimension modes.

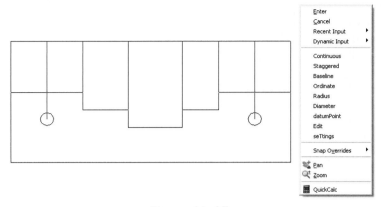

Figure 11–27

When the dimension line is identified, as in the following image, all continued dimensions are placed.

Figure 11–28

 Try It! – Open the drawing file 11_Qdim Continuous. Follow the command sequence below and previous images for performing this dimensioning task.

 Command: **QDIM**
Select geometry to dimension: *(Pick a point at "A")*
Specify opposite corner: *(Pick a point at "B")*
Select geometry to dimension: *(Press* ENTER *to continue)*
Specify dimension line position, or
[Continuous/Staggered/Baseline/Ordinate/Radius/Diameter/datumPoint/Edit/seTtings]
<Continuous>: *(Change to a different mode or locate the dimension line at "C")*

STAGGERED MODE

By default, the QDIM command places continued dimensions. Before locating the dimension line, you have the option of placing the staggered dimensions in the following image. The process with this style begins with adding dimensions to inside details and continuing outward until all features are dimensioned.

Try It! – Open the drawing file 11_Qdim Staggered. Activate the QDIM command and pick the lines labeled "A" through "H," as shown on the left in the following image. Change to the Staggered mode. Your display should appear similar to the illustration on the right in the following image.

Figure 11–29

BASELINE MODE

Another option of the QDIM command is the ability to place baseline dimensions. As with all baseline dimensions, an edge is used as the baseline, or datum. All dimensions are calculated from the left edge, as shown in the following image. The datumPoint option of the QDIM command would allow you to select the right edge of the object in the following image as the new datum or baseline.

 Try It! – Open the drawing file 11_Qdim Baseline. Activate the QDIM command and identify the same set of objects as in the first Qdim exercise. Change to the Baseline mode. Your display should appear similar to the illustration in the following image.

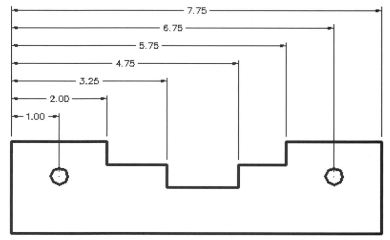

Figure 11–30

ORDINATE MODE

The Ordinate option of the QDIM command calculates dimensions from a known 0,0 corner. The main Ordinate dimensioning topic will be discussed in greater detail later in this chapter. A new User Coordinate System must first be created in the corner of the object. Then the QDIM command is activated along with the Ordinate options to display the results, as shown in the following image. Dimension lines are not used in ordinate dimensions, in order to simplify the reading of the drawing.

 Try It! – Open the drawing file 11_Qdim Ordinate. Activate the QDIM command and identify the same set of objects as in the first Qdim exercise. Change to the Ordinate mode. Your display should appear similar to the illustration in the following image.

Figure 11–31

 Tip: The QDIM command can also be used to edit an existing group of dimensions. Activate the command, select all existing dimensions with a crossing box, and enter an option to change the style of all dimensions.

DIAMETER AND RADIUS DIMENSIONING

Arcs and circles should be dimensioned in the view where their true shape is visible. The mark in the center of the circle or arc indicates its center point, as shown in the following image. You may place the dimension text either inside or outside the circle; you may also use grips to aid in the dimension text location of a diameter or radius dimension.

 Try It! – Open the drawing file 11_Dim Radial. Use the following command sequence and image for placing diameter and radius dimensions.

Command: **DDI** *(For DIMDIAMETER)*
Select arc or circle: *(Select the edge of the large circle)*
Dimension text = 2.50
Specify dimension line location or [Mtext/Text/Angle]: *(Pick a point to locate the diameter dimension)*

 Command: **DDI** *(For DIMDIAMETER)*
Select arc or circle: *(Select the edge of the small circle)*
Dimension text = 1.00
Specify dimension line location or [Mtext/Text/Angle]: *(Pick a point to locate the diameter dimension)*

 Command: **DRA** *(For DIMRADIUS)*
Select arc or circle: *(Select the edge of the arc)*
Dimension text = 1.50
Specify dimension line location or [Mtext/Text/Angle]: *(Pick a point to locate the radius dimension)*

Figure 11–32

USING QDIM FOR RADIUS AND DIAMETER DIMENSIONS

As with linear dimensions, the QDIM command has options that can be applied to radius and diameter dimensions. In the following image, all arcs and circles are selected. Activating either the Radius or Diameter option displays the results in the illustrations. When locating the radius or diameter dimensions, you can specify the angle of the leader. A predefined leader length is applied to all dimensions. Grips could be used to relocate the dimensions to better places.

 Try It! – Open the drawing file 11_Qdim Radius. Use the illustration on the left in the following image and the command sequence below for placing a series of radius dimensions using the QDIM command.

 Command: **QDIM**

Select geometry to dimension: *(Select the four arcs on the left in the following image)*
Select geometry to dimension: *(Press ENTER to continue)*
Specify dimension line position, or
[Continuous/Staggered/Baseline/Ordinate/Radius/Diameter/datumPoint/Edit]
<Continuous>: **R** *(For Radius)*

 Try It! – Open the drawing file 11_Qdim Diameter. Use the illustration on the right in the following image and the command sequence below for placing a series of diameter dimensions using the QDIM command.

 Command: **QDIM**

Select geometry to dimension: *(Select the six circles on the right in the following image)*
Select geometry to dimension: *(Press ENTER to continue)*
Specify dimension line position, or
[Continuous/Staggered/Baseline/Ordinate/Radius/Diameter/datumPoint/Edit]
<Continuous>: **D** *(For Diameter)*
Specify dimension line position, or
[Continuous/Staggered/Baseline/Ordinate/Radius/Diameter/datumPoint/Edit]
<Diameter>: *(Pick a point to locate the diameter dimension)*

Figure 11–33

ADDING JOGGED DIMENSIONS

At times, you need to add a radius dimension to a large arc. When creating the radius dimension, the center for the dimension is placed outside the boundaries of the drawing. To give better control over these situations, you can create a jog in the radius. This is represented by a zigzag appearance, or jog, in the leader holding the radius dimension. Use the following command sequence and image for creating a jogged radius dimension.

Command: **DJO** *(For DIMJOGGED)*
Select arc or circle: *(Select the arc in the following image)*
Specify center location override: *(Locate the center at "A")*
Dimension text = 23.1588
Specify dimension line location or [Mtext/Text/Angle]: *(Locate the dimension line)*
Specify jog location: *(Pick a point at "B")*

ADDING ARC DIMENSIONS

You can also dimension the length of an arc using the Dimension Arc command (DIMARC, or the shortcut DAR). After you select the arc, extension lines are created at the endpoints of the arc and a dimension arc is constructed parallel to the arc being dimensioned. The arc symbol is placed with the dimension text. This symbol can also be located above the dimension text. This technique will be covered in the next chapter.

Command: **DAR** *(For DIMARC)*
Select arc or polyline arc segment: *(Select the arc)*
Specify arc length dimension location, or [Mtext/Text/Angle/Partial]: *(Locate the dimension)*
Dimension text = 10.8547

Figure 11–34

LEADER LINES

A leader line is a thin, solid line leading from a note or dimension and ending with an arrowhead, illustrated at "A" in the following image. The arrowhead should always terminate at an object line such as the edge of a hole or an arc. A leader to a circle or arc should be radial; this means it is drawn so that if extended, it would pass through the center of the circle, as illustrated at "B." Leaders should cross as few object lines as possible and should never cross each other. The short horizontal shoulder of a leader should meet the dimension illustrated at "A." It is poor practice to underline the dimension with the horizontal shoulder, as illustrated at "C." Example "C" also illustrates a leader not lined up with the center or radial. This may affect the appearance of the leader. Again, check your company's standard practices to ensure that this example is acceptable.

Yet another function of a leader is to attach notes to a drawing, as illustrated at "D." Notice that the two notes attached to the view have different terminators: arrows and dots. It is considered good practice to adopt only one terminator for the duration of the drawing. The sequence for the LEADER command is as follows:

 Try It! – Open the drawing file 11_Dim Leader. Use the following image and command sequence for placing leaders.

Command: **LEADER**
Specify leader start point: **Nea**
to *(Select the edge of the circle at "A")*
Specify next point: *(Pick a point to locate the end of the leader)*
Specify next point or [Annotation/Format/Undo] <Annotation>: *(Press ENTER to accept this default)*
Enter first line of annotation text or <options>: *(Press ENTER to accept this default)*
Enter an annotation option [Tolerance/Copy/Block/None/Mtext] <Mtext>: *(Press ENTER to display the Multiline Text Editor dialog box. Enter the desired text to make up the leader note and click the OK button to place the leader)*

Figure 11–35

THE QLEADER COMMAND

The QLEADER command, or Quick Leader, provides numerous controls for placing leaders in your drawing. This command may appear to operate similarly to the LEADER command previously explained.

 Try It! – Open the drawing file 11_Qleader. Study the prompt sequence below and the following image for this command.

Command: **LE** *(For QLEADER)*
Specify first leader point, or [Settings]<Settings>: **Nea**
to *(Pick the point nearest at "A")*
Specify next point: *(Pick a point at "B")*
Specify next point: *(Press ENTER to continue)*
Specify text width <0.00>: *(Press ENTER to accept this default)*
Enter first line of annotation text <Mtext>: *(Press ENTER to display the Multiline Text Editor dialog box. Enter "R0.20" and "2 PLACES" in two separate text lines. Click the OK button to place the leader)*

Figure 11–36

Pressing ENTER at the first Quick Leader prompt displays the Leader Settings dialog box shown in the following image. Various controls are available to add greater functionality to your leaders through three tabs: Annotation, Leader Line & Arrow, and Attachment.

The Annotation tab deals with the object placed at the end of the leader. By default, the MText radio button is selected, allowing you to add a note through the Multiline

Text Editor dialog box. You could also copy an object at the end of the leader, have a geometric tolerancing symbol placed in the leader, have a pre-defined block placed in the leader, or leave the leader blank.

 Command: **LE** *(For QLEADER)*

Specify first leader point, or [Settings]<Settings>: *(Press* ENTER *to accept the default value and display the Leader Settings dialog box in the following image)*

Figure 11–37

The Leader Line & Arrow tab in the following image allows you to draw a leader line consisting of straight segments or in the form of a spline object. You can control the number of points used to define the leader; a maximum of three points will be more than enough to create your leader. You can even change the arrowhead through the list illustrated on the right in the following image.

Figure 11–38

The Attachment tab in the following image allows you to control how text is attached to the end of the leader through various justification modes. The settings in the figure are the default values.

Figure 11-39

DIMENSIONING ANGLES

Dimensioning angles requires two lines forming the angle in addition to the location of the vertex of the angle along with the dimension arc location. Before going any further, understand how the curved arc for the angular dimension is derived. At "A" in the following image, the dimension arc is struck with its center at the vertex of the object.

 Try It! – Open the drawing file 11_Dim Angle. Use the following command sequence and image for performing this operation on both angles.

 Command: **DAN** *(For DIMANGULAR)*
Select arc, circle, line, or <specify vertex>: *(Select line "A")*
Select second line: *(Select line "B")*
Specify dimension arc line location or [Mtext/Text/Angle]: *(Pick a point at "C" to locate the dimension)*
Dimension text = 53

Figure 11-40

DIMENSIONING SLOTS

For slots, first select the view where the slot is visible. Two methods of dimensioning the slot are illustrated in the following image. With the first method, you call out a slot by locating the center-to-center distance of the two semicircles, followed by a radius dimension to one of the semicircles; which radius dimension is selected depends on the available room to dimension. A second method involves the same center-to-center distance followed by an overall distance designating the width of the slot. This dimension is the same as the diameter of the semicircles. It is considered good practice to place this dimension inside the slot.

 Try It! – Open the drawing file 11_Dim Slot1. Use this file for practice in placing dimensions on the slot, as shown in the following image.

Figure 11–41

A more complex example in the following image involves slots formed by curves and angles. Here, the radius of the circular center arc is called out. Angles reference each other for accuracy. The overall width of the slot is dimensioned, which happens to be the diameter of the semicircles at opposite ends of the slot.

 Try It! – Open the drawing file 11_Dim Slot2. Use this file for practice in placing dimensions on the angular slot in the following image. Use the DIMALIGN command to place the .48 dimension. Use OSNAP-Nearest to identify the location at "A" and OSNAP-Perpendicular to identify the location at "B."

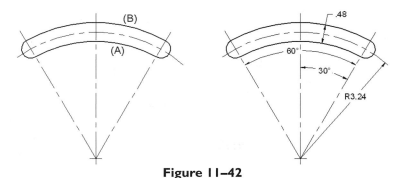

Figure 11–42

ADDING ORDINATE DIMENSIONS

The plate in the following image consists of numerous drill holes with a few slots, in addition to numerous 90° angle cuts along the perimeter. This object is not considered difficult to draw or make because it consists mainly of drill holes. However, conventional dimensioning techniques make the plate appear complex because a dimension is required for the location of every hole and slot in both the X and Y directions. Add standard dimension components such as extension lines, dimension lines, and arrowheads, and it is easy to get lost in the complexity of the dimensions even on this simple object.

Figure 11–43

A better dimensioning method, called ordinate, or datum, dimensioning, is illustrated in the following image. Here, dimension lines or arrowheads are not drawn; instead, one extension line is constructed from the selected feature to a location specified by you. A dimension is added to identify this feature in either the X or Y direction. It is important to understand that all dimension calculations occur in relation to the current User Coordinate System (UCS), or the current 0,0 origin. In the following image, with the 0,0 origin located in the lower left corner of the plate, all dimensions in the horizontal and vertical directions are calculated in relation to this 0,0 location. Holes and slots are called out with the DIMDIAMETER command. The following illustrates a typical ordinate dimensioning command sequence:

Command: **DOR** *(For DIMORDINATE)*
Specify feature location: *(Select a feature using an Osnap option)*
Specify leader endpoint or [Xdatum/Ydatum/Mtext/Text/Angle]: *(Locate a point outside of the object)*
Dimension text = Calculated value

Figure 11–44

To understand how to place ordinate dimensions, see the example in the following image and the prompt sequence below. Before you place any dimensions, a new User Coordinate System must be moved to a convenient location on the object with the UCS command and the Origin option. All ordinate dimensions will reference this new origin because it is located at coordinate 0,0. At the command prompt, enter DOR (for DIMORDINATE) to begin ordinate dimensioning. Select the quadrant of the arc at "A" as the feature. For the leader endpoint, pick a point at "B." Be sure Ortho mode is on. It is also helpful to snap to a convenient snap point for this and other dimensions along the direction. This helps in keeping all ordinate dimensions in line with one another.

 Try It! – Open the drawing file 11_Dim Ordinate. Follow the next series of figures and the following command sequences to place ordinate dimensions.

 Command: **DOR** *(For DIMORDINATE)*
Specify feature location: **Qua**
of *(Select the quadrant of the slot at "A")*
Specify leader endpoint or [Xdatum/Ydatum/Mtext/Text/Angle]: *(Locate a point at "B")*
Dimension text = 1.50

Figure 11–45

With the previous example highlighting horizontal ordinate dimensions, placing vertical ordinate dimensions is identical, as shown in the following image. With the UCS still located in the lower left corner of the object, select the feature at "A" using either the Endpoint or Quadrant mode. Pick a point at "B" in a convenient location on the drawing. Again, it is helpful if Ortho is on and you snap to a grid dot.

Command: **DOR** *(For DIMORDINATE)*
Specify feature location: **Qua**
of *(Select the quadrant of the slot at "A")*
Specify leader endpoint or [Xdatum/Ydatum/Mtext/Text/Angle]: *(Locate a point at "B")*
Dimension text = 3.00

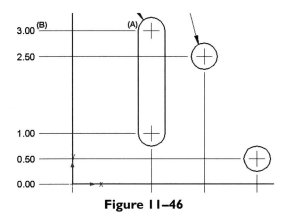

Figure 11–46

When spaces are tight to dimension to, two points not parallel to the X or Y axis will result in a "jog" being drawn, as shown in the following image. It is still helpful to snap to a grid dot when performing this operation; however, be sure Ortho is turned off.

Command: **DOR** *(For DIMORDINATE)*
Specify feature location: **End**
of *(Select the endpoint of the line at "A")*
Specify leader endpoint or [Xdatum/Ydatum/Mtext/Text/Angle]: *(Locate a point at "B")*
Dimension text = 2.00

Figure 11–47

Ordinate dimensioning provides a neat and easy way of organizing dimensions for machine tool applications. Only two points are required to place the dimension that references the current location of the UCS, as shown in the following image.

Figure 11–48

EDITING DIMENSIONS

Use the DIMEDIT command to add text to the dimension value, rotate the dimension text, rotate the extension lines for an oblique effect, or return the dimension text to its home position. The following image shows the effects of adding text to a dimension and rotating the dimension to a user-specified angle.

Try It! – Open the drawing file 11_Dim Dimedit. Follow the next series of figures and command prompt sequences for accomplishing this task.

 Command: **DED** *(For DIMEDIT)*

Enter type of dimension editing [Home/New/Rotate/Oblique] <Home>: **N** *(For New. This will display the Multiline Text Editor dialog box. Add the text "TYPICAL" on the other side of the <>. When finished, click the OK button)*
Select objects: *(Select the 5.00 dimension)*
Select objects: *(Press ENTER to perform the dimension edit operation)*

 Command: **DED** *(For DIMEDIT)*

Enter type of dimension editing [Home/New/Rotate/Oblique] <Home>: **R** *(For Rotate)*
Specify angle for dimension text: **10**
Select objects: *(Select the 5.00 dimension)*
Select objects: *(Press ENTER to perform the dimension edit operation)*

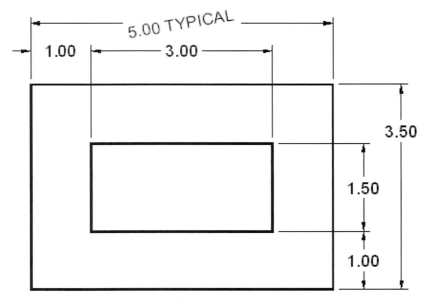

Figure 11–49

Regular AutoCAD modify commands can also affect dimensions. When you use the STRETCH command on the object on the left in the following image, points "A" and "B" identify a crossing window. Point "C" is the base point of displacement. The results are displayed on the right in the following image. Not only did the object lines stretch to the new position, but also the dimensions all updated themselves to new values.

Command: **S** *(For STRETCH)*
Select objects to stretch by crossing-window or crossing-polygon...
Select objects: *(Pick a point at "A")*
Specify opposite corner: *(Pick a point at "B" to activate the crossing window)*
Select objects: *(Press ENTER to continue)*
Specify base point or displacement: *(Pick a point at "C"; it could also be anywhere on the screen)*
Specify second point of displacement: **@0.75<180** *(Or use the Direct Distance mode; with ORTHO mode on, move your cursor to the left and type .75 to perform the stretch)*

Figure 11–50

One of the other options of the DIMEDIT command is the ability for you to move and rotate the dimension text and still have the text return to its original or home location. This is the purpose of the Home option. Selecting the 5.75 dimension will return it to its original position, as shown in the following image. However, you would have to use the New option of the DIMEDIT command to remove the text "TYPICAL."

Command: **DED** *(For DIMEDIT)*
Enter type of dimension editing [Home/New/Rotate/Oblique] <Home>: **H** *(Press ENTER to continue)*
Select objects: *(Select the 5.00 dimension)*
Select objects: *(Press ENTER to perform the dimension edit operation)*

Figure 11–51

The DIMEDIT command also has an Oblique option that allows you to enter an obliquing angle, which will rotate the extension lines and reposition the dimension line. This option is useful if you are interested in placing dimensions on isometric drawings.

 Try It! – Open the drawing file 11_Dim Oblique. Follow the next series of images and command prompt sequence for accomplishing this task.

 Command: **DED** (For DIMEDIT)
Enter type of dimension editing [Home/New/Rotate/Oblique] <Home>: **O** (For Oblique)
Select objects: (Select the 2.00 and 1.00 dimensions on the left in the following image)
Select objects: (Press ENTER to continue)
Enter obliquing angle (Press ENTER for none): **150**

The results are illustrated on the right in the following image, with both dimensions being repositioned with the Oblique option of the DIMEDIT command. Notice that the extension and dimension lines were affected; however, the dimension text remained the same.

Figure 11–52

Use the Oblique option to complete the editing of this drawing by rotating the dimension at "A" at an obliquing angle of 210°. An obliquing angle of −30° was used to rotate the dimensions at "B" and "C." The dimensions at "D" require an obliquing angle of 90°, as shown in the following image. This represents proper isometric dimensions, except for the orientation of the text.

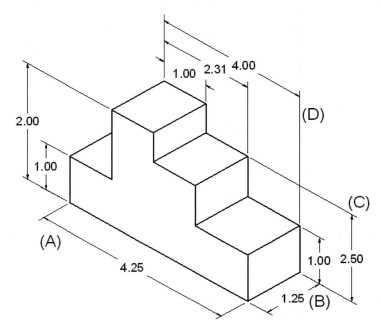

Figure 11–53

FLIPPING DIMENSION ARROWS

You can control the placement of dimension arrows by first selecting the dimensions and then right-clicking on one of the arrows to display the menu in the following image. Clicking on Flip Arrow from this menu will flip this arrowhead. The other arrowhead will remain normal. In the illustration on the right, the Flip Arrow mode was used twice to flip both arrowheads to the outside of the extension lines.

Figure 11–54

GEOMETRIC DIMENSIONING AND TOLERANCING (GDT)

In the object in the following image, approximately 1000 items need to be manufactured based on the dimensions of the drawing. Also, once constructed, all objects needed to be tested to be sure the 90° surface did not deviate over 0.005 units from its base. Unfortunately, the wrong base was selected as the reference for all dimensions. Which should have been the correct base feature? As all items were delivered, they were quickly returned because the long 8.80 surface drastically deviated from the required 0.005 unit deviation or zone. This is one simple example that demonstrates the need for using geometric dimensioning and tolerancing techniques. First, this method deals with setting tolerances to critical characteristics of a part. Of course, the function of the part must be totally understood by the designer in order for tolerances to be assigned. The problem with the object was the note, which did not really specify which base feature to choose as a reference for dimensioning.

Figure 11–55

The following image shows the same object complete with dimensioning and tolerancing symbols. The letter "A" inside the rectangle identifies the datum or reference surface. The tolerance symbol at the end of the long edge tells the individual making this part that the long edge cannot deviate more than 0.005 units using surface "A" as a reference. The datum triangles that touch the part are a form of dimension arrowhead.

Using geometric tolerancing symbols ensures more accurate parts with less error in interpreting the dimensions.

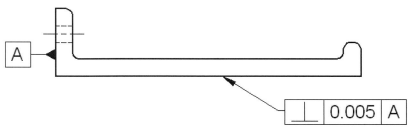

Figure 11–56

GDT SYMBOLS

Entering TOL at the command prompt brings up the main Geometric Tolerance dialog box, as shown in the following image. This box contains all tolerance zone boxes and datum identifier areas.

Figure 11–57

Clicking in one of the dark boxes under the Sym area displays the Symbol dialog box illustrated on the left in the following image. This dialog box contains all major geometric dimensioning and tolerancing symbols. Choose the desired symbol by clicking the specific symbol; this will return you to the main Geometric Tolerance dialog box.

The illustration on the right in the following image shows a chart outlining all geometric tolerancing symbols supported in the Symbol dialog box. Alongside each symbol is the characteristic controlled by the symbol, in addition to a tolerance type. Tolerances of form such as Flatness and Straightness can be applied to surfaces without a datum being referenced. On the other hand, tolerances of orientation such as Angularity and Perpendicularity require datums as reference.

Symbol	Purpose	Symbol	Purpose
⌗	Flatness	⊥	Perpendicularity
—	Straightness	//	Parallelism
○	Roundness	⊕	Position
⌭	Cylindricity	◎	Concentricity
⌒	Profile of a Line	=	Symmetry
⌓	Profile of a Surface	↗	Circular Runout
∠	Angularity	⌰	Total Runout

Figure 11–58

With the symbol placed inside this dialog box, you now assign such items as tolerance values, maximum material condition modifiers, and datums. In the following image, the tolerance of Parallelism is to be applied at a tolerance value of 0.005 units to Datum "A."

Figure 11-59

DIMENSION SYMBOLS

In today's global economy, the transfer of documents in the form of drawings is becoming a standard way of doing business. As drawings are shared with a subsidiary of an overseas company, two different forms of language may be needed to interpret the dimensions of the drawing. In the past, this has led to confusion in interpreting drawings, and as a result, a system of dimensioning symbols has been developed. It is hoped that a symbol will be easier to recognize and interpret than a note in a different language about the particular feature being dimensioned.

The following image shows some of the more popular dimensioning symbols in use today on drawings. Notice how the symbols are designed to make as clear and consistent an interpretation of the dimension as possible. As an example, the Deep or Depth symbol displays an arrow pointing down. This symbol is used to identify how far into a part a drill hole goes. The Arc Length symbol identifies the length of an arc, and so on.

Symbol	Description	Symbol	Description
⌒	Arc Length	2X	Number of Times
⌞x.xx⌟	Basic Dimension	R	Radius
▷	Conical Taper	(X.XX)	Reference Dimension
⌴	Counterbore/Spotface	SØ	Spherical Diameter
∨	Countersink	SR	Spherical Radius
↧	Deep or Depth	⌳	Slope
Ø	Diameter	□	Square
X.XX	Not to Scale		

Figure 11-60

CHARACTER MAPPING FOR DIMENSION SYMBOLS

Illustrated in the following image is a typical counterbore operation and the correct dimension layout. The LEADER command was used to begin the diameter dimension. The counterbore and depth symbols were generated with the MTEXT command along with the Unicode Character Mapping dialog box. In any case, the first step in using the dimension symbols is to first create a new text style, for which you can use any name.

Figure 11-61

These extra symbol characters can be found by right-clicking inside the Text Formatting edit box. From the menu that appears, click on Symbol followed by Other..., as shown in the following image.

Figure 11-62

Clicking Other... displays the Character Map dialog box, as shown on the left in the following image. Notice, in the upper left corner, that the current font is GDT, which holds all geometric tolerancing symbols along with the special dimensioning symbols such as counterbore, deep, and countersink. Be sure your current font is set to GDT. Once you identify a symbol, double-click it. A box will appear around the symbol. Also, the symbol will appear in the Characters to Copy area in the lower left corner of the dialog box. Click the Copy button to copy this symbol to the Windows Clipboard. Then close the Character Map. Return to the Multiline Text Editor dialog box and press CTRL+V, which performs a paste operation. You may also paste the symbol by right-clicking in the Multiline Text Editor dialog box and clicking Paste. Because the counterbore symbol was copied to the Windows Clipboard, it pastes into the dialog box, as shown on the right in the following image.

Figure 11–63

GRIPS AND DIMENSIONS

Grips have a tremendous amount of influence on dimensions. Grips allow dimensions to be moved to better locations; grips also allow the dimension text to be located at a better position along the dimension line.

Try It! – Open the drawing file 11_Dimgrip. In the example shown in the following image, notice the various unacceptable dimension placements. The 2.88 horizontal dimension lies almost on top of another extension line. To relocate this dimension to a better position, click the dimension. Notice that the grips appear and the entire dimension highlights. Now click the grip near "A," illustrated on the left in the following image. (This grip is located at the left end of the dimension line.) When the grip turns red, the Stretch mode is currently active. Stretch the dimension above the extension line but below the 3.50 dimension. Press the ESC key to turn off the grips. The same results can be accomplished with the 4.50 horizontal dimension as it is stretched closer to the 3.50 dimension, as shown on the right in the following image.

Figure 11–64

Notice that the two vertical dimensions shown on the left in the following image do not line up with each other; this would be considered poor practice. Pick both dimensions and notice the appearance of the grips, in addition to both dimensions being highlighted. Click the upper grip at "A" of the 1.25 dimension. When this grip turns red and places you in Stretch mode, select the grip at "B" of the opposite dimension. The result will be that both dimensions now line up with each other, as shown on the right in the following image.

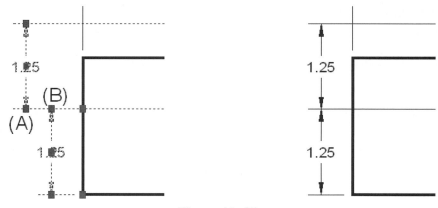

Figure 11–65

Illustrated on the left in the following image, the 2.50 dimension text is too close to the 2.00 vertical dimension on the right side of the object. Click the 2.50 dimension; the grips appear and the dimension highlights. Click in the grip representing the text location at "A." When this grip turns red, stretch the dimension text to a better location, as shown on the right in the following image.

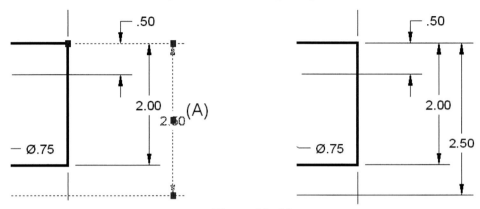

Figure 11–66

It is very easy to use grips to control the placement of diameter and radius dimensions. As shown on the left in the following image, click the diameter dimension; the grips appear, in addition to the diameter dimension being highlighted. Click the grip that locates the dimension text at "A." When this grip turns red, relocate the diameter dimension text to a better location using the Stretch mode, as shown on the right in the following image.

Figure 11–67

The completed object, with dimensions edited through grips, is displayed in the following image.

Figure 11–68

TUTORIAL EXERCISE: 11_FIXTURE.DWG

Figure 11–69

Purpose
The purpose of this tutorial is to add dimensions to the drawing of 11_Fixture.

System Settings
The drawing in the previous image is already constructed. Follow the steps in this tutorial for adding dimensions. Be sure the Endpoint and Intersection Object Snap modes are set.

Layers
Layers have already been created for this tutorial.

Suggested Commands
Use the DIMCENTER, DIMLINEAR, DIMCONTINUE, DIMBASELINE, DIMDIAMETER, and DIMRADIUS commands for placing dimensions throughout this tutorial.

STEP 1

Open the drawing file 11_Fixture, which can be found on the CD. A series of linear, baseline, continue, radius, and diameter dimensions will be added to this view. Before continuing, verify that running OSNAP is set to Endpoint and Intersection modes and that OSNAP is turned on.

A number of dimension styles have been created to control various dimension properties. You will learn more about dimension styles in chapter 12. For now, make the Center Mark dimension style current by clicking on its name in the Dimension Styles Control box, as shown in the following image.

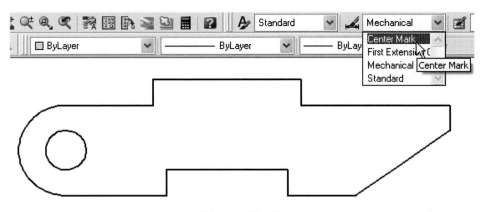

Figure 11–70

STEP 2

Using the Dimension Center command (DCE), add a center mark to identify the center of the circular features. Touch the edge of the arc to place the center mark.

The Center Mark dimension style already allows a small and long dash to be constructed. See the following image.

 Command: **DCE** *(For DIMCENTER)*
Select arc or circle: *(Pick the edge of the arc in the following image)*

Figure 11–71

STEP 3

Now that the center mark is properly placed, it is time to begin adding the horizontal and vertical dimensions. Before performing these tasks, first make the Mechanical dimension style current by clicking on its name in the Dimension Styles Control box in the following image.

Figure 11–72

STEP 4

Verify in the Status bar that OSNAP is turned on. Then place a linear dimension of 5.50 units from the intersection at "A" to the intersection at "B," as shown in the following image.

 Command: **DLI** *(For DIMLINEAR)*
Specify first extension line origin or <select object>: *(Pick the intersection at "A")*
Specify second extension line origin: *(Pick the intersection at "B")*
Specify dimension line location or
[Mtext/Text/Angle/Horizontal/Vertical/Rotated]: *(Locate the dimension in the following image)*
Dimension text = 5.50

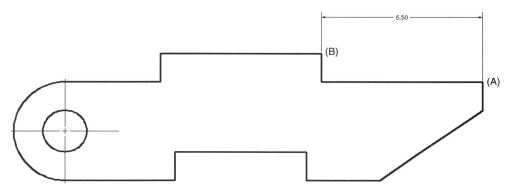

Figure 11–73

STEP 5

Place two Continue dimensions at intersection "A" and endpoint "B" in the following image. These Continue dimensions know to calculate the new dimension from the second extension line of the previous dimension.

 Command: **DCO** *(For DIMCONTINUE)*

Specify a second extension line origin or [Undo/Select] <Select>: *(Pick the intersection at "A")*
Dimension text = 5.50
Specify a second extension line origin or [Undo/Select] <Select>: *(Pick the endpoint at "B")*
Dimension text = 3.25
Specify a second extension line origin or [Undo/Select] <Select>: *(Press ENTER)*
Select continued dimension: *(Press ENTER)*

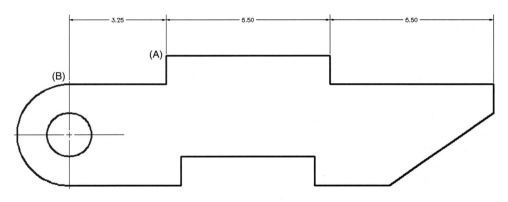

Figure 11–74

STEP 6

Place a linear dimension of 3.50 units from the intersection at "A" to the intersection at "B," as shown in the following image.

 Command: **DLI** *(For DIMLINEAR)*

Specify first extension line origin or <select object>: *(Pick the intersection at "A")*
Specify second extension line origin: *(Pick the intersection at "B")*
Specify dimension line location or
[Mtext/Text/Angle/Horizontal/Vertical/Rotated]: *(Locate the dimension in the following image)*
Dimension text = 3.50

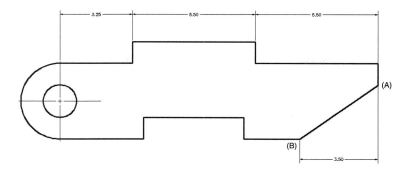

Figure 11–75

STEP 7

Add a series of baseline dimensions, as shown in the following image. All baseline dimensions are calculated from the first extension line of the previous dimension (the linear dimension measuring 3.50 units).

 Command: **DBA** *(For DIMBASELINE)*
Specify a second extension line origin or [Undo/Select] <Select>: *(Pick the intersection at "A")*
Dimension text = 6.00
Specify a second extension line origin or [Undo/Select] <Select>: *(Pick the intersection at "B")*
Dimension text = 10.50
Specify a second extension line origin or [Undo/Select] <Select>: *(Pick the endpoint at "C")*
Dimension text = 14.25
Specify a second extension line origin or [Undo/Select] <Select>: *(Press* ENTER*)*
Select base dimension: *(Press* ENTER*)*

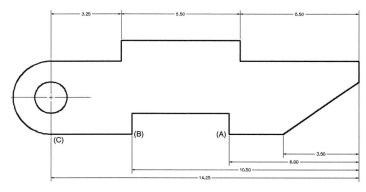

Figure 11–76

STEP 8

Add a diameter dimension to the circle using the DIMDIAMETER command (DDI). Then add a radius dimension to the arc using the DIMRADIUS command (DRA), as shown in the following image.

 Command: **DDI** *(For DIMDIAMETER)*
Select arc or circle: *(Pick the edge of the circle at "A")*
Dimension text = 1.50
Specify dimension line location or [Mtext/Text/Angle]: *(Locate the diameter dimension in the following image)*

 Command: **DRA** *(For DIMRADIUS)*
Select arc or circle: *(Pick the edge of the arc at "B")*
Dimension text = 1.75
Specify dimension line location or [Mtext/Text/Angle]: *(Locate the diameter dimension in the following image)*

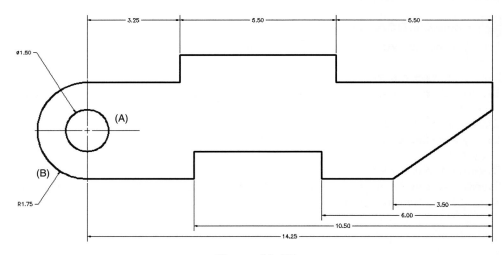

Figure 11–77

STEP 9

Place a vertical dimension of 1.00 units using the DIMLINEAR command (DLI). Pick the first extension line origin at the intersection at "A" and the second extension line origin at the intersection at "B." Then follow this by adding a baseline dimension using the DIMBASELINE command (DBA). See the following image.

 Command: **DLI** *(For DIMLINEAR)*

Specify first extension line origin or <select object>: *(Pick the intersection at "A")*
Specify second extension line origin: *(Pick the intersection at "B")*
Specify dimension line location or
[Mtext/Text/Angle/Horizontal/Vertical/Rotated]: *(Locate the dimension in the following image)*
Dimension text = 1.00

 Command: **DBA** *(For DIMBASELINE)*

Specify a second extension line origin or [Undo/Select] <Select>: *(Pick the intersection at "C")*
Dimension text = 3.50
Specify a second extension line origin or [Undo/Select] <Select>: *(Press ENTER)*
Select base dimension: *(Press ENTER)*

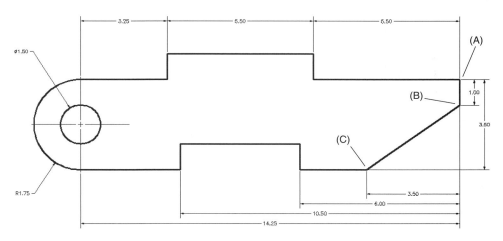

Figure 11–78

STEP 10

Add an angular dimension of 144° using the DIMANGULAR command (DAN), as shown in the following image.

Command: **DAN** *(For DIMANGULAR)*
Select arc, circle, line, or <specify vertex>: *(Pick the line at "A")*
Select second line: *(Pick the line at "B")*
Specify dimension arc line location or [Mtext/Text/Angle]: *(Locate the dimension in the following image)*
Dimension text = 144

Figure 11-79

STEP 11

The next series of dimensions involves placing linear dimensions so that one extension line is constructed by the dimension and the other is actually an object line that already exists. You need to turn off one extension line while leaving the other extension line turned on. This is accomplished by using an existing dimension style. Make the First Extension Off dimension style current by clicking on its name in the Dimension Styles Control box, as shown in the following image. This dimension style will suppress or turn off the first extension line while leaving the second extension line visible. This will prevent the extension line from being drawn on top of the object line.

Figure 11-80

STEP 12

Add the two linear dimensions to the slots by using the following command sequences and image as a guide.

| Command: **DLI** *(For DIMLINEAR)* | Command: **DLI** *(For DIMLINEAR)* |

Specify first extension line origin or
   <select object>:
   (Pick the intersection at "A")
Specify second extension line origin:
   (Pick the intersection at "B")
Specify dimension line location or
[Mtext/Text/Angle/Horizontal/Vertical/
   Rotated]: *(Locate the dimension in the
   following image)*
Dimension text = 1.00

Specify first extension line origin or
   <select object>:
   (Pick the intersection at "C")
Specify second extension line origin:
   (Pick the intersection at "D")
Specify dimension line location or
[Mtext/Text/Angle/Horizontal/Vertical/
   Rotated]: *(Locate the dimension in the
   following image)*
Dimension text = 1.00

Figure 11–81

The completed fixture drawing with all dimensions is shown in the following image.

Figure 11–82

TUTORIAL EXERCISE: 11_DIMENSION VIEWS.DWG

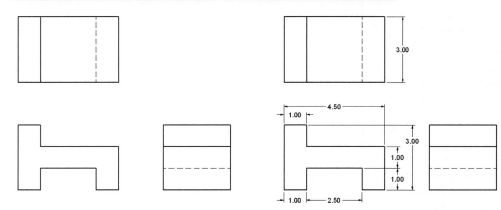

Figure 11–83

Purpose
The purpose of this tutorial is to add dimensions to the three-view drawing named 11_Dimension Views.dwg.

System Settings
The drawing in the previous image is already constructed. Dimensions must be added to various views to call out overall distances, in addition to features such as cuts and slots. Be sure the Object Snap modes Endpoint, Center, Intersect, and Extension are set.

Layers
Layers have already been created for this tutorial.

Suggested Commands
Use the DIMLINEAR command for horizontal and vertical on different views in this drawing.

STEP 1

Open the drawing file 11_Dimension Views.dwg. Linear dimensions are placed identifying the overall length, width, and depth dimensions. The DIMLINEAR command is used to perform this task, as shown in the following image.

Figure 11–84

STEP 2

Detail dimensions that identify cuts and slots are placed. Because these cuts are visible in the Front view, the dimensions are placed there, as shown in the following image. The Dimension Continue (DCO) command could be used for the second 1.00 dimension.

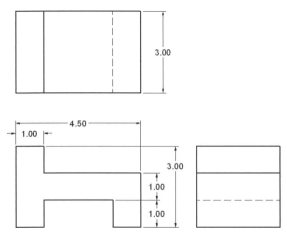

Figure 11–85

STEP 3

Once the spaces between the Front, Top, and Right Side views are used up by dimensions, the outer areas are used for placing additional dimensions such as the two horizontal dimensions in the following image. Again, use Dimension Continue (DCO) for the second dimension. Use grips to adjust dimension text locations if necessary.

Figure 11–86

PROBLEMS FOR CHAPTER 11

Problems 11–1 through 11–18

1. Open each drawing, which is located on the CD.
2. Use proper techniques to dimension each drawing.
3. Follow the steps involved in completing 11_Dimension Views.dwg.

Figure 11–87

Figure 11–88

Figure 11-89

CHAPTER 12

Managing Dimension Styles

Dimensions have different settings that could affect the group of dimensions. These settings include the control of the dimension text height, the size and type of arrowhead used, and whether the dimension text is centered in the dimension line or placed above the dimension line. These are but a few of the numerous settings available to you. In fact, some settings are used mainly for architectural applications, while other settings are only for mechanical uses. As a means of managing these settings, dimension styles are used to group a series of dimension settings under a unique name to determine the appearance of dimensions. This chapter will cover in detail the Dimension Style Manager dialog box and the following tabs associated with it: Lines, Symbols and Arrows; Text; Fit; Primary Units; Alternate Units; and Tolerances. Additional topics include overriding a dimension style and modifying the dimension style of an object.

THE DIMENSION STYLE MANAGER DIALOG BOX

Begin the process of creating a dimension style by choosing Style... from the Dimension pull-down menu, as shown on the left in the following image. This activates the Dimension Style Manager dialog box, shown on the bottom right in the following image. You can also launch this dialog by picking the appropriate button from the Dimension or Standard toolbars, as shown on the top right in the following image. Entering the keyboard command DDIM or D can also activate this dialog box, as can choosing Dimension Style from the Format pull-down menu. The current dimension style is listed as Standard. This style is automatically available when you create any new drawing. In the middle of the dialog box is an image icon that displays how some dimensions will appear based on the current value of the dimension settings. Various buttons are also available to set a dimension style current, create a new dimension style, make modifications to an existing dimension style, display dimension overrides, and compare two dimension styles regarding their differences and similarities.

Figure 12-1

To create a new dimension style, click on the New… button. This activates the Create New Dimension Style dialog box, as shown at the top of the following image. Enter a new name such as Mechanical in the New Style Name area. Then click the Continue button. This takes you to the New Dimension Style: Mechanical dialog box, as shown at the bottom of the following image, where a number of tabs hold all the settings needed in dimensioning.

Refer to the following table for a brief description of each tab located in the Dimension Style Manager dialog box:

Tab	Description
Lines	The Lines tab deals with settings that control dimension and extension lines.
Symbols and Arrows	The Lines and Arrows tab controls arrowheads and center marks.
Text	The Text tab contains the settings that control the appearance, placement, and alignment of dimension text.
Fit	The Fit tab contains various fit options for placing dimension text and arrows, especially in narrow places.
Primary Units	You control the units used in dimensioning through the Primary Units tab.
Alternate Units	If you need to display primary and secondary units in the same drawing, the Alternate Units tab is used.
Tolerances	Finally, for mechanical applications, various ways to show tolerances are controlled in the Tolerances tab.

When you make changes to any of the settings under the tabs, the preview image will update to show these changes. This provides a quick way of previewing how your dimensions will appear in your drawing. All these areas will be explained in greater detail throughout this chapter.

Figure 12–2

When finished making changes, click the OK button. This returns you to the Dimension Style Manager dialog box, as shown in the following image. Notice that the Mechanical dimension style has been added to the list of styles. Also, any changes made in the tabs are automatically saved to the dimension style you are creating or modifying. Click on Mechanical and then pick the Set Current button to make this style the current dimension style. Clicking the Close button returns you to your drawing. Let's get a closer look at all the tabs and their settings. Type the letter D to reopen the Dimension Style Manager. Make sure Mechanical is highlighted and click on the Modify… button.

Figure 12–3

THE LINES TAB

The Modify Dimension Style: Mechanical dialog box in the following image displays the Lines tab, which will now be discussed in greater detail. This tab consists of four main areas dealing with dimension lines and extension lines.

Figure 12–4

DIMENSION LINE SETTINGS

Use the Dimension lines area, shown in the following image, to control the color, lineweight, visibility, and spacing of the dimension line.

Figure 12–5

The Baseline spacing setting in the Dimension lines area controls the spacing of baseline dimensions, because they are placed at a distance from each other similar to the illustration on the left in the following image. This value affects the DIMBASELINE command (or DBA for short).

By default, dimension line suppression is turned off. To turn on suppression of dimension lines, place a check in the Dim line 1 or Dim line 2 box next to Suppress. This operation turns off the display of dimension lines for all dimensions placed under this dimension style. See the illustration on the right in the following image. This may be beneficial where tight spaces require that only the dimension text be placed.

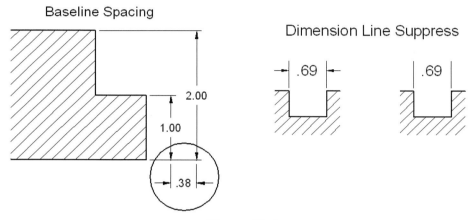

Figure 12–6

EXTENSION LINES

The Extension lines area controls the color, the distance the extension extends past the arrowhead, the distance from the object to the end of the extension line, and the visibility of extension lines, as shown in the following image. Additional controls allow you to set a different linetype for extension lines and provide the ability to use a fixed length value for all extension lines.

Figure 12–7

The Extend beyond dim lines setting controls how far the extension extends past the arrowhead or dimension line, as shown on the left in the following image. By default, a value of 0.18 is assigned to this setting.

The Offset from origin setting controls how far away from the object the extension line will start, as shown on the right in the following image. By default, a value of 0.06 is assigned to this setting.

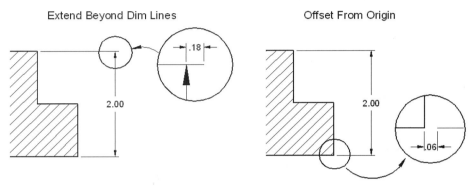

Figure 12–8

The Suppress Ext line 1 and Ext line 2 check boxes control the visibility of extension lines. They are useful when you dimension to an object line and for avoiding placing the extension line on top of the object line. Placing a check in the Ext line 1 box of Suppress turns off the first extension line. Similarly, when you place a check in the Ext line 2 box of Suppress, the second extension line is turned off. Suppressing extension lines by checking these boxes will suppress all extension lines for dimensions placed under this dimension style. Study the examples in the following image to get a better idea about how suppression of extension lines operates.

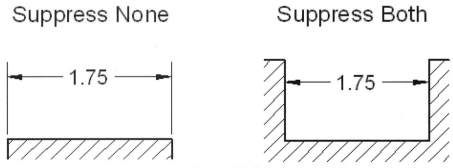

Figure 12–9

EXTENSION LINETYPES

You can assign linetypes presently loaded in your drawing, as shown in the following image. In this example, the Center linetype is assigned to extension lines. This provides an efficient means of applying custom linetypes to dimensions where appropriate, as in the dimensioning of the circular features in the plate.

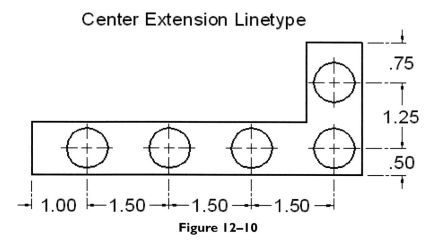

Figure 12–10

FIXED LENGTH EXTENSION LINES

Fixed length extension lines is a setting found under the Extension lines area of the Lines tab. It sets a user-defined length for all extension lines. In the illustration on the left in the following image, extension lines of varying lengths are displayed. All extension lines begin by default .0625 units away from the object. Illustrated on the right is another example of how extension lines are displayed at a fixed length of .25 units. Notice that all extension lines are the same length in this example. This value is calculated from where the endpoint of the arrow intersects with the extension to the bottom of the extension line. As you can see, this results in very short extension lines that are all the same length.

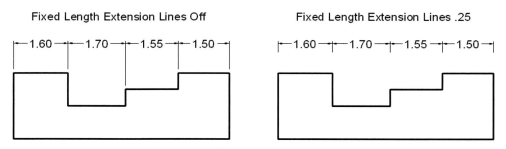

Figure 12–11

THE SYMBOLS AND ARROWS TAB

The Symbols and Arrows tab controls four main areas, as shown in the following image, namely, the size and type of arrowheads, the size and type of center marks, the style of the arc length symbol, and the angle formed when placing a radius jog dimension.

Figure 12–12

ARROWHEAD SETTINGS

Use the Arrowheads area to control the type of arrowhead terminator used for dimension lines and leaders, as shown in the following image. This dialog box also controls the size of the arrowhead.

Figure 12–13

Clicking in the First box displays a number of arrowhead terminators. Choose the desired terminator from the list, as shown on the left in the following image. When you choose the desired arrowhead from the First box, the Second box automatically updates to the selection made in the First box. If you choose an arrowhead from the Second edit box, the first and second arrowheads may be different at opposite ends of the dimension line; this is desired in some applications. Choosing a terminator in the Leader box displays the arrowhead that will be used whenever you place a leader.

Illustrated on the right in the following image is the complete set of arrowheads, along with their names. The last arrow type in the image on the left is a User Arrow, which allows you to create your own custom arrowhead.

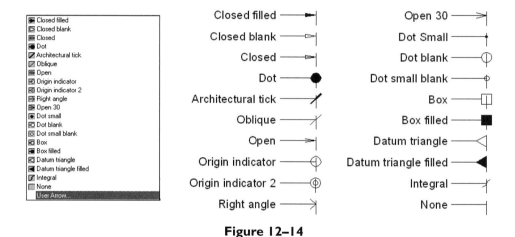

Figure 12–14

Use the Arrow size setting to control the size of the arrowhead terminator. By default, the arrow size is set to .18 units, as shown in the following image.

Figure 12–15

CENTER MARK SETTINGS

The Center marks for circles area allows you to control the type of center marker used when identifying the centers of circles and arcs. You can make changes to these settings by clicking on the three center mark modes: None, Mark, or Line, as shown on the left in the following image. The Size box controls the size of the small plus mark (+). If Mark or Line is chosen, these lines will show up when placing radius and diameter dimensions.

The three types of center marks are illustrated on the right in the following image. The Mark option places a plus mark in the center of the circle. The Line option places the plus mark and extends a line past the edge of the circle. The None option displays a circle with no center mark.

 Tip: It is considered good practice to add center marks to all circles and arcs before adding radius and diameter dimensions. To do this, first pick the Center Line type and use the DCE (Dimension Center) command to place the center marks. When finished, pick the None type in the center mark type area. Now all diameter and radius dimensions can be added without duplicating the center marks.

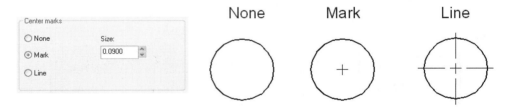

Figure 12–16

CONTROLLING DIMENSION ARC LENGTH SYMBOL

When placing an arc length dimension, you can control where the arc symbol is placed. In the following image, notice the three settings; namely, the arc symbol preceding the dimension text, the arc symbol placed above the dimension text, or no arc symbol displayed.

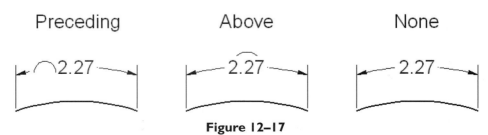

Figure 12–17

CONTROLLING THE RADIUS DIMENSION JOG ANGLE

You can control the angle of a jog dimension, as shown in the following image. By default, a 45° angle defines the jog, as shown on the left. Notice the effects of entering a 30° or 70° angle in the other examples in this image.

Figure 12–18

THE TEXT TAB

Use the Text tab, shown in the following image, to change the text appearance, for example, height; the text placement, for example, centered vertically and horizontally; and the text alignment, for example, always horizontal or parallel with the dimension line.

Figure 12–19

TEXT APPEARANCE

The Text height setting controls the size of the dimension text, as shown on the left in the following image. By default, a value of 0.18 is assigned to this setting.

Placing a check in the Draw frame around text box will draw a rectangular box around all dimensions, as shown on the right in the following image. This can be done to emphasize the dimension text. It also is a way of calling out a basic dimension.

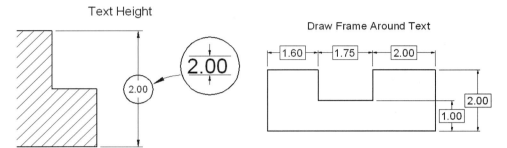

Figure 12–20

If the primary dimension units are set to architectural or fractional, the Fraction height scale activates. Changing this value affects the height of fractions that appear in the dimension. In the following image, a fractional height scale of 0.5000 will make the fractions as tall as the primary dimension number in the preview window.

Figure 12–21

TEXT PLACEMENT

The Text placement area, shown in the following image, allows you to control the vertical and horizontal placement of dimension text. You can also set an offset distance from the dimension line for the dimension text.

Figure 12–22

VERTICAL TEXT PLACEMENT

The Vertical area of Text placement controls the vertical justification of dimension text. Clicking in the drop-down edit box, as shown on the left in the following image, allows you to set vertical justification modes. By default, dimension text is centered vertically in the dimension line. Other modes include justifying vertically above the dimension line, justifying vertically outside the dimension line, and using the JIS (Japan International Standard) for placing text vertically.

Illustrated in the middle in the following image is the result of setting the Vertical justification to Centered. The dimension line will automatically be broken to accept the dimension text.

Illustrated on the right in the following image is the result of setting the Vertical justification to Above. Here the text is placed directly above a continuous dimension line. This mode is very popular for architectural applications.

Figure 12–23

The following image illustrates the result of setting the Vertical justification mode to outside. All text, including that contained in angular and radial dimensions, will be placed outside the dimension lines and leaders.

Figure 12–24

HORIZONTAL TEXT PLACEMENT

At times, dimension text needs to be better located in the horizontal direction; this is the purpose of the Horizontal justification area. Illustrated in the following image are the five modes of justifying text horizontally. By default, the horizontal text justification is centered in the dimension.

Figure 12–25

Clicking on the Centered option of the Horizontal justification area displays the dimension text, as shown on the left in the following image. This is the default setting because it is the most commonly used text justification mode in dimensioning.

Clicking on the At Ext Line 1 option displays the dimension text as shown in the middle in the following image, where the dimension text slides close to the first extension line. Use this option to position the text out of the way of other dimensions. Notice the corresponding option to have the text positioned nearer to the second extension line.

Clicking on the Over Ext Line 1 option displays the dimension text parallel to and over the first extension line, as shown on the right in the following image. Notice the corresponding option to position dimension text over the second extension line.

Figure 12–26

The last item of the Text placement area deals with setting an offset distance from the dimension line. When you place dimensions, a gap is established between the inside ends of the dimension lines and the dimension text. Entering different values depending on the desired results can control this gap. Study the examples in the middle and on the right in the following image, which have different text offset settings. Entering a value of zero (0) will force the dimension lines to touch the edge of the dimension text. Negative values are not supported.

Figure 12–27

TEXT ALIGNMENT

Use the Text alignment area, shown on the left in the following image, to control the alignment of text. Dimension text can be placed either horizontally or parallel (aligned) to the edge of the object being dimensioned. An ISO (International Standards Organization) standard is also available for metric drawings. Click in the appropriate radio button to turn the desired text alignment mode on.

If the Horizontal radio button is clicked, all text will be read horizontally, as shown in the middle in the following image. This includes text located inside and outside the extension lines.

Clicking on the Aligned with dimension line radio button displays the alignment results shown on the right in the following image. Here all text is read parallel to the edge being dimensioned. Not only will vertical dimensions align the text vertically, but the 2.06 dimension used to dimension the incline is also parallel to the edge.

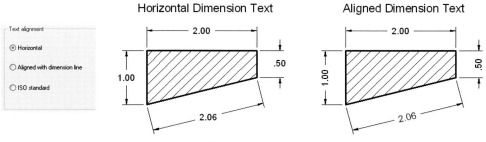

Figure 12–28

THE FIT TAB

Use the Fit tab in the following image to control how text and/or arrows are displayed if there isn't enough room to place both inside the extension lines. You could place the text over the dimension with or without a leader line. The Scale for dimension features area is very important when you dimension in Model Space or when you scale dimensions to Paper Space units. You can even fine-tune the placement of text manually and force the dimension line to be drawn between extension lines.

Figure 12-29

FIT OPTIONS

The Fit options area, shown on the left in the following image, has the radio button set for Either text or arrows (best fit). AutoCAD will decide to move either text or arrows outside extension lines. It will determine the item that fits the best. This setting is illustrated in the preview area on the right in the following image. It so happens that the preview image is identical when you click on the radio button for Arrows.

This tells AutoCAD to move the arrowheads outside the extension lines if there isn't enough room to fit both text and arrows.

Figure 12–30

Clicking on the Text radio button of the Fit options area updates the preview image, as shown in the following image. Here you are moving the dimension text outside the extension lines if the text and arrows do not fit. Since the value 1.0159 is the only dimension that does not fit, it is placed outside the extension lines, but the arrows are drawn inside the extension lines.

Figure 12–31

Clicking on the Both text and arrows radio button updates the preview image, as shown in the following image, where the 1.0159 dimension text and arrows are both placed outside the extension lines.

Figure 12–32

If you click on the radio button for Always keep text between ext lines, the result is illustrated in the preview image in the following image. Here all dimension text, including the radius dimension, is placed between the extension lines.

Figure 12–33

If you click on the radio button for Either the text or the arrows (best fit) and you also place a check in the box for Suppress arrows if they don't fit inside the extension lines, the dimension line is turned off only for dimensions that cannot fit the dimension text and arrows. This is illustrated in the following image.

Figure 12–34

TEXT PLACEMENT

You control the placement of the text if it is not in the default position. Your choices, shown in the following image, are Beside the dimension line, which is the default, Over dimension line, with leader, or Over dimension line, without leader.

If you click on the radio button for Text in the Fit options area and you click on the radio button for Over dimension line, with leader, the result is illustrated in the middle in the following image. For the 1.0159 dimension that does not fit, the text is placed outside the dimension line with the text connected to the dimension line with a leader.

If you click on the radio button for Text in the Fit options area and you click on the radio button for Over dimension line, without leader, the result is illustrated on the right in the following image. For the 1.0159 dimension that does not fit, the text is placed outside the dimension. No leader is used.

Figure 12–35

SCALE FOR DIMENSION FEATURES

This setting acts as a multiplier and globally affects all current dimension settings that are specified by sizes or distances. This means that if the current overall scale value is 1.00, other settings that require values will remain unchanged. There is also a radio button allowing you to have dimensions automatically scaled to Paper Space units inside a layout, as shown on the left in the following image. The scale of the viewport will control the scale of the dimensions.

The illustrations in the following image show the effects of overall scale factors of 1.00 and 2.00. The dimension text, arrows, origin offset, and extension beyond the arrow have all doubled in size.

Figure 12–36

 Try It! – Open the drawing file 12_Dimscale. A simple floor plan is displayed on the left in the following image. This floor plan is designed to be plotted at a scale of 1/2″ = 1′0″. Also displayed in this floor plan are dimensions in the magenta color. However, the dimensions are too small to be viewed. Set the Scale for dimension features found under the Fit tab of the Dimension Style dialog box to 24 (1′ [12″] divided by 1/2 is 24). The dimension values and tick marks should now be visible, as shown on the right in the following image.

Figure 12–37

FINE-TUNING

You have two options to add further control of the fitting of dimension text in the Fine tuning area, as shown on the left in the following image. You can have total control for horizontally justifying dimension text if you place a check in the box for Place text manually when dimensioning. You can also force the dimension line to be drawn between extension lines by placing a check in this box. The results are displayed on the right in the following image.

Figure 12–38

THE PRIMARY UNITS TAB

Use the Primary Units tab, shown in the following image, to control settings affecting the primary units. This includes the type of units the dimensions will be constructed in (decimal, engineering, architectural, and so on) and whether the dimension text requires a prefix or suffix.

Figure 12–39

LINEAR AND ANGULAR DIMENSION UNITS

The Linear dimensions area of the Primary Units tab, shown in the previous image, has various settings that deal with primary dimension units. A few of the settings deal with the format when working with fractions. This area activates only if you are working in architectural or fractional units. Even though you may be drawing in architectural units, the dimension units are set by default to decimal. You can also designate the decimal separator as a Period, Comma, or Space.

Clicking in the box for Unit format, as shown on the left in the following image, displays the types of units you can apply to dimensions. You also control the precision of the primary units by clicking in the Precision box.

In a similar way, clicking in the box for Units format in the Angular dimensions category, as shown on the right in the following image, displays the various formats angles can be displayed in. The precision of these angle units can also be controlled by clicking in the Precision box.

Figure 12–40

ROUNDING OFF DIMENSION VALUES

Use a Round off value to round off all dimension distances to the nearest unit based on the round off value. With a round off value of 0, the dimension text reflects the actual distance being dimensioned, as shown on the left in the following image. With a round off value set to .25, the dimension text reflects the next .25 increment, namely 2.50, as shown on the right in the following image.

Figure 12–41

APPLYING A DIMENSION PREFIX OR SUFFIX

A prefix allows you to specify text that will be placed in front of the dimension text whenever you place a dimension. Use the Suffix box to control the placement of a character string immediately after the dimension value. Both Prefix and Suffix control boxes are illustrated on the left in the following image. In the illustration on the right, the suffix "mm" was added to the dimension, signifying millimeters. Other common suffix letters include "M" for meters, "MI" for miles, and "in" for inches, to name just a few.

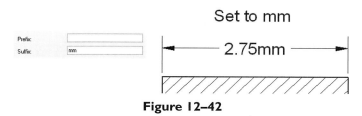

Figure 12–42

MEASUREMENT SCALE

The Measurement scale area, shown in the following image, acts as a multiplier for all linear dimension distances, including radius and diameter dimensions. When a dimension distance is calculated, the current value set in the Scale factor edit box is multiplied by the dimension to arrive at a new dimension value.

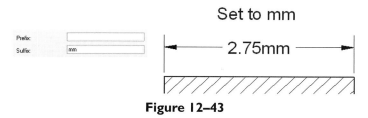

Figure 12–43

Illustrated on the left in the following image, and with a linear scale value of 1.00, the dimension distances are taken at their default values.

Illustrated on the right in the following image, the linear scale value has been changed to 2.00 units. This means 2.00 will multiply every dimension distance; the result is that the previous 3.00 and 2.00 dimensions are changed to 6.00 and 4.00, respectively. In a similar fashion, having a linear scale value set to 0.50 will reduce all dimension values to half their original values.

Figure 12–44

THE ALTERNATE UNITS TAB

Use the Alternate Units tab, shown in the following image, to enable alternate units, set the units and precision of the alternate units, set a multiplier for all units, use a round-off distance, and set a prefix and suffix for these units. You also have two placement modes for displaying these units. By default, alternate units are placed beside primary units. The alternate units are enclosed in square brackets. With two sets of units being displayed, your drawing could tend to become very busy. An application of using Alternate Units would be to display English and metric dimension values, since some design firms require both.

Figure 12–45

ALTERNATE UNITS

Once alternate units are enabled, all items in the Alternate Units area become active, as shown on the left in the following image. By default, the alternate unit value is placed in brackets alongside the calculated dimension value, as shown on the right in the following image. This value depends on the current setting in the Multiplier for all units edit box. This factor, set to 25.40 in the following image, is used as a multiplier for all calculated alternate dimension values.

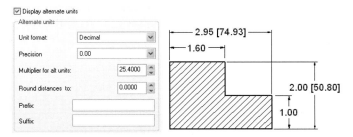

Figure 12–46

You could also click on the radio button next to Below primary value, as shown on the left in the following image. This places the primary dimension above the dimension line and the alternate dimension below it, as shown on the right in the following image.

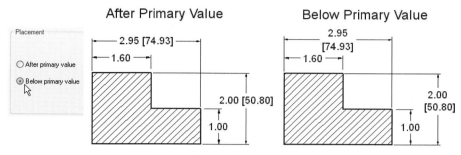

Figure 12–47

THE TOLERANCES TAB

The Tolerances tab, shown in the following image, consists of various edit boxes used to control the five types of tolerance settings: None, Symmetrical, Deviation, Limits, and Basic. Depending on the type of tolerance being constructed, an Upper value and Lower value may be set to call out the current tolerance variance. The Vertical position setting allows you to determine where the tolerance will be drawn in relation to the location of the body text. The Scaling for height setting controls the text size of the tolerance. Usually this value is smaller than the body text size.

Figure 12–48

TOLERANCE FORMAT

The five tolerance types available in the drop-down list are illustrated in the following image. A tolerance setting of None uses the calculated dimension value without applying any tolerances. The Symmetrical tolerance uses the same value set in the Upper and Lower value. The Deviation tolerance setting will have a value set in the Upper value and an entirely different value set in the Lower value. The Limits tolerance will use the Upper and Lower values and place the results with the larger limit dimension placed above the smaller limit dimension. The Basic tolerance setting does not add any tolerance value; instead, a box is drawn around the dimension value.

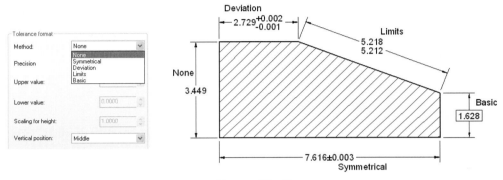

Figure 12–49

CONTROLLING THE ASSOCIATIVITY OF DIMENSIONS

THE DIMASSOC SYSTEM VARIABLE

The associativity of dimensions is controlled by the DIMASSOC variable. By default, this value is set to 2 for new drawings. This means that the dimension is associated with the object being dimensioned. Associativity means that if the object changes size or if some element of an object changes location, the dimension associated with the object will change as well.

You may find that for existing drawings, this variable is set to 1. This is called a non-associative dimension. This dimension is not associated with the object being dimensioned; however, it is possible to have the dimension value automatically updated using conventional AutoCAD editing commands.

The DIMASSOC variable can also be set to 0, which will create exploded dimensions. There is no association between the various elements of the dimension. All dimension lines, extension lines, arrowheads, and dimension text are drawn as separate objects. This means that grips will not have any effect on this dimension. You cannot stretch an object and have this dimension update to the object's new length. You cannot make a change in the Dimension Styles Manager dialog box and have this dimension affected. In other words, never set this variable to 0. The same effects can be achieved by using the EXPLODE command on a dimension; this is not recommended and is considered poor practice.

 Try It! – Open the drawing 12_Dimassoc2. The object on the left in the following image has dimensions placed with the DIMASSOC variable set to 2. Another polyline shape on the right has various dimensions placed on its outside. Click the polyline on the left to activate its grips. Click the grip at "A" and stretch the polyline vertex up and to the left. Notice that all dimension locations, orientations, and values are automatically updated. Click the polyline vertex at "B" and stretch this vertex up and to the right. The same results occur, with the dimension elements being updated based on the new location of the object. Use grips to realign any dimension that is not updated to a new location.

Figure 12–50

 Tip: When you make changes to the DIMASSOC variable, these changes are stored in the drawing file.

REASSOCIATING DIMENSIONS

It is possible to change a non-associative dimension into an associative dimension through the process called Reassociation. This command will allow you to pick the non-associative dimension and reestablish its endpoints with new endpoints located on the object being dimensioned. The following Try It! will demonstrate this capability.

Try It! – Open the drawing 12_Dim Reassoc. When the corner of the object at "B" stretches, the vertical 1.60 dimension does not readjust to the stretch operation. This is because the dimension may have been placed in a previous version of AutoCAD with DIMASSOC set to 1. Activate the DIMREASSOCIATE command from the Dimension pull-down menu, as shown on the left in the following image. Pick the 1.60 dimension as the object to reassociate. When the blue X appears, pick endpoints on the object to perform the reassociation, as shown in the middle of the following image.

Command: **DIMREASSOCIATE**
Select dimensions to reassociate ...
Select objects: *(Pick the vertical 1.60 dimension)*
Select objects: *(Press ENTER to continue)*
Specify first extension line origin or [Select object] <next>: *(When the blue X appears at the end of the proper extension line, pick the endpoint of the object at "B")*
Specify second extension line origin <next>: *(The blue X will move to the opposite endpoint of the dimension. Pick the endpoint of the object at "C" in order to complete the dimension reassociation)*

You can test to see whether this dimension is associative by selecting the object, picking either grip at "B" or "C," and stretching the object. The dimension should now update to the changes in the object's shape.

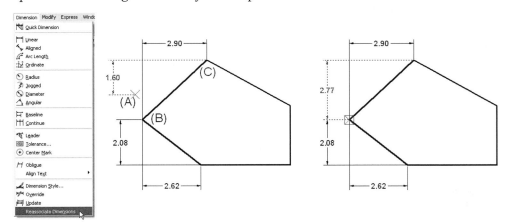

Figure 12–51

You will be able to identify associated dimensions by clicking on the dimension while in the REASSOCIATE command. The familiar X is surrounded by a blue box, as shown on the right in the previous image. If you do not see the blue box, the dimension is not associative.

Reassociating works only on non-associative dimensions that have their DIMASSOC value set to 1. The process of Reassociation does not have any effect on exploded dimensions (DIMASSOC set to 0).

While AutoCAD gives you this ability to reassociate dimensions, it would be very time-consuming to perform this task on hundreds of dimensions in a drawing. Reassociation is ideal when dealing with only a few dimensions.

USING DIMENSION TYPES IN DIMENSION STYLES

In addition to creating dimension styles, you can assign dimension types to dimension styles. The purpose of using dimension types is to reduce the number of dimension styles defined in a drawing. For example, the object in the following image consists of linear dimensions with three-decimal-place accuracy, a radius dimension with one-decimal-place accuracy, and an angle dimension with a box surrounding the number. Normally you would have to create three separate dimension styles to create this effect. However, the linear, radius, and angular dimensions consist of what are called dimension types.

Figure 12–52

Dimension types appear similar to the following image when assigned to the Mechanical dimension style. Before creating an angular dimension, you create the dimension type and make changes in various tabs located in the Modify Dimension Style dialog box. These changes will apply only to the angular dimension type. To expose the dimension types, click on the New button in the main Dimension Style Manager dialog box.

Figure 12–53

When the Create New Dimension Style dialog box appears, as shown in the following image, click in the Use for box. This box usually displays All dimensions. Notice all dimension types appearing. You can make changes to dimension settings that will apply only for linear, angular, radius, diameter, and ordinate dimensions. Also, a dimension type for Leaders and Tolerances is available. Highlight this option and then click the Continue button.

Figure 12–54

This takes you to the New Dimension Style: Mechanical: Leader dialog box, shown in the following image. Any changes you make in the tabs will apply only whenever you place a leader dimension. The use of dimension types is an efficient means of keeping the number of dimension styles down to a manageable level.

Figure 12–55

 Try It! – Open the drawing file 12_Dimension Types. A series of Dimension Types have already been created. Use the illustration in the following image and add all dimensions to this object.

Figure 12–56

OVERRIDING A DIMENSION STYLE

 Try It! – Open the drawing file 12_Dimension Override and activate the Dimension Style Manager dialog box. For special cases in which you need to change the settings of one dimension, a dimension override would be used. First launch the Dimension Style Manager dialog box. Clicking on the Override button displays the Override Current Style: Mechanical dialog box. Under the Lines tab, check the Ext line 1 and Ext line 2 boxes in the Extension lines area for suppression, as shown in the following image. One dimension needs to be constructed without displaying extension lines.

Figure 12-57

Clicking the OK button returns you to the main Dimension Style Manager dialog box, as shown in the following image. Notice that, under the Mechanical style, a new dimension type has been created called <style overrides> (Angular, Diameter, Linear, and Radial were already existing in this example). Also notice that the image in the Preview box shows sample dimensions displayed without extension lines. The <style overrides> dimension style is also the current style. Click the Close button to return to your drawing.

 Note: If the Preview box does not show the correct sample image when you create the new <style overrides>, close and immediately reopen the Dimension Style Manager dialog box. The Preview box will show the correct sample image.

Figure 12-58

In the illustration on the right in the following image, a linear dimension is placed identifying the vertical distance of 1.250. To avoid the mistake of placing extension lines on top of existing object lines, the extension lines are not drawn due to the style override.

Figure 12–59

Unfortunately, if you place other linear dimensions, these will also lack extension lines. The Dimension Style Manager dialog box is once again activated. Clicking on an existing dimension style such as Mechanical and then clicking on the Set Current button displays the AutoCAD Alert box shown in the following image. If you click OK, the style override will disappear from the listing of dimension types. If you would like to save the overrides under a name, click the Cancel button, right-click on the <style overrides> listing, and rename this style to a new name. This will preserve the settings under this new name.

Figure 12–60

MODIFYING THE DIMENSION STYLE OF AN OBJECT

Existing dimensions in a drawing can easily be modified through a number of methods that will be outlined in this section. The following image illustrates the first of these methods: the use of the Dim Style Control box, which is part of the Dimension toolbar. Be sure to have this toolbar displayed on your screen before continuing.

Try It! – Open the drawing file 12_Dimension Edit. First click on the dimension shown in the following image and notice that the dimension highlights and the grips appear. The current dimension style is also displayed in the Dim Style Control box. Click in the control box to display all other styles currently defined in the drawing. Click on the style name TOLERANCE to change the highlighted dimension to that style. Press ESC to turn off the grips.

Tip: Be careful when docking this toolbar along the side of your screen. In lower screen resolutions, the Dim Style Control box will not display.

Figure 12–61

The second method of modifying dimensions is illustrated in the following image. In this example, select the dimension shown in this image; it highlights and grips appear. Right-clicking the mouse button displays the cursor menu. Choosing the Dim Style heading displays the cascading menu of all dimension styles defined in the drawing. Click on the TOLERANCE style to change the highlighted dimension to the new style.

Managing Dimension Styles 645

Figure 12–62

The third method of modifying dimensions begins with selecting the overall length dimension shown in the following image. Again, the dimension highlights and grips appear. Clicking on the Properties button displays the Properties Window, shown in the following image. Click on the Alphabetic tab to display all dimension settings and current values. If you click in the name box next to Dim style, all dimension styles will appear in the edit box. Click on the OVERALL style to change the highlighted dimension to the new style. Close the Properties Palette and press the ESC key to turn off the grips.

Figure 12–63

TUTORIAL EXERCISE: 12_DIMEX.DWG

Figure 12–64

Purpose
The purpose of this tutorial is to place dimensions on the drawing of the two-view object illustrated in the previous image.

System Settings
No special system settings need to be made for this drawing file.

Layers
The drawing file 12_Dimex.Dwg has the following layers already created for this tutorial.

Name	Color	Linetype
Object	Magenta	Continuous
Hidden	Red	Hidden
Center	Yellow	Center
Dim	Yellow	Continuous

Suggested Commands
Open the drawing called 12_Dimex.Dwg. The following dimension commands will be used: DIMLINEAR, DIMCONTINUE, DIMBASELINE, DIMCENTER, DIMRADIUS, DIMDIAMETER, DIMANGULAR, and LEADER. All dimension commands may be chosen from the Dimension toolbar or the Dimension pull-down menu, or entered from the keyboard. Use the ZOOM command to get a closer look at details and features that are being dimensioned.

STEP 1

To prepare for the dimensioning of the drawing, type D to activate the Dimension Style Manager dialog box. Click on the New button, which activates the Create New Dimension Style dialog box, shown in the following image. In the New Style Name area, enter MECHANICAL. Click the Continue button to create the style, as shown in the following image.

Figure 12–65

STEP 2

When the New Dimension Style: MECHANICAL dialog box appears, make the following changes in the Lines tab, as shown on the left in the following image: Change the size of the Extend beyond dim lines from a value of 0.18 to a new value of 0.07. Switch to the Symbols and Arrows tab and change the Center marks to Line, as shown on the right in the following image.

Figure 12–66

STEP 3

Next, click on the Text tab and change the Text height from a value of 0.18 to a new value of 0.12, as shown on the left in the following image. Then click on the Primary Units tab and change the number of decimal places from 4 to 2. Also place a check in the box next to Leading in the Zero suppression area, as shown on the right in the following image. This will turn off the leading zero for dimension values under 1 unit. When finished, click the OK button to return to the main Dimension Style Manager dialog box.

Figure 12–67

STEP 4

You will now create a dimension type dealing with all diameter dimensions and make a change to the center mark settings for this type. Click the New button to display the Create New Dimension Style dialog box. In the Use for box, click on Diameter dimensions, as shown on the left in the following image, and then click the Continue button. Click on the Symbols and Arrows tab and change the Center marks setting to None, as shown on the right in the following image. This will prevent center marks from being displayed when placing diameter dimensions. When finished, click the OK button to return to the main Dimension Styles Manager dialog box.

Figure 12–68

STEP 5

In the Dimension Style Manager dialog box, click on MECHANICAL in the Styles area. Then click on the Set Current button to make MECHANICAL the current dimension style. Your display should appear similar to the following image. Click the Close button to save all changes and return to the drawing.

Figure 12–69

STEP 6

Begin placing center markers to identify the centers of all circular features in the Top view. Use the DIMCENTER command (or the shortcut DCE) to perform this operation on circles "A" through "E," as shown in the following image.

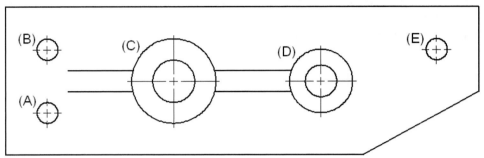

Figure 12–70

STEP 7

Use the QDIM command to place a string of baseline dimensions. First select the individual lines labeled "A" through "F," as shown in the following image. When the group of dimensions previews as continued dimensions, change this grouping to baseline and click a location to place the baseline group of dimensions, as shown in the following image. Use grips to place the text as shown.

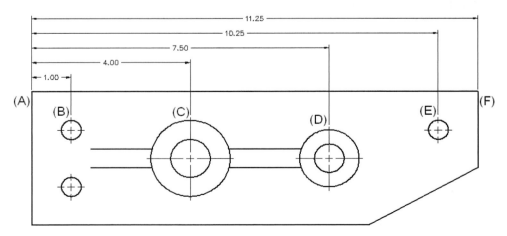

Figure 12–71

STEP 8

Magnify the left side of the Top view using the ZOOM command. Then use the DIMLINEAR command (or DLI shortcut) to place the .75 vertical dimension, as shown on the left in the following image. Next, use the DIMCONTINUE command (or DCO shortcut) to place the next dimension in line with the previous dimension, as shown in the middle in the following image. Then use the DIMLINEAR command (or DLI shortcut) to place the 1.75 vertical dimension, as shown on the right in the following image. Use grips to place the text as shown.

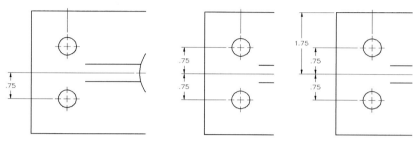

Figure 12-72

STEP 9

Use the PAN command to slide over to the right side of the Top view while keeping the same zoom percentage. Then use the QDIM command and select the lines "A" through "C" in the following image. When the group of dimensions previews as continued dimensions, change this grouping to baseline and click a location to place the baseline group of dimensions, as shown in the following image. You may have to identify a new base point (datumPoint) in order for your dimension to match the illustration. Then use the DIMANGULAR command (or DAN shortcut) to place the 61° dimension in the following image.

Figure 12-73

STEP 10

Use the ZOOM command and the Extents option to display both the Front and Top views. Then use the DIMDIAMETER command (or DDI shortcut) to place two diameter dimensions as shown in the following image.

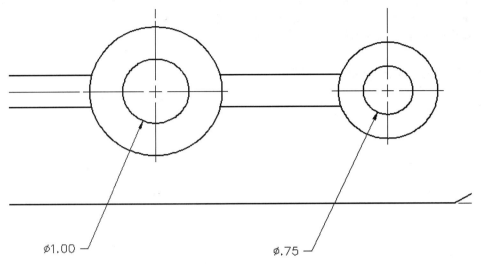

Figure 12–74

STEP 11

Place a diameter dimension using the DIMDIAMETER command on the circle as shown in the following image.

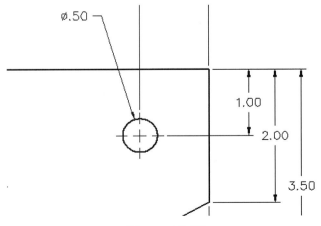

Figure 12–75

STEP 12

Since the two other smaller holes share the same diameter value, use the DDEDIT command (or ED shortcut) to edit this dimension value. Clicking on the diameter value activates the Text Formatting dialog box. Begin by typing 3X to signify three holes of the same diameter, as shown on the left in the following image. Click the OK button to return to your drawing. The results are illustrated on the right in the following image.

Figure 12–76

STEP 13

One more dimension needs to be placed in the Top view: the .50 width of the rib. Unfortunately, because of the placement of this dimension, extension lines will be drawn on top of the object's lines. This is considered poor practice. To remedy this, a dimension override will be created. To do this, activate the Dimension Style Manager dialog box and click on the Override button, as shown on the left in the following image. This displays the Override Current Style: MECHANICAL dialog box. In the Lines tab, place checks in the boxes to Suppress (turn off) Ext line 1 and Ext line 2 in the Extension lines area, as shown in the middle in the following image.

Click the OK button and notice the new <style overrides> listing under MECHANICAL, as shown on the right in the following image. Notice also that the preview image lacks extension lines. Click the Close button to return to the drawing.

Figure 12–77

STEP 14

Pan to the middle area of the Top view and place the .50 rib-width dimension while in the dimension style override, as shown in the following image. This dimension should be placed without having any extension lines visible. Also, use grips to drag the dimension text inside the visible object lines (temporarily turn off OSNAP to better accomplish this task).

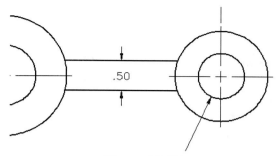

Figure 12–78

STEP 15

The completed Top view, including dimensions, is illustrated in the following image. Use this figure to check that all dimensions have been placed and all features such as holes and fillets have been properly identified.

Figure 12–79

STEP 16

The Front view in the following image will be the focus for the next series of dimensioning steps. Again use the ZOOM and PAN commands whenever you need to magnify or slide to a better drawing view position.

Figure 12–80

STEP 17

Before continuing, activate the Dimension Style Manager dialog box, click on the MECHANICAL style, and then click on the Set Current button. An AutoCAD Alert box displays, as shown in the following image. Making MECHANICAL current will discard any style overrides. Click the OK button to discard the changes because you need to return to having extension lines visible in linear dimensions.

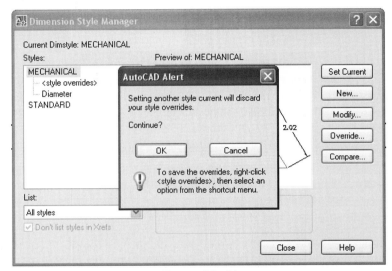

Figure 12–81

STEP 18

Place two horizontal linear dimensions that signify the diameters of the two cylinders using the DIMLINEAR command, as shown in the following image. Also add the 1.50 vertical dimension to the side of the Front view, as shown in the following image.

Figure 12–82

STEP 19

Use the QDIM command to place the vertical baseline dimensions as shown in the following image. Also place the 45° angular dimension in this image

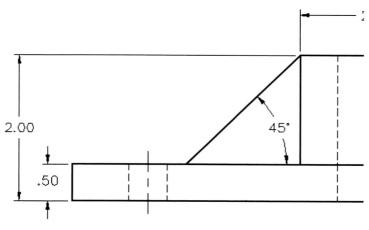

Figure 12–83

STEP 20

The remaining step is to add the diameter symbol to the 2.00 and 1.50 dimensions since the dimension is placed on a cylindrical object. Use the Text Formatting dialog box as shown in the following image to accomplish this task.

Figure 12–84

Once you have completed all dimensioning steps, your drawing should appear similar to the following image. Use grips to close the gaps between the extension and center lines.

Figure 12–85

TUTORIAL EXERCISE: 12_ARCHITECTURAL DIMENSION.DWG

Figure 12–86

Purpose

This tutorial is designed to develop and apply dimension styles to an architectural drawing project, shown in the previous image. The dimensions will be applied to a Model Space environment.

System Settings

No special settings are needed for this project.

Layers

Layers have already been created for this drawing.

Suggested Commands

An architectural dimension style will be created to reflect the settings required for an architectural drawing. Various horizontal and vertical dimensions will be placed around the floor plan to call out distances between windows, doors, and walls.

STEP 1

Open the 12_Architectural Dimension.dwg file, shown in the following image. Check to make sure that the dimension layer is the current layer.

Figure 12–87

STEP 2

The Dimension Style Manager will be used to set up characteristics of the dimensions used in architectural applications. Activate the Dimension Style Manager dialog box by clicking on Dimension Style, located in the Format pull-down menu. Selecting the New... button will launch the Create New Dimension Style dialog box, as shown in the following image. Replace Copy of Standard in the New Style Name edit box with Architectural. This dimension style will initially be based on the current standard dimension style and applied to all dimensions that are placed in the drawing.

Figure 12–88

STEP 3

Note that the name Architectural is now displayed in the title bar of the New Dimension Style dialog box, as shown in the following image. In the Symbols and Arrows section, change the arrowhead style to Architectural tick. When you change the First arrowhead, it also changes the Second arrowhead. Also change the arrow size to 1/8" or .125, as shown in the following image.

Figure 12–89

STEP 4

Select the Text tab and make changes to the following items, as shown in the following image. Change the Text style from Standard to Architectural (this text style already exists in the drawing). As with the arrowheads, change the Text height to 1/8″. Next, change the Vertical Text placement from Centered to Above. This will place dimension text above the dimension line, which is the architectural standard. Finally, in the Text alignment area, click on the radio button next to Aligned with dimension line. This will force the dimension text to display parallel to the dimension line. This also means that for vertical dimensions the dimension must be read from the side, which represents another architectural standard.

Figure 12–90

STEP 5

Next, select the Fit tab and change the overall scale of the dimension. The overall scale is based on the plotting scale when plotting is done from the model tab. The anticipated plotting scale for this drawing is ¼″ = 1′. Because ¼″ divides 48 times into 1′, type 48 in the Use overall scale of edit box, as shown in the following image.

Figure 12-91

STEP 6

Next, choose the Primary Units tab. In the Linear dimensions section, change the Unit format option to Architectural. Also, change the precision to 0´-0˝. Your display should appear similar to the following image.

Figure 12-92

STEP 7

Clicking the OK button will return you to the main Dimension Style Manager dialog box. Notice the appearance of the new dimension style, namely Architectural. As this dimension style is created, it is not current. Double-clicking on Architectural or picking the Set Current button will make this the current dimension style, as shown in the following image. When finished, click the Close button to return to the drawing editor.

Figure 12–93

STEP 8

Place a horizontal dimension using the LINEAR DIMENSION command (DLI), as shown in the following image. Be sure that OSNAP Endpoint and Intersection modes are active.

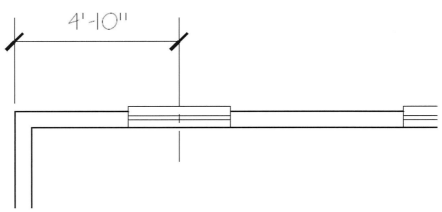

Figure 12–94

STEP 9

Use the CONTINUE DIMENSION command (DCO) to apply dimensions in succession to dimension the window locations, as shown in the following image.

Figure 12–95

STEP 10

Next add an overall length dimension using the LINEAR DIMENSION command (DLI), as shown in the following image.

Figure 12–96

STEP 11

Continue adding other dimensions to the floor plan, including those defining vertical and interior dimensions of the building, as shown in the following image.

Figure 12–97

PROBLEMS FOR CHAPTER 12

Directions for Problems 12–1 through 12–18

1. Open each drawing, which is located on the CD that is supplied with this book.
2. Use proper techniques to dimension each drawing.

Directions for Problems 12–19 through 12–24

1. Use the isometric drawings provided to develop orthographic view drawings, showing as many views as necessary to communicate the design.
2. Fully dimension your drawings.

PROBLEM 12–19

PROBLEM 12–20

PROBLEM 12–21

PROBLEM 12–22

PROBLEM 12–23

PROBLEM 12–24

CHAPTER 13

Analyzing 2D Drawings

This chapter will show how a series of commands may be used to calculate distances and angles of selected objects. It will also show how surface areas may be calculated on complex geometric shapes. The following pages highlight all of AutoCAD's inquiry commands and show how you can use them to display useful information on an object or group of objects.

USING INQUIRY COMMANDS

You can invoke AutoCAD's inquiry commands by using the Inquiry toolbar, by entering the commands at the keyboard, or by choosing them from the Tools menu, as illustrated in the following image.

Figure 13–1

The following is a list of the inquiry commands with a short description of each:

Button	Tool	Shortcut	Description
	Distance	DI	Calculates the distance between two points. Also provides the delta X,Y,Z coordinate values, the angle in the XY plane, and the angle from the XY plane.
	Area	AREA	Calculates the surface area after you specify a series of points or select a polyline or circle. You can add or subtract multiple objects to calculate the area with holes and cutouts.
	Mass Properties	MASSPROP	Used for determining the mass properties of a region or 3D solid model.
	List	LI	Used to list important information on an object.
	Locate	ID	Used to determine the X, Y, and Z coordinates of a point.
	Time	TIME	Displays the time spent in the drawing editor.
	Status	STATUS	Displays important information on the current drawing.
	Set Variable	SETVAR	Used for making changes to one of the many system variables internal to AutoCAD.

The following pages give a detailed description of how these commands are used in interaction with different AutoCAD objects.

FINDING THE AREA OF AN ENCLOSED SHAPE

The AREA command is used to calculate the area through the selection of a series of points. Select the endpoints of all vertices in the following image with the OSNAP-Endpoint option. Once you have selected the first point along with the remaining points in either a clockwise or counterclockwise pattern, respond to the prompt "Next point:" by pressing ENTER to calculate the area of the shape. Along with the area is a calculation for the perimeter.

Try It! – Open the drawing file 13_Area1. Use the following image and the prompt sequence below for finding the area by identifying a series of points.

Command: **AA** *(For AREA)*
Specify first corner point or [Object/Add/Subtract]: *(Select the endpoint at "A")*
Specify next corner point or press ENTER for total: *(Select the endpoint at "B")*
Specify next corner point or press ENTER for total: *(Select the endpoint at "C")*
Specify next corner point or press ENTER for total: *(Select the endpoint at "D")*
Specify next corner point or press ENTER for total: *(Select the endpoint at "E")*
Specify next corner point or press ENTER for total: *(Select the endpoint at "A")*
Specify next corner point or press ENTER for total: *(Press ENTER to calculate the area)*
Area = 25.25, Perimeter = 20.35

Figure 13–2

FINDING THE AREA OF AN ENCLOSED POLYLINE OR CIRCLE

The previous example showed how to find the area of an enclosed shape using the AREA command and identifying the corners and intersections of the enclosed area by a series of points. For a complex area, this could be a very tedious operation. As a result, the AREA command has a built-in Object option that will calculate the area and perimeter of a polyline and the area and circumference of a circle. Finding the area of a polyline can only be accomplished if one of the following conditions is satisfied:

- The shape must have already been constructed through the PLINE command.
- The shape must have already been converted to a polyline through the PEDIT command if originally constructed from individual objects.

Try It! – Open the drawing file 13_Area2. Use the following image and the prompt sequence below for finding the area of both shapes.

Command: **AA** *(For AREA)*
Specify first corner point or [Object/Add/Subtract]: **O** *(For Object)*
Select objects: *(Select the polyline at "A")*
Area = 24.88, Perimeter = 19.51

Command: **AA** *(For AREA)*
Specify first corner point or [Object/Add/Subtract]: **O** *(For Object)*
Select objects: *(Select the circle at "B")*
Area = 7.07, Circumference = 9.42

Figure 13–3

 Try It! – Open the drawing file 13_Extrude1. In the following image, either convert all line segments into a single polyline object or use the BOUNDARY command to trace a polyline on the top of all the line segments. In either case, answer Question 1 regarding 13_Extrude1.

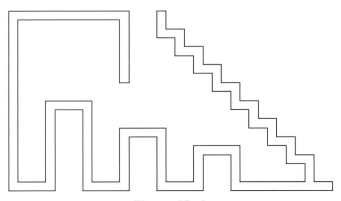

Figure 13–4

Question 1
What is the total surface area of 13_Extrude1?

(A) 9.1242

(B) 9.1246

(C) 9.1250

(D) 9.1254

(E) 9.1258

The total surface area of 13_Extrude1 is "C," <u>9.1250</u>.

 Try It! – Open the drawing file 13_Extrude2. In the following image, either convert all line segments into a single polyline object or use the BOUNDARY command to trace a polyline on the top of all the line segments. In either case, answer Question 1 regarding 13_Extrude2.

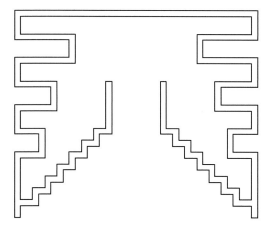

Figure 13–5

Question 1
What is the total surface area of 13_Extrude2?

(A) 28.9362

(B) 28.9366

(C) 28.9370

(D) 28.9374

(E) 28.9378

The total surface area of 13_Extrude2 is "A," 28.9362.

FINDING THE AREA OF A SHAPE BY SUBTRACTION

You may wish to calculate the area of a shape with holes cut through it. The steps you use to calculate the total surface area are: (1) calculate the area of the outline and (2) subtract the objects inside the outline. All individual objects, except circles, must first be converted to polylines through the PEDIT command. Next, find the overall area and add it to the database using the Add mode of the AREA command. Exit the Add mode and remove the inner objects using the Subtract mode of the AREA command. Remember that all objects must be in the form of a circle or polyline. This means that the inner shape at "B" in the following image must also be converted to a polyline through the PEDIT command before the area is calculated. Care must be taken when selecting the objects to subtract. If an object is selected twice, it is subtracted twice and may yield an inaccurate area in the final calculation.

In the following image, the total area with the circle and rectangle removed is 30.4314.

 Try It! – Open the drawing file 13_Area3. Use the following image and the prompt sequence to verify this area calculation.

 Command: **AA** *(For AREA)*
Specify first corner point or [Object/Add/Subtract]: **A** *(For Add)*
Specify first corner point or [Object/Subtract]: **O** *(For Object)*
(ADD mode) Select objects: *(Select the polyline at "A")*
Area = 47.5000, Perimeter = 32.0000
Total area = 47.5000
(ADD mode) Select objects: *(Press ENTER to exit ADD mode)*
Specify first corner point or [Object/Subtract]: **S** *(For Subtract)*
Specify first corner point or [Object/Add]: **O** *(For Object)*
(SUBTRACT mode) Select objects: *(Select the polyline at "B")*
Area = 10.0000, Perimeter = 13.0000
Total area = 37.5000
(SUBTRACT mode) Select objects: *(Select the circle at "C")*
Area = 7.0686, Circumference = 9.4248
Total area = 30.4314
(SUBTRACT mode) Select objects: *(Press ENTER to exit SUBTRACT mode)*
Specify first corner point or [Object/Add]: *(Press ENTER to exit this command)*

Figure 13–6

 Try It! – Open the drawing file 13_Shield. In the following image, use the BOUNDARY command to trace a polyline on the top of all the line segments. Subtract all four-sided shapes from the main shape using the AREA command. Answer Question I regarding 13_Shield.

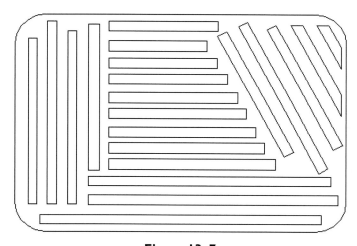

Figure 13–7

Question 1
What is the total surface area of 13_Shield with all inner four-sided shapes removed?

    (A) 44.2242

    (B) 44.2246

    (C) 44.2250

    (D) 44.2254

    (E) 44.2258

The total surface area of 13_Shield is "B," 44.2246.

USING FIELDS IN CALCULATIONS

Fields can be of great use when performing area calculations on geometric shapes. In the following image, the area of the room of the partial office plan will be calculated using fields. When the geometric shape changes, the field can be updated to display the new area calculation. Begin this process by first adding the following multiline text object, (AREA=). Then pick the Insert Field button from the Text Formatting toolbar.

Figure 13–8

This action will activate the Field dialog box, as shown in the following image. Locate Object under the Field Names area. Under the Object Type heading, click on the Select object button.

Figure 13–9

Clicking on the Select object button in the previous image will return you to the drawing editor. Once you are there, the rectangular polyline shape is selected, as shown in the following image.

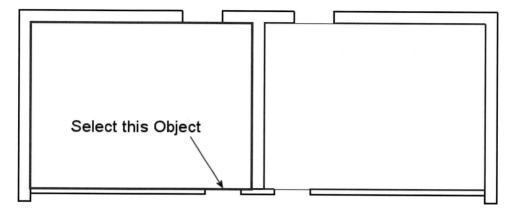

Figure 13–10

This will return you back to the Field dialog box, as shown in the following image. Notice that the property changed to Area and the Format changed to Architectural. The Preview area shows the current area calculation. Click the OK button to exit the Field dialog box.

Figure 13–11

The results are illustrated in the following image. A field consisting of the square footage information is added to the end of the multiline text. Notice also that the field is identified by the text with the gray background.

Figure 13–12

In the following image, the room was stretched 6´ to the left. Regenerating the drawing screen will update the field to reflect the change in the square footage calculation, as shown in the following image.

Figure 13–13

THE DIST (DISTANCE) COMMAND

The DIST command calculates the linear distance between two points on an object, whether it be the distance of a line, the distance between two points, or the distance from the quadrant of one circle to the quadrant of another circle. The following information is also supplied when you use the DIST command: the angle in the XY plane, the angle from the XY plane, and the delta X,Y,Z coordinate values. The angle in the XY plane is given in the current angular mode set by the Drawing Units dialog box. The delta X,Y,Z coordinate is a relative coordinate value taken from the first point identified by the DIST command to the second point.

 Try It! – Open the drawing file 13_Distance. Use the following image and the prompt sequences below for the DIST command.

 Command: **DI** *(For DIST)*
Specify first point: *(Select the endpoint at "A")*
Specify second point: *(Select the endpoint at "B")*
Distance = 6.36, Angle in XY Plane = 45.00, Angle from XY Plane = 0.00
Delta X = 4.50, Delta Y = 4.50, Delta Z = 0.00

 Command: **DI** *(For DIST)*
Specify first point: *(Select the endpoint at "C")*
Specify second point: *(Select the endpoint at "D")*
Distance = 9.14, Angle in XY Plane = 192.75, Angle from XY Plane = 0.00
Delta X = -8.91, Delta Y = -2.02, Delta Z = 0.00

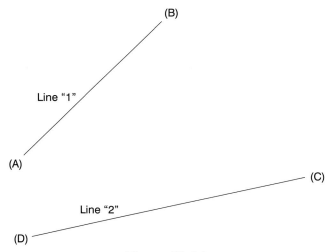

Figure 13–14

INTERPRETATION OF ANGLES USING THE DIST COMMAND

Previously, it was noted that the DIST command yields information regarding distance, delta X,Y coordinate values, and angle information. Of particular interest is the angle in the XY plane formed between two points. Picking the endpoint of the line segment at "A" as the first point followed by the endpoint of the line segment at "B" as the second point displays an angle of 42°, as shown on the left in the following image. This angle is formed from an imaginary horizontal line drawn from the endpoint of the line segment at "A" in the zero direction.

Take care when using the DIST command to find an angle on an identical line segment, illustrated on the right in the following image, compared to the example on the left. However, notice that the two points for identifying the angle are selected differently. With the DIST command, you select the endpoint of the line segment at "B" as the first point, followed by the endpoint of the segment at "A" for the second point. A new angle in the XY plane of 222° is formed. On the right in the following image, the angle is calculated by the construction of a horizontal line from the endpoint at "B," the new first point of the DIST command. This horizontal line is also drawn in the zero direction. Notice the relationship of the line segment to the horizontal baseline. In other words, be careful when identifying the order of line segment endpoints for extracting angular information.

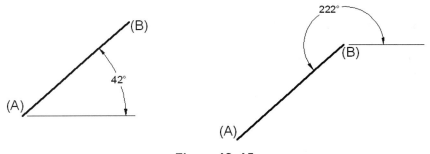

Figure 13–15

THE ID (IDENTIFY) COMMAND

The ID command is probably one of the more straightforward of the inquiry commands. ID can stand for "Identify" or "Locate Point" and allows you to obtain the current absolute coordinate listing of a point along or near an object.

In the following image, the coordinate value of the center of the circle at "A" was found through the use of ID and the OSNAP-Center mode. The coordinate value of the starting point of text string "B" was found with ID and the OSNAP-Insert mode. The coordinate value of the endpoint of line segment "C" was found with ID and the OSNAP-Endpoint mode. The coordinate value of the midpoint of line segment at

"CD" was found with ID and the OSNAP-Midpoint mode. Finally, the coordinate value of the current position of point "E" was found with ID and the OSNAP-Node mode.

 Try It! – Open the drawing file 13_ID. Follow the prompt sequences below for calculating the X,Y,Z coordinate point of the objects in the following image.

 Command: **ID**
Specify point: **Cen**
of *(Select the edge of circle "A")*
X = 2.00 Y = 7.00 Z = 0.00

 Command: **ID**
Specify point: **Ins**
of *(Select the text at "B")*
X = 5.54 Y = 7.67 Z = 0.00

 Command: **ID**
Specify point: **End**
of *(Select the line at "C")*
X = 8.63 Y = 4.83 Z = 0.00

 Command: **ID**
Specify point: **Mid**
of *(Select line "CD")*
X = 5.13 Y = 3.08 Z = 0.00

Command: **ID**
Specify point: **Nod**
of *(Select the point at "E")*
X = 9.98 Y = 1.98 Z = 0.00

Figure 13–16

THE LIST COMMAND

Use the LIST command to obtain information about an object or group of objects. In the following image, two rectangles are displayed along with a circle, but are the rectangles made up of individual line segments or a polyline object? Using the LIST command on each object informs you that the first rectangle at "A" is a polyline, the circle lists as a circle, and the second rectangle is actually a block reference. In addition to the object type, you also can obtain key information such as the layer that the object resides on, area and perimeter information for polylines, and circumference information for circles.

 Try It! – Open the drawing file 13_List. Study the prompt sequence below for using the LIST command.

Command: **LI** *(For LIST)*
Select objects: *(Select the objects at "A", "B", and "C" in order)*
Select objects: *(Press ENTER to list the information on each object)*
LWPOLYLINE   Layer: "Object"
Space: Model space
Handle = 2B
Closed
Constant width   0.0000
area   3.3835
perimeter   7.6235
at point X=   4.7002 Y=   4.1846 Z=   0.0000
at point X=   7.1049 Y=   4.1846 Z=   0.0000
at point X=   7.1049 Y=   5.5916 Z=   0.0000
at point X=   4.7002 Y=   5.5916 Z=   0.0000
CIRCLE   Layer: "Object"
Space: Model space
Handle = 2B
center point, X=   7.6879 Y=   4.8881 Z=   0.0000
radius   1.4719
circumference   9.2479
area   6.8058
BLOCK REFERENCE Layer: "Object"
Space: Model space
Handle = 34
Block Name "rec"
at point, X=   10.6756 Y=   4.1846 Z=   0.0000
X scale factor   -1.0000
Y scale factor   1.0000
rotation angle   0
Z scale factor   1.0000

Insunits: Inches

Unit conversion 1.0000
Scale uniformity: no
Allow exploding: yes

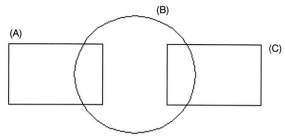

Figure 13–17

TUTORIAL EXERCISE: C-LEVER.DWG

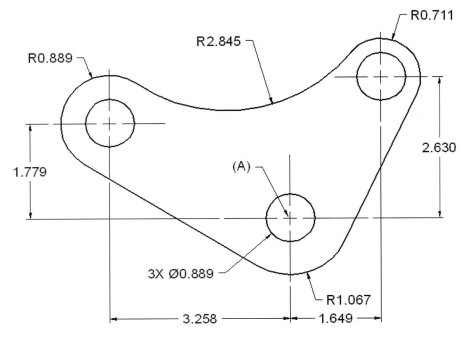

Figure 13-18

Purpose
This tutorial is designed to show you various methods of constructing the C-Lever object in the previous image. Numerous questions will be asked about the object, requiring the use of the AREA, DIST, ID, and LIST commands.

System Settings
Use the Drawing Units dialog box and change the number of decimal places past the zero from four units to three units. Keep the default drawing limits at (0.000,0.000) for the lower left corner and (12.000,9.000) for the upper right corner. Check to see that the following Object Snap modes are already set: Endpoint, Extension, Intersection, Center.

Layers
Create the following layers with the format:

Name	Color	Linetype
Boundary	Magenta	Continuous
Object	Yellow	Continuous

Suggested Commands
Begin drawing the C-Lever with point "A" illustrated in the previous image at absolute coordinate 7.000,3.375. Begin laying out all circles. Then draw tangent lines and arcs. Use the TRIM command to clean up unnecessary objects. To prepare to answer the AREA command question, convert the profile of the C-Lever to a polyline using the BOUNDARY command. Other questions pertaining to distances, angles, and point identifications follow. Do not dimension this drawing.

STEP 1

Make the Object layer current. Then construct one circle of 0.889 diameter with the center of the circle at absolute coordinate 7.000,3.375, as shown in the following image. Construct the remaining circles of the same diameter by using the COPY command. Use of the @ symbol for the base point in the COPY command identifies the last known point, which in this case is the center of the first circle drawn at coordinate 7.000,3.375.

 Command: **C** *(For CIRCLE)*
Specify center point for circle or [3P/2P/Ttr (tan tan radius)]: **7.000,3.375**
Specify radius of circle or [Diameter]: **D** *(For Diameter)*
Specify diameter of circle: **0.889**

Command: **CP** *(For COPY)*
Select objects: **L** *(For Last)*
Select objects: *(Press ENTER to continue)*
Specify base point or [Displacement] <Displacement>: **@**
Specify second point or <use first point as displacement>: **@1.649,2.630**
Specify second point or [Exit/Undo] <Exit>: **@-3.258,1.779**
Specify second point or [Exit/Undo] <Exit>: *(Press ENTER to exit this command)*

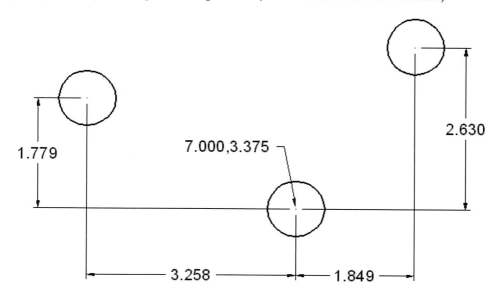

Figure 13–19

STEP 2

Construct three more circles, as shown in the following image. Even though these objects actually represent arcs, circles will be drawn now and trimmed later to form the arcs.

Command: **C** *(For CIRCLE)*
Specify center point for circle or [3P/2P/Ttr (tan tan radius)]: *(Select the edge of the circle at "A" to snap to its center)*
Specify radius of circle or [Diameter] <0.445>: **1.067**

Command: **C** *(For CIRCLE)*
Specify center point for circle or [3P/2P/Ttr (tan tan radius)]: *(Select the edge of the circle at "B" to snap to its center)*
Specify radius of circle or [Diameter] <1.067>: **0.889**

Command: **C** *(For CIRCLE)*
Specify center point for circle or [3P/2P/Ttr (tan tan radius)]: *(Select the edge of the circle at "C" to snap to its center)*
Specify radius of circle or [Diameter] <0.889>: **0.711**

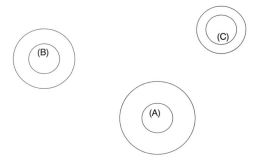

Figure 13–20

STEP 3

Construct lines tangent to the three outer circles, as shown in the following image.

Command: **L** *(For LINE)*
Specify first point: **Tan**
to *(Select the outer circle near "A")*
Specify next point or [Undo]: **Tan**
to *(Select the outer circle near "B")*
Specify next point or [Undo]: *(Press* ENTER *to exit this command)*

Command: **L** *(For LINE)*
Specify first point: **Tan**
to *(Select the outer circle near "C")*
Specify next point or [Undo]: **Tan**
to *(Select the outer circle near "D")*
Specify next point or [Undo]: *(Press* ENTER *to exit this command)*

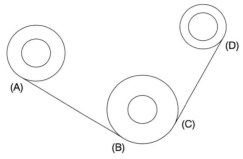

Figure 13–21

STEP 4

Construct a circle tangent to the two circles in the following image, using the CIRCLE command with the Tangent-Tangent-Radius option (TTR).

 Command: **C** *(For CIRCLE)*
Specify center point for circle or [3P/2P/Ttr (tan tan radius)]: **TTR**
Specify point on object for first tangent of circle: *(Select the outer circle near "A")*
Specify point on object for second tangent of circle: *(Select the outer circle near "B")*
Specify radius of circle <0.711>: **2.845**

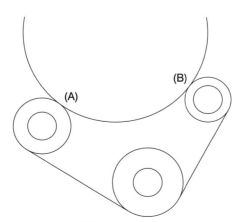

Figure 13–22

STEP 5

Use the TRIM command to clean up and form the finished drawing. Pressing ENTER at the "Select objects:" prompt will select all objects as cutting edges although they will not highlight, as shown in the following image. Study the following prompts for selecting the objects to trim.

Command: **TR** *(For TRIM)*
Current settings: Projection=UCS Edge=None
Select cutting edges ...
Select objects or <select all>: *(Press ENTER)*
Select object to trim or shift-select to extend or
    [Fence/Crossing/Project/Edge/eRase/Undo]: *(Select the circle at "A")*
Select object to trim or shift-select to extend or
    [Fence/Crossing/Project/Edge/eRase/Undo]: *(Select the circle at "B")*
Select object to trim or shift-select to extend or
    [Fence/Crossing/Project/Edge/eRase/Undo]: *(Select the circle at "C")*
Select object to trim or shift-select to extend or
    [Fence/Crossing/Project/Edge/eRase/Undo]: *(Select the circle at "D")*
Select object to trim or shift-select to extend or
    [Fence/Crossing/Project/Edge/eRase/Undo]: *(Press ENTER to exit this command)*

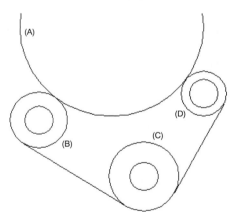

Figure 13–23

CHECKING THE ACCURACY OF C-LEVER.DWG

Once the C-Lever has been constructed, answer the following questions to determine the accuracy of this drawing. Use the following image to assist in answering the questions.

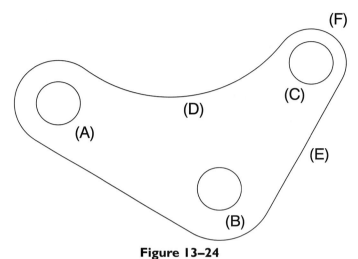

Figure 13–24

1. What is the total area of the C-Lever with all three holes removed?

   (A) 13.744
   (B) 13.749
   (C) 13.754
   (D) 13.759
   (E) 13.764

2. What is the total distance from the center of circle "A" to the center of circle "B"?

   (A) 3.692
   (B) 3.697
   (C) 3.702
   (D) 3.707
   (E) 3.712

3. What is the angle formed in the XY plane from the center of circle "C" to the center of circle "B"?

   (A) 223°
   (B) 228°
   (C) 233°
   (D) 238°
   (E) 243°

4. What is the delta X,Y distance from the center of circle "C" to the center of circle "A"?

   (A) -4.907,-0.851
   (B) -4.907,-0.856
   (C) -4.907,-0.861
   (D) -4.907,-0.866
   (E) -4.907,-0.871

5. What is the absolute coordinate value of the center of arc "D"?

   (A) 5.869,8.218
   (B) 5.869,8.223
   (C) 5.869,8.228
   (D) 5.869,8.233
   (E) 5.869,8.238

6. What is the total length of line "E"?

   (A) 3.074
   (B) 3.079
   (C) 3.084
   (D) 3.089
   (E) 3.094

7. What is the total length of arc "F"?

   (A) 2.051
   (B) 2.056
   (C) 2.061
   (D) 2.066
   (E) 2.071

A solution for each question follows, complete with the method used to arrive at the answer. Apply these methods to any type of drawing that requires the use of inquiry commands.

Tip: Before performing an area calculation that requires you to remove numerous objects, turn the BLIPMODE variable on. This will place a small mark on the object you select. In this way, you can keep better track of the objects you select. When finished, turn BLIPMODE off.

SOLUTIONS TO THE QUESTIONS ON C-LEVER

Question 1

What is the total area of the C-Lever with all three holes removed?

(A) 13.744   (D) 13.759
(B) 13.749   (E) 13.764
(C) 13.754

First make the Boundary layer current. Then use the BOUNDARY command and pick a point inside the object at "A" in the following image. This will trace a polyline around all closed objects on the Boundary layer.

Command: **BO** *(For BOUNDARY)*
(When the Boundary Creation dialog box appears, click on the Pick points button.)
Select internal point: *(Pick a point inside of the object at "Y")*
Selecting everything...
Selecting everything visible...
Analyzing the selected data...
Analyzing internal islands...
Select internal point: *(Press* ENTER *to create the boundaries)*
BOUNDARY created 4 polylines

Next, turn off the Object layer. All objects on the Boundary layer should be visible. Then use the AREA command to add and subtract objects to arrive at the final area of the object.

Command: **AA** *(For AREA)*
Specify first corner point or [Object/Add/Subtract]: **A** *(For Add)*
Specify first corner point or [Object/Subtract]: **O** *(For Object)*
(ADD mode) Select objects: *(Select the edge of the shape near "X")*
Area = 15.611, Perimeter = 17.771
Total area = 15.611
(ADD mode) Select objects: *(Press ENTER to continue)*
Specify first corner point or [Object/Subtract]: **S** *(For Subtract)*
Specify first corner point or [Object/Add]: **O** *(For Object)*
(SUBTRACT mode) Select objects: *(Select circle "A")*
Area = 0.621, Perimeter = 2.793
Total area = 14.991
(SUBTRACT mode) Select objects: *(Select circle "B")*
Area = 0.621, Perimeter = 2.793
Total area = 14.370
(SUBTRACT mode) Select objects: *(Select circle "C")*
Area = 0.621, Perimeter = 2.793
Total area = 13.749
(SUBTRACT mode) Select objects: *(Press ENTER to continue)*
Specify first corner point or [Object/Add]: *(Press ENTER to exit this command)*

The total area of the C-Lever with all three holes removed is (B), 13.749.

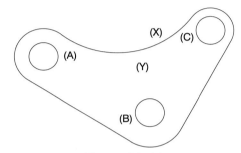

Figure 13–25

Question 2

What is the total distance from the center of circle "A" to the center of circle "B"?

(A) 3.692 (D) 3.707
(B) 3.697 (E) 3.712
(C) 3.702

Use the DIST (Distance) command to calculate the distance from the center of circle "A" to the center of circle "B" in the following image. Be sure to use the OSNAP-Center mode for locating the centers of all circles.

 Command: **DI** *(For DIST)*
Specify first point: *(Select the edge of the circle at "A")*
Specify second point: *(Select the edge of the circle at "B")*
Distance = 3.712, Angle in XY Plane = 331, Angle from XY Plane = 0
Delta X = 3.258, Delta Y = -1.779, Delta Z = 0.000

The total distance from the center of circle "A" to the center of circle "B" is (E), 3.712.

Figure 13–26

Question 3

What is the angle formed in the XY plane from the center of circle "C" to the center of circle "B"?

(A) 223° (D) 238°
(B) 228° (E) 243°
(C) 233°

Use the DIST (Distance) command to calculate the angle from the center of circle "C" to the center of circle "B" in the following image. Be sure to use the OSNAP-Center mode for locating the centers of all circles.

 Command: **DI** *(For DIST)*
Specify first point: *(Select the edge of the circle at "C")*
Specify second point: *(Select the edge of the circle at "B")*
Distance = 3.104, Angle in XY Plane = 238, Angle from XY Plane = 0
Delta X = -1.649, Delta Y = -2.630, Delta Z = 0.000

The angle formed in the XY plane from the center of circle "C" to the center of circle "B" is (D), <u>238°</u>.

Figure 13–27

Question 4

What is the delta X,Y distance from the center of circle "C" to the center of circle "A"?

(A) -4.907,-0.851 (D) -4.907,-0.866
(B) -4.907,-0.856 (E) -4.907,-0.871
(C) -4.907,-0.861

Use the DIST (Distance) command to calculate the delta X,Y distance from the center of circle "C" to the center of circle "A" in the following image. Be sure to use the OSNAP-Center mode. Notice that additional information is given when you use the DIST command. For the purpose of this question, we will be looking only for the delta X,Y distance. The DIST command will display the relative X,Y,Z distances. Since this is a 2D problem, only the X and Y values will be used.

 Command: **DI** *(For DIST)*
Specify first point: *(Select the edge of the circle at "C")*
Specify second point: *(Select the edge of the circle at "A")*
Distance = 4.980, Angle in XY Plane = 190, Angle from XY Plane = 0
Delta X = -4.907, Delta Y = -0.851, Delta Z = 0.000

The delta X,Y distance from the center of circle "C" to the center of circle "A" is (A), <u>-4.907,-0.851</u>.

Figure 13–28

Question 5

What is the absolute coordinate value of the center of arc "D"?

(A) 5.869,8.218   (D) 5.869,8.233
(B) 5.869,8.223   (E) 5.869,8.238
(C) 5.869,8.228

The ID command is used to get the current absolute coordinate information on a desired point, as shown in the following image. This command will display the X,Y,Z coordinate values. Since this is a 2D problem, only the X and Y values will be used.

 Command: **ID**
Specify point: **Cen**
of *(Select the edge of the arc at "D")*
X = 5.869   Y = 8.223   Z = 0.000

The absolute coordinate value of the center of arc "D" is (B), <u>5.869,8.223</u>.

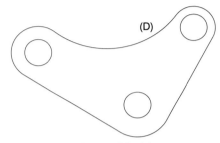

Figure 13–29

Question 6

What is the total length of line "E"?

(A) 3.074   (D) 3.089
(B) 3.079   (E) 3.094
(C) 3.084

Use the DIST (Distance) command to find the total length of line "E" in the following image. Be sure to use the OSNAP-Endpoint mode. Notice that additional information is given when you use the DIST command. For the purpose of this question, we will be looking only for the distance.

 Command: **DI** *(For DIST)*
Specify first point: *(Select the endpoint of the line at "X")*
Specify second point: *(Select the endpoint of the line at "Y")*

Distance = 3.084, Angle in XY Plane = 64, Angle from XY Plane = 0
Delta X = 1.328, Delta Y = 2.783, Delta Z = 0.000

The total length of line "E" is (C), <u>3.084</u>.

Figure 13–30

Question 7

What is the total length of arc "F"?

(A) 2.051   (D) 2.066

(B) 2.056   (E) 2.071

(C) 2.061

The LIST command is used to calculate the lengths of arcs. However, a little preparation is needed before you perform this operation. If arc "F" is selected, as shown on the left in the following image, notice that the entire outline is selected because it is a polyline. Use the EXPLODE command to break the outline into individual objects. Use the LIST command to get a listing of the arc length, as shown on the right in the following image.

Command: **X** *(For EXPLODE)*

Select objects: *(Select the edge of the dashed polyline in the following image)*
Select objects: *(Press ENTER to perform the explode operation)*

Command: **LI** *(For LIST)*

Select objects: *(Select the edge of the arc at "F" in the following image)*
Select objects: *(Press ENTER to continue)*
ARC   Layer: "Boundary"
Space: Model space
Handle = 94
center point, X= 8.649 Y= 6.005 Z= 0.000
radius   0.711
start angle   334
end angle   141
length   2.071

The total length of arc "F" is (E), <u>2.071</u>.

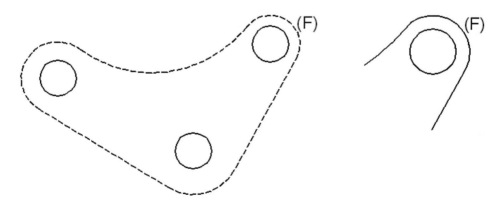

Figure 13–31

TUTORIAL EXERCISE: 13_FIELD_CALC.DWG

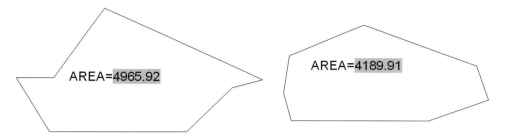

Figure 13-32

Purpose
This tutorial is designed to utilize the Field function of AutoCAD to calculate the area of an object.

System Settings
This drawing has drawing settings already created

Layers
Layers already exist in this drawing.

Suggested Commands
Open the drawing 13_Field_Calc.Dwg and activate the multiline text command. Set the text height to 12 and add a text header dealing with the Area. Then activate the Field dialog box and select a polyline object in the drawing. This will retrieve the area information and add this text calculation to the Text Formatting dialog box. Once the field is added, stretching the original polyline shape and regenerating the drawing will recalculate the area.

STEP 1

Open the drawing 13_Field_Calc.Dwg. Activate the Text Formatting dialog box, change the text height to 5, and add the text (AREA=), as shown in the following image. Continue by clicking on the Insert Field button.

Figure 13–33

STEP 2

When the Field dialog box appears, select the Object Field name, as shown in the following image. This will prepare the dialog box to display field information based on the object that you select. Continue by clicking on the Select object button as shown in the following image.

Figure 13–34

STEP 3

Clicking the Select object button will return you to the drawing editor, where you can select the polyline shape of the object. This action will again return you to the Field dialog box. Verify the following changes in this dialog box: under the Field Names category, pick Object; under the Property category, pick Area; under the Format category, pick Decimal. Then click the OK button to exit the Field dialog box.

Figure 13–35

STEP 4

The results are illustrated in the following image. A field consisting of the square footage information is added to the end of the multiline text.

Figure 13–36

STEP 5

Now click on the polyline shape to display grips, choose a number of grips, and stretch the vertices of the polyline to new locations as shown in the following image. Your display may not match the following image exactly. You will notice that the square footage does not automatically update itself.

Figure 13-37

STEP 6

To update the field information and change the area for the new polyline shape, click on the Regen command found under the View pull-down menu, as shown in the following image.

Figure 13-38

STEP 7

The results are illustrated in the following image. Notice that the area field has changed to reflect the new size of the closed polyline shape.

Figure 13–39

PROBLEMS FOR CHAPTER 13

PROBLEM 13–1 ANGLEBLK.DWG

Directions for Angleblk.Dwg

Use the Drawing Units dialog box to set the units to decimal. Set the precision to two. Be sure the system of angle measure is set to decimal degrees and the number of decimal places for the display of angles is zero.

Keep all remaining default unit values. Keep the default settings for the drawing limits.

Begin the drawing shown in the following image by locating the lower-left corner of Angleblk, identified by "X" at coordinate (2.35,3.17).

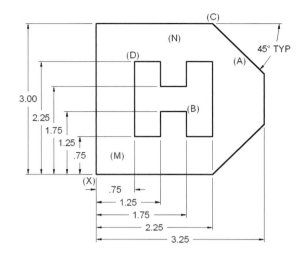

Refer to the drawing of the Angleblk and answer the following questions.

1. What is the total surface area of the Angleblk with the inner "H" shape removed?

   (A) 6.66 (C) 6.89 (E) 7.11
   (B) 6.77 (D) 7.00

2. What is the total area of the inner "H" shape?

   (A) 1.75 (C) 1.87 (E) 1.99
   (B) 1.81 (D) 1.93

3. What is the total length of line "A"?

   (A) 1.29 (C) 1.41 (E) 1.53
   (B) 1.35 (D) 1.47

4. What is the absolute coordinate value of the endpoint of the line at "B"?

   (A) 4.04,4.42 (D) 4.22,4.42
   (B) 4.10,4.42 (E) 4.28,4.42
   (C) 4.16,4.42

5. What is the absolute coordinate value of the endpoint of the line at "C"?

   (A) 4.60,6.11 (D) 4.60,6.29
   (B) 4.60,6.17 (E) 4.60,6.35
   (C) 4.60,6.23

6. Use the stretch command and extend the inner "H" shape a distance of 0.37 units in the 180° direction. Use "N" as the first corner of the crossing window and "M" as the other corner. Use the endpoint of "D" as the base point of the stretching operation. What is the new surface area of Angleblk with the inner "H" shape removed?

   (A) 6.63 (C) 6.75 (E) 6.87
   (B) 6.69 (D) 6.82

7. Use the scale command with the endpoint of the line at "D" as the base point. Reduce the size of just the inner "H" using a scale factor of 0.77. What is the new surface area of Angleblk with the inner "H" removed?

   (A) 7.48 (C) 7.60 (E) 7.72
   (B) 7.54 (D) 7.66

PROBLEM 13–2 LEVER2.DWG

Directions for Lever2.Dwg

Use the Drawing Units dialog box to set the units to decimal. Keep the precision at four places.

Be sure the system of angle measure is set to decimal degrees and the number of decimal places for the display of angles is zero.

Keep the remaining default unit values.

Begin the drawing shown in the following image by locating the center of the 1.0000-diameter circle at absolute coordinate (2.2500,4.0000).

Refer to the drawing of Lever2 and answer the following questions.

1. What is the total area of Lever2 with the inner irregular shape and both holes removed?
   (A) 17.6813   (D) 17.6831
   (B) 17.6819   (E) 17.6837
   (C) 17.6825

2. What is the absolute coordinate value of the center of the 4.5000-radius arc "C"?
   (A) 4.8944,-0.0822
   (B) 4.8944,-0.8226
   (C) 4.8950,-0.8232
   (D) 4.8956,-0.8238
   (E) 4.8962,-0.8244

3. What is the absolute coordinate value of the center of the 6.0000-radius arc "D"?
   (A) 6.0828,0.7893
   (B) 6.0834,0.7899
   (C) 6.0840,0.7905
   (D) 6.0846,0.7911
   (E) 6.0852,0.7917

4. What is the total length of arc "C"?
   (A) 5.3583   (D) 5.3601
   (B) 5.3589   (E) 5.3607
   (C) 5.3595

5. What is the distance from the center of the 1.0000-diameter circle "A" to the intersection of the circle and centerline at "B"?
   (A) 6.8456   (D) 6.8480
   (B) 6.8462   (E) 6.8486
   (C) 6.8474

6. What is the angle formed in the XY plane from the upper quadrant of arc "D" to the center of the 1.5000-circle "E"?
   (A) 313°   (C) 317°   (E) 321°
   (B) 315°   (D) 319°

7. What is the delta X,Y distance from the upper quadrant of arc "C" to the center of the 1.0000-hole "A"?
   (A) -2.6444,0.3220   (C) -2.6444,0.3232
   (B) -2.6444,0.3226   (D) -2.6444,0.3238
   (E) -2.6444,0.3244

PROBLEM 13–3 MAIN ROCKER.DWG

Directions for Main Rocker.Dwg

Begin the construction of Main Rocker by keeping the default units set to decimal but changing the number of decimal places past the zero from four to two. Be sure the system of angle measure is set to decimal degrees and the number of decimal places for the display of angles is zero. Keep the remaining default unit values. Use the LIMITS command to change the drawing limits to (0.00,0.00) for the lower left corner and (15.00,12.00) for the upper right corner

Begin the drawing by placing the center of the regular hexagon and 2.25-radius arc at coordinate (6.25,6.50).

Refer to the drawing of Main Rocker and answer the following questions.

1. What is the distance from the center of the 1.00-diameter hole "A" to the center of the other 1.00-diameter hole "B"?
   (A) 8.39   (C) 8.46   (E) 8.52
   (B) 8.42   (D) 8.49

2. What is the absolute coordinate value of the center of the 3.00-radius arc "C"?
   (A) 7.79,1.48   (D) 7.76,1.51
   (B) 7.79,1.51   (E) 7.76,1.48
   (C) 7.79,1.54

3. What is the total length of line "G"?
   (A) 4.14   (C) 4.24   (E) 4.34
   (B) 4.19   (D) 4.29

4. What is the length of the 2.25-radius arc segment "E"?
   (A) 2.10   (C) 2.16   (E) 2.22
   (B) 2.13   (D) 2.19

5. What is the angle formed in the XY plane from the center of the 1.00-diameter circle "B" to the center of the 2.25-radius arc "D"?
   (A) 203°   (C) 207°   (E) 211°
   (B) 205°   (D) 209°

6. What is the total surface area of Main Rocker with the hexagon and both 1.00-diameter holes removed?
   (A) 20.97   (D) 21.06
   (B) 21.00   (E) 21.09
   (C) 21.03

7. What is the total length of arc "D"?
   (A) 2.31   (D) 2.46
   (B) 2.36   (E) 2.51
   (C) 2.41

PROBLEM 13–4 PATTERN4.DWG

Directions for Pattern4.Dwg

Start a new drawing called Pattern4. Even though this is a metric drawing, no special limits need be set. Keep the default setting of decimal units precision to zero. Be sure the system of angle measure is set to decimal degrees and the number of decimal places for the display of angles is zero. Keep all remaining default unit values.

AB=94
BC=40
CD=35
DE=57
EF=82
FG=61
GH=38
HJ=85
JK=53

Begin the drawing, as shown in the following image, by constructing Pattern4 with vertex "A" at absolute coordinate (50,30).

Refer to the drawing of Pattern4 and answer the following questions.

1. What is the total distance from the intersection of vertex "K" to the intersection of vertex "A"?

    (A) 33   (C) 35   (E) 37
    (B) 34   (D) 36

2. What is the total area of Pattern4 with the slot removed?

    (A) 14493   (C) 14529   (E) 14620
    (B) 14500   (D) 14539

3. What is the perimeter of the outline of Pattern4?

    (A) 570   (C) 586   (E) 602
    (B) 578   (D) 594

4. What is the distance from the intersection of vertex "A" to the intersection of vertex "E"?

    (A) 186   (C) 194   (E) 202
    (B) 190   (D) 198

5. What is the absolute coordinate value of the intersection at vertex "G"?

    (A) 104,117   (D) 107,120
    (B) 105,118   (E) 108,121
    (C) 106,119

6. What is the total length of arc "L"?

    (A) 20   (C) 22   (E) 24
    (B) 21   (D) 23

7. Stretch the portion of Pattern4 around the vicinity of angle "E." Use "Y" as the first corner of the crossing box. Use "X" as the other corner. Use the endpoint of "E" as the base point of the stretching operation. For the new point, enter a polar coordinate value of 26 units in the 40° direction. What is the new degree value of the angle formed at vertex "E"?

    (A) 78°   (D) 81°
    (B) 79°   (E) 82°
    (C) 80°

PROBLEM 13–5 ARCHITECTURAL APPLICATION—APARTMENT.DWG

Directions for Apartment.Dwg

Change from decimal units to architectural units. Keep all remaining default values. The thickness of all walls measures 4″.

Refer to the drawing of Apartment and answer the following questions.

1. What is the total area in square feet of all bedrooms? (Pick the closest value.)
   (A) 296 sq. ft.   (D) 326 sq. ft.
   (B) 306 sq. ft.   (E) 336 sq. ft.
   (C) 316 sq. ft.

2. What is the distance from the intersection of the corner at "A" to the intersection of the corner at "B"? (Pick the closest value.)
   (A) 25′-2″   (C) 26′-0″   (E) 26′-10″
   (B) 25′-7″   (D) 26′-5″

3. What is the total area in square feet of all closets (C1, C2, and C3)? (Pick the closest value.)
   (A) 46 sq. ft.   (D) 52 sq. ft.
   (B) 48 sq. ft.   (E) 54 sq. ft.
   (C) 50 sq. ft.

4. What is the total area in square feet of the laundry and bathroom? (Pick the closest value.)
   (A) 91 sq. ft.   (D) 97 sq. ft.
   (B) 93 sq. ft.   (E) 99 sq. ft.
   (C) 95 sq. ft.

5. What is the distance from the intersection of the corner at "C" to the intersection of the corner at "D"? (Pick the closest value.)
   (A) 31′-0″   (D) 32′-3″
   (B) 31′-5″   (E) 32′-8″
   (C) 31′-10″

6. What is the angle in the XY plane from the intersection of the wall corner at "A" to the intersection of the wall corner at "D"? (Pick the closest value.)
   (A) 1°   (D) 7°
   (B) 3°   (E) 9°
   (C) 5°

7. Stretch the floor plan up at a distance of 2′-5″. Use "E" and "F" as the locations of the crossing box. What is the new total area of Bedroom #1, Bedroom #2, C1, and C2? (Pick the closest value.)
   (A) 315 sq. ft.   (D) 327 sq. ft.
   (B) 319 sq. ft.   (E) 331 sq. ft.
   (C) 323 sq. ft.

Intermediate-Level Drawings

PROBLEM 13–6 GASKET2.DWG

Directions for Gasket2.Dwg

Use the Drawing Units dialog box to set the units to decimal. Set the precision to two. Be sure the system of angle measure is set to decimal degrees and the number of decimal places for the display of angles is zero.

Keep the remaining default unit values.

Begin the drawing by locating the center of the 6.00 × 3.00 rectangle at absolute coordinate (6.00,4.75).

Refer to the drawing of Gasket2 and answer the following questions.

1. What is the total surface area of Gasket2 with the rectangle and all ten holes removed?
   (A) 21.46 (C) 21.50 (E) 21.54
   (B) 21.48 (D) 21.52

2. What is the distance from the center of arc "A" to the center of arc "B"?
   (A) 6.63 (C) 6.71 (E) 6.79
   (B) 6.67 (D) 6.75

3. What is the length of arc segment "C"?
   (A) 3.44 (C) 3.50 (E) 3.56
   (B) 3.47 (D) 3.53

4. What is the absolute coordinate value of the center of the 0.75-radius arc "D"?
   (A) 4.83,2.50 (D) 4.80,2.50
   (B) 4.83,2.47 (E) 4.83,2.56
   (C) 4.83,2.53

5. What is the angle formed in the XY plane from the center of the 0.75-radius arc "D" to the center of the 0.75-radius arc "A"?
   (A) 116° (C) 120° (E) 124°
   (B) 118° (D) 122°

6. What is the delta X,Y distance from the intersection at "E" to the midpoint of the line at "F"?
   (A) -4.50,3.65 (D) -4.50,3.75
   (B) -4.50,-3.65 (E) -4.50,-3.75
   (C) -4.50,-3.70

7. Use the SCALE command to reduce the size of the inner rectangle. Use the midpoint of the line at "F" as the base point. Use a scale factor of 0.83 units. What is the new total surface area with the rectangle and all ten holes removed?
   (A) 26.99 (D) 27.14
   (B) 27.04 (E) 27.19
   (C) 27.09

PROBLEM 13–7 HANGER.DWG

Directions for Hanger.Dwg

Use the Drawing Units dialog box to set units to decimal. Set precision to three. Be sure the system of angle measure is set to decimal degrees and the number of decimal places for the display of angles is zero. Keep all remaining default unit values. Use the LIMITS command and set the upper right corner of the screen area to a value of (250.000,150.000). Perform a Zoom All operation.

Begin drawing the hanger by locating the center of the 40.000-radius arc at absolute coordinate (55.000,85.000).

Refer to the drawing of the hanger and answer the following questions.

1. What is the area of the outer profile of the hanger?
   (A) 9970.567   (D) 9975.005
   (B) 9965.567   (E) 9980.347
   (C) 9975.567

2. What is the area of the hanger with the polygon, circle, and irregular shapes removed?
   (A) 7304.089 (D) 7304.000
   (B) 7305.000 (E) 7306.098
   (C) 7303.890

3. What is the absolute coordinate value of the center of the 28.000-radius arc "A"?
   (A) 120.000,69.000   (D) 121.082,69.000
   (B) 121.082,68.964   (E) 121.082,66.937
   (C) 121.520,68.237

4. What is the absolute coordinate value of the center of the 20.000-diameter circle "B"?
   (A) 154.000,35.000   (D) 155.000,35.000
   (B) 156.000,36.000   (E) 156.000,37.000
   (C) 156.147,35.256

5. What is the angle formed in the XY plane from the center of the 5.000-radius arc "C" to the center of the 40.000-radius arc "D"?
   (A) 219°   (C) 223°   (E) 227°
   (B) 221°   (D) 225°

6. What is the total area of irregular shape "E"?
   (A) 271.613   (D) 271.801
   (B) 271.723   (E) 271.822
   (C) 271.784

7. What is the total area of irregular shape "F"?
   (A) 698.511   (D) 699.856
   (B) 698.621   (E) 699.891
   (C) 699.817

PROBLEM 13–8 HOUSING1.DWG

Directions for Housing1.Dwg

Begin constructing Housing1 by keeping the default units set to decimal but changing the precision to three. Be sure the system of angle measure is set to decimal degrees and the number of decimal places for the display of angles is zero. Place the center of the 1.500-radius circular centerline at absolute coordinate (6.500,5.250).

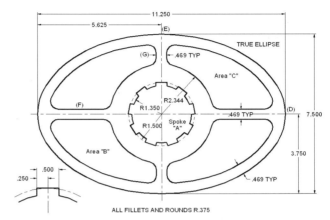

Before constructing the outer ellipse, set the PELLIPSE command to a value of 1. This will draw the ellipse as a polyline object.

Refer to the drawing of Housing1 and answer the following questions.

1. What is the perimeter of Spoke "A"?
   (A) 11.564   (D) 11.582
   (B) 11.570   (E) 11.588
   (C) 11.576

2. What is the perimeter of Area "B"?
   (A) 12.513   (D) 12.531
   (B) 12.519   (E) 12.537
   (C) 12.525

3. What is the total area of Area "C"?
   (A) 7.901   (C) 7.913   (E) 7.928
   (B) 7.907   (D) 7.919

4. What is the absolute coordinate value of the intersection of the ellipse and centerline at "D"?
   (A) 12.125,5.244   (D) 12.125,5.262
   (B) 12.125,5.250   (E) 12.125,5.268
   (C) 12.125,5.256

5. What is the total surface area of Housing1 with the spoke and all slots removed?
   (A) 27.095   (D) 28.113
   (B) 28.101   (E) 28.119
   (C) 28.107

6. What is the distance from the midpoint of the horizontal line segment at "F" to the midpoint of the vertical line segment at "G"?
   (A) 4.235   (C) 4.247   (E) 4.259
   (B) 4.241   (D) 4.253

7. Increase Spoke "A" in size using the SCALE command. Use the center of the 1.500-radius arc as the base point. Use a scale factor of 1.115 units. What is the new total area of Housing1 with the spoke and all slots removed?
   (A) 26.535   (D) 26.553
   (B) 26.541   (E) 26.559
   (C) 26.547

PROBLEM 13-9 ROTOR.DWG

Directions for Rotor.Dwg

Start a new drawing called Rotor, as shown in the following image. Keep the default setting of decimal units precision to three.

Be sure the system of angle measure is set to decimal degrees and the number of decimal places for the display of angles is zero. Keep all remaining default unit values. Begin the drawing by constructing the center of the 6.250-unit-diameter circle at "A" at absolute coordinate (5.500,5.000).

Refer to the drawing of the rotor and answer the following questions.

1. What is the absolute coordinate value of the center of the 0.625-diameter circle "B"?
   (A) 4.294,2.943
   (B) 4.300,2.943
   (C) 4.306,2.943
   (D) 4.312,2.943
   (E) 4.318,2.943

2. What is the total area of the rotor with all eight holes and the center slot removed?
   (A) 21.206
   (B) 21.210
   (C) 21.214
   (D) 21.218
   (E) 21.222

3. What is the total length of arc "F"?
   (A) 3.260
   (B) 3.264
   (C) 3.268
   (D) 3.272
   (E) 3.276

4. What is the distance from the center of the 0.625 circle "C" to the center of the 0.625 circle "D"?
   (A) 3.355
   (B) 3.359
   (C) 3.363
   (D) 3.367
   (E) 3.371

5. What is the angle formed in the XY plane from the center of the 0.625 circle "B" to the center of the 0.625 circle "E"?
   (A) 11°
   (B) 13°
   (C) 15°
   (D) 17°
   (E) 19°

6. What is the delta X,Y distance from the intersection at "H" to the intersection at "I"?
   (A) -3.827,-3.827
   (B) -3.827,3.827
   (C) -3.834,3.820
   (D) -3.841,3.813
   (E) -3.848,-3.806

7. Use the SCALE command to increase the size of just the center slot "G." Use the center of arc "F" as the base point. Use a scale factor of 1.500 units. What is the new surface area of the rotor with the center slot and all eight holes removed?
   (A) 17.861
   (B) 17.868
   (C) 17.875
   (D) 17.882
   (E) 17.889

PROBLEM 13–10 PATTERN5.DWG

Directions for Pattern5.Dwg

Begin the construction of Pattern5 by keeping the default units set to decimal but changing the precision to zero. Be sure the system of angle measure is set to decimal degrees and the number of decimal places for the display of angles is zero. Begin the drawing by placing Vertex "A" at absolute coordinate (190,30).

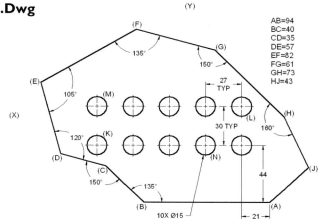

Refer to the drawing of Pattern5 and answer the following questions.

1. What is the distance from the intersection of vertex "J" to the intersection of vertex "A"?
   (A) 38   (D) 41
   (B) 39   (E) 42
   (C) 40

2. What is the perimeter of Pattern5?
   (A) 523   (D) 526
   (B) 524   (E) 527
   (C) 525

3. What is the total area of Pattern5 with all 10 holes removed?
   (A) 16369   (D) 16372
   (B) 16370   (E) 16373
   (C) 16371

4. What is the distance from the center of the 15-diameter hole "K" to the center of the 15-diameter hole "L"?
   (A) 109   (D) 112
   (B) 110   (E) 113
   (C) 111

5. What is the angle formed in the XY plane from the center of the 15-diameter hole "M" to the center of the 15-diameter hole "N"?
   (A) 340°   (D) 346°
   (B) 342°   (E) 348°
   (C) 344°

6. What is the absolute coordinate value of the intersection at "F"?
   (A) 90,163   (D) 96,167
   (B) 92,163   (E) 98,169
   (C) 94,165

7. Use the STRETCH command to lengthen Pattern5. Use "Y" as the first point of the stretch crossing box. Use "X" as the other corner. Pick the intersection at "F" as the base point and stretch Pattern5 a total of 23 units in the 180° direction. What is the new total area of Pattern5 with all ten holes removed?
   (A) 17746   (D) 17767
   (B) 17753   (E) 17774
   (C) 17760

PROBLEM 13–11 RATCHET.DWG

Directions for Ratchet.Dwg

Use the Drawing Units dialog box to change the precision to two. Be sure the system of angle measure is set to decimal degrees and the number of decimal places for the display of angles is zero. Keep the remaining default unit values. Begin by drawing the center of the 1.00-radius arc of the ratchet at absolute coordinate (6.00,4.50), as shown in the following image.

Refer to the drawing of the ratchet and answer the following questions.

1. What is the total length of the short line segment "A"?
   (A) 0.22   (D) 0.28
   (B) 0.24   (E) 0.30
   (C) 0.26

2. What is the total length of line "B"?
   (A) 1.06   (D) 1.45
   (B) 1.19   (E) 1.57
   (C) 1.33

3. What is the total length of arc "C"?
   (A) 1.81   (D) 2.19
   (B) 1.93   (E) 2.31
   (C) 2.08

4. What is the perimeter of the 1.00-radius arc "D" with the 0.25 × 0.12 keyway?
   (A) 6.52   (D) 6.89
   (B) 6.63   (E) 7.02
   (C) 6.77

5. What is the total surface area of the ratchet with all 4 slots, the two 1.00-diameter holes, and the 1.00-radius arc with the keyway removed?
   (A) 42.98   (D) 43.16
   (B) 43.04   (E) 43.22
   (C) 43.10

6. What is the absolute coordinate value of the endpoint at "F"?
   (A) 2.01,5.10   (D) 2.20,5.95
   (B) 2.01,5.63   (E) 2.37,5.95
   (C) 2.01,5.95

7. What is the angle formed in the XY plane from the endpoint of the line at "F" to the center of the 1.00-diameter hole "G"?
   (A) 304°   (D) 310°
   (B) 306°   (E) 312°
   (C) 308°

PROBLEM 13–12 GENEVA.DWG

Directions for Geneva.Dwg

Start a new drawing called Geneva. Keep the default settings of decimal units, and precision set to two. Be sure the system of angle measure is set to decimal degrees and the number of decimal places for the display of angles is zero. Begin the drawing by constructing the 1.50-diameter arc at absolute coordinate (7.50,5.50).

Refer to the drawing of the Geneva and answer the following questions.

1. What is the total length of arc "A"?
   (A) 3.00   (C) 3.21   (E) 3.40
   (B) 3.10   (D) 3.30

2. What is the angle formed in the XY plane from the intersection at "B" to the center of arc "C"?
   (A) 11°   (C) 15°   (E) 19°
   (B) 13°   (D) 17°

3. What is the absolute coordinate value of the midpoint of line "D"?
   (A) 5.27,7.13   (D) 5.31,7.13
   (B) 5.27,7.17   (E) 5.31,7.09
   (C) 5.23,7.13

4. What is the total area of the Geneva with the 1.50-diameter hole and keyway removed?
   (A) 27.40   (D) 27.52
   (B) 27.44   (E) 27.66
   (C) 27.48

5. What is the total distance from the midpoint of arc "F" to the center of arc "A"?
   (A) 8.24   (D) 8.39
   (B) 8.29   (E) 8.44
   (C) 8.34

6. What is the delta X,Y distance from the intersection at "E" to the center of arc "C"?
   (A) 3.93,-3.75   (D) 3.75,3.93
   (B) 3.93,3.75   (E) 3.75,-3.93
   (C) 3.75,3.80

7. Use the SCALE command to reduce the Geneva in size. Use 7.50,5.50 as the base point; use a scale factor of 0.83 units. What is the absolute coordinate value of the intersection at "E"?
   (A) 8.12,8.71   (D) 8.12,8.86
   (B) 8.12,8.76   (E) 8.12,8.91
   (C) 8.12,8.81

PROBLEM 13–13 STRUCTURAL APPLICATION—GUSSET.DWG

Directions for Gusset.Dwg

Begin the construction of the Gusset by keeping the default units set to decimal, but changing the precision to two.

Refer to the drawing of the Gusset and answer the following questions.

1. What is the total area of the gusset plate with all 35 rivet holes removed?
   (A) 1045.46
   (B) 1050.67
   (C) 1055.21
   (D) 1060.40
   (E) 1065.88

2. What is the delta X,Y distance from the center of hole "A" to the center of hole "B"?
   (A) 9.07,-12.19
   (B) 9.07,-12.24
   (C) 9.07,-12.29
   (D) 9.07,-12.34
   (E) 9.07,-12.39

3. What is the angle formed in the XY plane from the center of hole "C" to the center of hole "D"?
   (A) 285°
   (B) 288°
   (C) 291°
   (D) 294°
   (E) 297°

4. What is the angle formed in the XY plane from the center of hole "E" to the center of hole "F"?
   (A) 165°
   (B) 168°
   (C) 171°
   (D) 174°
   (E) 177°

5. What is the total length of line "J"?
   (A) 13.15
   (B) 13.20
   (C) 13.25
   (D) 13.30
   (E) 13.35

6. What is the total length of line "K"?
   (A) 44.02
   (B) 44.07
   (C) 44.12
   (D) 44.17
   (E) 44.22

7. Stretch the gusset plate directly to the left using a crossing box from "H" to "G" and at a distance of 3.00 units. What is the new area of the gusset plate with all 35 rivet holes removed?
   (A) 1121.98
   (B) 1126.72
   (C) 1131.11
   (D) 1136.59
   (E) 1141.34

CHAPTER 14

Working with Drawing Layouts

INTRODUCING DRAWING LAYOUTS

Up to this point in the book, all drawings have been constructed in Model Space. This consisted of drawings composed of geometry such as arcs, circles, lines, and even dimensions. You can even plot your drawing from Model Space; this gets a little tricky, especially when you attempt to plot the drawing at a known scale, or even trickier if multiple details need to be arranged on a single sheet at different scales.

You can elect to work in two separate spaces on your drawing; namely Model Space and Layout mode, sometimes referred to as Paper Space. Part of the geometry and dimensions is still drawn in Model Space. However, items such as notes, annotations, and title blocks are laid out separately in a drawing layout, which is designed to simulate an actual sheet of paper. To arrange a single view of a drawing or multiple views of different drawings, you arrange a series of viewports in the drawing layout to view the images. These viewports are mainly rectangular in shape and can be made any size. You can even create circular and polygonal viewports. An option of the ZOOM command allows you to scale the image inside the viewport. In this way, a series of images may be scaled differently in order for the drawing to be plotted out at a scale of 1:1. A dedicated toolbar containing standard scales is also available to perform this task.

This chapter introduces you to the controls used in a drawing layout to manage information contained in a viewport. A tutorial exercise is provided to help you practice creating drawing layouts.

MODEL SPACE

Before starting on the topic of Paper Space, you must first understand the environment in which all drawings are originally constructed, namely Model Space. It is here that the drawing is drawn full size or in real-world units. Model Space is easily identified by the appearance of the User Coordinate System icon located in the lower left corner of the active drawing area in the following image. This icon is associated with the Model Space environment. Past and present users of AutoCAD have felt that it clutters the display screen and gets in the way of geometry, and they have used the UCSICON command to turn the icon off. While these points may seem true, the icon should be present at all times to identify the space you are working in. Another indicator that you are in Model Space is the presence of the Model tab, located just below the User Coordinate System icon, also in the following image.

Figure 14–1

TILED VIEWPORTS IN MODEL SPACE

An advantage of constructing in the Model Space environment is the ability to divide the display screen into a number of smaller screens, as illustrated in the following image. (Drawing file 14_V-Step is provided and can be opened to practice creating viewports in Model Space.) In the illustration, the display screen is divided into four smaller screens or viewports. These viewports are further arranged in a certain pattern, or configuration. You do not have much control over this viewport arrangement. For this reason these are called tiled viewports, because they resemble the tile pattern commonly used in floors. Do not be confused that you have four separate files open in the illustration. Rather, it is the same drawing file, which enables you to view separate parts of the drawing in each viewport. Again, notice the appearance of the familiar User Coordinate System icon present in the lower left corner of each viewport. This signifies you are in Model Space in each viewport. Also in the illustration, notice how the lower right viewport has a thick border around it along with the presence of the AutoCAD cursor. This happens to be the active viewport. Only one viewport can be active at one time. If a viewport is inactive, the appearance of an arrow instead of a cursor is present. To activate another viewport, simply move the arrow to the viewport and click inside the viewport; this activates a new current viewport.

Figure 14–2

CREATING VIEWPORTS IN MODEL MODE

Viewports are created through the VPORTS command. You can access this command by choosing Viewports from the View pull-down menu. Choosing New Viewports…, as shown in the following image, activates the Viewports dialog box. A toolbar dedicated to viewports is also illustrated in the following image. Numerous layout examples and configurations for viewports are available. Clicking on the desired viewport in the Standard viewports area previews the viewport layout at the right of the dialog box. These changes will be made to the display after you click the OK button at the bottom of the dialog box.

Figure 14–3

MODEL SPACE AND PAPER SPACE

You have already seen that Model Space is the environment set aside for constructing your drawing in real-world units or full size, as shown on the left in the following image. Model space can be easily recognized by the familiar User Coordinate System icon found in the lower left corner of every model space screen.

Clicking on the Layout1 tab found in the lower left corner of the display screen will activate the Paper Space environment. Paper Space is considered an area used to lay out your drawing before it is plotted. Paper Space is also used to place title block information and notes associated with the drawing. The drawing illustrated on the right in the following image has been laid out in Paper Space and shows a sheet of paper with the drawing surrounded by a viewport. The dashed lines along the outer perimeter of the sheet are referred to as margins. Anything inside the margins will plot; for this reason, this is called the printable area. Notice also at the bottom of the screen that the Layout1 tab is activated. Both tabs at the bottom of the screen can be used to easily display your drawing in either Model Space or Paper Space. Also notice the icon in the lower left corner of the illustration; this icon is in the form of a triangle and is used to quickly identify the Paper Space environment.

Figure 14–4

PAPER SPACE FEATURES

When you draw and plot from Model Space, the limits of the drawing have to be properly set depending on the scale of the drawing and the sheet of paper the drawing is planned for. Also, title block information has to be scaled depending on the scale of the drawing. The text height that made up the notes in a drawing also has to be scaled depending on the scale of the drawing. Finally, the scale of the drawing is applied to the Plot Settings tab of the Plot dialog box, which scales the drawing for placement on a certain sheet of paper. There can be many problems when you use this method. For instance, a certain design office assigned an individual to plot all drawings produced by the engineers and designers. It was this individual's responsibility to plot the drawings at the proper scale. However, with numerous drawings to plot, it was easy to get the plot scales confused. As a result, drawings were plotted at the wrong scale. This not only cost paper but valuable time. This scenario is the reason for Paper Space; all drawings, no matter how many details are arranged, can be plotted at 1 = 1 with little or no error encountered, through the use of Paper Space. Plotting becomes the fundamental reason for using the Paper Space environment. Here are a few other reasons for arranging a drawing in Paper Space:

- Paper Space is based on the actual sheet size. If you are plotting a drawing on a D-size sheet of paper, you use the actual size (36 × 24) in Paper Space.
- Title blocks and the text used for notes do not have to be scaled as in the past with Model Space.
- Viewports created in Paper Space are user-defined. Viewports created in Model Space are dependent on a configuration set by AutoCAD. In Paper Space, the viewports can be different sizes and shapes, depending on the information contained in the viewport.
- Multiple viewports can be created in Paper Space, as in Model space. However, the images assigned to Paper Space viewports can be at different scales. Also, the control of layers in Paper Space is viewport-dependent. In other words, layers turned on in one viewport can be turned off or frozen in another viewport.
- All drawings, no matter how many viewports are created or details arranged, are plotted out at a scale of 1 = 1.

SETTING UP A PAGE

Right-clicking on Layout1 and picking Page Setup Manager, as shown on the left in the following image, will display the Page Setup Manager dialog box, shown on the right in the following image. The purpose of this initial dialog box is to create a series of page setups under unique names. These page setup names can hold information such as plotter name, page sizes, and plot style tables used for pen assignments when plotting. To get a better idea of what these page settings are, click on the Modify button, as shown on the right in the following image.

Figure 14–5

A larger, more comprehensive, Page Setup dialog box will display, as shown in the following image. The following areas will now be explained:

Printer/plotter—Area A: Use this area to choose the plotter you will be outputting to. Even though you may not be ready to plot, you can still designate a plotter. In this image the DWF6 ePlot.pc3 file will be used.

Paper size—Area B: This area holds all paper sizes. It is very important to know that these paper sizes are based on the plotter you selected back in Area A.

Plot style table (pen assignments)—Area C: This area is used to communicate with AutoCAD. Plot style will be assigned to this page for plotting purposes.

Another important area to take note of is the Plot scale. This is the plotted scale that the layout will be based on. Since the main purpose of creating a layout is to plot at a scale of 1 to 1, the scale 1:1 reflects this practice. This also means that the information contained in the layout has already been scaled. This will be explained in greater detail in this chapter.

Figure 14–6

A more detailed look at the available paper sizes is illustrated on the left in the following image. Again, these paper sizes are a direct result of the printer/plotter you are using. Notice in this image a number of paper sizes that have the word "expand" associated with them. These paper sizes have a larger printable area as opposed to the paper sizes that are not identified with "expand."

Illustrated on the right in the following image are the default Plot style tables that are available when AutoCAD is loaded. Plot style tables control the appearance of a printed drawing, which can include color, linetype, and even lineweight. For example, the acad.ctb table is designed for a color plot. Whatever objects are red will plot out in the color red (provided you are using a color printer). The monochrome.ctb plot style will take all colors and plot them out as black objects on a white sheet of paper. Notice other color tables that will plot out objects in different shades of gray (Grayscale.ctb) and a number of screening plot styles. These are useful when you want to fade away a group of objects. A screening factor of 25% would have the objects faded more than a screen factor of 50%, and so on.

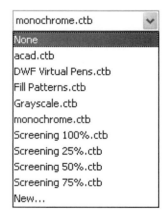

Figure 14–7

When you have finished making changes in the Plot Setup dialog box, click the OK button to return to the main Page Setup Manager dialog box. Click OK here to return to your drawing.

Note: When you make a number of changes in the Page Setup dialog box and then save your drawing, these settings are automatically saved inside the drawing.

FLOATING MODEL SPACE

When a viewport is created in Paper Space, the drawing image automatically fills up the viewport, as shown in the following image. Notice also the appearance of the drafting triangle icon in the lower left corner of the display screen. This signifies that you are presently in the Paper Space environment.

Figure 14–8

While in Paper Space, operations such as adding a border, title block, and general notes are usually performed. Paper Space is also the area in which you plot your drawing.

Other operations need to be performed in Model Space without leaving the Layout environment. This is referred to as floating Model Space. To activate floating Model Space, double-click inside a viewport. The UCS icon normally will display, as shown in the following image.

Figure 14–9

Note: If you need to switch back to the Layout mode, double-click outside the viewport. The Paper Space drafting triangle will reappear.

SCALING VIEWPORT IMAGES

One of the operations performed while in a floating Model Space viewport is the scaling of the image to Paper Space. It was previously mentioned that by default, images that are brought into viewports displayed in a ZOOMED-Extents appearance. To scale an image to a viewport, use the next series of steps and refer to the following image.

- Have the Viewports toolbar present somewhere on your screen.
- Activate floating Model Space by double-clicking inside the viewport to scale.
- In the Viewports toolbar, click in the scale box to display other scales.
- Select various scales until the image fits in the viewport.

Figure 14–10

Note: You can also apply a scale to a viewport by simply picking on an edge of the viewport while in Layout mode and setting the proper scale.

If you find that the image is very close to fitting but is cut off by one edge of a viewport, there are two options to resolve the problem:

- Stretch the viewport (using grips or the STRETCH command)
- Pan the image into position.

A combination of both methods usually proves to be the most successful.

To pan the image, double-click inside the viewport to start floating Model Space. You will notice the thick borders on the viewport. Using the PAN command or holding down the wheel on a wheel mouse, move the image to fit the viewport. Once the image is panned into position, double-click outside the viewport on the drawing surface (usually the area surrounding the drawing sheet) to return from the floating Model Space back to the layout space.

CONTROLLING THE LIST OF SCALES

You will notice that when setting the scale of the image inside the Paper Space viewport, the list of scales is very long. You may find yourself using only certain scales and ignoring others. You could also be confronted with assigning a scale that is not present on the list. All these situations can be easily handled through the use of the Scale List dialog box, which is activated by picking Scale List from the Format pull-down menu, as shown in the following image.

Figure 14–11

Various buttons are available that allow you to add, edit, or delete scales from the list. Clicking on the Add button will display the Add Scale dialog box, as shown on the left in the following image. In this dialog box, a new scale of 1:1000 is being created since it is not available in the default scale list. Clicking on the Edit button will display the Edit Scale dialog box, as shown on the right in the following image. Here an existing scale is being edited to reflect the new scale of 1:500.

Figure 14–12

Clicking on the Delete button will remove a scale from this list. You can also rearrange scales so that your most popular scales will always appear at the top of the list. Simply click the Move Up button to move the selected scale up the list or the Move Down button to arrange the selected scale at the bottom of the list.

USING THE ZOOM XP OPTION

The ZOOM command can also be used to scale images inside viewports to Paper Space. This is accomplished using the XP option, which stands for "Times Paper." For example, to size a floating Model Space image to the scale of 1/8″ = 1′-0″, you would enter the following at the ZOOM command:

Command: **Z** *(For ZOOM)*
Specify corner of window, enter a scale factor (nX or nXP), or
[All/Center/Dynamic/Extents/Previous/Scale/Window] <real time>: **1/96XP**

Below is a list of commonly used architectural scales that utilize the ZOOM XP option:

1/8″ = 1′-0″	Zoom 1/96XP
1/4″ = 1′-0″	Zoom 1/48XP
3/8″ = 1′-0″	Zoom 1/32XP
1/2″ = 1′-0″	Zoom 1/24XP
3/4″ = 1′-0″	Zoom 1/16XP
1″ = 1′-0″	Zoom 1/12XP
1 1/2″ = 1′-0″	Zoom 1/8XP

LOCKING VIEWPORTS

When you have scaled the contents of a floating Model Space viewport, it is very easy to accidentally change this scale by rolling the wheel of the mouse or performing another ZOOM operation. To prevent accidental panning and zooming of a viewport image once it has been scaled, a popular technique is to lock the viewport. To perform this task, use these steps and refer to the following image:

Figure 14–13

- Be sure you are in Layout mode (look for the drafting triangle in the lower left corner).
- Select the viewport to lock. It will highlight and grips will appear.
- Right-click on your screen. A menu will appear.
- Select Display Locked followed by "Yes."

Note: When you lock a viewport, your scale will be grayed out inside the Viewports toolbar. If you need to change the scale of an image inside a viewport, you must first unlock the viewport.

MAXIMIZING A VIEWPORT

The VPMAX command is designed to maximize the size of a viewport in a layout. This is especially helpful when editing drawings with small viewports. It also eliminates the need to constantly switch between Model Space and a Layout. The VPMAX command can be entered at the keyboard. You could also click on the Maximize Viewport button located in the status bar at the bottom of the display screen, as shown in the following image.

Note: This button is visible only when you are in a Layout; it is not displayed in Model Space. Also, if numerous viewports are created, arrow buttons appear, allowing you to move between viewports. Click on the Maximize Viewport button a second time to minimize the viewport and return to the Layout environment.

Figure 14–14

The following image displays a drawing that has its viewport maximized. To return to Paper Space or Layout mode, double-click on the red outline.

Figure 14–15

HIDING LAYOUT AND MODEL TABS

As a means of regaining valuable screen space, you can hide the layout and model tabs. To do this, right-click on one of the tabs and choose Hide Layout and Model tabs from the menu, as shown in the following image.

Figure 14–16

When clicking on this menu item in the previous image, all tabs are removed from the screen. The tabs are replaced with two buttons that are located in the status bar, as shown in the following image. Here you can easily switch back and forth from Model to Layout by clicking on the appropriate button. If multiple layouts are available, these can be displayed by clicking on the double arrow buttons, as shown in the following image.

Figure 14–17

To redisplay the tabs on the screen, right-click on the Model or Layout buttons, and pick Display Layout and Model Tabs, as shown in the following image.

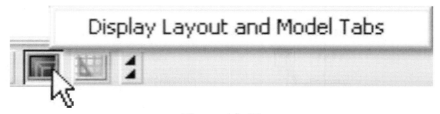

Figure 14–18

CREATING A PAPER SPACE LAYOUT

 Try It! – Open the drawing file 14_Gasket1. The illustration on the left in the following image shows the drawing originally created in Model Space. This drawing is required to be laid out in Paper Space.

As you click on the Layout1 tab at the bottom of the display screen, a layout is automatically created, as shown on the right in the following image.

Figure 14–19

Before creating the layout in Paper Space, it is considered good practice to rename the Layout1 tab located at the bottom of your screen. Do this by first right-clicking on the Layout1 tab. When the menu appears, click Rename. Since this drawing consists of a single view, the Layout1 tab is renamed One View Drawing, as shown in the following image.

Figure 14–20

Next, right-click on the One View Drawing tab and pick Page Setup Manager from the menu, as shown on the left in the following image. This will launch the Page Setup Manager, illustrated on the right in the following image. Click the Modify button.

Figure 14–21

When the Page Setup dialog box appears, as shown in the following image, click in the Printer/plotter area and change the name of the plot device to DWF6 ePlot.

Figure 14–22

Next, click in the Paper size area and select ANSI expand B (17.00 × 11.00 Inches) from the available list.

Figure 14–23

Because you increased the size of the sheet, the image of the gasket in the viewport did not grow in size. The size of the viewport will depend on the scale applied in this next step.

Rather than activating floating Model Space, simply click on the edge of the viewport, as shown on the left in the following image. A special toolbar is available to control viewports. To display it, first right-click any button on your display screen. This displays the toolbar pop-up menu. Clicking on Viewports displays the Viewports toolbar, as shown on the right in the following image. The number inside the edit box in this image is the current scale of the image inside the viewport. Whenever you create a drawing layout, the image is automatically zoomed to the extents of the viewport. For this reason, the desired scale will never be correctly displayed in this area.

To have the image properly scaled inside the viewport, click the down arrow in the Viewports toolbar. This displays the drop-down list of all standard scales. These include standard engineering and architectural scales. The image is assigned a scale of 1:1, which is applied to the viewport.

Figure 14–24

You can see on the left in the following image that the gasket has grown inside the viewport. However, it is properly scaled. The viewport needs to be increased in size to accommodate the scaled image of the gasket. Clicking once on the viewport displays grips located at the four corners, as shown on the left in the following image. Use these grips to size the viewport to the image. Accomplish this by clicking on one of the grips in the corner and stretching the viewport to a new location, as shown on the right in the following image.

Figure 14–25

The results of this operation are displayed in the following image. Press ESC to remove the object highlight and the grips.

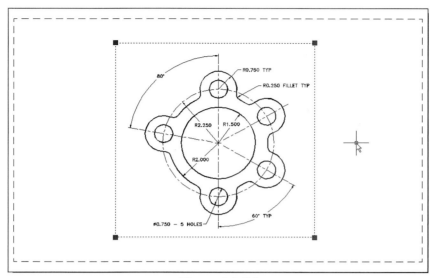

Figure 14–26

Before the drawing is plotted, a title block containing information such as drawing scale, date, title, company name, and designer is inserted in the Paper Space environment. Accomplish this by first choosing Block… from the Insert pull-down menu, as shown on the left in the following image. This activates the Insert dialog box, shown on the right in the following image. This feature of AutoCAD will be covered in greater detail in chapter 16. Be sure the block ANSI B title block is listed in the Name edit box.

Figure 14–27

Click the OK button to place the title block in the following image. Be sure the title block displays completely inside the paper margins; otherwise, part of it will not plot.

Figure 14–28

At this point, the MOVE command is used to move the viewport containing the drawing image to a better location based on the title block. Unfortunately, if the drawing were to be plotted out at this point, the viewport would also plot. It is considered good practice to assign a layer to the viewport and then turn that layer off before plotting. You could also set the viewport layer to No Plot. This would allow the viewport to be visible in drawing mode yet not plot out. In the following image, the viewport is first selected and grips appear. Click in the Layer Control box and click in the Viewport layer. Then click the lightbulb icon in the Viewport layer row to turn it off.

Figure 14–29

The drawing display will appear similar to the following image, with the drawing laid out and properly scaled in Paper Space.

Figure 14–30

USING A WIZARD TO CREATE A LAYOUT

The AutoCAD Create Layout wizard can be especially helpful in laying out a drawing in the Paper Space environment.

Clicking on Create Layout… displays the Create Layout-Begin dialog box, as shown on the right in the following image. You cycle through the different categories, and when you are finished, the drawing layout is displayed, complete with title block and viewport.

Figure 14–31

Functions of the Create Layout wizard dialog box are briefly described in the following table:

Create Layout Wizard

Category	Purpose
Begin	Change the name of the layout here.
Printer	Use the Printer category to select a configured plotter from the list provided.
Paper Size	The Paper Size category displays all available paper sizes supported by the currently configured plotter.
Orientation	Use the Orientation category to designate whether to plot the drawing in Landscape or Portrait mode.
Title Block	Depending on the paper size, choose a corresponding title block to automatically be inserted in the layout sheet.
Define Viewports	The Define Viewports category is used to either create a viewport or leave the drawing layout empty of viewports.
Pick Location	The Pick Location category creates the viewport to hold the image in Paper Space. If you click the Next> button, the viewport will be constructed to match the margins of the paper size. If you click the Select location< button, you return to the drawing and pick two diagonal points to define the viewport.
Finish	The Finish category alerts you to a new layout name that will be created. Once it is created, modifications can be made through the Page Setup dialog box.

ARRANGING ARCHITECTURAL DRAWINGS IN PAPER SPACE

Architectural drawings pose certain challenges when you lay them out in Paper Space based on the scale of the drawing. For example, the following image shows a drawing of a stair detail originally created in Model Space. The scale of this drawing is 3/8″ = 1′-0″. This scale will be referred to later on in this segment on laying out architectural drawings in Paper Space.

Figure 14–32

Try It! – Open the drawing file 14_Stair Detail. Click on the Layout1 tab or use the Create Layout wizard to arrange the stair detail in Paper Space so that your drawing appears similar to the following image. In the figure, we see the drawing sheet complete with viewport that holds the image and a title block that has been inserted. Use ARCH expand C as the paper size and insert the ANSI C title block. Because the illustration in the following image is now visible in Paper Space, it is not necessarily scaled to 3/8″ = 1′-0″. Double-clicking inside the viewport places you in floating Model Space (notice the appearance of the User Coordinate System icon only in the viewport). Activate the Viewports toolbar and scale the image in floating Model Space to the proper scale. Notice, in the figure, a complete listing of the more commonly used architectural scales. You could also have used the ZOOM command and entered an XP value of 1/32 (found by dividing 1′ [12″] by 3/8″).

Figure 14–33

Changing the viewport to a different layer and then turning that layer off displays the completed layout of the stair detail in the following image. Since the purpose of Paper Space is to lay out a drawing and scale the image inside the viewport, this drawing will be plotted at a scale of 1:1.

 Note: Another common practice is to set the layer that contains all viewports to a No Plot state in the Layer Properties Manager dialog box. The viewport will always remain visible in your layout. However, when previewing your image before plotting, the viewport will disappear. Exiting Plot Preview will make the viewport reappear.

Figure 14–34

CREATING MULTIPLE DRAWING LAYOUTS

The methods explained so far have dealt with the arrangement of a single layout in Paper Space. AutoCAD provides for greater flexibility when working in Paper Space by enabling you to create multiple layouts of the same drawing. This will be explained through the following image, which shows a floor plan complete with dimensions. The drawing appears very busy, with a foundation plan with its dimensions and an electrical plan with its notes all overlaid on the floor plan. Because it is difficult to read each plan, individual layouts will be created to display separate images of the floor, foundation, and electrical plans. Because multiple layouts are used, a special feature of layers will allow certain layers frozen in one layout to be visible in another layout.

Figure 14–35

In the following image, a Floor Plan layout is already created. However, you can still view foundation and electrical information, which makes the image very busy. You want to see only the floor plan and its dimensions inside this viewport. First, double-click inside the viewport in order to switch to floating Model Space. Activating the Layer Properties Manager dialog box displays the layer information, as shown in the following image. When you create a new drawing layout and use this dialog box, two additional layer modes are added. These are the ability to freeze layers only in the active viewport and the ability to freeze layers in new viewports. Freezing layers in all viewports is not an effective means of controlling layers when you create multiple layouts. You need to be very familiar with the layers created in order to perform this task. To display only the floor plan information, notice that the following layers have been frozen under the Current VP Freeze heading, as shown in the following image: Construction, Electrical_Symbols, Electrical_Wiring, Foundation, Foundation_Centers, Foundation_Dimensions, Piers.

All other layers, such as Appliances, First_Floor, Floor_Dimensions, Title Block, Title Text, Viewport, and Windows, remain visible in this viewport.

Figure 14–36

The resulting image is displayed in the following image with only the floor plan information visible in the viewport. The next series of steps involve the creation of an additional layout to display the foundation plan.

Figure 14–37

To create an additional layout, right-click on the Floor Plan tab at the bottom of the display screen, as shown on the left in the following image. This displays the menu also shown on the left. Clicking on Move or Copy… displays the Move or Copy dialog box, shown on the right in the following image. You have the opportunity to shuffle drawing layouts around to change their order. You can also click in the check box to create a copy of the existing layout based on the floor plan. This new layout will be moved to the end; it will be placed after the floor plan.

Figure 14–38

Clicking the OK button displays the new layout. When you copy a layout, all information inside the viewport, such as title block, viewport, and information, is copied. Begin the conversion of the new layout to the foundation plan by right-clicking on the Floor Plan (2) tab at the bottom of the display screen. This displays the menu, as shown on the left in the following image. Click on the Rename option to activate the Rename Layout dialog box, as shown on the right in the following image. Rename this layout to Foundation Plan and click the OK button.

Figure 14–39

Be sure you are in floating Model Space by double-clicking inside the viewport. Then activate the Layer Properties Manager dialog box, as shown in the following image. Because this layout will hold all information that pertains to the Foundation Plan, only the following layers need to be frozen in the current viewport: Appliances, Construction, Electrical_Symbols, Electrical_Wiring, Floor_Dimensions.

The remaining layers, thawed in the current viewport, will display the Foundation Plan while freezing all other layers in this viewport.

Figure 14–40

The resulting image is displayed in the following image, which shows the information needed to describe the foundation plan.

Figure 14–41

TRANSLATING DISTANCES BETWEEN SPACES

Suppose you want to label the rooms located in the floor plan illustrated in the following image. It is considered good practice to add this type of text while still inside Model Space. This will allow the text to be selected with the floor plan if the drawing needs to move to a new location. Unfortunately, you must first determine the plotted text size. You then multiply plotted text height by the plotted scale of the drawing. This will give you the text height while inside the model.

A better way to accomplish this task is to use the Convert distance between spaces tool located in the Text toolbar illustrated in the following image. The actual command name is SPACETRANS, or Translate between model space and paper space.

Try It! – Open the drawing 14_SpaceTrans, as shown in the following image. The floor plan is currently displayed in floating Model Space. Also, the scale of the floor plan inside the viewport is ¼" = 1" (The scale is grayed out in the figure because the viewport is locked.) You want the text height of all rooms to plot out at a height of 0.25 units. Instead of calculating this height manually based on the viewport scale, the SPACETRANS command will be used.

Figure 14–42

Activate the MTEXT command and pick a point to specify the first corner of the multiline text rectangle. However, do not pick the second rectangle point. You will be prompted to enter a number of text properties; one property is Height. After you enter Height from the keyboard, click on the Convert distance between spaces button located in the previous image. Follow the prompts and enter the desired paper space distance of 0.25 (remember, this is the plotted text height). The MTEXT command resumes and a new text height of 12 is applied. This value was automatically calculated by the SPACE- TRANS command by multiplying 0.25 by 48 (the viewport scale). Picking the opposite corner of the mtext rectangle, as shown in the following image, and entering the text SECOND FLOOR completes the command sequence.

Command: **MT** *(For MTEXT)*
Current text style: "STANDARD" Text height: 7 3/16"
Specify first corner: *(Pick a first corner on your screen…do not pick the opposite corner yet)*
Specify opposite corner or [Height/Justify/Line spacing/Rotation/Style/Width]: **H** *(For Height)*
Specify height <7 3/16">: **'_spacetrans** *(Pick the SpaceTrans button from the Text toolbar)*
>>Specify paper space distance <1">: **0.25** *(This represents the plotted text height)*
Resuming MTEXT command.
Specify height <7 3/16">: **12.00000000000000** *(This is the text height calculated by the SPACETRANS command based on the viewport scale of ¼" = 1'-0")*
Specify opposite corner or [Height/Justify/Line spacing/Rotation/Style/Width]: *(Now pick the opposite corner to define the width of the multiline text, as shown in the following image. When the Multiline Text Editor appears, enter the text SECOND FLOOR)*

Figure 14–43

Now try the SPACETRANS command again by placing the text BEDROOM 1, as shown in the following image. The plotted text height of the room names will be 0.15 units. Follow the MTEXT prompt for completing this task using the SPACETRANS command.

[A] Command: **MT** *(For MTEXT)*
Current text style: "STANDARD" Text height: 1´
Specify first corner: *(Pick a first corner on your screen...do not pick the opposite corner yet)*
Specify opposite corner or [Height/Justify/Line spacing/Rotation/Style/Width]: **H** *(For Height)*
Specify height <1´>: **'spacetrans** *(Pick the SpaceTrans button from the Text toolbar)*
>>Specify paper space distance <1/4"˃: **0.15** *(This represents the plotted text height)*
Resuming MTEXT command.
Specify height <1´>: **7.20000000000001** *(This is the new text height calculated by the SPACETRANS command based on the viewport scale of ¼" = 1´-0")*
Specify opposite corner or [Height/Justify/Line spacing/Rotation/Style/Width]: *(Now pick the opposite corner to define the width of the multiline text. When the Multiline Text Editor appears, enter the text BEDROOM 1)*

Figure 14–44

ASSOCIATIVE DIMENSIONS AND LAYOUTS

It was once considered good practice to dimension all objects while inside Model Space. The main reason was because there was no associativity with the Model Space objects and dimensions placed in a Layout or Paper Space. This has changed. Once you scale the Model Space objects inside a layout, the scale of the layout has a direct bearing on the dimensions being placed. This is accomplished only if the DIMASSOC variable is set to 2.

Try It! – Open the drawing file 14_Inlay. Two viewports are arranged in a single layout called Plan Views, as shown in the following image. The images inside these viewports are scaled differently; the main inlay pattern in the left viewport is scaled to 3/4″ = 1′-0″. The detail image in the right viewport has been scaled 3″ = 1′-0″. The DIMASSOC variable is currently set to 2. This will allow you to dimension the objects at two different scales and still have the correct dimensions appear.

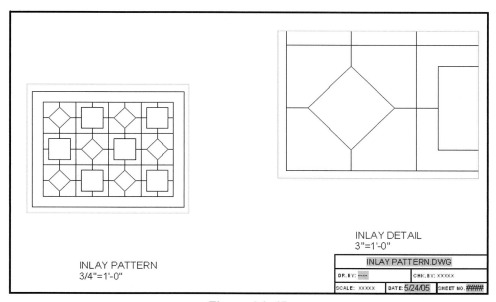

Figure 14–45

Use the DIMLINEAR command and place a few linear dimensions on the main inlay plan in the left viewport. Now switch and add a few linear dimensions in the right viewport. Zoom in to a few of the dimensions and see that they reflect the Model Space distances. Place more dimensions on the main inlay pattern and detail using a combination of dimension commands and observe the correct values being placed.

Figure 14–46

Next, pick and drag each viewport to a more convenient location on the drawing screen. Notice that the dimensions move along with the viewports.

Figure 14–47

 Note: If you drag viewports around and the dimensions do not keep up with the viewports, try entering the DIMREGEN command from the keyboard. This should update all dimensions to the new viewport positions.

HATCH SCALING RELATIVE TO PAPER SPACE

While in a layout, you have the opportunity to scale a hatch pattern relative to the current Paper Space scale in a viewport. In this way, you can easily have AutoCAD calculate the hatch pattern scale, since this is determined by the scale of the image inside a viewport. Open the file 14_Valve Gasket. This drawing consists of a small gasket that does not have crosshatching applied to the thin inner border. If you hatched this shape using the default hatch settings while in Model Space, the gasket would appear similar to the illustration in the following image. Notice that the hatch pattern spacing is too large. You would need to experiment with various hatch scales in order to achieve the desired hatch results.

Figure 14–48

A better way to approach this problem would be to enter Paper Space, activate floating Model Space, and produce the hatch pattern from there. While inside a floating Model Space viewport, activate the Hatch and Gradient dialog box. Notice the Relative to paper space check box. Placing a check in this box activates this feature, as shown in the following image. This feature is grayed out if you try to hatch in Model Space.

Figure 14–49

The finished gasket with hatching applied is illustrated in the following image. Notice how the hatch scale is based on the viewport scale.

Figure 14–50

 Try It! – Open the drawing file 14_Hatch Partial Plan. The inner walls of the object in the following image need to be crosshatched. Apply the technique of making the hatch scale relative through the Hatch and Gradient dialog box.

Figure 14–51

The finished partial plan is illustrated in the following image.

Figure 14–52

NOTES TO THE STUDENT AND INSTRUCTOR

A tutorial exercise has been designed around this introductory topic of laying out drawings in Paper Space. As with all tutorials, you will tend to follow the steps very closely, taking care not to make a mistake. This is to be expected. However, most individuals rush through tutorials to get the correct solution, only to forget the steps used to complete the tutorial. This is also to be expected.

Of all tutorial exercises in this book, the Drawing Layout tutorial will probably pose the greatest challenge. Certain operations have to be performed in Paper Space while other operations require floating Model Space. For example, it was already illustrated that user-defined viewports are constructed in Paper Space; scaling an image to Paper Space units through the Viewports toolbar or through the XP option of the ZOOM command must be accomplished in floating Model Space, and so on.

It is recommended to both student and instructor that all tutorial exercises, and especially this tutorial that deals with Paper Space, be performed two or even three times. Completing the tutorial the first time will give you the confidence that it can be done; however, you may not understand all the steps involved. Completing the tutorial a second time will allow you to focus on where certain operations are performed and why things behave the way they do. This still may not be enough. Completing the tutorial exercise a third time will allow you to anticipate each step and have a better idea of which operation to perform in each step.

This recommendation should be exercised with all tutorials; however, it should especially be practiced when you work on the following Paper Space tutorial.

TUTORIAL EXERCISE: 14_HVAC.DWG

Figure 14–53

Purpose
This tutorial is designed to create multiple layouts of the HVAC drawing in the previous image in Paper Space.

System Settings
All unit, limit, and plotter settings have already been made in this drawing.

Layers
Layers have already been created for this exercise.

Suggested Commands
Begin by opening the drawing file 14_Hvac.Dwg. Click on Layout1 to begin the first drawing layout, which consists of the grid lines. When the layout is created, enter floating Model space and scale the image to 1/8″ = 1′-0″. Also, freeze the layers that pertain to the floor and HVAC plans only in the current viewport. Press ENTER to return to Paper Space and insert an architectural title block and change the viewport to the Viewports layer. From this layout, create another layout called Floor Plan. While in floating Model Space, freeze the layers that pertain to the grid lines and HVAC plans. From this layout, create another layout called Hvac Plan. While in floating Model Space, freeze the layers that pertain to the grid lines. Turn off all viewports and add drawing titles to each layout.

STEP 1

Illustrated in the following image is a layout of an HVAC plan (heating, venting, and air conditioning). A viewport already exists and the image inside the viewport has been scaled to a value of 1/8″ = 1′-0″. A border also exists in this layout. The goal is to create two extra layouts that show different aspects of the HVAC plan. One layout will display only the grid pattern used to lay out the plan. Another layout will show just the floor plan information. The third layout will show the HVAC and floor plans together.

Figure 14–54

STEP 2

Before you start to create the new layouts, the viewport present in the existing layout needs to be locked. This will prevent any accidental zooming in and out while inside floating Model Space. To lock a viewport, click on the edge of the viewport, right-click, and choose Display Locked followed by Yes from the menu, as shown in the following image. This step is considered good practice to perform before creating any extra layouts. Now that this viewport is locked, the copied viewports created in other layouts will also be locked.

Figure 14–55

STEP 3

Next, right-click on Layout1, located in the lower left corner of the display screen. From the menu, pick Rename and change the name to Grid Lines in the Rename Layout dialog box, as shown in the following image. It is always considered good practice to give your layouts meaningful names.

Figure 14–56

STEP 4

Click on the drawing title (HVAC PLAN); the title should highlight and grips will appear. Then right-click the mouse and pick Properties at the bottom of the menu to activate the Properties palette. Change HVAC PLAN to GRID LINES PLAN, as shown in the following image; pressing ENTER automatically updates the text to the new value. Dismiss this dialog box when finished. Press ESC to remove the object highlight and grips.

Figure 14–57

STEP 5

Prepare to create the second layout by right-clicking on the Grid Lines tab to display the cursor menu, as shown on the left in the following image. Click on the Move or Copy... option to make a copy of the current layout. With the Move or Copy dialog box displayed, as shown on the right in the following image, place a check in the box to make a copy. Also highlight the area to move the new layout to the end. When finished, click the OK button to create the new layout.

Figure 14–58

STEP 6

Before continuing, right-click on the new layout Grid Lines (2) and click on the Rename option in the menu, as shown on the left in the following image. In the Rename Layout dialog box, shown on the right in the following image, change the current layout name, Grid Lines (2), to the new name Floor Plan. Click the OK button to dismiss the dialog box and make the change.

Figure 14–59

STEP 7

While in this new layout, right-click on the Floor Plan tab to display the menu, as shown on the left in the following image. Click on the Move or Copy... mode to make one more copy of the current layout. With the Move or Copy dialog box displayed, as shown on the right in the following image, place a check in the box to create a copy. Also, highlight the area to move the new layout to the end. When finished, click the OK button to create the new layout.

Figure 14–60

STEP 8

Before continuing, right-click on the new layout Floor Plan (2) and click on the Rename option in the menu, as shown on the left in the following image. Change the current layout name, Floor Plan (2), to the new name HVAC Plan, as shown on the right in the following image. Click the OK button to dismiss the dialog box and make the change.

Figure 14–61

When you are finished creating and renaming all the layouts, the lower left corner of your display should appear similar to the image on the right.

Figure 14–62

STEP 9

Click on the Grid Lines tab. Double-click anywhere inside the viewport, as shown in the following image; this places you in floating Model Space. You will need to turn off all layers that deal with the floor plan and HVAC, which leave only the layers with the grid lines visible.

Activate the Layer Properties Manager dialog box, as shown in the following image, and freeze the layers DOORS, FLOOR, HVAC DIM, and HVAC SUP in the current viewport, which pertain to the floor plan and HVAC plan. Pick one of the layers and hold down the CTRL key as you pick the other three. This allows you to freeze all four layers at once, as shown in the following image.

Figure 14-63

Clicking the OK button in the Layer Properties Manager dialog box returns you to the drawing editor. Double-click anywhere outside the viewport to return to Paper Space. Your drawing should appear similar to the following image. Notice that only the columns and grid lines appear visible in this layout.

Figure 14-64

STEP 10

Click on the Floor Plan tab. Double-click anywhere inside the viewport, as shown in the following image; this places you in floating Model Space. You will need to turn off all layers that deal with the grid lines and HVAC.

Activate the Layer Properties Manager dialog box, shown in the following image, and freeze the layers in the current viewport that pertain to the grid lines plan and HVAC plan (GRID, GRID DIM, GRID ID, GRID_LINES, HVAC DIM, and HVAC SUP). Pick the GRID layer and hold down the SHIFT key as you pick HVAC SUP. This allows you to freeze all six layers at once, as shown in the following image.

Figure 14–65

Clicking the OK button returns you to the drawing editor. Double-click anywhere outside the viewport to return to Paper Space. Use the Properties Palette to change the drawing title from GRID LINES PLAN, as shown in the following image, to FLOOR PLAN. Do this by first selecting GRID LINES PLAN. With this object highlighted, right-click and pick Properties from the bottom of the menu to activate the Properties Palette and change the text. When finished, press ESC to remove the object highlight and grips.

Figure 14–66

Your drawing now displays only floor plan information, as shown in the following image, while the HVAC and GRID layers are frozen only in this viewport.

Figure 14–67

STEP 11

Click on the HVAC Plan tab. Double-click anywhere inside the viewport, as shown in the following image; this places you in floating Model Space. You will need to turn off all layers that deal with the grid lines plan.

Activate the Layer Properties Manager dialog box, as shown in the following image, and freeze the layers in the current viewport that pertain to the grid lines plan (COLUMNS, GRID, GRID DIM, GRID ID, and GRID_LINES).

Figure 14–68

Clicking the OK button returns you to the drawing editor. Double-click anywhere outside the viewport to return to Paper Space. Use the Properties Palette to change the drawing title from GRID LINES PLAN, as shown in the following image, to HVAC PLAN. When finished, press ESC to remove the object highlight and grips.

Figure 14–69

Your drawing should appear similar to the following image. In this viewport, you see the floor plan and HVAC ductwork but no grid lines or columns.

Figure 14–70

STEP 12

Turn off the Viewports layer. The completed drawing is displayed in the following image.

This exercise illustrated how to create multiple layouts. It also illustrated how to freeze layers in one viewport and have the same layers visible in other viewports. If the Viewports layer is set to No Plot through the Layer Properties Manager dialog box, performing a plot preview will not display the Viewports layer.

Figure 14–71

Plotting Your Drawings

This chapter discusses plotting through a series of tutorial exercises designed to perform the following tasks:

- Configure a new plotter
- Plot from a drawing layout (Paper Space)
- Control lineweights
- Use the Plotter Configuration Editor to filter available paper sizes
- Create a Color Dependent plot style table
- Create a web page consisting of various drawing layouts for viewing over the Internet

CONFIGURING A PLOTTER

Before plotting, you must first establish communication between AutoCAD and the plotter. This is called configuring. From a list of supported plotting devices, you choose the device that matches the model of plotter you own. This plotter becomes part of the software database, which allows you to choose this plotter many times. If you have more than one output device, each device must be configured before being used. This section discusses the configuration process used in AutoCAD.

STEP I

Begin the plotter configuration process by choosing Plotter Manager from the File pull-down menu, as shown on the left in the following image. This activates the Plotters program group, as shown on the right in the following image, which lists all valid plotters that are currently configured. The listing in this image displays the default plotters configured after the software is loaded. Except for the DWF devices, which allow you to publish a drawing for viewing over the Internet, a plotter has not yet been configured. Double-click on the Add-A-Plotter Wizard icon to continue with the configuration process.

Figure 15–1

STEP 2

Double-clicking on the Add-A-Plotter Wizard icon displays the Add Plotter-Introduction Page dialog box, as shown in the following image. This dialog box states that you are about to configure a Windows or non-Windows system plotter. This configuration information will be saved in a file with the extension .PC3. Click the Next> button to continue on to the next dialog box.

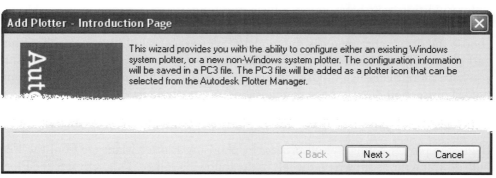

Figure 15–2

STEP 3

In the Add Plotter-Begin dialog box, shown on the left in the following image, decide how the plotter will be controlled: by the computer you are currently using, by a network plot server, or by an existing system printer where changes can be made specifically for AutoCAD. Click on the radio button next to My Computer. Then click on the Next> button to continue on to the next dialog box.

Use the Add Plotter-Plotter Model dialog box, as shown on the right in the following image, to associate your plotter model with AutoCAD. You would first choose the appropriate plotter manufacturer from the list provided. Once this is done, all models supported by the manufacturer appear to the right. If your plotter model is not listed, you are told to consult the plotter documentation for a compatible plotter. For the purposes of this tutorial, click on Hewlett-Packard in the list of Manufacturers. Click on the DesignJet 750C Plus C3195A for the plotter model. A Driver Info dialog box may appear giving you more directions regarding the type of HP DeskJet plotter selected. Click the Continue button to move on to the next dialog box used in the plotter configuration process.

Figure 15–3

STEP 4

PCP and PC2 files have been in existence for many years. They were designed to hold plotting information such as pen assignments. In this way, you used the PCP or PC2 files to control pen settings instead of constantly making pen assignments every time you performed a plot; at least this is how pen assignments were performed in past versions of AutoCAD. The Add Plotter-Import Pcp or Pc2 dialog box, shown on the left in the following image, allows you to import those files for use in AutoCAD in a PC3 format. If you will not be using any PCP or PC2 files from previous versions of AutoCAD, click on the Next> button to move on to the next dialog box.

In the Add Plotter-Ports dialog box, shown on the right in the following image, click on the port used for communication between your computer and the plotter. The LPT1 port will be used for the purposes of this tutorial. Place a check in its box and then click on the Next> button to continue on to the next dialog box.

Figure 15–4

STEP 5

In the Add Plotter-Plotter Name dialog box, shown on the left in the following image, you have the option of giving the plotter a name other than the name displayed in the dialog box. For the purpose of this tutorial, accept the name that is given. This name will be displayed whenever you use the Page Setup and Plot dialog boxes.

The last dialog box is displayed on the right in the following image. In the Add Plotter-Finish dialog box, you can modify the default settings of the plotter you just configured. You can also test and calibrate the plotter if desired. Click the Finish button to dismiss the Add Plotter-Finish dialog box.

Plotter Name Dialog

The model name you selected is the default plotter configuration name. You can accept the default name, or enter a new name to identify the new PC3 file you have created. The name you apply will be displayed in the Page Setup and Plot dialog boxes.

Plotter Name:
DesignJet 750C C3195A

Note: If you enter a name that is exactly the same as a System Printer's name, you will not see the System Printer listed in the AutoCAD 2006 Plot or Page Setup dialog boxes.

Finish Dialog

The plotter DesignJet 750C C3195A has been installed with its default configuration settings. To modify the default settings, choose Edit Plotter Configuration.

Edit Plotter Configuration...

Optionally, to perform a plot calibration test on the newly configured plotter, and verify that your drawing measurements plot accurately, choose Calibrate Plotter.

Calibrate Plotter...

Figure 15–5

STEP 6

Exiting the Add Plotter dialog box returns you to the Plotters program group, shown in the following image. Notice that the icon for the DesignJet 750C Plus plotter has been added to this list. This completes the steps used to configure the DesignJet 750C Plus plotter. Follow these same steps if you need to configure another plotter.

Figure 15–6

PLOTTING FROM PAPER SPACE

Open the drawing 15_Center Guide to step through the process of plotting the drawing from Layout mode, or Paper Space.

STEP 1

Open the drawing 15_Center_Guide.Dwg. This drawing should already be laid out in Paper Space. A layout called Four Views should be present at the bottom of the screen next to the Model tab.

STEP 2

Begin the process of plotting this drawing by choosing Plot… from the File pull-down menu, as shown in the following image. You could also choose the plot icon from the Standard toolbar. You can also type the word Plot or hold down the CTRL key and type the letter P.

Figure 15–7

This activates the Plot dialog box, shown in the following image. For the purposes of this tutorial, the DWF6 ePlot is being used as the output device. You can opt to create additional copies of your plot; by default this value is set to 1. The DWF6 ePlot can plot only one copy, so this value is grayed out. Next, make sure the paper size is currently set to ANSI expand C (22.00 × 17.00 Inches). In the Drawing orientation area, make sure that the radio button adjacent to Landscape is selected. You could also plot the drawing out in Portrait mode, where the short edge of the paper is the top of the page. For special plots, you could plot the drawing upside down.

In the Plot area, the Layout radio button is selected. Since you created a layout, this is the obvious choice. The Extents mode allows you to plot the drawing based on all objects that make up the drawing. Plotting the Display will plot your current drawing view, but be careful: If you are currently zoomed in to your drawing, plotting the Display will plot only this view. In this case, it would be more predictable to use Layout or Extents to plot. When you plot a layout in Paper Space, the Plot scale will be set to 1:1. Since you pre-scaled the drawing to the Paper Space viewport using the Viewports toolbar, all drawings in Paper Space are designed to be plotted at this scale. The Plot offset is designed to move or shift the location of your plot on the paper if it appears off center. In the Plot options area, you have more control over plots by applying lineweights, using existing plot styles, plotting Paper Space last (Model Space first), or even hiding objects, on a 3D solid model.

One other area to change is in the Plot style table area, located in the upper right corner of the dialog box. This area controls the appearance of the plot, for instance, a colored plot versus a monochromatic plot (black lines on white paper). You can even create your own plot style, which will be discussed later in this chapter. For the purposes of this tutorial, monochrome will be used, as shown in the following image.

Figure 15–8

STEP 3

One of the more efficient features of plotting is that you can preview your plot before sending the plot information to the plotter. In this way, you can determine whether the entire drawing will plot based on the sheet size (this includes the border and title block). Clicking on the Preview… button activates the image shown in the following image. The sheet size is shown along with the border and four-view drawing. Right-clicking anywhere on this preview image displays the cursor menu, allowing you to perform various display functions such as ZOOM and PAN to assist with the verification process. If everything appears satisfactory, click on the Plot option to send the drawing information to the plotter. Clicking on the Exit option returns you to the Plot dialog box, where you can make changes in the Plot Device or Plot Settings tabs.

Figure 15–9

 Note: You can also plot from Model Space. However, this is more involved because you must bring borders, title blocks, and notes into Model Space. Also, the sizes of notes and dimensions in Model Space must be properly sized to plot correctly.

ENHANCING YOUR PLOTS WITH LINEWEIGHTS

This section on plotting will describe the process of assigning lineweights to objects and then having the lineweights appear in the finished plot. Open the drawing called 15_V_Step.Dwg, shown in the following image, and notice that you are currently in

Model Space (the Model tab is current at the bottom of the screen). Follow the next series of steps to assign lineweights to a drawing before it is plotted.

Figure 15–10

STEP 1

From the illustration of the drawing in the previous image, all lines on the Object layer need to be assigned a lineweight of 0.60 mm. All objects on the Hidden layer need to be assigned a lineweight of 0.30 mm. There is also a title block that will be used with this drawing. The Title Block layer needs a lineweight assignment of 0.80 mm. Click in the Layer Properties Manager dialog box, shown in the following image, and make these lineweight assignments. When you are finished, click OK to save the lineweight assignments and return to Model Space.

Figure 15–11

STEP 2

Click on the LWT button in the Status bar to display the lineweights, as shown in the following image.

Figure 15-12

You can control the lineweight scale to have the linetypes give off a more pleasing appearance in your drawing. Clicking on Lineweight in the Format pull-down menu displays the Lineweight Settings dialog box, as shown in the following image. Notice the slider bar in the area called Adjust Display Scale. This controls only the way lineweights display. Use this slider bar to reduce or increase the scale of the lineweights when being viewed in your drawing. In the following image, the slider bar has been adjusted toward the minimum side of the scale. This should make the lineweights better to read in the drawing. Click the OK button to return to the drawing and observe the results.

Figure 15-13

All hidden and object lines have been reduced in scale in the following image. This action adds to the clarity and appearance of the drawing. If after making changes to the lineweights you decide to increase the scale, return to the Lineweight Settings dialog box and readjust the slider bar until you achieve the desired results.

Figure 15–14

STEP 3

Clicking on the Orthographic Views tab switches you to Paper Space, as shown in the following image. The lineweights will not appear in Paper Space at first glance. Zooming in to the drawing will display all lineweights at their proper widths.

Figure 15–15

STEP 4

Activate the Plot dialog box, as shown in the following image. Verify that the current plot device is the HP DWF6 ePlot. If this is not the device, click on the Plot Device tab to activate this plotter. In the Plot options area of the main Plot dialog box, be sure to check the Plot object lineweights box. This will ensure that the linetypes will plot. If this box appears to be inactive, remove the check from the Plot with plot styles area.

Figure 15–16

STEP 5

Click on the Preview… button; your display should appear similar to the following image. Notice that the viewport is not present in the plot preview. The Viewports layer was either turned off or set to a non-plot state inside the Layer Properties Manager dialog box. To view the lineweights in Preview mode, zoom in to segments of your drawing.

Figure 15–17

CREATING A COLOR-DEPENDENT PLOT STYLE TABLE

This section of the chapter is devoted to the creation of a Color Dependent plot style table. Once the table is created, it will be applied to a drawing. From there, the drawing will be previewed to see how this type of plot style table affects the final plot. Open the drawing 15_Color_R-Guide.Dwg. Your display should appear similar to the image in the following illustration. A two-view drawing together with an isometric view is arranged in a layout called Orthographic Views.

Figure 15–18

The drawing is also organized by layer names and color assignments. The object is to create a Color Dependent plot style table where all layers will plot out black. Also, through the Color Dependent plot style table, the hidden lines will be assigned a lineweight of 0.30 mm, object lines 0.70 mm, and the title block 0.80 mm, as shown in the following table. Follow the next series of steps to perform this task.

Color	Layer	Lineweight
Red	Hidden	0.30
Yellow	Center	Default
Green	Viewports	Default
Cyan	Text	Default
Blue	Title Block	0.80
Magenta	Dimensions	Default
Black	Object	0.70

STEP 1

Begin the process of creating a Color Dependent plot style table by choosing Plot Style Manager from the File pull-down menu, as shown on the left in the following image. This activates the Plot Styles dialog box, as shown on the right in the following image. Various Color Dependent and Named plot styles already exist in this dialog box. To create a new plot style, double-click on the Add-A-Plot Style Table Wizard.

Figure 15–19

STEP 2

A partial illustration of the Add Plot Style Table dialog box appears, as shown in the following image, and introduces you to the process of creating plot style tables. Plot styles contain plot definitions for color, lineweight, linetype, end capping, fill patterns, and screening. You will be presented with various choices in creating a plot style from scratch, using the parameters in an existing plot style, or importing pen assignments information from a PCP, PC2, or CFG file. You will also have the choice of saving this plot style information in a CTB (Color Dependent) or STB (Named) plot style. This chapter will discuss only the creation of a Color Dependent plot style. A Next> button is displayed at the bottom of all Plot Style Wizard dialog boxes. Even though they are not displayed in these images, click it on your display to move on to the next dialog.

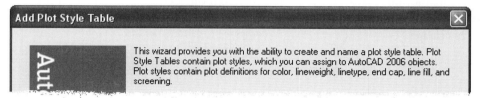

Figure 15–20

STEP 3

In the partial illustration of the Add Plot Style Table-Begin dialog box, shown on the left in the following image, four options are available for you to choose, depending on how you want to create the plot style table. Click on the Start from scratch radio button to create this plot style from scratch. If you have made pen assignments from previous releases of AutoCAD, you can import them through this dialog box. Click the Next> button to display the next dialog box. This Plot Style Table will be started from scratch.

In the partial illustration of the Add Plot Style Table-Pick Plot Style Table dialog box, shown on the right in the following image, click on the Color-Dependent Plot Style Table radio button to make this the type of plot style you will create. Click the Next> button to display the next dialog box.

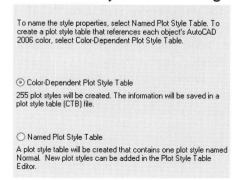

Figure 15–21

STEP 4

Use the Add Plot Style Table-File Name dialog box, shown as a partial image on the left in the following image, to assign a name to the plot style table. Enter the name Ortho_Drawings in the File name area. The extension CTB will automatically be added to this file name. Click the Next> button to display the next dialog box.

The Finish dialog box, shown on the right in the following image, alerts you that a plot style called Ortho_Drawings.ctb has been created. However, you want to have all colors plot out black and you need to assign different lineweights to a few of the layers. To accomplish this, click on the Plot Style Table Editor… button as shown in this image.

Figure 15–22

STEP 5

Clicking on the Plot Style Table Editor… button displays the Plot Style Table Editor dialog box, as shown in the following image. Notice the name of the plot style table present at the top of the dialog box. Also, three tabs are available for making changes to the current plot style table (Ortho-Drawings.ctb). The first tab is General and displays file information about the current plot style table being edited. It is considered good practice to add a description to further document the purpose of this plot style table. It must be pointed out at this time that this plot style table will be used on other drawings besides the current one. Rather, if layers are standard across projects, the same plot style dialog box can be used. This is typical information that can be entered in the Description area.

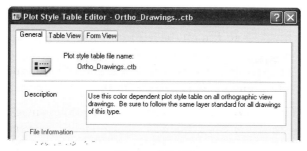

Figure 15–23

STEP 6

Clicking on the Table View tab activates the dialog box, as shown on the left in the following image. You can use the horizontal scroll bar to get a listing of all 255 colors along with special properties that can be changed. This information is presented in a spreadsheet format. You can click in any of the categories under a specific color and make changes, which will be applied to the current plot style table. The color and lineweight changes will be made through the next tab.

Clicking on the Form View tab displays the dialog box shown on the right in the following image. Here the colors are arranged vertically with the properties displayed on the right.

Click on Color 1 (Red) and change the Color property to Black. Whatever is red in your drawing will plot out in the color black. Since red is used to identify hidden lines, click in the Lineweight area and set the lineweight for all red lines to 0.30 mm.

For Color 2 (Yellow), Color 3 (Green), Color 4 (Cyan), and Color 6 (Magenta), change the color to Black in the Properties area. All colors can be selected at one time by holding down CTRL while picking each of them. Changes can then be made to all colors simultaneously. Whatever is yellow, green, cyan, or magenta will plot out in the color black. No other changes need to be made in the dialog box for these colors.

Figure 15–24

STEP 7

Click on Color 5 (Blue), as shown on the left in the following image, and change the Color property to Black. Whatever is blue in your drawing will plot out in the color black. Since blue is used to identify the title block lines, click in the Lineweight area and set the lineweight for all blue lines to 0.80 mm.

Click on Color 7 (Black), as shown on the right in the following image, and change the Color property to Black. Since black is used to identify the object lines, click in the Lineweight area and set the lineweight for all object lines to 0.70 mm.

This completes the editing process of the current plot style table. Click on the Save & Close button; this returns you to the Finish dialog box. Clicking on the Finish button will return you to the Plot Styles dialog box. Close this box to display your drawing.

Figure 15–25

STEP 8

Activate the Plot dialog box in the Plot device tab, verify that the plotter is the DWF6 ePlot plotter (this plotter has been used in all plotting tutorials throughout this chapter). In the Plot style table (pen assignments) area, make the current plot style Ortho_Drawings.ctb, as shown on the left in the following image. Notice at the top of the dialog box that the plot style table will be saved to this layout. This means that if you need to plot this drawing again in the future, you will not have to look for the desired plot style table.

Click on the Plot Settings tab in the Plot dialog box and verify in the lower right corner of the dialog box under Plot options that you will be plotting with plot styles, as shown on the right in the following image. Click on the Preview button to display the results. Unless a plotter is attached to your computer, it is not necessary to click the OK button.

Figure 15–26

STEP 9

The results of performing a plot preview are illustrated in the following image. Notice that all lines are black even though they appear in color in the drawing file.

Figure 15–27

Notice that when you zoom in on a part of the preview, different lineweights appear, as in the following image, even though they all appear the same in the drawing file.

This is the result of using a Color Dependent plot style table on this drawing. This file can also be attached to other drawings that share the same layer names and colors.

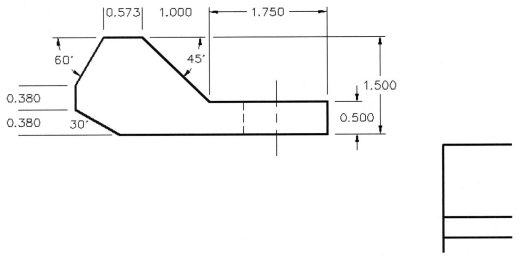

Figure 15–28

CONVERTING DRAWINGS TO DIFFERENT PLOT STYLES

You have seen how to create a Color Dependent plot style (CTB) for the purpose of controlling the plotted color, lineweight, linetype, and other properties when printing out a drawing. Another plot style type is available; it is called a Named plot style. These files have the extension STB and do not depend on color. For instance, you may want to assign different lineweights to the same color in a drawing. Also, these colors are assigned to different layers. You would create a Named plot style and make the lineweight assignments to the individual layers in the Layer Properties Manager dialog box.

In the event that you encounter a drawing that has a Named plot style (STB) attached, you can easily convert this drawing back to a Color Dependent plot style (CTB) by using the CONVERTPSTYLES command. This command must be entered from the keyboard at the command prompt. Follow the next series of steps, which illustrate this important process.

STEP 1

Open the drawing file 15_Convert from STB, as shown in the following image.

Figure 15–29

STEP 2

To convert Named Plot Styles, launch the Layer Properties Manager dialog box and notice that the Plot Style heading is active and assigned the Normal plot style for all layers, as shown on the left in the following image. Also, activate the Page Setup dialog box and notice that the extension of all plot styles is STB, for Named plot style, as shown on the right in the following image. This means that you are not plotting by color, which is the plotting method preferred by most AutoCAD users. Unfortunately, there is no control in the Page Setup dialog box that allows you to easily switch to the Color Dependent plot style (CTB). Click OK to proceed to the next step.

Figure 15–30

STEP 3

Begin the conversion process from an STB to a CTB by entering the following command at the prompt:

Command: **CONVERTPSTYLES**

Pressing ENTER after this command will display the AutoCAD alert box illustrated in the following image. The alert box informs you that all plot style names will be removed from objects and that any Named plot style tables will be detached from the drawing. Click OK to initiate the conversion.

Figure 15–31

STEP 4

Notice that after the conversion process is complete, your drawing will not appear to have changed. Again activate the Layer Properties Manager dialog box, illustrated on the left in the following image, and the Page Setup dialog box, illustrated on the right. Notice in the list of layer states that the Plot Style tab is inactive and all color assignments under this heading are grayed out. Notice also in the Page Setup dialog box that all of the Plot style table file extensions are CTB. You can now assign a Color Dependent plot style table to your drawing and control the output of your plot by these colors.

Figure 15–32

Note: A drawing that has a Color Dependent plot style (CTB) attached can be changed to the Named plot style (STB) by using the CONVERTCTB command. A dialog box will display, allowing you to select the Color Dependent plot style to convert. Then another dialog box will appear, allowing you to select an existing Named plot style from the list or create a new Named plot style.

PUBLISHING TO THE WEB

Yet another way of applying electronic plots is through the Publish to the Web utility that is provided with AutoCAD. This feature creates a project that consists of a formatted HTML page and your drawing content in either DWF, JPG, or PNG image formats. Through the Publish to Web wizard, you can select how your layout will look from a number of preformatted designs. Once the HTML page is created, you can post the page to an Internet location through the wizard. Follow the next series of steps, which demonstrate how easy it is to publish drawings to the web.

STEP 1

Open the drawing file 15_Publish to Web. Notice that four layouts exist, namely Four Views, Front View, Top View, and Right Side View. All four of these layouts will be arranged in a single web page to demonstrate how easy it is to perform this task. This example demonstrates how you can publish various layouts of the same drawing to the web; you can also publish different drawings as well.

Begin the process of publishing to the web by clicking on the File pull-down menu followed by Publish to Web…, as shown on the left in the following image. This will launch the Publish to Web wizard as shown on the right in the following image. In this first dialog box (called Begin), be sure the radio button next to Create New Web Page is selected. If you already have an existing web page that needs updating or editing, you could click the radio button next to Edit Existing Web Page. When finished, click the Next> button and continue on to the next step of this wizard.

Figure 15–33

STEP 2

In the Create Web Page dialog box, shown in the following image, add the name of the web page as Four Views with Details; then specify the location of this web page. It is also considered good practice to add a description of the web page in the event that others will be manipulating your web page. After completing the description, click the Next> button to go to the next step.

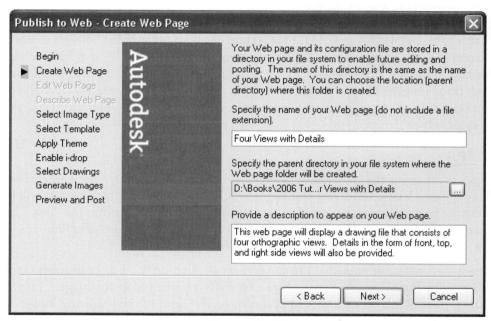

Figure 15–34

STEP 3

When the Select Image Type dialog box appears, you will have the opportunity to pick the type of image that the drawings will be generated in. By default, your web page will be created in DWF format, as shown on the left in the following image. These are vector-based images of your drawing that are viewed in Autodesk DWF Viewer, which is a downloadable free viewer. Another image type option is illustrated in the middle in the following image, in which the AutoCAD drawings consist of JPG images. JPG files consist of raster images that do not perform very well if your drawing contains a lot of text. However, for this tutorial exercise, JPG will be the image format used. Yet another image format is displayed on the right in the following image. The PNG formats are also raster-based representations of your AutoCAD drawing. They produce high-quality images when compared with JPG files. Be sure that the image type reads JPG and click the Next> button when finished.

Figure 15–35

STEP 4

The next step in the process of publishing to the web is to select a web template. The Select Template dialog box in the following image displays four different templates for you to choose from. Click on each one to preview each template at the right of the dialog box. For this tutorial, click on Array of Thumbnails and click the Next> button when finished.

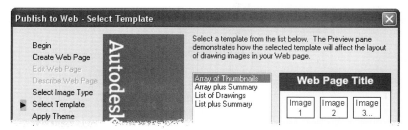

Figure 15–36

STEP 5

The next step when publishing to the web is the selection of a theme. The illustration on the left in the following image shows seven possible themes that you can apply. Click each one to preview the contents of each. For this tutorial, the Classic theme will be used, which is illustrated on the right in the following image. When finished, click the Next> button to continue.

Figure 15–37

STEP 6

The next step displays the Enable i-drop dialog box, shown in the following image. Placing a check in the Enable i-drop box allows those who visit your web page to drag and drop drawing files into their session of AutoCAD. While this is a very powerful feature, we will not be demonstrating it during this tutorial. Click the Next> button to continue.

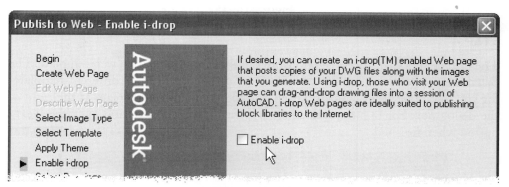

Figure 15–38

STEP 7

The next dialog box allows you to add the drawings that will make up the web page. As illustrated on the left in the following image, the drawing name is listed along with layout and label information. The label can be any name or series of names that you wish to appear in your web page. After filling in a description of this drawing, click the Add-> button to add the label to the image list. Under the Layout heading, shown on the right in the following image, click on Front View as the new layout, add a description, and click the Add-> button to add this label to the image list.

Figure 15–39

Do the same for the Right Side View and Top View. After adding the four layouts to the image list, your dialog should appear similar to the following image. Click the Next> button to continue.

Figure 15–40

STEP 8

The Generate Images dialog box, shown on the left in the following image, will create the web page in the folder that you specified earlier. Clicking on the Next> button of this dialog box will pause your system while all the layouts are plotted. This may take time depending on the number and complexity of drawings being published. Be sure the radio button next to "Regenerate images for drawings that have changed" is selected.

After the regeneration of all images, the Preview and Post dialog box appears, as shown on the right in the following image. You can either preview your results or post the web page to a web site at a later time. Click the Preview button.

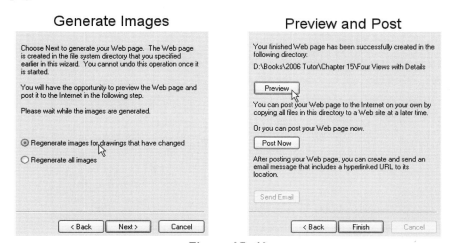

Figure 15–41

STEP 9

Clicking on the Preview button in the previous step will launch Microsoft Internet Explorer and display your web page, as shown in the following image.

Figure 15–42

To view each image separately, click on the thumbnail to enlarge the image and show more detail (see the image of the four views in the following image). Dismissing Internet Explorer takes you back to the Preview and Post dialog box. Pick the Finish button to end this task.

Figure 15–43

CHAPTER 16

Working with Blocks

This chapter begins the study of how blocks are created and merged into drawing files. This is a major productivity enhancement and is considered an electronic template similar to those methods used with block templates in manual drafting practices. The first segment of this chapter will discuss what blocks are and how they are created. Blocks are typically inserted in the current drawing but can be inserted in any drawing by utilizing the proper commands and techniques. The chapter continues by discussing other topics such as redefining blocks and the effects blocks have on table objects. Dynamic blocks provide even greater control of block objects. A number of dynamic block techniques will be utilized. Next, this chapter continues with a discussion about using the Insert dialog box and the DesignCenter to bring blocks into drawings. The DesignCenter is a special feature that allows blocks to be inserted in drawings with drag and drop techniques. In addition, blocks found in other drawings can easily be inserted in the current drawing from the DesignCenter. Yet another feature, the Tool Palette, will allow you to organize blocks and hatch patterns in one convenient area. These object types can then be shared with the current drawing through drag and drop techniques. Finally, the chapter will discuss the use of MDE (Multiple Design Environment). This feature allows the opening of multiple drawings within a single AutoCAD session and provides a convenient method of exchanging data, such as blocks, between one drawing and another. As an added bonus to AutoCAD users, a series of block libraries is supplied with the package. These block libraries include such application areas as mechanical, architectural, electrical, piping, and welding, to name just a few.

WHAT ARE BLOCKS?

Blocks usually consist of smaller components of a larger drawing. Typical examples include doors and windows for floor plans, nuts and bolts for mechanical assemblies, and resistors and transistors for electrical schematics. In the following image, which shows an electrical schematic, all resistors, capacitors, tetrodes, and diodes are considered blocks that make up the total drawing of the electrical schematic. The capacitor is highlighted as one of these components.

Figure 16–1

Blocks are created and then inserted in a drawing. When creating the block, you must first provide a name for the block. The capacitor illustrated on the left in the following image was assigned the name CAPACITOR. Also, when you create a block, an insertion point is required. This acts as a reference point from which the block will be inserted. In the illustration on the left, the insertion point of the block is the left end of the line.

At times, blocks have to be rotated into position. In the illustration on the right in the following image, notice that one capacitor is rotated at a 45° angle while the other capacitor is rotated 270°. In this way the same block can be used numerous times even though it is positioned differently in the drawing.

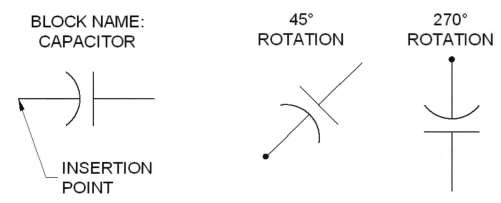

Figure 16–2

METHODS OF SELECTING BLOCK COMMANDS

Commands that deal with blocks can be selected from one of two menu areas. The first one is illustrated on the left in the following image; selecting Block from the Draw pull-down menu exposes Make (the Block Definition dialog box) and Base, which defines a new base or insertion point for a block.

The second area that contains block-related commands is found on the Draw toolbar, as shown on the right in the following image.

Figure 16–3

You can also enter block-related commands from the keyboard, either using their entire name or through command aliases, as in the following examples. These commands will be discussed in greater detail in the following pages.

Enter B to activate the Block Definition dialog box (BLOCK).

Enter I to activate the Insert dialog box (INSERT).

Enter W to activate the Write Block dialog box (WBLOCK).

Enter ADC to activate the DesignCenter (ADCENTER).

CREATING A LOCAL BLOCK USING THE BLOCK COMMAND

The illustration on the left in the following image is a drawing of a hex head bolt. This drawing consists of one polygon, representing the hexagon, a circle, indicating that the hexagon is circumscribed about the circle, and two centerlines. Rather than copy these individual objects numerous times throughout the drawing, you can create a block using the BLOCK command. Selecting Block from the Draw pull-down menu and then selecting Make activates the Block Definition dialog box, illustrated on the right in the following image. Entering the letter B from the keyboard also activates the Block Definition dialog box.

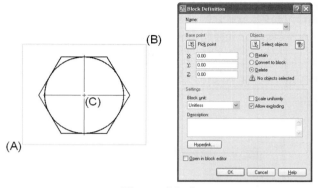

Figure 16–4

When you create a block, the numerous objects that make up the block are all considered a single object. This dialog box allows for the newly created block to be merged into the current drawing file. This means that a block is available only in the drawing it was created in; it cannot be shared directly with other drawings. (The WBLOCK command is used to create global blocks that can be inserted in any drawing file. This command will be discussed later in this chapter.)

To create a block through the Block Definition dialog box, enter the name of the block, such as "Hexbolt," in the Name edit box, as shown in the following image. You can use up to 255 alphanumeric characters when naming a block.

Next, click on the Select objects button; this will return you to the drawing editor. Create a window from "A" to "B" around the entire hex bolt to select all objects, as shown in the previous image. When finished, press ENTER. This will return you to the Block Definition dialog box, where a previewed image of the block you are about to create is displayed in the upper right corner of the dialog box, as shown in the following image. Whenever you create a block, you can elect to allow the original objects that made up the hex bolt to remain on the screen (select Retain button), to be replaced with an instance of the block (select Convert to Block button), or to be removed after the block is created (select Delete button). Select the Delete radio button to erase the original objects after the block is created. If the original objects that made up the hex bolt are unintentionally removed from the screen during this creation process, the OOPS command can be used to retrieve all original objects to the screen.

The next step is to create a base point or insertion point for the block. This is considered a point of reference and should be identified at a key location along the block. By default, the Base point in the Block Definition dialog box is located at 0,0,0. To enter a more appropriate base-point location, click on the Pick point button; this will return you to the drawing editor and allow you to pick the center of the hex bolt at "C," as shown in the previous image. Once this point is selected, this returns you to the Block Definition dialog box; notice how the Base point information now reflects the key location along the block, as shown in the following image.

Figure 16–5

Yet another feature of the Block Definition dialog box allows you to add a description for the block. Many times, the name of the block hides the real meaning of the block, especially if the block name consists of only a few letters. Click in the Description edit box and add the following statement: "This is a hexagonal head bolt." This will allow you to refer to the description in case the block name does not indicate the true intended purpose for the block, as shown in the following image. Specifying the Block unit determines the type of units utilized for scaling the block when it is inserted from the DesignCenter.

Figure 16–6

As the block is written to the database of the current drawing, the individual objects used to create the block automatically disappear from the screen (due to our earlier selection of the Delete radio button). When the block is later inserted in a drawing, it will be placed in relation to the insertion point.

CREATING A DRAWING FILE USING THE WBLOCK COMMAND

In the previous example of using the BLOCK command, several objects were grouped into a single object for insertion in a drawing. As this type of block is created, its intended purpose is for insertion in the drawing it was created in. Using the WBLOCK (Write block) command creates a block by a method that is similar to the process used in the BLOCK command. However, this will write the objects to a separate file, which allows the block to be inserted in any drawing. Entering WBLOCK or just W at the command prompt activates the Write Block dialog box, illustrated on the left in the following image.

In this image, the source for the write (global) block is Objects (the Objects radio button is selected). Follow the same steps demonstrated earlier for creating blocks by selecting the objects that will make up the global block. The Block radio button (grayed out because no blocks exist in the current drawing) would allow you to select an existing block to create the global block. The Entire drawing radio button would allow you to select the complete current drawing for creation of the block file. Next, click on the Base point area's Pick point button to select an appropriate insertion point for the global block.

It may be unclear where the block file is actually being written. Be sure to carefully provide the destination information for the Write Block dialog box. Provide an appropriate file name in the edit box. To give a location for the block, click on the Browse button and choose a suitable drive and folder in which to store the file, as shown on the right in the following image. Clicking on the OK button in the Write Block dialog box writes the objects to the drive location with the file name specified by the user.

Figure 16-7

INSERTING BLOCKS

Once blocks are created through the BLOCK and WBLOCK commands, they are merged or inserted in the drawing through the INSERT command. Enter INSERT or I from the keyboard or choose the command from the Insert pull-down menu, the Insert toolbar, or the Draw toolbar, as shown in the following image, to activate the Insert dialog box, also shown in the following image. This dialog box is used to dynamically insert blocks.

Figure 16–8

First, by clicking on the Name drop-down list box, select a block from the current drawing (clicking on the Browse button locates global blocks or drawing files). After you identify the name of the block to insert, the point where the block will be inserted must be specified, along with its scale and rotation angle. By default, the Insertion point area's Specify On-screen box is checked. This means you will be prompted for the insertion point at the command prompt area of the drawing editor. The default values for the scale and rotation insert the block at the original size and orientation. It was already mentioned that a block consists of several objects combined into a single object. The Explode box determines whether the block is inserted as one object or the block is inserted and then exploded back to its individual objects. Once the name of the block, such as Hexbolt, is selected, click on the OK button; the following prompts complete the block insertion operation:

Specify insertion point or [Basepoint/Scale/X/Y/Z/Rotate/PScale/PX/PY/PZ/PRotate]:
  (Mark a point at "A" in the following image to insert the block)

If the Specify On-screen boxes are also checked for Scale and Rotation, the following prompts will complete the block insertion operation.

Specify insertion point or [Basepoint/Scale/X/Y/Z/Rotate/PScale/PX/PY/PZ/PRotate]: *(Mark a point at "A," as shown in the following image, to insert the block)*
Enter X scale factor, specify opposite corner, or [Corner/XYZ] <1>: *(Press* ENTER *to accept default X scale factor)*
Enter Y scale factor <use X scale factor>:): *(Press* ENTER *to accept default)*
Specify rotation angle <0>: *(Press* ENTER *to accept the default rotation angle and insert the block, as shown in the following image)*

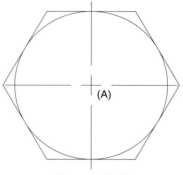

Figure 16–9

If blocks are defined as part of the database of the current drawing, they may be selected from the Name drop-down list box, as shown in the following image.

Figure 16–10

For inserting global blocks in a drawing, select the Insert dialog box Browse button, which displays the Select Drawing File dialog box illustrated in the following image. This is the same dialog box associated with opening drawing files. In fact, you will be unable to distinguish between global blocks and any other AutoCAD drawing file. Global blocks are simply drawing files created in a unique way, and either can be inserted in the current drawing. Select the desired folder where the global block or drawing file is located, followed by the name of the drawing. This will return you to the main Insert dialog box with the file now available in the Name drop-down list box and ready for insertion.

Figure 16–11

APPLICATIONS OF BLOCK INSERTIONS

The following image shows the results of entering different scale factors and rotation angles when blocks are inserted in a drawing file. The image at "A" shows the block inserted in a drawing with its default scale and rotation angle values. The image at "B" shows the result of inserting the block with a scale factor of 0.50 and a rotation angle of 0°. The image appears half its normal size. The image at "C" shows the result of inserting the block with a scale factor of 1.75 and a rotation angle of 0°. In this image, the scale factor increases the block in size while keeping the same proportions. The image at "D" shows the result of inserting the block with different X and Y scale factors. In this image, the X scale factor is 0.50 while the Y scale factor is 2.00 units. Notice how out of proportion the block appears. There are certain applications where different scale factors are required to produce the desired effect. The image at "E" shows the result of inserting the block with the default scale factor and a rotation angle of 30°. As with all rotations, a positive angle rotates the block in the counterclockwise direction; negative angles rotate in the clockwise direction.

Figure 16–12

The following image shows the results of entering different negative scale factors when inserting blocks in a drawing file. The image at "A" shows the block inserted in a drawing with its default scale and rotation angle values. The image at "B" shows the result of inserting the block with an X scale factor of −1.00 units; notice how the image flips in the X direction. This result is similar to using the MIRROR command with the Y axis as the axis to mirror about. The image at "C" shows the result of inserting the block with a Y scale factor of −1.00 units; this time, notice how the image flips in the Y direction. The image at "D" shows the result of inserting the block with both X and Y scale factors of −1.00 units. Here, the object is flipped in both the X and Y directions.

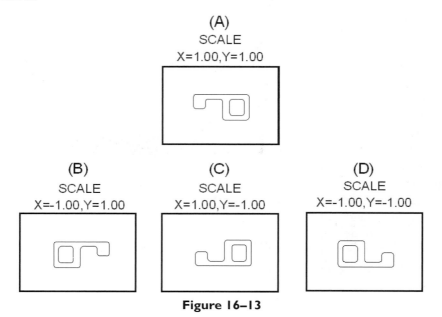

Figure 16–13

TRIMMING TO BLOCK OBJECTS

The ability to trim to a block object is possible through the TRIM command. As you select cutting edges on the block to trim to, the command isolates the cutting edges from the remainder of the objects that make up the block. In the following image, the outside edges of the bolt act as cutting edges.

 Try It! – To test this feature, open the drawing 16_Trim Plates and use the following image as a guide. The bolt is a block. Activate the TRIM command, pick the edges of the bolt at "A" and "B" as cutting edges, and pick the lines at "C" through "E" as the objects to trim.

The result of using the TRIM command is illustrated on the right in the following image.

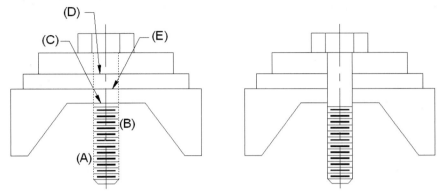

Figure 16–14

EXTENDING TO BLOCK OBJECTS

As with the TRIM command, objects can be extended to boundary edges that are part of a block with the EXTEND command. Selecting boundary edges that consist of block components isolates these objects. This enables you to extend objects to these boundary edges.

 Try It! – To test this feature, open the drawing 16_Extend Plates and use the following image as a guide. Again the bolt is a block. Activate the EXTEND command, pick the edges of the bolt at "A" and "B" as boundary edges, and pick the lines at "C" through "H" as the objects to extend.

The result of using the EXTEND command on block objects as cutting edges is illustrated on the right in the following image.

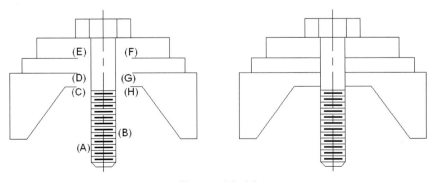

Figure 16–15

EXPLODING BLOCKS

It has already been mentioned that as blocks are inserted in a drawing file, they are considered one object even though they consist of numerous individual objects. At times it is necessary to break a block up into its individual parts. The EXPLODE command is used for this.

Illustrated on the left in the following image is a block that has been selected. Notice that the entire block highlights and the insertion point of the block is identified by the presence of the grip. Using the EXPLODE command on a block will result in the illustration on the right in the following image. Here, when one of the lines is selected, only that object highlights. Exploding a block breaks the block back up into its individual objects. As a result, you must determine when it is appropriate to explode a block.

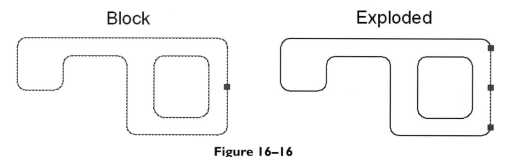

Figure 16–16

MANAGING UNUSED BLOCK DATA WITH THE PURGE DIALOG BOX

AutoCAD stores named objects (blocks, dimension styles, layers, linetypes, multiline styles, plot styles, shapes, table styles, and text styles) with the drawing. When the drawing is opened, AutoCAD determines whether other objects in the drawing reference each named object. If a named object is unused and not referenced, you can remove the definition of the named object from the drawing by using the PURGE command. This is a very important productivity technique used for compressing or cleaning up the database of the drawing. Picking File from the pull-down menu, followed by Drawing Utilities, displays Purge…, as shown on the left in the following image. Clicking on Purge… displays the Purge dialog box, as shown on the right in the following image.

Figure 16–17

The Blocks category is expanded by clicking on the "+." This produces the list of all the items that are currently unused in the drawing that can be removed, as shown on the left in the following image. Clicking on the item SOFA2, right-clicking, and then picking Purge from the shortcut menu or by simply selecting the button at the bottom of the dialog box will display the Confirm Purge dialog box, as shown in the middle in the following image. Click the Yes button and the block is removed from the listed items, as shown on the right in the following image.

Figure 16–18

Other controls are available in the Purge dialog box, such as the ability to view items that cannot be purged and the capability of purging nested items. An example of purging a nested item would be purging a block definition that lies inside another block definition. Entering -PURGE or PU at the Command prompt also activates the Purge dialog box.

Tip: The layer "0," Standard text and dimension styles, and the Continuous linetype cannot be purged from a drawing.

Try It! – Open the drawing file 16_Purge, as shown on the left in the following image. Activate the Purge dialog box and click the Purge All button. You will be prompted to remove all items individually through the Confirm Purge dialog box, as shown on the right in the following image. Answer Yes to each item you want purged or Yes to All, until all items, including blocks, layers, linetypes, text styles, dimension styles, and multiline styles, have been removed from the database of the drawing.

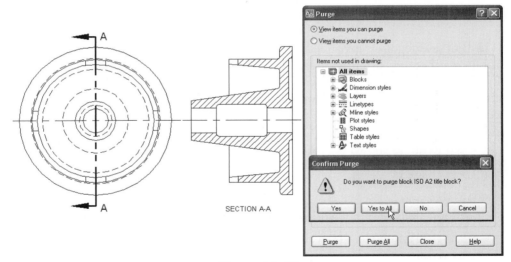

Figure 16–19

REDEFINING BLOCKS

At times, a block needs to be edited. Rather than erase all occurrences of the block in a drawing, you can redefine it. This is considered a major productivity technique because all blocks that share the same name as the block being redefined will automatically update to the latest changes. The illustration on the left in the following image shows various blocks inserted in a drawing. The block name is Step Guide and the insertion point is at the lower left corner of the object.

Try It! – The drawing file 16_Block Redefine has been provided on CD for you to experiment with redefining blocks. When the drawing appears on your screen, as shown on the left in the following image, double-click on any block in the drawing. This will launch the Edit Block Definitions dialog box, as shown on the right in the following image. A list of all blocks defined in the drawing will appear. Click on Step Guide from the list and click the OK button.

Figure 16–20

After you click on the OK button, the Block Editor area will appear, as shown on the left in the following image. A majority of the tools displayed in the palette on the left are designed for creating dynamic blocks, which will be explained in detail later in this chapter. For now, use this environment to make changes to the geometry that makes up the Step Guide. You can add new objects, change the properties of objects, or make modifications to this object. For this example, erase the circle and stretch the upper right corner of the object down by 2 units, as shown on the right in the following image.

To return to your drawing, you must first save the changes made to the object by clicking on the Save Block Definition button, also shown on the right in the following image. To exit the Block Editor area, click on the Close Block Editor button.

Figure 16–21

When you return to the drawing editor, notice that all Step Guide blocks have been updated to the changes made in the Block Editor. This provides a very productive method of making changes to all blocks in a drawing. This method works only if blocks have not been exploded in a drawing. It is for this reason that you must exercise caution when exploding blocks in a drawing.

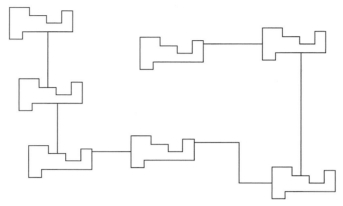

Figure 16–22

BLOCKS AND THE DIVIDE COMMAND

The DIVIDE command was already covered in chapter 5. It allows you to select an object, give the number of segments, and place point objects at equally spaced distances depending on the number of segments. The DIVIDE command has a Block option that allows you to place blocks at equally spaced distances.

 Try It! – Open the drawing file 16_Speaker. In the following image, a counterbore hole and centerline need to be copied 12 times around the elliptical centerline so that each hole is equally spaced from others. (The illustration of the block CBORE is displayed at twice its normal size.) Because the ARRAY command is used to copy objects in a rectangular or circular pattern, that command cannot be used for an ellipse. The DIVIDE command's Block option allows you to specify the name of the block and the number of segments. In the following image, the elliptical centerline is identified as the object to divide. Follow the command sequence to place the block CBORE in the elliptical pattern:

Command: **DIV** *(For DIVIDE)*
Select object to divide: *(Select the elliptical centerline)*
Enter the number of segments or [Block]: **B** *(For Block)*
Enter name or block to insert: **CBORE**
Align block with object? [Yes/No] <Y>:*(Press* ENTER *to accept)*
Enter the number of segments: **12**

The results are illustrated on the right in the following image. Notice how the elliptical centerline is divided into 12 equal segments by 12 blocks called CBORE. In this way, any object may be divided through the use of blocks.

Figure 16–23

BLOCKS AND THE MEASURE COMMAND

As with the DIVIDE command, the MEASURE command offers increased productivity when you measure an object and insert blocks at the same time.

 Try It! – Open the drawing file 16_Chain. In the following image, a polyline path will be divided into 0.50-length segments using the block CHAIN2. The perimeter of the polyline was calculated by the LIST command to be 22.00 units, which is evenly divisible by 0.50 and will allow the insertion of 44 blocks. Follow the command prompt sequence below for placing a series of blocks called CHAIN2 around the polyline path to create a linked chain.

Figure 16–24

Command: **ME** *(For MEASURE)*
Select object to measure: *(Pick the polyline path)*
Specify length of segment or [Block]: **B** *(For Block)*
Enter name of block to insert: **CHAIN2**
Align block with object? [Yes/No] <Y> *(Press ENTER to accept)*
Specify length of segment: **0.50**

The result is illustrated on the left in the following image, with all chain links being measured along the polyline path at increments of 0.50 units.

Answering No to the prompt "Align block with object? [Yes/No] <Y>" displays the results, as shown on the right in the following image. Here all blocks are inserted horizontally and travel in 0.50 increments. While the polyline path has been successfully measured, the results are not acceptable for creating the chain.

Figure 16–25

RENAMING BLOCKS

Blocks can be renamed to make their meanings more clear through the RENAME command. This command can be found in the Format pull-down menu, as shown on the left in the following image, and when selected will display a dialog box similar to the one illustrated on the right in the following image. Clicking on Blocks in the dialog box lists all blocks defined in the current drawing. One block with the name REF1 was abbreviated, and we wish to give it a full name. Clicking on the name REF1 will paste it in the Old Name edit box. Type the desired full name REFRIGERATOR in the Rename To edit box and click on the Rename To button to rename the block.

Figure 16–26

TABLES AND BLOCKS

In the following image, electrical symbols and their descriptions are arranged in the legend to call out the symbols in a table. To add blocks to a table, click inside the cell that will hold the block and right-click to display the menu, then click Insert Block...

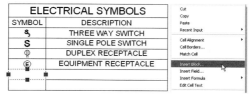

Figure 16–27

When the Insert a Block in a Table Cell dialog appears, locate the name of the block to insert, as shown in the following image.

Figure 16–28

The results are illustrated in the following image, where the Ground Fault block is inserted into the highlighted cell of the table. Because the AutoFit function is checked in the previous figure, the block will be scaled to fit inside the cell, no matter how large or small the block is. The title of the symbol was also added to the table by double-clicking inside the cell to launch the Text Formatting toolbar.

ELECTRICAL SYMBOLS	
SYMBOL	DESCRIPTION
S₃	THREE WAY SWITCH
S	SINGLE POLE SWITCH
⌽	DUPLEX RECEPTACLE
Ⓔ	EQUIPMENT RECEPTACLE
⏚	GROUND FAULT

Figure 16–29

 Note: You can type the TINSERT command at the command prompt to insert a block into a table cell. After selecting the cell in which to insert the block, the Insert Block in Table dialog box will appear. Select a block from the list to be inserted into the selected table cell.

ADDITIONAL TIPS FOR WORKING WITH BLOCKS

CREATE BLOCKS ON LAYER 0

Blocks are best controlled, when dealing with layer colors, linetypes, and lineweights, by being drawn on Layer 0, because it is considered a neutral layer. By default, Layer 0 is assigned the color White and the Continuous linetype. Objects drawn on Layer 0 and then converted to blocks will take on the properties of the current layer when inserted in the drawing. The current layer will control color, linetype, and lineweight.

CREATE BLOCKS FULL SIZE IF APPLICABLE

The illustration on the left in the following image shows a drawing of a refrigerator, complete with dimensions. In keeping with the concept of drawing in real-world units in CAD or at full size, individual blocks must also be drawn at full size in order for them to be inserted in the drawing at the correct proportions. For this block, construct a rectangle 28 units in the X direction and 24 units in the Y direction. Create a block called Refrigerator by picking the rectangle. When testing out this block, be sure to be in a drawing that is set to the proper units based on a scale such as ½″ = 1′-0″ or ¼″ = 1′-0″. These are typical architectural scales, and the block of the refrigerator will be in the correct proportions on these types of sheet sizes.

One exception to the full-size rule is illustrated on the right in the following image. Rather than create each door block separately to account for different door sizes, create the door so as to fit into a 1-unit by 1-unit square. The purpose of drawing the door block inside a 1-unit square is to create only one block of the door and insert it at a scale factor matching the required door size. For example, for a 2′-8″ door, enter 2′ 8 or 32 when prompted for the X and Y scale factors. For a 3′-0″ door, enter 3′ or 36 when prompted for the scale factors. Numerous doors of different types can be inserted in a drawing using only one block of the door. Also try using a negative scale factor to mirror the door as it is inserted.

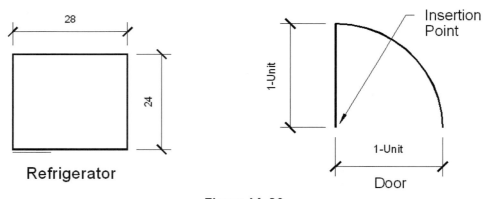

Figure 16–30

USE GRID WHEN PROPORTIONALITY, BUT NOT SCALE, IS IMPORTANT

Sometimes blocks represent drawings in which the scale of each block is not important. In the previous example of the refrigerator, scale was very important in order for the refrigerator to be drawn according to its full-size dimensions. This is not the case, as shown on the left in the following image, of the drawing of the resistor block. Electrical schematic blocks are generally not drawn to any specific scale; however, it is important that all blocks are proportional to one another. Setting up a grid is considered good practice in keeping all blocks at the same proportions. Whatever the size of the grid, all blocks are designed around the same grid size. The result is shown on the right in the following image; with four blocks being drawn with the same grid, their proportions look acceptable.

Figure 16–31

USING DESIGNCENTER

A revolutionary tool called the DesignCenter has been developed as a simple, efficient device for sharing drawing data. It provides a means of inserting blocks and drawings even more efficiently than through the Insert dialog box. This feature has the distinct advantage of inserting specific blocks internal to one drawing into another drawing. If the DesignCenter is not present, you can load it by choosing Palettes and then DesignCenter from the Tools pull-down menu, as shown on the left in the following image, or by selecting the DesignCenter button from the Standard toolbar, as shown on the right in the following image.

Figure 16–32

When used for the first time, the DesignCenter loads in the middle of the screen. An Auto-hide feature for the DesignCenter allows it to hide (collapse) once you move the cursor outside the DesignCenter window. This helps clear the drawing area when the DesignCenter is not being used. To expand it again, simply move the cursor over the DesignCenter title bar. A shortcut menu allows you to turn the Auto-hide and docking features on or off as desired, as shown in the following image. If you prefer that the DesignCenter remain on the screen, you can still dock it to the left or right side of the AutoCAD drawing screen. The DesignCenter can be resized as the user requires. You can unload it by clicking the X in the title bar, selecting DesignCenter from the Tools pull-down menu, clicking on the DesignCenter button on the Standard toolbar, or entering the ADCCLOSE command at the keyboard.

Figure 16–33

Tip: The DesignCenter can also be launched by pressing and holding down the CTRL key and typing 2 (CTRL+2).

Block libraries may be prepared in different formats in order for them to be used through the DesignCenter. One method is to place all global blocks in one folder. The DesignCenter identifies this folder and graphically lists all drawing files to be inserted. Another method of organizing blocks is to create one drawing containing all local blocks. When this drawing is identified through the DesignCenter, all blocks internal to this drawing display in the DesignCenter palette area.

DESIGNCENTER COMPONENTS

The DesignCenter is isolated in the following image. The following components of the dialog box are identified below and in the figure: Control buttons, Tree View or Navigation Pane, Palette or Content Pane, and Shortcut Menu.

Figure 16–34

A more detailed illustration of the DesignCenter Control buttons is found in the following image. These buttons are identified as: Load, Back, Forward, Up, Search, Favorites, Home, Tree View Toggle, Preview, Descriptions, and Views. It may be necessary to resize the DesignCenter to see all the buttons.

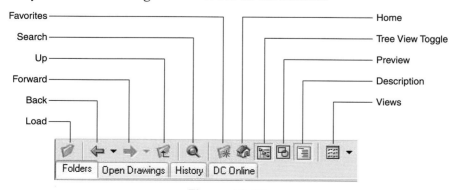

Figure 16–35

USING THE TREE VIEW

As mentioned earlier, clicking the Tree View button expands the DesignCenter to look similar to the illustration in the following image. The DesignCenter divides into two major areas; on the right is the familiar Palette area; on the left is the Tree View (a layout of network and local drive folders). Clicking on a folder displays its contents in the Palette area, as shown in the following image.

Figure 16–36

Clicking on a drawing in the Tree View displays the drawing objects, as shown in the following image, that can be shared through the DesignCenter (Blocks, Dimstyles, Layers, Layouts, Linetypes, Tablestyles, Textstyles, and Xrefs).

Figure 16–37

Double-click the Blocks icon (or click the "+" symbol next to the drawing name and choose Blocks in the Tree View) to display the blocks available in the selected drawing, as shown in the following image.

Figure 16–38

INSERTING BLOCKS THROUGH THE DESIGNCENTER

The following image displays a typical floor plan drawing along with the Design-Center (docked to the left side of the screen), showing the blocks identified in the current drawing. If no blocks are found internal to the drawing, the Palette area will be empty.

Figure 16–39

The DesignCenter operates on the "drag and drop" principle. Select the desired block located in the Palette area of the DesignCenter, drag it out, and drop it into the desired location of the drawing. The following image shows a queen-size bed dragged and dropped into the bedroom area of the floor plan.

Figure 16–40

When performing a basic drag and drop operation with the left mouse button, all you have to do is identify where the block is located and drop it into that location (the use of running object snaps can ensure that the blocks are dropped in a specific location). What if the block needs to be scaled or rotated? If the drag and drop method is performed with the right mouse button, a shortcut menu is provided, as shown in the following image. Selecting Insert Block from the shortcut menu will display the Insert dialog box, covered earlier in this chapter, which allows you to specify different scale and rotation values. The following image shows a rocking chair inserted and rotated into position in the corner of the room. Instead of dragging the block with the right mouse button, double-click on the block in the Palette area and the Insert dialog box will be provided immediately.

Figure 16–41

INSERTING BLOCKS USING THE TOOL PALETTE

Generally, tool palettes allow you to organize blocks and hatch patterns for insertion into your drawing. This feature is somewhat similar to the DesignCenter in its ability to drag and drop blocks, layers, dimension styles, text styles, and hatch patterns into a drawing. The Tool Palette, however, is specific to blocks, commands, and hatch patterns. You can organize and customize the Tool Palette to meet your individual drawing needs.

By default when AutoCAD is first loaded, the Tool Palette is positioned in the upper right corner of your display screen. If the Tool Palette is not visible, it can be activated by selecting it from the Tools pull-down menu or from the Standard toolbar, as shown on the left and in the middle of the following image.

 Tip: The Tool Palette can also be activated by pressing down the CTRL key and typing 3 (CTRL+3).

Once displayed, the Tool Palette provides a number of sample tabs for you to experiment with, as shown on the right in the following image. Notice in this figure how block icons are displayed with a lightning bolt graphic. This means that the block is considered dynamic. This feature of creating and using dynamic blocks will be covered later in this chapter.

Figure 16-42

Additional tool palettes are available to assist in your design capabilities. To access these extra palettes, click on the area located in the lower left corner of any tool palette, as shown in the following image. A long list activates alongside the existing palette.

A number of the palettes consist of blocks that can be dragged and dropped into a drawing. Other palettes such as Lighting and Cameras deal with the 3D rendering module contained in AutoCAD. Notice also a number of material palettes. These also deal with 3D rendering.

Figure 16–43

Illustrated in the following image is an example of placing hatch patterns. In this example, a concrete pattern is dragged from the Tool Palette, as shown on the right in the following image, and dropped into a closed area of your drawing, as in the fireplace hearth illustrated in the figure. The results are shown in the following image, with the hatching pattern applied to the closed area.

Figure 16–44

The Tool Palette has a number of very powerful features to automate its operation. For example, in the following image, a Vehicles symbol is selected. Right-clicking on this block activates a shortcut menu. The following options are available:

Cut—Cutting the block from the Tool Palette to the Windows clipboard

Copy—Copying the block to the Windows clipboard

Delete—Deleting this block from the Tool Palette

Rename—Changing the name of this block in the Tool Palette

Properties—Changing the properties of this block in the Tool Palette

The Properties option allows you to modify the object's properties to suit your specific needs. Select the Properties option, as shown on the left in the following image. This activates the Tool Properties dialog box, as shown in the middle of the following image. All the information in the fields can be changed and applied to this symbol. For example, suppose you need to change the insertion scale of this block for a number of drawings. Changing the scale in the Tool Properties dialog box will change the block's scale as it is inserted into the drawing. This feature is also available for hatch patterns when using the Tool Palette. Scrolling down this dialog box will display a heading for Custom Properties. In the illustration on the right in the following image, numerous versions of the block are displayed. This is because the item selected was constructed as a dynamic block.

Figure 16–45

CREATING A NEW TOOL PALETTE

The process for creating new tool palettes is very easy and straightforward. First, right-click anywhere inside the Tool Palette to display the menu, illustrated on the left in the following image. Options of this Tool Palette menu include:

Allow Docking—Allows the Tool Palette to be docked to the sides of your display screen. Removing the check disables this feature.

Auto-Hide—When checked, this feature collapses the Tool Palette so only the thin blue strip is displayed. When you move your cursor over the blue strip, the Tool Palette redisplays.

Transparency—Activates a Transparency dialog box, which controls the opaqueness of the Tool Palette.

View Options—Activates the View Options dialog box, which controls the size of the hatch and block icons and whether the icon is labeled or not.

Delete Tool Palette—Deletes the Tool Palette. A warning dialog box appears asking whether you really want to perform this operation.

Rename Tool Palette—Renames the Tool Palette.

Customize—Activates the Customize dialog box, which allows you to create a new Tool Palette.

New Palette—Creates a blank Tool Palette.

Click on the New Palette option, as shown on the left in the following image. This automatically creates a blank Tool Palette. As illustrated on the right in the following image, a new Tool Palette name, Electrical, has been entered in the edit box.

Figure 16–46

Once the Tool Palette name is entered, a tab is created for this palette, as shown on the right in the following image. To add blocks and hatch patterns to this new palette, activate the DesignCenter, search for the folder that contains the symbols you wish to place in the Tool Palette, and drag and drop these blocks or hatch patterns from the DesignCenter into the Tool Palette, as shown on the left in the following image.

Figure 16–47

If you want to create a new Tool Palette from one whole drawing in the DesignCenter, activate the DesignCenter and go to the drawing that contains the blocks. Right-clicking on this drawing displays the menu, as shown on the left in the following image. Clicking on the Create Tool Palette option will create the Tool Palette using the same name as the DesignCenter drawing. The new Tool Palette will contain all blocks from this drawing, as shown on the right in the following image.

 Tip: If you right-click a folder and select Create Tool Palette of Blocks, a new Tool Palette will be created with the name of the folder and will contain the drawings from that folder.

Figure 16–48

TRANSPARENCY CONTROLS FOR PALETTES

The Transparency feature of the menu, illustrated on the left in the following image, allows you to control the opaqueness of the Tool Palette. Picking Transparency activates the Transparency dialog box, illustrated in the middle in the following image. By default, the slider bar is set to the Less position. This means that if the Tool Palette is positioned on top of your drawing, you will not be able to view what is underneath it. Changing the slider bar to the More position, as shown on the right in the following image, will make the Tool Palette transparent to the point that you will be able to see objects under the Tool Palette.

Figure 16–49

The results of setting the Transparency dialog box to More are illustrated in the following image. You can see both the Tool Palette and the drawing at the same time. This is a very powerful option for displaying more information in your drawing.

Figure 16–50

WORKING WITH MULTIPLE DRAWINGS

The Multiple Design Environment allows users to open multiple drawings within a single session of AutoCAD, as shown in the following image. This feature, like DesignCenter, allows sharing of data between drawings. You can easily copy and move objects, such as blocks, from one drawing to another.

Figure 16–51

OPENING MULTIPLE DRAWINGS

Repeat the OPEN command as many times as necessary to open all drawings you will need. In fact, you can select multiple drawings in the Select File dialog box by holding down CTRL or SHIFT as you select the files, as shown in the following image.

Figure 16–52

Once the drawings are open, use CTRL+F6 or CTRL+TAB to switch back and forth between the drawings. To efficiently work between drawings, you may wish to tile or cascade the drawing windows utilizing the Window pull-down menu, as shown in the following image. A list of all the open drawings is displayed at the bottom of the Window pull-down menu. Selecting one of the file names is another convenient way to switch between drawings. Remember to use the CLOSE command (Window or File pull-down menu) to individually close any drawings that are not being used. To close all drawings in a single operation, the CLOSEALL command (Window pull-down menu) can be used. If changes were made to any of the drawings, you will be prompted to save those changes before the drawing closes.

Figure 16–53

WORKING BETWEEN DRAWINGS

Once your drawings are opened and arranged on the screen, you are now ready to cut and paste, copy and paste, or drag and drop objects between drawings. The first step is to cut or copy objects from a drawing. The object information is stored on the Windows clipboard until you are ready for the second step, which is to paste the objects into that same drawing or any other open drawing. These operations are not limited to AutoCAD. In fact, you can cut, copy, and paste between different Windows applications.

Use one of the following commands to cut and copy your objects:

CUTCLIP—To remove selected objects from a drawing and store them on the clipboard

COPYCLIP—To copy selected objects from a drawing and store them on the clipboard

COPYBASE—Similar to the COPYCLIP command but allows the selection of a base point for locating your objects when they are pasted

Use one of the following commands to paste your objects:

PASTECLIP—Pastes the objects at the location selected

PASTEBLOCK—Similar to PASTECLIP command but objects are inserted as a block and an arbitrary block name is assigned

The commands listed can be typed at the keyboard, selected from the Edit pull-down menu, as shown on the left in the following image, or selected by right-clicking on the display screen when a command is not in progress, as shown in the middle in the following image.

Objects may also be copied between drawings with drag and drop operations. After selecting the objects, place the cursor over the objects (without selecting a grip) and then drag and drop the objects in the new location. Dragging with the right mouse button depressed will provide a shortcut menu allowing additional control over pasting operations, as shown on the right in the following image.

Figure 16–54

ADVANCED BLOCK TECHNIQUES—CREATING DYNAMIC BLOCKS

Dynamic blocks are blocks that change in appearance whenever they are edited through grips or through custom tables embedded inside the block. In the following image, a dynamic block called Drawing Title was located under the Architectural tab of the Tool Palettes, as shown on the right in the following image. This block was then dragged and dropped into the drawing. Clicking on this block will display the normal insertion point grip. However, notice a second grip, which appears as an arrow pointing in the right direction. This grip identifies a stretching action associated with the dynamic block. When you click on this arrow and move your cursor, that portion of the dynamic block will change, giving the block a different appearance without redefining or exploding the block. In this example, the arrow grip is designed to stretch the line. Notice also that dynamic blocks are identified in the Tool Palette by the appearance of a lightning bolt icon.

Figure 16-55

The creation process begins with the Block Editor. This can be accessed by double-clicking on the block, which will display the Edit Block Definitions dialog box. This is the same dialog box used when redefining a block. You select the block name from the list and click the OK button. At this point, you enter the Block Editor environment, as shown in the following image. It is here that you make assignments to the dynamic block. The first step is to assign a parameter to the block. In the following image, a Linear Parameter was selected from the list of parameters in the toolbar and assigned to the geometry, in this case a line segment. This parameter usually is named Distance by default. In this example, the default name was renamed to a term with more meaning, Title Line Length. The Properties Palette is used for this renaming task. Once a parameter is present in the Block Editor, an action item such as Stretch is linked to the parameter. It is the action item that allows the block to change when edited back in the drawing editor. In this example, two action items are present, namely, Stretch and Move. The Stretch action will allow the line to be stretched to different lengths. The Move action is required in order to move the VPSCALE attribute along with the line as it is being stretched.

Figure 16-56

When you are satisfied with the parameter and action assignments and wish to test the features out on the block, you first click the Save button to save the changes to the block name and then pick the Close Block Editor button, as shown in the following image, to return to the drawing editor.

Figure 16–57

The following table gives a brief description of all buttons found in the toolbar of the Block Editor:

Button	Description
	Switches back to the Edit Block Definition dialog box
	Saves any changes to the current block name
	Used to save any changes under a different block name
	Toggles the Block Authoring Palettes on or off
	Used for assigning parameters to the block
	Used for assigning action items to the block
	Launches the Attribute Definition dialog box for assigning attributes. Attributes are covered in chapter 17
	Used for updating the parameter and action item text size
	Used for learning more about dynamic blocks

The following image illustrates another, more powerful example of using dynamic blocks. Normally you would have to create four different blocks in order to show the various door swings. Through the use of Visibility States, all four doors pictured in the following image belong to a single block name. You simply pick the desired door opening from a list that displays with the block. Another feature of dynamic blocks illustrated in the following image is the ability to flip the door to different locations. Two Flip Grips allow you to flip the door along horizontal or vertical hinges. Notice also a Stretch Grip. This is present to stretch the door based on different wall thicknesses. As you can see, dynamic blocks can easily become a major productivity tool used to reduce the overall number of differently named blocks in your drawing.

Figure 16–58

Yet another example of dynamic blocks is illustrated in the following image. Through the use of Visibility States, you have the ability to consolidate a number of blocks under a single name. In this image, notice the block name Trees – Imperial, as shown on the right in the Tool Palette. This single block name actually contains 12 different tree blocks. When you insert one of these dynamic tree blocks, the Visibility States grip will appear. Clicking on the grip will activate the list of trees. You pick the tree from the list and the previous tree block changes based on what you select from the list.

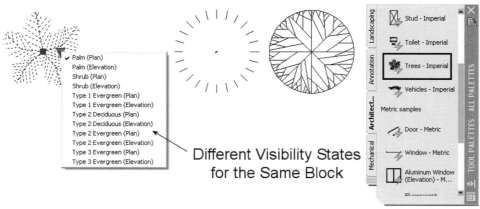

Figure 16–59

The following table gives a brief description of each dynamic block grip.

Dynamic Block Grip	Name	Description
■	Insertion Grip	The original grip that displays at the insertion point of a block; can also be used for moving the block to a new location
▶ ▼	Lengthen	Allows the block to be lengthened or shortened by stretching, scaling, or arraying
⬠	Alignment	Positions or aligns the block based on an object; the positioning occurs when you move the block near the object
⬅ ⬇	Flip	Allows the entire block or items inside the block to be flipped
●	Rotate	Allows the entire block or items inside the block to be rotated
▼	List	Allows you to choose from a list of items

Before you begin the process of designing dynamic blocks, here are a few items to consider:

- For what intended purpose are you designing this dynamic block?
- How do you want this block to change when it is being edited?
- What parameters are needed in order to create changes in the block?
- What actions need to be assigned to the parameters?
- Do you need the block to contain various size values in order to make incremental changes?
- Do you want to organize various blocks under a single name and control what is displayed through Visibility States?
- Do you want to create a table consisting of different values and change the size of a block through the table?

The next series of exercises will allow you to experiment with various capabilities of dynamic blocks.

WORKING WITH PARAMETERS AND ACTIONS

This exercise is designed to familiarize you with the basics of assigning parameters and actions to a block and testing its dynamic nature. You will assign a Parameter and Action item to a block and then test the results by stretching the block to new sizes without exploding the block.

Try It! – Open the drawing file 16_Dynamic_Basics. An existing block called Table01 is already created. The size of this table is 6´ by 4´. Double-click on the existing block Table01 to launch the Edit Block Definition dialog box, as shown on the right in the following image.

Figure 16–60

Clicking the OK button in the Edit Block Definition dialog box will launch the Dynamic Block Editor, as shown in the following image. You can make changes to the geometry of the existing block; or you can assign parameters and actions, making the block dynamic. These assignments are made through the Block Authoring Palettes, as shown in the following image.

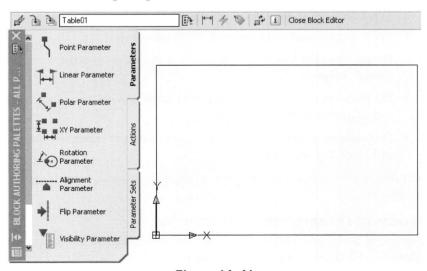

Figure 16–61

We want the ability to change the overall width of this table block. To do this, first make sure the Parameters tab is active and click on the Linear Parameter tool, as shown in the following image. You will be directed through a series of prompts, which can be found in the command prompt area.

Just as in dimensioning, you pick a starting point and endpoint for the parameter, namely, the endpoints of the bottom line of the rectangle. You also specify the label location, as shown in the following image.

Figure 16–62

Adding a parameter is the first step in creating a dynamic block. An action must now be associated with this parameter. Select the Actions tab on the palette. Since you want the ability to change the width of the table, select the Stretch action, which will be used to accomplish this task. As with adding the parameter, the Stretch action comes with a lengthy series of command prompts.

When initiating the Stretch action, first pick the existing Distance parameter. Since the parameter was created with two endpoints, you specify the endpoint to associate the Stretch action with. Select the rightmost parameter endpoint. Next, you create a crossing box around the area to Stretch, as shown in the following image. Then you select the rectangle as the object to stretch. Finally, you specify a location for the icon that signifies the Stretch action.

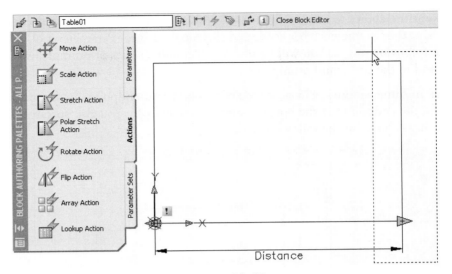

Figure 16–63

The following image illustrates all the components that make up the dynamic block. The Stretch icon should be positioned near the side of the object the action occurs at. Notice also the Alert icon. One more action needs to be created in order for the Alert icon to disappear. For the purposes of this exercise, we will not need an additional action because we will be stretching this rectangle only to the right. When you are finished assigning parameters and actions, click the Save Block Definition button to save these changes to the block. To return to the drawing editor, click on the Close Block Editor button.

Figure 16–64

While back in the drawing, add a dimension to the existing block. Then click on the rectangle and notice the two arrows appear. These arrows represent the parameter. Clicking on the right arrow will stretch the block to the right or left (be sure Osnap is turned off). Notice also that the dimension changes to the new value. If you attempt to stretch the other side of the rectangle, the arrow will move but the rectangle will not. This is because the Stretch action was not assigned to this side.

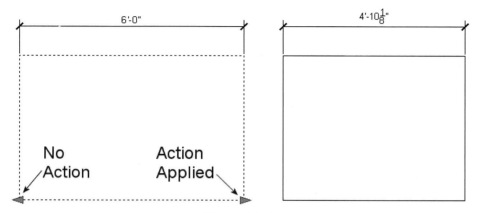

Figure 16–65

WORKING WITH VALUE SET PROPERTIES FOR PARAMETERS

Value sets are associated with parameters and are used to define custom property values for block references. Three types of value sets can be defined in a block reference; they are List, Increment, and None. This exercise will concentrate on using the Increment value set.

 Try It! – Open the drawing file 16_Dynamic_Value_Sets. We will be using the same table block from the previous Try It! exercise. However, instead of stretching the table to the right at random lengths, we want to better control the stretching operation. We want to stretch this table in increments of 6″. We also do not want the table to get narrower than 4′ or wider than 8′. We will be able to control these items through the Value Set associated with the Distance parameter.

Double-click on the block to activate the Edit Block Definition dialog box and click the OK button to enter the Block Authoring environment. Use the Properties command to activate the Properties Palette, and click on the Distance parameter. Change the parameter name from Distance to Table Width, as shown in the following image. This gives the parameter a more meaningful name. Actions can also be renamed through the Properties Palette.

Figure 16–66

With the Properties Palette active and the Table Width parameter selected, scroll down to the bottom of the Properties Palette and notice the Value Set category. It is here that you activate one of the three Value Set types. For this example, make the Increment type active. This will also display other information related to this type. Change the Distance Increment field to 6″, change the Distance Minimum field to 4′, and change the Distance Maximum field to 8′, as shown in the following image. As you make these changes, notice the appearance of tick marks signifying these three values.

Figure 16–67

When you have finished creating these Value Sets for the Table Width parameter, save these changes to the current block name and then close the Block Editor. This will return you to your drawing. Click on the block and then on the rightmost parameter arrow. Notice the appearance of the two parameters and tick marks, as shown on the left in the following image. The tick marks are in increments of 6″. The tick mark on the far left begins at the 4′ distance of the table. The tick mark on the far right ends at the 8′ distance. Slide your cursor to the right or left and notice your cursor snapping to the increment tick marks. Clicking on one of the tick marks will adjust the table to that distance, as shown on the right in the following image.

Figure 16–68

BASE PARAMETER APPLICATIONS

When stretching dynamic blocks, it is possible to stretch the block off its base position. To prevent this from happening, you add a Base Parameter to the dynamic block. Then the base point of the block will always maintain its position relative to some of the geometry in the block.

 Try It! – Open the drawing file 16_Dynamic_Base. This block already has a parameter and two Stretch actions assigned to it. Double-click on the block to activate the Edit Block Definition dialog box, and click the OK button to enter the Block Authoring environment. Notice the existence of the Table Width parameter in addition to one Stretch action to the right and the other to the left; these were all renamed through the Properties Palette. Notice also the position of the parameter arrows near the middle of the block. These arrows were moved straight up using the MOVE command with Polar mode on. Click the Close Block Editor button to return to the drawing editor, click on the block, pick the left parameter arrow, and stretch the block to the right. (This can still be accomplished, even though the action is named Left.) Notice the appearance of the lone grip off the table block and to the left. This represents the insertion point of the block. We want this point to always maintain its relative position, even when Stretch actions occur. A Base Point Parameter will be applied to this corner. Undo this Stretch operation to return the block to its original shape.

Figure 16–69

Activate the Block Editor, and in the Block Authoring Palette, identify the Base Point Parameter (you may have to scroll down in the palette window to see this parameter). Click on the Base Point Parameter tool and pick the lower left corner of the table block as the new base point, as shown in the following image.

Figure 16–70

Unfortunately, if you return to the drawing editor to test this new block, the base point will not follow any stretching operations. This is because the Base Point Parameter was placed into the block after the Stretch Left action was already created. Double-clicking on the Stretch Left action will activate a number of action redefine prompts. First re-create the crossing box, being careful to include the Base Point Parameter in the selection. Next, you will have to pick the Base Point Parameter in order to add it to the selection set. When finished redefining the Stretch Left action, save the block and return to the drawing editor. Test your block by stretching the left side of the table and notice that the base point now follows along with the stretch.

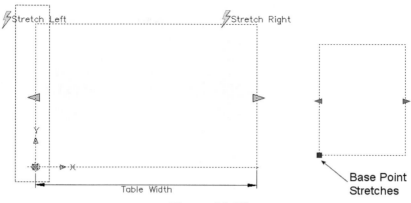

Figure 16–71

ALIGNMENT PARAMETER APPLICATIONS

Alignment Parameters allow you to drag a block object over an object and have the block automatically rotate and orientate itself relative to the geometry your cursor is over. This makes the Alignment Parameter one of the easiest modes of adding to a dynamic block. Two types of Alignment Parameters are available: Perpendicular and Tangent. When creating an Alignment Parameter, you specify an alignment point and alignment angle.

 Try It! – Open the drawing file 16_Dynamic_Alignment. A block of a sink is already defined in this drawing. Double-click on the block to activate the Edit Block Definition dialog box, and click the OK button to enter the Block Authoring environment, as shown in the following image.

Before continuing, make the following preparations: turn on Polar, turn on and set Osnap mode to Quadrant, and turn on Otrack. Then click the Alignment Parameter tool in the Block Authoring Palette.

Figure 16–72

For the base point of the alignment, move your cursor over the top of the ellipse, track straight up, and enter a value of 2, as shown on the left in the following image. For the alignment direction, move your cursor to the right of the last known point and pick a blank part of your screen, as shown in the middle of the following image. The Alignment Parameter will appear, as shown on the right in the following image. The Alignment Parameter being used for this application is Perpendicular.

Figure 16–73

Saving this sink block and exiting the Block Editor will return you to the drawing editor. Clicking on the sink block, as shown on the left in the following image, will expose the Perpendicular Alignment Parameter. Click on this Alignment Parameter, move your cursor to any of the inner walls of the bathroom plan, and notice the sink automatically rotating at an angle perpendicular to the wall. When you are satisfied with the desired orientation, click to place the block, as shown on the right in the following image.

Figure 16–74

MULTIPLE ALIGNMENT POINT APPLICATIONS

You can assign multiple alignment points to a single block. Then, when you insert the block, you can cycle through the grip points by pressing the CTRL key while dragging the block into position until the desired insertion point is found. The next Try It! exercise will illustrate this capability of dynamic blocks.

 Try It! – Open the drawing file 16_Dynamic_Insertion_Cycling. A block of a window is already defined in this drawing. Double-click on the block to activate the Edit Block Definition dialog box and click the OK button to enter the Block Authoring environment, as shown in the following image. Activate the Alignment Parameter and create the three parameters shown in the following image. Use OSNAP endpoint for the alignment location. For the alignment angle, pick to the right or left of the endpoint, as shown in the following image. The three icons identify the Perpendicular alignment tool being used.

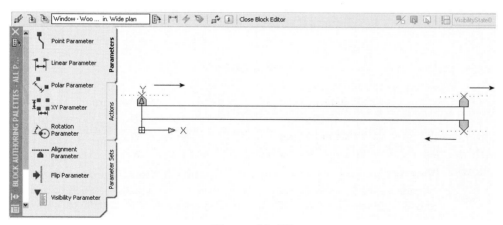

Figure 16–75

Save this window block and exit the Block Editor to return to the drawing editor. Clicking on the existing block will activate the three alignment points just created. The grip located in the lower left corner of the block is the original insertion point. Now experiment by inserting the window and pressing the CTRL key to cycle through all insertion points. You should be able to place all windows in the following image without the need to move, copy, or rotate the block into position. Since the Perpendicular Alignment Parameter was used, the window will align itself perpendicular (at a 90° angle) with each wall.

Figure 16–76

SCALE AND ANGLE OFFSET APPLICATIONS

This application of dynamic blocks allows you to increase or decrease the angle direction of a part of the block relative to a parameter.

 Try It! – Open the drawing file 16_Dynamic_Offsets. Two door blocks are already defined in this drawing. Double-click on the left door block to activate the Edit Block Definition dialog box and click the OK button to enter the Block Authoring environment. Activate the Linear Parameter. For the Start Point, click the bottom endpoint of the line, and for the End Point, click the bottom endpoint of the arc. Specify the label location below the door block; the label should read Distance.

Click on the Distance Parameter you just created and activate the Properties Palette. In the palette, change the Property Label from Distance to Door Width. Under the Value Set area, change the Distance Type to Increment, change the Distance Increment to 2″, change the Distance Minimum to 1′, and change the Distance Maximum to 3′. Refer to the illustration on the left in the following image for these changes.

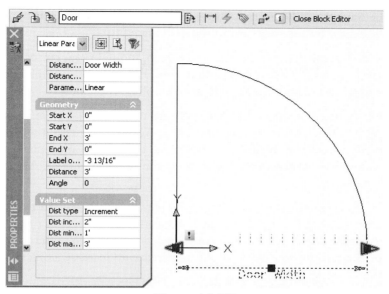

Figure 16–77

Next, activate the Scale Action, select Door Width as the parameter, and select the line and arc as the objects to scale. Press ENTER to continue. Specify the action location, as shown on the right in the following image.

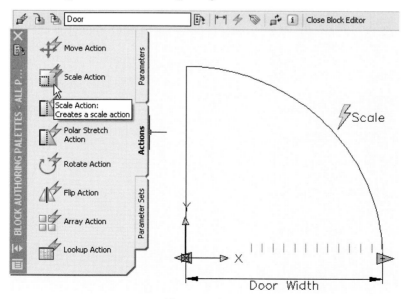

Figure 16–78

Click on the Save Block Definition button to save these changes, and then click the Close Block Editor button to return to the drawing editor.

Click on the door symbol, select the right arrow grip, and slide your cursor to one of the increment hash marks. Notice how the door symbol reduces in size in 2˝ increments. Notice also how the arc and line scale to accommodate these new distances.

Double-click on the right door block to activate the Edit Block Definition dialog box, and click the OK button to enter the Block Authoring environment. This door symbol actually shows the thickness of the door. The same Linear Parameter used for the previous door symbol has already been applied to this door. Also, the Scale Action already exists.

However, in this example, the Scale Action has been applied only to the arc. As the door increases or decreases in size, the arc will be affected by these actions. However, you do not want the door thickness to increase or decrease. You want the door thickness to remain unchanged but the door height to increase or decrease along with the arc.

To have the door height be affected by changes in the Door Width parameter, first click on the Stretch Action. Select the Door Width parameter. In order for the parameter point to be associated with the action, type S for second point. This should select the point at the bottom of the arc. Next, erect a crossing box, as shown on the left in the following image. Select the two vertical edges of the door along with the short horizontal top of the door.

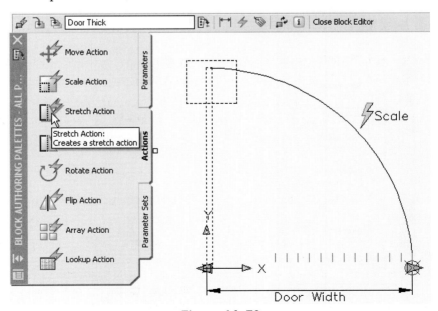

Figure 16–79

Press ENTER. The following prompt will display:

Specify action location or [Multiplier/Offset]:

Type O for Offset, and at the next prompt, type 90 for the offset angle. This will stretch the top of the door straight up, or in the 90° direction, when the door symbol increases or decreases in size. Finally, locate the Stretch Action icon near the top of the door, as shown on the right in the following image.

Figure 16–80

Click on the Save Block Definition button to save these changes, and then click the Close Block Editor button to return to the drawing editor.

Click on the door symbol, select the right arrow grip, and slide your cursor to one of the increment hash marks. Notice how the door symbol reduces in size in 2˝ increments. Notice also how the arc scales and door thickness stretches to accommodate these new distances.

CENTERING OBJECTS

This feature of dynamic blocks allows for certain parts of geometry to remain centered in the block even as the block increases or decreases in size.

 Try It! – Open the drawing file 16_Dynamic_Centered. A rectangular block with a circle is already defined in this drawing. The idea is to keep the circle centered as the rectangle is stretched to longer or shorter distances. Double-click on the rectangle to activate the Edit Block Definition dialog box and click the OK button to enter the Block Authoring environment.

Notice in the following image that a Linear Parameter (Distance) and Stretch Action (Stretch Right) are already defined in your drawing. In order to have the circle change in location with the rectangle, a Move Action needs to be applied to the circle. To accomplish this, first click the Move Action from the palette list, as shown on the left in the following image. Select the Distance Parameter. In order for the parameter point to associate with the action, type S for second point and then pick the edge of the circle as the object to move. Locate the Move icon somewhere near the edge of the circle, as shown in the following image. In this image, the Move Action was renamed to Move Circle through the Properties Palette, which was explained in a previous Try It! exercise.

Figure 16–81

Unfortunately, attaching the Move Action to the circle is not enough to center the circle in the rectangle. A special distance property must be changed to keep the circle centered. To accomplish this, first click on the Move Circle icon and click on the Properties button to launch the Properties Palette. Notice near the bottom of the palette a condition called Overrides. By default, the Distance Multiplier value is set to 1. Because we want this value to be exactly half of whatever the rectangle distance is, change this value to .50, as shown on the left in the following image.

Click on the Save Block Definition button to save these changes, and then click the Close Block Editor button to return to the drawing editor.

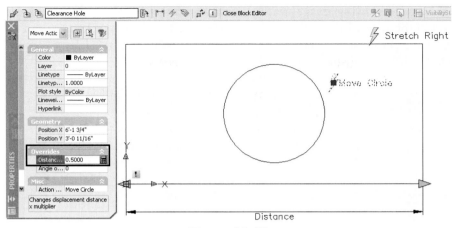

Figure 16–82

When you are returned to the drawing editor, experiment with the rectangle to see whether the circle will center itself to any changes in the rectangle. These changes are affected only when the rectangle is stretched to the right; this is how the parameter was set up originally. Click on the edge of the rectangle, pick the arrow grip on the right side of the rectangle, and stretch the rectangle to the right. Notice that the circle should remain centered in the rectangle. The following image illustrates a few examples of this action.

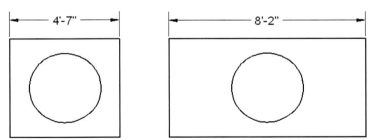

Figure 16–83

STRETCHING A BLOCK EQUALLY IN OPPOSITE DIRECTIONS

This exercise will illustrate how to stretch a dynamic block equally in opposite directions. An example of an I beam will be used to illustrate this concept.

 Try It! – Open the drawing file 16_Dynamic_I-Beam. A block representing an I beam is already defined in this drawing. Double-click on the I beam to activate the Edit Block Definition dialog box and click the OK button to enter the Block Authoring environment.

Notice in the following image that a Linear Parameter (I-Beam Width) is already defined in the Block Editor. Click on this parameter, activate the Properties Palette, and Change the Base location to Midpoint. This will allow the I beam to remain centered as it is being stretched.

Figure 16–84

Next, activate the Stretch Action and select the I-Beam Width parameter. Accept the base point for the stretch operation located in the upper right corner of the I beam. Create a crossing box and select the six items, as shown in the following image. Locate the Stretch icon in a convenient location.

Figure 16–85

Notice in the following image that the first Stretch Action was renamed to a more meaningful term (Stretch I-Beam Right). A second Stretch Action needs to be created similar to the first one in the previous image. This time, after accepting the base point for the stretch again in the upper right corner, create a crossing box and select the six objects, as shown in the following image. Locate this icon in a convenient location of the screen. This icon could also be renamed to Stretch I-Beam Left.

Figure 16–86

One more setting needs to be changed before you can test the I-beam. Click on the Stretch I-Beam Left icon and activate the Properties Palette. Scroll down to the bottom of the palette and change the Angle Override to 180°. This will ensure that as the right side of the I beam stretches to the right or left, the left side will stretch in the opposite direction using the same distance.

Click on the Save Block Definition button to save these changes, and then click the Close Block Editor button to return to the drawing editor.

Figure 16–87

When you are returned to the drawing editor, experiment with the I beam to see whether it stretches equally in opposite directions. In the following image, two blocks have been stretched to different widths. Notice how the middle rib remains centered in both blocks due to this type of dynamic block technique.

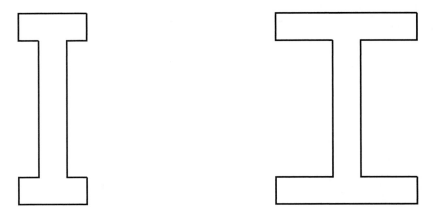

Figure 16–88

ARRAY APPLICATIONS

Linking an Array Action to a dynamic block can have some interesting effects. As the dynamic block increases in width, items associated with the block can be added depending on the distance of the width. This application will utilize a table block consisting of four chairs positioned around a 4´ table. As the table increases in width, extra chairs need to be added to the block.

 Try It! – Open the drawing file 16_Dynamic_Array. A block consisting of four chairs arranged around a 4´ table is displayed in the drawing editor. This exercise will illustrate how to stretch the table in 4´ increments and have extra columns of chairs created. Double-click on the table block to activate the Edit Block Definition dialog box and click the OK button to enter the Block Authoring environment.

Click on the Table Width parameter and launch the Properties Palette. Make changes under the Value Set area, as shown in the following image. Set the Distance type to Increment, the Distance increment to 4´, the Distance minimum to 4´, and the Distance maximum to 16´.

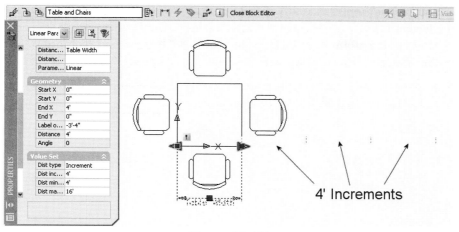

Figure 16–89

Next, assign a Stretch Action to the right end of the table and rightmost positioned chair, as shown in the following image. After activating the Stretch Action, pick the Table Width parameter. In order for the parameter point to associate with the action, type S for second point. Create a crossing box to surround the rightmost chair and right end of the table, and select the table and rightmost chair as the objects to stretch. Locate the Stretch Action icon in a convenient location of your screen.

Figure 16–90

If you experiment by stretching this new block in the drawing editor, you will find that the table and rightmost chair stretch; however, the two chairs located in the 90° and 270° positions remain in their original positions. We want these two chairs to be copied or arrayed at a specified distance depending on how long the table stretches. To perform this operation, click on the Array Action button back in the Block Authoring editor. Select the Table Width parameter, select both chairs, as shown in the following image, and enter a Column Width distance of 3´ 11˝ as the distance from one chair to another. Locate the Array Action icon in a convenient screen location.

Click on the Save Block Definition button to save these changes, and then click the Close Block Editor button to return to the drawing editor.

Figure 16–91

When your return to the drawing editor, click on the block of the table and chairs, pick the right arrow grip, as shown on the left in the following image, and begin stretching the table. When you stretch the table to a new length of 8´, two new columns of chairs are added. Stretch to a new length of 16´, and your screen should appear similar to the illustration on the right in the following image.

Figure 16–92

WORKING WITH VISIBILITY STATES

Using a visibility state on a block allows you to define various versions of the block under one unique name. For example, rather than create four separate blocks to represent the fittings of a gate valve, as shown in the following image, one master block consisting of all geometry native to all versions will be created. You create visibility states and turn off the geometry you do not want to see. This is an extremely productive way to keep the number of block names down in a drawing file without sacrificing the various types of blocks normally created individually.

Figure 16–93

 Try It! – Open the drawing file 16_Dynamic_Visibility. A block consisting of a gate valve is displayed in the drawing editor. This exercise will illustrate how to create various visibility states using the gate valve block provided. Double-click on the gate valve block to activate the Edit Block Definition dialog box and click the OK button to enter the Block Authoring environment.

First, create a Visibility Parameter and locate the icon, as shown in the following image.

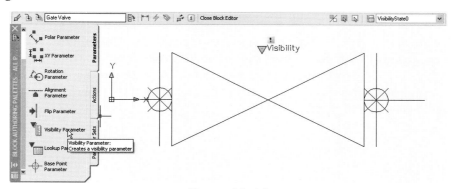

Figure 16–94

Next, click on the Manage Visibility States button to launch the Visibility States dialog box, as shown in the following image. Immediately rename the default state (VisibilityState0) to Main Gate Valve. This is the main block that contains all geometry associated with all the similar but different gate valve blocks.

Figure 16–95

The following table gives a brief description of the visibility tools present in the Block Editor:

Button	Description
	Toggles between all visibility modes, visible or invisible
	Used for making objects in a visibility state visible
	Used for making objects in a visibility state invisible
	Launches the Visibility States dialog box used for managing visibility states

Next, click on the New button and begin creating additional visibility states. When the New Visibility State dialog box appears, enter Soldered, as shown in the following image. Keep all remaining default options for new states unchanged. Click the OK button when finished. Use the same method to create three other visibility states: Flanged, Screwed, and Welded.

Figure 16–96

Next, make the Flanged visibility state current by clicking on its name, located in the upper right corner of the Block Editor dialog box. Select all the geometry that you do not want to be displayed when in the Flanged state. In the following image, both circles and four diagonal lines are highlighted. Click on the Make Invisible button to turn this geometry off. The result will be the gate valve symbol displaying only the geometry necessary to define a Flanged gate valve.

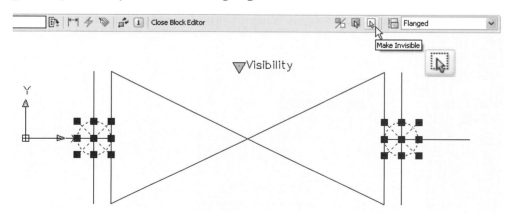

Figure 16–97

Continue by making the Screwed Visibility state current, noticing that all the geometry reappears in the block. For a screwed gate valve, select both circles, four diagonal lines, and two vertical line segments, as shown in the following image. Click on the Make Invisible button, and the remaining geometry defines a screwed gate valve. Repeat this procedure for the Welded and Soldered visibility states. The Welded state should consist of the gate valve and the four diagonal lines. The Soldered state should consist of the gate valve and both circles.

Click on the Save Block Definition button to save these changes, and then click the Close Block Editor button to return to the drawing editor.

Figure 16–98

When you return to the drawing editor, create various copies of the gate valve. Then click on an individual block and notice the visibility state grip. Click on the grip and pick the desired gate valve fitting, as shown in the following image. This method provides a quick and efficient means of organizing blocks that are similar but different in appearance.

Figure 16–99

Note: Another technique of creating blocks through visibility states is to open an existing block in the Block Editor and rename the default visibility state. You then create a new visibility state, click on the Hide All Objects button, and then close the Visibility States dialog box. Since the Block Editor will remain empty, insert a block and then explode it at the origin of the Block Editor. Display the Visibility States dialog box and repeat this procedure for as many blocks as desired.

USING LOOKUP PARAMETERS AND TABLES

A Lookup Parameter allows you to create a lookup property whose value is controlled by a table of custom properties located in the block. While inside this table, you add a series of different input properties that are tied directly to a single parameter or multiple parameters assigned to the block. You also create lookup properties that are tied to the input properties. These lookup properties will appear as a list inside the block once you return to the drawing editor. You then pick different lookup properties from the list in order to make changes to the block. This concept of creating lookup properties will be demonstrated in the next Try It! exercise.

Try It! – Open the drawing file 16_Dynamic_Lookup. A rectangular block with diagonal lines is already defined in this drawing. The name of this block is Basic Stud and represents a 2 × 4 construction object. This exercise will illustrate how to create a lookup table. After this is done, you will be able to choose various sizes of this basic stud ranging in size from the standard 2 × 4 to a 2 × 12. Begin by double-clicking on the 2 × 4 Basic Stud block to activate the Edit Block Definition dialog box, and click the OK button to enter the Block Authoring environment.

Notice in the following image that a Linear Parameter (Stud Width) and Stretch Action (Stretch Right) are already defined in your drawing. The grips at both ends of the parameter name, normally visible, have been turned off through the Properties Palette. You will be controlling the sizes of the stud through the lookup list.

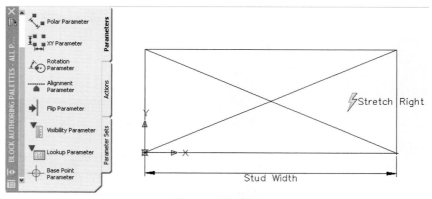

Figure 16–100

Begin by clicking on Lookup Parameter located under the Parameters tab, as shown in the following image. Locate this parameter in a convenient location of the screen and use the Properties Palette to rename the parameter to Basic Stud Sizes.

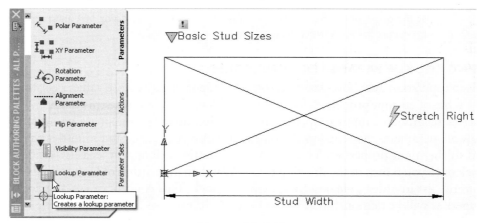

Figure 16–101

Create a Lookup Action item by selecting this item from the Block Authoring Palette, as shown on the left in the following image. Associate this action with a parameter by picking the Basic Stud Sizes Lookup Parameter that you just created in the previous step. Locate the Lookup Action next to the Basic Stud Sizes Lookup Parameter. After you select a screen position for the action location, the Property Lookup Table dialog box will appear, as shown on the right in the following image. Click the Add Properties button.

Figure 16–102

When the Add Parameter Properties dialog box appears, as shown on the left in the following image, notice the linear parameter listed. For this example, this is the only parameter that will be used. In more complicated Lookup Actions, you may have numerous parameter names to choose from. For now, click the OK button to accept this parameter name.

This will return you to the Property Lookup Table dialog box. Notice the column for Stud Width. Start entering the values for various stud widths, such as 3.5, 5.5, 7.5, 9.5, and 11.5, as shown on the left side of this dialog box.

You also need to fill in the right column of the dialog box under the heading Lookup Properties Basic Stud Sizes. These are the listings that will be visible when you return to the drawing editor and test the Basic Stud block for its various sizes. Under this column, enter the following sizes; 2 × 4, 2 × 6, 2 × 8, 2 × 10, and 2 × 12. Be certain these correspond with the Stud Width values on the left side of the dialog box. When finished, your dialog box should appear similar to the illustration on the right in the following image.

Figure 16–103

One other very important operation needs to be performed inside this dialog box. On the right side of the dialog box, notice Custom. Click on this item if it appears grayed out. This will activate the Read only mode listed just below it. Click on Read only, pick the down arrow, and choose Allow reverse lookup, as shown in the following image. This will activate the lookup table when you click on the block once back in the drawing editor. If this is not selected, the table will not display in the block when you return to the drawing editor.

Input Properties		Lookup Properties	
	Stud Width		Basic Stud Sizes
	3.5000		2 x 4
	5.5000		2 x 6
	7.5000		2 x 8
	9.5000		2 x 10
	11.5000		2 x 12
	<Unmatched>		Custom
▶			Allow reverse lookup
			Allow reverse lookup
			Read only

Figure 16–104

When you return to the Block Editor screen, change the name of the Lookup Action item to Basic Stud Table. Save the block and close the Block Editor to return to the drawing editor. Click on the Block and notice the appearance of the Lookup Parameter. Click the down arrow in the Lookup grip and test the various Basic Stud Sizes by picking them from the list, as shown in the following image.

Figure 16–105

TUTORIAL EXERCISE: ELECTRICAL SCHEMATIC

Figure 16–106

Purpose
This tutorial is designed to lay out electrical blocks such as resistors, transistors, diodes, and switches to form an electrical schematic, as shown in the previous image.

System Settings
Create a new drawing called ELECT_SYMB to hold the electrical blocks. Keep all default units and limits settings. Set the GRID and SNAP commands to 0.0750 units. This will be used to assist in the layout of the blocks. Once this drawing is finished, create another new drawing called ELECTRICAL_SCHEM1. This drawing will show the layout of an electrical schematic. Keep the default units but use the LIMITS command to set the upper right corner of the display screen to (17.0000,11.0000). Grid and snap do not have to be set for this drawing.

Layers
Create the following layers for ELECTRICAL_SCHEM1 with the format:

Name	Color	Linetype
Border	White	Continuous
Blocks	Red	Continuous
Wires	Blue	Continuous

Suggested Commands
This tutorial begins by creating a new drawing called ELECT_SYMB, which will hold all electrical blocks. You will use the image below as a guide in drawing all electrical blocks. A grid with spacing 0.0750 would provide further assistance with the drawing of the blocks. Once all blocks are drawn, the BLOCK command is used to create blocks out of the individual blocks. You will save this

drawing and create a new drawing file called ELECTRICAL_SCHEM1. The DesignCenter will be used to drag and drop the internal blocks from the drawing ELECT_SYMB into the new drawing, ELECTRICAL_SCHEM1.

Finally, you will connect all blocks with lines that represent wires and electrical connections, add block identifiers, and save the drawing.

STEP 1

Begin a new drawing and call it ELECT_SYMB. Be sure to save this drawing to a convenient location. It will be used along with the DesignCenter to bring the internal blocks into another drawing file.

STEP 2

Using a grid/snap of 0.075 units, construct each block using the following image as a guide. The grid/snap will help keep all blocks proportional to each other. Create all the blocks on the neutral layer 0. Also, the X located on each block signifies its insertion point (do not draw the X).

Figure 16–107

STEP 3

For a more detailed approach to creating the blocks, the resistor block will be created. Using the illustration on the left in the following image as a guide, first construct the resistor with the LINE command while using the grid/snap to connect points until the resistor is created. Be sure Snap mode is active for this procedure. Then issue the BLOCK command, which will bring up the dialog box illustrated on the right in the following image. This will be an internal block to the drawing ELECT_SYMB; name the block RESISTOR. In the Select objects area of the dialog box, select all lines that represent the resistor (if you labeled the symbols, do not select the text as part of the block). Select the Retain radio button so that the symbol will remain visible in your drawing. Returning to the dialog box will show the objects selected in a small image icon on the right of the dialog box. Next, identify the insertion point of the resistor at the X. It is very important to use Snap or an Object Snap mode when picking the insertion point. Finally, add a description for the block. When finished with this dialog box, click on the OK button to create the internal block. Repeat this procedure for the remainder of the electrical symbols. When complete, save the drawing as ELECT_SYMB.

Figure 16–108

STEP 4

Begin another new drawing called ELECTRICAL_SCHEM1. Activate the DesignCenter. Click on the Tree View icon and expand the DesignCenter to display the folders on your hard drive. Locate the correct folder in which the drawing ELECT_SYMB was saved and click on it. Double-click on ELECT_SYMB and select Blocks to display the internal blocks, as illustrated in the following image. These internal blocks can now be inserted in any AutoCAD drawing file.

Figure 16–109

STEP 5

Make the Blocks layer current and begin dragging and dropping the blocks into the drawing, as shown in the following image. Double-clicking an image of a block located in the DesignCenter will launch the Insert dialog box should you need to change the scale or rotation angle of the block. In the case of the electrical schematic, it is not critical to place the blocks exactly, because they are moved to better locations when connected to lines that represent wires.

Figure 16–110

STEP 6

Make the Wires layer current and begin connecting all the blocks to form the schematic, as shown in the following image. Some type of Object Snap mode such as Endpoint or Insert must be used for the wire lines to connect exactly with the endpoints of the blocks. If spaces get tight or if the blocks look too crowded, move the block to a better location and continue drawing lines to form the schematic. Hint: The STRETCH command is a fast and easy way to reposition the blocks without having to reconnect the wires.

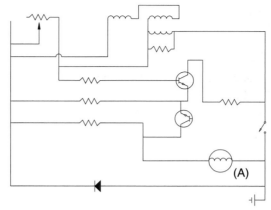

Figure 16–111

STEP 7

Add text (place it on the Blocks layer) identifying all resistors, transistors, diodes, and switches by number. The text height used was 0.12; however, this could vary depending on the overall size of your schematic. Also, certain areas of the schematic show connections with the presence of a dot at the intersection of the connection. Use the DONUT command with an inside diameter of 0.00 and an outside diameter of 0.075. Place donuts on the Wires layer in all locations, as shown in the following image.

Figure 16–112

TUTORIAL EXERCISE: 16_DYNAMIC_SWITCH_PLATE.DWG

Purpose
This tutorial is designed to link an Array action to a dynamic block. As the dynamic block increases in width, items associated with the block can be added, depending on the width.

System Settings
No special system settings need to be made for this exercise.

Layers
No special layers need to be created for this exercise.

STEP 1

Open the drawing file 16_Dynamic_Switch_Plate. A block consisting of a single-opening switch plate is displayed in the drawing editor. Dimensions have been added to the switch plate to calculate the incremental distances. This exercise will illustrate how to stretch the switch plate in 45 mm increments and have additional switch place openings created. Double click on the switch-plate block to activate the Edit Block Definition dialog box and click the OK button to enter the Block Authoring environment.

Suggested Commands
This application will utilize a switch plate consisting of a basic rectangular shape and three openings: two circular openings that represent screw holes and a rectangular opening used to accept the switch. As the switch plate increases in width, extra openings will automatically be added to the block.

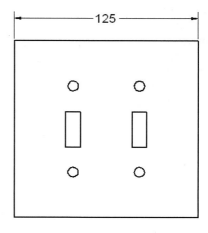

Figure 16–113

STEP 2

Add a Linear Parameter to the top of the switch plate. Change the name of the Distance parameter to Plate Width, as shown in the following image.

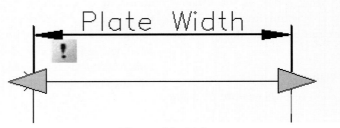

Figure 16–114

STEP 3

Next, click on the Plate Width parameter and launch the Properties Palette. Make changes under the Value Set area, as shown in the following image; Set the Distance type to Increment, the Distance increment to 45, the Distance minimum to 80, and the Distance maximum to 305, as shown in the following image. When the switch plate is at its minimum distance of 80 mm, a single opening will be displayed. When the switch plate is at its maximum distance of 305 mm, six openings will be displayed.

Figure 16–115

STEP 4

Next, assign a Stretch Action to the right side of the switch plate, as shown on the left in the following image. After activating the Stretch Action, pick the Plate Width parameter, select the base point located in the upper right corner of the table, create a crossing box to surround the right side of the switch plate, and select the entire right side of the switch plate as the objects to stretch. Locate the Stretch Action icon in a convenient location of your screen, as shown on the right in the following image. Use the Properties window to rename the Stretch action to Stretch Right.

Figure 16–116

STEP 5

Now assign a Stretch Action to the left side of the switch plate using the same steps performed in the previous step. After activating the Stretch Action, pick the Plate Width parameter, select the base point located in the upper left corner of the switch plate, create a crossing box to surround the left side of the switch plate, and select the entire left side of the switch plate as the objects to stretch. Locate the Stretch Action icon in a convenient location of your screen, as shown in the following image. Use the Properties window to rename the Stretch action to Stretch Left.

Figure 16–117

STEP 6

If you experiment by stretching this new block back in the drawing editor, you will find that the switch plate will stretch; however, the screw holes and switch opening will remain in their original positions. We want these openings to be copied or arrayed at a specified distance, depending on how much the switch plate stretches. To perform this operation, click on the Array Action button back in the Block Authoring editor. Select the Plate Width parameter, select all three openings, as shown in the following image, and enter a Column Width distance of 45 mm as the distance from one set of openings to another. Locate the Array Action icon in a convenient screen location.

Click on the Save Block Definition button to save these changes, and then click the Close Block Editor button to return to the drawing editor.

Figure 16–118

STEP 7

When your return to the drawing editor, click on the block of the switch plate, pick the right arrow grip, and begin stretching the switch plate. When you stretch the switch plate to a new length of 305 mm, which represents the maximum distance, a series of six switch openings are created, as shown on the right in the following image. This completes this tutorial exercise on creating a dynamic block of a switch plate.

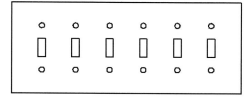

Figure 16–119

TUTORIAL EXERCISE: 16_DYNAMIC_ELECTRICAL_LIST.DWG

Purpose
This tutorial is designed to create visibility states to have numerous electrical symbols contained in a single block name.

System Settings
No special system settings need to be made for this tutorial.

Layers
No special layers need to be created for this tutorial.

Suggested Commands
The concept of creating visibility states in dynamic blocks begins with the creation of a block using the Edit Block Definition dialog box. You then create a visibility state while inside the Block Authoring environment. During this process, you insert an electrical symbol and immediately explode it into individual objects. You then enter the Visibility States dialog box and change the name of the current state to match that of the electrical symbol. You then create a new visibility state, name this state the next electrical symbol, and hide all existing objects in this new state. When you return to the Block Authoring environment, you insert the next electrical symbol, explode it, and follow the same process for arranging a number of electrical symbols in the same visibility state, as shown in the following image.

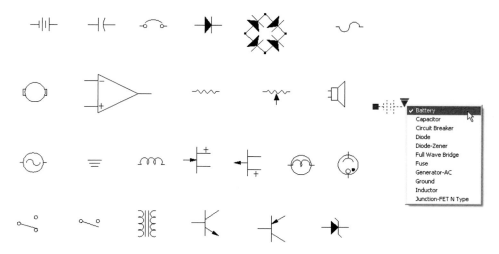

Figure 16–120

STEP 1

Open the drawing file 16_Dynamic_Electrical_List. Before creating visibility states, you must create a new general block name that will hold all individual electrical symbols. You must also enter the block editor to create the visibility states. First, click on the Block Editor button located in the Standard toolbar, as shown on the left in the following image. When the Edit Block Definition dialog box appears, as shown on the right in the following image, enter Electrical Symbols as the block name and click the OK button. This action will open the new block name in the Block Editor.

Figure 16–121

STEP 2

Insert the first block, called Battery, in the Block Editor, as shown in the following image. Use an insertion point of 0,0. In fact, all individual electrical blocks will be inserted at 0,0 for consistency. Once the block appears in the Block Editor, use the EXPLODE command to break up the block into individual objects. Then, locate the Visibility Parameter in the Parameters tab of the Block Authoring Palette, and place the marker to the right and below the battery symbol, as shown in the following image.

Figure 16–122

STEP 3

Notice in the upper right corner of the Block Authoring Palette the appearance of a toolbar. Use this toolbar for managing visibility states. First, click on the Manage Visibility States button, as shown in the following image.

Figure 16–123

STEP 4

This will launch the Visibility States dialog box. When you use this dialog box for the first time, a default state called VisibilityState0 is already created. Use the Rename button to change the name of this visibility state to Battery. This state will coincide with the object currently displayed in the Block Authoring screen. Next, a new visibility state needs to be created by clicking on the New button in the Visibility States dialog box.

Figure 16–124

STEP 5

This will launch the New Visibility State dialog box, as shown in the following image. In the Visibility state name edit box, enter the name Capacitor. Also, be sure to check the radio button next to Hide all existing objects in new state. This will hide all objects when you return to the Block Editor. Click the OK button to continue.

Figure 16–125

STEP 6

When you return to the Block Editor, insert the block Capacitor at an insertion point of 0,0 and explode the block back into its individual objects. The objects that remain in the Block Editor form the Capacitor visibility state.

Launch the Manage Visibility States dialog box again. Create another new visibility state called Circuit Breaker, and make sure Hide all existing objects in new state is selected. Insert the Circuit Breaker block at an insertion point of 0,0 and explode the block back into its individual objects.

Repeat the procedure for creating more visibility states based on the other electrical block symbols. Close the Block Editor and save the changes to the block.

Insert a Battery block in the drawing and test the visibility states by selecting them from the list, as shown in the following image. If you created visibility states of all electrical symbols, your list will be longer.

Figure 16-126

PROBLEMS FOR CHAPTER 16

PROBLEM 16–1

Directions for Problem 16–1

Construct this battery pack using the block library supplied with AutoCAD: Basic Electronics.dwg. This drawing is not to scale.

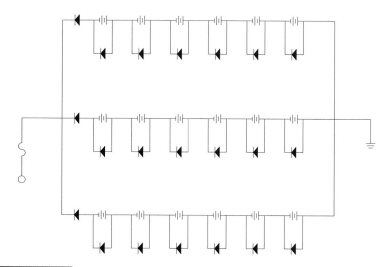

PROBLEM 16–2

Directions for Problem 16–2

Construct this electrical schematic using the block library supplied with AutoCAD: Basic Electronics.dwg. This drawing is not to scale.

PROBLEM 16-3

Directions for Problem 16-3

Construct this electrical schematic using the block library supplied with AutoCAD: Basic Electronics.dwg. This drawing is not to scale.

PROBLEM 16-4

Directions for Problem 16-4

Construct this logic gate schematic using the following block libraries supplied with AutoCAD: CMOS Integrated Circuits.dwg and Basic Electronics.dwg. This drawing is not to scale.

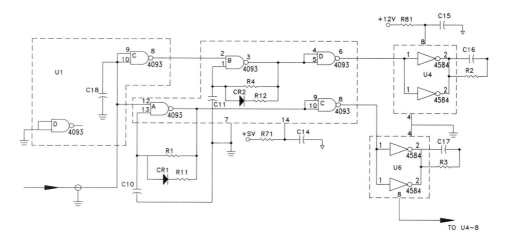

Intermediate Level Drawings

PROBLEM 16–5

Directions for Problem 16–5

Construct this instrumentation pipe diagram using the following block library supplied with AutoCAD: Pipe Fittings.dwg. This drawing is not to scale.

PROBLEM 16-6

Directions for Problem 16-6

Construct the floor plan of the house using the dimensions provided. Create the appropriate layers to separate the walls from the dimensions, and so on. Once the floor plan is completed, create an interior plan consisting of furniture and appliances. Place the blocks in their appropriate spaces using the designated room titles. Use the following block libraries supplied with AutoCAD: House Designer.dwg, Home – Space Planner.dwg, and Kitchens.dwg.

Use the following suggestions for the creation of doors and windows:

All windows measure 2'-8" wide.
All bedroom doors measure 2'-6" wide.
The bathroom door measures 2'-0" wide.
The main entrance door measures 3'-0" wide.
The kitchen and laundry doors each measure 2'-8" wide.
All closet openings measure 4'-0" wide.
All interior walls measure 4".
The exterior walls with brick veneer measures 5".

FLOOR PLAN
1/4" = 1'-0"

CHAPTER 17

Working with Attributes

This chapter begins the study of how to create, display, edit, and extract attributes. Attributes consist of intelligent text data that is attached to a block. The data could consist of part description, part number, catalog number, and price. Whenever the block is inserted in a drawing, you are prompted for information that, once entered, becomes attribute data. Attributes could also be associated with a title block for entering such items as drawing name; who created, checked, revised, and approved the drawing; drawing date; and drawing scale. Once included in the drawing, the attributes can be extracted and shared with other programs.

ABOUT ATTRIBUTES

An attribute may be considered a label that is attached to a block. This label is called a tag and can contain any type of information that you desire. Examples of attribute tags are illustrated on the left in the following image. The tags are RESISTANCE, PART_NAME, WATTAGE, and TOLERANCE and relate to the particular symbol they are attached to, in this case an electrical symbol of a resistor. Attribute tags are placed in the drawing with the symbol. When the block, which includes the tags and symbol, is inserted, AutoCAD requests the values for the attributes. The same resistor with attribute values is illustrated on the right in the following image. When you create the attribute tags, you determine what information is requested, the actual prompts, and the default values for the information requested. Once the values are provided and inserted in the drawing, the information contained in the attributes can be displayed, extracted, and even shared with spreadsheet or database programs.

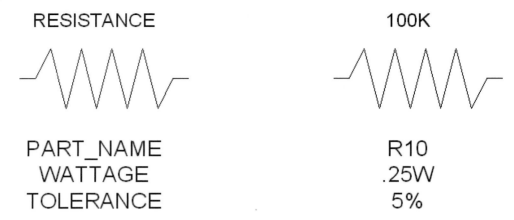

Figure 17–1

Once attributes are inserted in the drawing, the following additional tasks can be performed on attributes:

1. Attributes can be turned on, turned off, or displayed normally through the use of the ATTDISP command.
2. The individual attributes can be changed through the Attribute Edit dialog box, which is activated through the ATTEDIT command.
3. Attributes can be globally edited using the -ATTEDIT command.
4. Once edited, attributes can be extracted into various text file modes through the Extract Attributes dialog box, which is activated by the ATTEXT command.
5. The text file created by the attribute extraction process can be exported to a spreadsheet or database program.

The following commands, which will be used throughout this chapter, assist in the creation and manipulation of attributes:

ATTDEF—Activates the Attribute Definition dialog box used for the creation of attributes

ATTDISP—Used to control the visibility of attributes in a drawing

ATTEDIT—Activates a dialog box used to edit attribute values

-ATTEDIT—Used to edit attributes singly or globally; the hyphen (-) in front of the command activates prompts from the command line

ATTEXT—Activates the Attribute Extraction dialog box, used for extracting attributes in CDF, SDF, or text file formats

The following additional commands are available in AutoCAD to better assist with managing and manipulating attributes:

> BATTMAN—Activates the Block Attribute Manager dialog box, used to edit attribute properties of a block definition
>
> EATTEDIT—Activates the Enhanced Attribute Editor dialog box, used to edit the attributes of a block
>
> EATTEXT—Activates the Attribute Extraction wizard, used to extract attributes

CREATING ATTRIBUTES THROUGH THE ATTRIBUTE DEFINITION DIALOG BOX

Selecting Define Attributes from the Block cascading menu on the Draw pull-down menu, shown on the left in the following image, activates the Attribute Definition dialog box, illustrated on the right in the following image. The following components of this dialog box will be explained in this section: Attribute Tag, Attribute Prompt, Attribute Value, and Attribute Mode.

Figure 17–2

ATTRIBUTE TAG

A tag is the name given to an attribute. Typical attribute tags could include PART_NAME, CATALOG_NUMBER, PRICE, DRAWING_NAME, and SCALE. The underscore is used to separate words because spaces are not allowed in tag names.

ATTRIBUTE PROMPT

The attribute prompt is the text that appears on the text line when the block containing the attribute is inserted in the drawing. If you want the prompt to be the same as the tag name, enter a null response by leaving the Prompt edit box blank. If the Constant mode is specified for the attribute, the prompt area is not available.

ATTRIBUTE VALUE

The attribute value is the default value displayed when the attribute is inserted in the drawing. This value can be accepted or a new value entered as desired. The attribute value is handled differently if the attribute mode selected is Constant or Preset.

ATTRIBUTE MODES

Invisible—This mode is used to determine whether the label is invisible when the block containing the attribute is inserted in the drawing. If you later want to make the attribute visible, you can use the ATTDISP command.

Constant—Use this mode to give every attribute the same value. This might be very useful when the attribute value is not subject to change. However, if you designate an attribute to contain a constant value, you cannot change it later in the design process.

Verify—Use this mode to verify that every value is correct. This is accomplished by prompting the user twice for the attribute value.

Preset—This allows for the presetting of values that can be changed. However, you are not prompted to enter the attribute value when inserting a block. The attribute values are automatically set to their preset values. The preset option is not available if an attribute is entered in the Value edit box of the Attribute Definition dialog box.

Note: The effects caused by invoking the Verify and Preset modes will be apparent only when the ATTDIA system variable is off (the Enter Attributes dialog box is not displayed). When entering data at the command prompt, you will be asked twice for data that is to be verified and you will not be asked at all to supply data for attributes that are preset.

TUTORIAL EXERCISE: B TITLE BLOCK.DWG

Figure 17–3

Purpose
This tutorial is designed to assign attributes to a title block. The completed drawing is illustrated in the previous image.

System Settings
Because the drawings B Title Block and 17_Pulley Section are provided on the CD, all units and drawing limits have already been set.

Layers
All layers have already been created for this drawing.

Suggested Commands
Open the drawing B Title Block. Using the Attribute Definition dialog box (activated through the ATTDEF command), assign attributes consisting of the following tag names: DRAWING_NAME, DRAWN_BY, CHECKED_BY, APPROVED_BY, DATE, and SCALE. Save this drawing file with its default name. Open the drawing 17_Pulley Section and insert the title block in this drawing and answer the questions designed to complete the title block information.

STEP 1

Open the drawing B Title Block and observe the title block area, shown in the following image. Various point objects are present to guide you in the placement of the attribute information. All points are located on the layer Points.

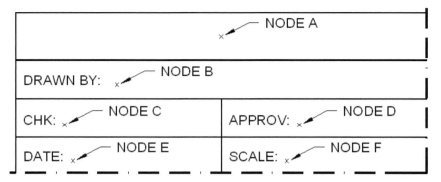

Figure 17–4

STEP 2

Activate the Attribute Definition dialog box, as shown on the left in the following image. Leave all items unchecked in the Mode area of this dialog box. In the Attribute area, make the following changes: Enter DRAWING_NAME in the Tag edit box. In the Prompt edit box, enter What is the name of the drawing? In the Value edit box, enter UNNAMED. In the Text Options area, change the Justification to Middle and the Height to 0.25 units. When finished, click the OK button. This will return you to your drawing. Using OSNAP-Node, pick the point at "A," as shown on the upper right in the following image. The DRAWING_NAME tag is added to the title block area, as shown on the lower right in the following image.

Figure 17–5

STEP 3

To define the next attribute, activate the Attribute Definition dialog box, as shown on the left in the following image. Leave all items unchecked in the Mode area of this dialog box. In the Attribute area, make the following changes: Enter DRAWN_BY in the Tag edit box. In the Prompt edit box, enter Who created this drawing? In the Value edit box, enter UNNAMED. In the Text Options area, verify that Justification is set to Left, and change the Height to 0.12 units. When finished, pick the OK button. This will return you to your drawing. Using OSNAP-Node, pick the point at "B," as shown on the upper right in the following image. The DRAWN_BY tag is added to the title block area, as shown on the lower right in the following image.

Figure 17–6

STEP 4

Activate the Attribute Definition dialog box, as shown on the left in the following image. Leave all items unchecked in the Mode area of this dialog box. In the Attribute area, make the following changes: Enter CHECKED_BY in the Tag edit box. In the Prompt edit box, enter Who will be checking this drawing? In the Value edit box, enter CHIEF DESIGNER. In the Text Options area, verify that Justification is set to Left and the Height is 0.12 units. When finished, click the OK button. This will return you to your drawing. Using OSNAP-Node, pick the point at "C," as shown on the upper right in the following image. The CHECKED_BY tag is added to the title block area, as shown on the lower right in the following image.

Figure 17–7

STEP 5

Activate the Attribute Definition dialog box, as shown on the left in the following image. Leave all items unchecked in the Mode area of this dialog box. In the Attribute area, make the following changes: Enter APPROVED_BY in the Tag edit box. In the Prompt edit box, enter Who will be approving this drawing? In the Value edit box, enter CHIEF ENGINEER. In the Text Options area, verify that Justification is set to Left and the Height is 0.12 units. When finished, click the OK button. This will return you to your drawing. Using OSNAP-Node, pick the point at "D," as shown on the upper right in the following image. The APPROVED_BY tag is added to the title block area, as shown on the lower right in the following image.

Figure 17–8

STEP 6

Activate the Attribute Definition dialog box, as shown on the left in the following image. Leave all items unchecked in the Mode area of this dialog box. In the Attribute area, make the following changes: Enter DATE in the Tag edit box. In the Prompt edit box, enter When was this drawing completed? In the Value edit box, enter UNDATED. In the Text Options area, verify that Justification is set to Left and the Height is 0.12 units. When finished, click the OK button. This will return you to your drawing. Using OSNAP-Node, pick the point at "E," as shown on the upper right in the following image. The DATE tag is added to the title block area, as shown on the lower right in the following image.

Figure 17–9

STEP 7

Activate the Attribute Definition dialog box, as shown on the left in the following image. Leave all items unchecked in the Mode area of this dialog box. In the Attribute area, make the following changes: Enter SCALE in the Tag edit box. In the Prompt edit box, enter What is the scale of this drawing? In the Value edit box, enter 1 = 1. In the Text Options area, verify that Justification is set to Left and the Height is 0.12 units. When finished, click the OK button. This will return you to your drawing. Using OSNAP-Node, pick the point at "F," as shown on the upper right in the following image. The SCALE tag is added to the title block area, as shown on the lower right in the following image.

You have now completed creating the attributes for the title block; turn off the Points layer. Close and save the changes to this drawing with its default name.

Figure 17–10

STEP 8

To test the attributes and see how they function in a drawing, first open the drawing file 17_Pulley Section. Notice that this image is viewed from inside a Layout (or Paper Space). The current page setup is based on the DWF6-ePlot.pc3 file, with the current sheet size as ANSI Expand B (17.00 × 11.00 inches). The viewport that holds the two views of the Pulley has the No Plot state assigned in the Layer Properties Manager dialog box. You could turn the Viewports layer off to hide the viewport if you prefer. You will now insert the title block with attributes in this drawing.

Verify that the system variable ATTDIA is set to 1 (This can be typed in from the keyboard). This will allow you to enter your attribute values through the use of a dialog box. Make the 0 layer current and activate the Insert Block dialog box. Browse for the Title Block drawing just created. Clear the Specify On-screen check box in the Insertion point area. This will automatically place the title block at the 0,0,0 location of the layout, in the lower left corner of the printable area indicators. Click OK. Before the title block can be placed, you must first fill in the boxes in the Edit Attributes dialog box, illustrated in the following image (if this dialog box does not appear and you are prompted for values at the command line, it is because ATTDIA is set to 0). Complete all boxes—enter appropriate names and initials as directed.

 Note: Pressing the TAB key is a quick way of moving from one box to another while inside any dialog box.

Figure 17–11

The completed drawing with title block and attributes inserted is illustrated in the following image.

Figure 17–12

FIELDS AND ATTRIBUTES

When defining attributes for a specific drawing object such as a title block, the attribute value can be converted into a field, as shown in the following image. In this figure, the attribute tag (PROJECT_NAME) and prompt (What is the project name?) are created using conventional methods. Before entering a default value, click on the Insert field button. This will display the Field dialog box. In this figure, the SheetSet field category is being used to assign a custom property called Project Name.

Figure 17–13

When you work with attribute values as fields, the actual default entry is illustrated, as shown in the following image. In this example, #### was inserted into this cell from the Field dialog box in the previous figure. When a title block that contains this attribute is inserted into a drawing, you will be prompted to enter a new value for the PROJECT_NAME tag.

Figure 17–14

CONTROLLING THE DISPLAY OF ATTRIBUTES

You don't always want attribute values to be visible in the entire drawing. The ATTDISP command is used to determine the visibility of the attribute values. This command can be entered from the keyboard or can be selected from the View pull-down menu under Display > Attribute Display, as shown in the following image.

Command: **ATTDISP**
Enter attribute visibility setting [Normal/ON/OFF] <Normal>:
  (Enter the desired option)

The following three modes are used to control the display of attributes:

VISIBILITY NORMAL

This setting displays attributes based on the mode set through the Attribute Definition dialog box. If some attributes were created with the Invisible mode turned on, these attributes will not be displayed through this setting, which makes this setting popular for displaying certain attributes and hiding others.

VISIBILITY ON

Use this setting to force all attribute values to be displayed on the screen. This affects even attribute values with the Invisible mode turned on.

VISIBILITY OFF

Use this setting to force all attribute values to be turned off on the screen. This is especially helpful in busy drawings that contain lots of detail and text.

All three settings are illustrated in the following image. With ATTDISP set to Normal, attribute values defined as Invisible will not be displayed. This is the case for the WATTAGE and TOLERANCE tags. The Invisible mode was turned on inside the Attribute Definition dialog box when they were created. As a result, these values will not be visible. With ATTDISP set to on, all resistor values are forced to be visible. When set to off, this command makes all resistor attribute values invisible.

Figure 17–15

EDITING ATTRIBUTE PROPERTIES

After attributes have been placed in a drawing, changes to properties such as attribute position, text height, and angle can be made through the -ATTEDIT command.

Command: **-ATTEDIT** *(The hyphen (-) in front of the command activates prompts from the command line)*
Edit attributes one at a time? [Yes/No] <Y>

The response determines the string of options, which follow:

Entering Yes selects attributes individually (one by one) for editing. The attributes that are currently visible on the screen can be edited. The attributes to be edited can be further restricted by object selection, block names, tags, or values of the attributes to be edited.

Entering No prepares the attributes for global editing (all attributes). You can also restrict the editing to block names, tags, values, and on-screen visibility. Global editing will be discussed in greater detail later in this chapter.

After choosing to edit attributes one at a time, you are next asked to select the block name to be edited. You can choose the blocks to be edited by entering a specific block name (specify more than one by separating the names with a comma), by using the global wildcard (*), or by using the (?) symbol to replace common characters.

Enter block name specification <*>: *(Type a specific block name or press* ENTER *to affect all blocks in the drawing)*

AutoCAD prompts for the parts of the attributes to be edited:

Enter attribute tag specification <*>: *(Type a specific tag or press* ENTER *to affect all attribute tags)*
Enter attribute value specification <*>: *(Type a specific value or press* ENTER *to affect all attribute values)*

Your reply to each prompt targets the specific attributes to be edited.

You then select the attributes. Only those attributes that satisfy the criteria specified can be selected.

Select Attributes: *(Select individually or by window)*

After you have selected each attribute to be edited with the object selection process, an X marks the first attribute value that can be edited, as illustrated in the following image. Notice also that the attribute value is currently highlighted as an additional means of identifying the attribute value being modified.

Figure 17–16

The X mark stays with the current attribute to be edited until you enter the Next option (or press ENTER) and a new attribute is marked and highlighted in the following image. You are then prompted for the following options:

Enter an option [Value/Position/Height/Angle/Style/Layer/Color/Next] <N>:
   (Enter the desired option)

Figure 17–17

These are the various properties of an attribute that can be modified. The first letter can be used to select the appropriate option, or press ENTER to move to the next attribute to edit. A brief description of each property is listed below:

> **Value**—Enter a new attribute value.
>
> **Position**—Locate a new position for the attribute text.
>
> **Height**—Change the attribute text height.
>
> **Angle**—Change the attribute text angle.
>
> **Style**—Change the text style of the attribute text.
>
> **Layer**—Change the layer of the attribute text.
>
> **Color**—Change the attribute text color.
>
> **Next**—Move to the next attribute to be edited.

The following image illustrates a number of the properties that can be changed when attribute values are edited individually.

Figure 17–18

CHANGING OR REPLACING VALUES

When you enter the Value option, the following prompt appears:

Enter type of value modification [Change/Replace] <R>:

The Change option is used to modify a few characters in the attribute text, as in a misspelling. If you choose C, the following prompt appears:

Enter string to change: *(Enter those letters that make up the string to change)*
Enter new string: *(Enter the new letters to replace the old string)*

You should respond to the first prompt with the string of characters to be changed and to the second prompt with the string you want it replaced with.

The Replace option is used to change the entire attribute value. You are prompted:

Enter new Attribute value: *(Enter the new attribute value)*

EDITING ATTRIBUTES GLOBALLY

Global editing is used to edit multiple attributes at one time. As usual, the criteria you specify will limit the set of attributes selected for editing.

You can choose global editing by picking this command from the Modify pull-down menu, as shown in the following image.

Figure 17–19

Clicking on Global in the previous image will activate the -ATTEDIT command. This command is driven completely from a series of prompts and must be followed very carefully.

Command: -ATTEDIT
Edit Attributes one at a time? [Yes/No] <Y>: **N** *(For No)*

Entering No to this prompt will result in the following comment:

Performing global editing of attribute values.

The next prompt asks whether you will be editing only attributes visible on the screen.

Edit only attributes visible on screen? [Yes/No] <Y>:

You are then prompted:

Enter block name specification <*>:
Enter attribute tag specification <*>:
Enter attribute value specification <*>:

The next prompt asks you to select the attributes to globally edit:

Select Attributes:

Use the standard object selection process to choose the group of attributes to edit. The objects selected will be highlighted.

Respond to the next prompts with the string you wish to change and the changes you wish to make.

Enter string to change:
Enter new string:

Let's examine how this process of globally editing attributes works. In the following image, suppose the tolerance value of 5% needs to be changed to 10% on all blocks containing this attribute value. You could accomplish this through the Edit Attributes dialog box. However, this would allow you only to edit each attribute value one at a time. A more productive way would be to use the -ATTEDIT command and edit the group of attributes globally. Follow the examples and prompt sequence below for accomplishing this task.

Figure 17–20

Command: **-ATTEDIT**
Edit attributes one at a time? [Yes/No] <Y>: **N**
Performing global editing of attribute values.
Edit only attributes visible on screen? [Yes/No] <Y>: *(Press* ENTER *to accept the default)*
Enter block name specification <*>: **RESISTOR**
Enter attribute tag specification <*>: **TOLERANCE**
Enter attribute value specification <*>: *(Press* ENTER *to affect all value specifications)*
Select Attributes: *(Pick a point at "A" in the following image)*
Specify opposite corner: *(Pick a point at "B")*
Select Attributes: *(Press* ENTER *to continue)*
3 attributes selected.
Enter string to change: **5%**
Enter new string: **10%**

Figure 17–21

The result is shown in the following image, where all attributes were changed in a single command.

Figure 17–22

TUTORIAL EXERCISE: 17_COMPUTERS.DWG

Figure 17–23

Purpose

This tutorial is designed to globally edit attribute values using the -ATTEDIT command on the group of computers in the previous image.

System Settings

Open the drawing 17_Computers. All units and drawing limits have already been set. Attributes have also been assigned to various computer components.

Layers

All layers have already been created for this drawing.

Suggested Commands

Activate the -ATTEDIT command. Switch to global attribute editing mode and remove all dollar signs from the COST attribute values. This editing operation would be necessary if you needed to change the data in the COST box from character to numeric values for an extraction operation (extracting attribute data will be discussed in more detail later in this chapter).

STEP 1

Open the drawing 17_Computers. The following image shows an enlarged view of four computer workstations. For each workstation, $400 needs to be changed to 400, and each $2400 needs to be changed to 2400. In other words, the dollar sign needs to be removed from all values for each workstation.

Figure 17-24

STEP 2

Use the -ATTEDIT command and following image to globally edit all attribute values and remove all dollar signs from all values. Pick this command from the Modify pull-down menu by clicking on Object followed by Attribute and then Global.

Command: **-ATTEDIT**
Edit attributes one at a time? [Yes/No] <Y>: **N** *(For No)*
Performing global editing of attribute values.
Edit only attributes visible on screen? [Yes/No] <Y>: *(Press ENTER to accept this default)*
Enter block name specification <*>: *(Press ENTER to accept this default)*
Enter attribute tag specification <*>: **COST**
Enter attribute value specification <*>: *(Press ENTER to accept this default)*
Select Attributes: *(Pick a point at "A" the following image)*
Specify opposite corner: *(Pick a point at "B")*
Select Attributes: *(Press ENTER)*
32 attributes selected.
Enter string to change: **$**
Enter new string: *(Press ENTER; this will remove the dollar sign from all attribute values)*

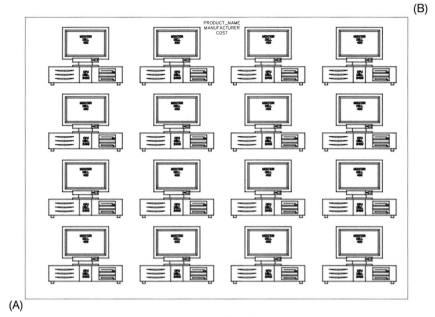

(A) (B)

Figure 17–25

The results are illustrated in the following image, with all dollar signs removed from all attribute values. The attributes can now be prepared for extraction.

Figure 17–26

EXTRACTING ATTRIBUTES

Extracting attributes is great for keeping records of your inserted blocks. You could maintain a database on furniture; model and cost figures on parts used in a design; or the number, type, and cost of windows in a plan. You could then print out all these items in a report or transfer this data to a database or spreadsheet program.

The ATTEXT command is used to extract attribute data from the drawing in a specified form. Entering the command activates the Attribute Extraction dialog box, as shown in the following image.

Command: **ATTEXT**

Figure 17–27

CREATING TEMPLATE FILES

To extract attribute information, you must first create a template file. This file controls the content and the display format for the information that will be extracted from the drawing. A text editor can be used to prepare the template file. Each line of the template represents one field to be listed in the attribute output file. You can also specify the width of the field (in characters) and the number of decimal places to be displayed in numerical fields. Each field will be listed in the order shown in the template file. The following image shows an example of an attribute template file created in Notepad.

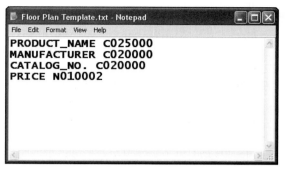

Figure 17–28

The following image shows a breakdown of what is contained in a template file. The entries of NAME, DEPARTMENT, TITLE, and PHONE represent the attribute tags. The template file will search for these exact names. As a result, spelling is important. If you misspell an attribute tag in the template file, AutoCAD will be unable to locate the attribute information and complete the extraction process.

The second column in the template file contains formatting information used for extracting and displaying the data. In the following image, next to the NAME tag is the entry C010000. The leading C stands for Character. This tells AutoCAD to search for character data during the extraction process (character data can consist of numbers, letters, and symbols). In a similar way, if the leading entry in the template file is N, this instructs AutoCAD to search for numeric data. Letters and symbols (other than a decimal point) are not allowed. If you define the leading character as N but your first attribute value begins with a symbol such as the $ sign, AutoCAD ignores this entry and no extraction takes place.

Figure 17–29

The next series of numbers next to the C in the NAME tag is 010. This tells the extraction process to allocate 10 spaces in the output file to hold the attribute value, which is used mainly by the SDF format. This is illustrated in the following image, in which four columns identified by the attribute tag names show the number of characters allotted to create a column effect. The last series, consisting of three zeros, 000, deals with the number of decimal places. While this is not appropriate for a character attribute, it is important for a numeric attribute value. You could have an attribute tag called PRICE along with the following information: N020002. The 002 tells AutoCAD to add two decimal places with every numeric value it extracts under the tag of PRICE.

Figure 17–30

Each line of the template file specifies one field to be written to the attribute output file. This includes the name of the field (the attribute tag), the character width, and the numeric precision. Each record of the attribute output file contains all information given in the order of the template file.

REDEFINING ATTRIBUTES

If more sweeping changes need to be made to attributes, a mechanism exists that allows you to redefine the attribute tag information globally. You follow a process similar to redefining a block, and the attribute values are also affected. A few examples of why you would want to redefine attributes might be to change their mode status (Invisible, Constant, etc.), change the name of a tag, reword a prompt, change a value to something completely different, add a new attribute tag, or delete an existing tag entirely.

Any new attributes assigned to existing block references will use their default values. Old attributes in the new block definition retain their old values. If you delete an attribute tag, AutoCAD deletes any old attributes that are not included in the new block definition.

EXPLODING A BLOCK WITH ATTRIBUTES

Before redefining an attribute, you might first have to copy an existing block with attributes and explode it. This will return the block to its individual objects and return the attribute values to their original tag information. The following image shows a kitchen sink with attribute values on the left and, on the right, the same sink but this time with attribute tags. The EXPLODE command was used on the right block to return the attribute values to their tags.

Figure 17-31

USING THE PROPERTIES PALETTE TO EDIT TAGS

A useful tool in making changes to attribute information is the Properties Palette. For example, clicking the CATALOG_NO. tag and then activating the Properties Palette, shown in the following image, allows you to make changes to the attribute prompt and values. You could even replace the attribute tag name with something completely different without having to use the Attribute Definition dialog box, as shown in the middle in the following image. Scrolling down the Properties Palette, as shown on the right in the following image, exposes the four attribute modes. If an attribute was originally created to be visible, you could make it invisible by changing the Invisible modifier here from No to Yes.

Figure 17-32

You could also double-click on the attribute tag in order to make any changes. Double-clicking on the CATALOG_NO. attribute tag, illustrated in the following image, displays the Edit Attribute Definition dialog box (DDEDIT command). Use this dialog box to make changes to the attribute tag, prompt, and default.

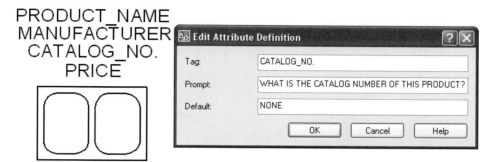

Figure 17-33

REDEFINING THE BLOCK WITH ATTRIBUTES

Once you have made changes to the attribute tags, you are ready to redefine the block along with the attributes with the ATTREDEF command. Use the following prompt sequence and illustration for accomplishing this task.

Command: **ATTREDEF**
Enter name of the block you wish to redefine: **SINK**
Select objects for new Block...
Select objects: *(Pick a point at "A" in the following image)*
Specify opposite corner: *(Pick a point at "B")*
Select objects: *(Press ENTER to continue)*
Specify insertion base point of new Block: **MID**
of *(Pick the midpoint of the sink at "C")*

Figure 17-34

TUTORIAL EXERCISE: 17_RESISTORS.DWG

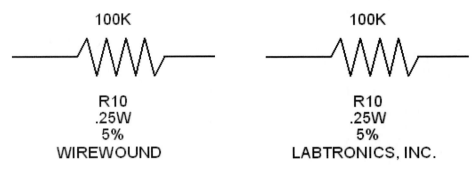

Figure 17-35

Purpose
This tutorial is designed to redefine attribute tags and update existing attribute values, as shown in the previous image.

System Settings
All units and drawing limits have already been set. Attributes have also been assigned to computer components.

Layers
All layers have already been created for this drawing.

Suggested Commands
The PROPERTIES and DDEDIT commands will be used to edit various characteristics of attributes.

STEP 1

Open the drawing file 17_Resistors. In the following image a series of resistors is arranged in a partial circuit design. A number of changes need to be made to the original attribute definitions. Once the changes are made, the block and attributes will be redefined and all the blocks will automatically be updated.

Figure 17–36

STEP 2

Copy one of the resistor symbols to a blank part of your screen. Then explode this block. This should break the block down into individual objects and return the attribute values to their tags, as shown on the left in the following image.

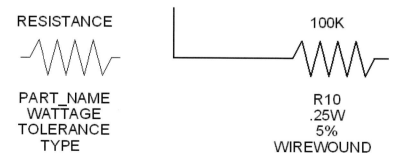

Figure 17–37

STEP 3

Let's examine what needs to be accomplished. The ATTDISP command is set to Normal; however, all the attribute values are displayed. Some of this information will be made invisible to simplify the drawing. Once invisible, this data could still be extracted with the ATTEXT command or viewed by setting the ATTDISP command to on. In this step, the tags WATTAGE, TOLERANCE, and TYPE will have their attribute mode changed to Invisible.

First, select the WATTAGE, TOLERANCE, and TYPE tags. These items will be highlighted, along with grips being displayed. Activate the Properties Palette and observe at the top that three Attributes are currently selected. Scroll down to the bottom of this window and under the Misc section, take note of the Invisible attribute mode. Click No, pick the down arrow in this edit box, and set the Invisible mode to Yes, as shown in the following image. Move the cursor into the drawing area and press ESC to remove the object highlight and the grips. These three tags have now been changed to Invisible.

Figure 17–38

STEP 4

In this step, the tag TYPE needs to be replaced with a new tag. This also means creating a new prompt and default value. Double-click on the TYPE tag to display the Edit Attribute Definition dialog box, as shown in the following image, and change the Tag from TYPE to SUP-PLIER. In the Prompt edit box, change the existing prompt to Supplier Name? In the Default edit box, change WIRE-WOUND to LABTRONICS, INC, as shown in the following image. When finished, click the OK button to accept the changes and dismiss the dialog box.

Figure 17-39

STEP 5

After these changes are made to the exploded block, the ATTREDEF command will be used to redefine the block and automatically update all resistor blocks to their new values and states. Activate the ATTREDEF command. Enter RESISTOR as the name of the block to redefine. Select the block and attributes, as shown on the left in the following image. Pick the new insertion point at "C"; as the drawing regenerates, all blocks are updated to the new attribute values, as shown on the right in the following image.

Command: **ATTREDEF**
Enter name of the block you wish to redefine: **RESISTOR**
Select objects for new Block...
Select objects: *(Pick a point at "A" in the following image)*
Specify opposite corner: *(Pick a point at "B")*
16 found
Select objects: *(Press ENTER to continue)*
Specify insertion base point of new Block: **End** *(For Object Snap Endpoint mode)*
of *(Pick the endpoint of the resistor at "C")*

Figure 17–40

STEP 6

To see if the SUPPLIER tag replaced the TYPE tag, activate the ATTDISP command and turn all attribute values on. The results are displayed in the following image. This action forces all attributes to be turned on.

Command: **ATTDISP**
Enter attribute visibility setting [Normal/ON/OFF] <Normal>: **ON**

Figure 17–41

ENHANCED ATTRIBUTE EDIT AND EXTRACTION TOOLS

A number of extra attribute tools in the form of dialog boxes are available to better assist in the modification and extraction of attributes. These are the Enhanced Attribute Editor dialog box, the Block Attribute Manager dialog box, and the Attribute Extraction wizard. These tools can all be selected from the Modify II toolbar, as shown in the following image.

The Enhanced Attribute Editor dialog box can also be displayed by selecting Object from the Modify pull-down menu, followed by Attribute, and then Single, as shown in the following image. The Block Attribute Manager dialog box can also be selected from this area of the Object > Attribute cascading menu. Activate the Attribute Extraction wizard by selecting Attribute Extraction from the Tools pull-down menu, also shown in the following image. All three of these tools will be explained in greater detail.

Figure 17–42

The following table gives a brief description of the Enhanced Attribute Extraction tools available.

Button	Tool	Command	Function
	Enhanced Attribute Editor	EATTEDIT	Launches a dialog box designed to edit the attributes in a block reference
	Block Attribute Manager	BATTMAN	Launches a dialog box used for managing attributes for blocks in the current drawing
	Synchronize Attributes		Updates all instances of the selected block with the currently defined attribute properties
	Attribute Extraction Wizard	EATTEXT	Exports the attributes found in a block reference out to a table or external file

THE ENHANCED ATTRIBUTE EDITOR DIALOG BOX

Click the Edit Attribute toolbar button to display the Enhanced Attribute Editor dialog box, as shown in the following image. This dialog box can also be activated by entering EATTEDIT at the command prompt. In the image, the Dryer block was selected. Notice that the attribute value is based on the selected tag. The value can then be modified in the Value edit box. To change to a different tag, select it with your mouse cursor. Notice that this dialog box has three tabs. The Attribute tab allows you to select the attribute tag that will be edited.

Figure 17–43

 Note: Double-clicking on a block containing attributes values will also launch the Enhanced Attribute Editor dialog box, illustrated in the previous image.

Clicking the Text Options tab displays the dialog box illustrated on the left in the following image. Use this area to change the properties of the text associated with the attribute tag selected, such as text style, justification, height, rotation, and so on. Once changes are made, click the Apply button.

Clicking the Properties tab displays the dialog box illustrated on the right in the following image. Use this area to change such properties as layer, color, lineweight, and so on. When you have completed the changes to the attribute, click the Apply button.

All three tabs of the Enhanced Attribute Editor dialog box have a Select block button visible in the upper right corner of each dialog box. Once you have edited a block, use this button to select a different block for editing, if desired.

Figure 17–44

THE BLOCK ATTRIBUTE MANAGER

Clicking the Block Attribute Manager toolbar button or entering BATTMAN at the command prompt displays the Block Attribute Manager dialog box, as shown in the following image. This dialog box is displayed as long as attributes are defined in your drawing. It does not require you to pick any blocks because it searches the database of the drawing and automatically lists all blocks with attributes. These are illustrated in the drop-down list, also shown in the following image. As will be shown, the Block Attribute Manager dialog box allows you to edit attributes globally.

Figure 17–45

Clicking the Edit button displays the Edit Attribute dialog box, as shown in the following image. Three tabs similar to those in the Enhanced Attribute Editor dialog box are displayed. Use these tabs to edit the attribute characteristics, the text options (style, justification, etc.), or the properties (layer, etc.) of the attribute. Notice at the bottom of the dialog box that, as changes are made, you will automatically see these changes because the Auto preview changes check box is selected.

Figure 17–46

Clicking the Settings button in the main Block Attribute Manager dialog box displays the Settings dialog box, as shown on the left in the following image. If you click one of the properties such as Height and click the OK button in the figure, this property will be displayed in the main Block Attribute Manager dialog box, as shown on the right in the following image.

Figure 17–47

TUTORIAL EXERCISE: 17_BLOCK ATTRIB MGR.DWG

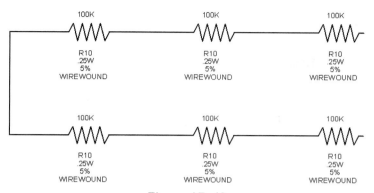

Figure 17-48

Purpose
This tutorial is designed to modify the properties of all attributes with the Block Attribute Manager dialog box, using the drawing illustrated in the previous image.

System Settings
All units and drawing limits have already been set. Attributes have also been assigned to the resistor block symbols.

Layers
All layers have already been created for this drawing.

Suggested Commands
The Block Attribute Manager dialog box will be used to turn the Invisible mode on for a number of attribute values. These changes will be automatically seen as they are made.

STEP 1
Open the drawing file 17_Block Attrib Mgr, as shown in the following image. The following attribute tags need to have their Invisible modifier turned on: TYPE, TOLERANCE, and WATTAGE. Activate the Block Attribute Manager dialog box (BATTMAN command), as shown in the following image, and select the TYPE tag. Then select the Edit button.

Figure 17-49

STEP 2

This takes you to the Edit Attribute dialog box, as shown in the following image. In the Attribute tab, place a check next to the Invisible mode and click the OK button. Notice that the changes are global and take place automatically. The attribute value WIREWOUND is no longer visible for any of the resistor blocks.

Figure 17–50

STEP 3

Turn the Invisible mode on for the TOLERANCE and WATTAGE tags. Your display should appear similar to the following image, with all attributed tags invisible, except for PART_NAME and RESISTANCE.

Figure 17–51

THE ATTRIBUTE EXTRACTION WIZARD

Clicking the Attribute Extraction wizard toolbar button or entering EATTEXT at the command prompt activates the Attribute Extraction wizard, shown in the following image. Use this wizard to step you through the attribute extraction process. You will be prompted to pick a new type of template with the BLK extension. If one is not already available, you can create it here in this dialog box. You will be able to individually select the attributes as well as specific block information to extract. You can preview the extraction file, save the template, and export the results in either TXT (tab-delimited file), CSV (comma delimited), or XLS (Microsoft Excel) format or extract the results into an AutoCAD table object.

Figure 17-52

The following table gives a brief description of each page used for extracting attributes using the Attribute Extraction wizard.

Attribute Extraction Wizard

Page and Title	Description
Page 1—Begin	Create a table or external file from scratch or use a template.
Page 2—Select Drawings	Select the current drawing or browse for other drawings.
Page 3—Select Attributes	Select the blocks and attributes to include as row and column headers.
Page 4—Finalize Output	Select to output to a table or to an external file.
Page 5—Table Style	Select a table style.
Page 6—Finish	Choose to save the settings out to a template or finish to produce the extracted information.

TUTORIAL EXERCISE: 17_CAD DEPT MODULES.DWG

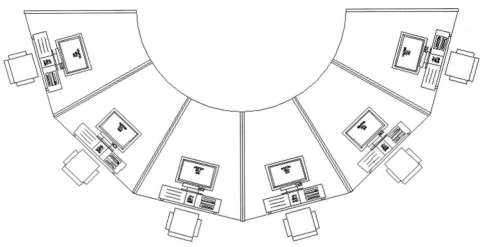

Figure 17–53

Purpose
This tutorial is designed to extract attribute data through the use of the Attribute Extraction wizard for the drawing 17_CAD Dept Modules, illustrated in the previous image.

System Settings
All units and drawing limits have already been set. Attributes have also been assigned to computer components.

Layers
All layers have already been created for this drawing.

Suggested Commands
This drawing appears inside a layout that has already been created. A table consisting of the extracted attributes will be inserted into this layout. Follow the steps provided by the Attribute Extraction wizard to extract the attributes out to an AutoCAD table.

STEP 1

Open the drawing file 17_CAD Dept Modules.dwg. Activate the Attribute Extraction wizard, as shown in the following image, by clicking on Attribute Extraction from the Tools pull-down menu, as shown on the left. You need to decide to create the attribute extraction from scratch or from a template.

If using a template, the Attribute Extraction wizard requires a block template with the BLK extension. Don't worry; you can still extract attributes through the wizard without a template. If you will be performing extraction operations numerous times to gain the same information, it would be useful to create and save your own block template. However, you cannot create the block template in such applications as Notepad. You will create this type of template with the help of the wizard in a later step.

This example will extract the attributes from scratch. When finished, click on the Next> button. (The next series of extraction dialog boxes is partially displayed. Each contains a Next> button at the bottom to proceed to the next step.)

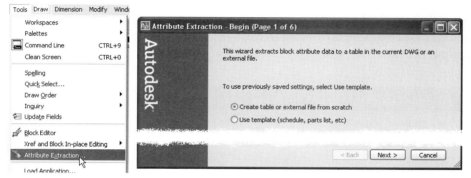

Figure 17–54

STEP 2

In this next step, you pick the data source from which the attributes will be extracted. Three types of data sources are available, as shown in the following image. You can select individual objects that have attributes, select all attributes from the current drawing, or select external drawings or sheet sets. For this exercise, verify that the radio button is set on Current drawing, as shown in the following image. Click the Next> button to continue.

Figure 17–55

STEP 3

Use the following dialog box to select the blocks and attributes to include as the final output. Two columns of information are available in the following image. On the left side, you click the blocks to perform the extraction on. In order to fine-tune your search, a check box is available to Exclude blocks without attributes. In this example, the extraction will be performed on two blocks, namely, CPU and Monitor. Illustrated on the right side of this dialog box are all attribute tags and their values. You could extract other information relative to the block. For this example, only attributes will be extracted. This is the reason for placing a check inside the box next to Exclude general block properties.

Figure 17–56

STEP 4

The next dialog box gives a preview of how the extracted attributes will be organized. In the following image, the Full Preview button was picked and a second dialog box previews the results. Notice that the preview resembles a table; this is because the attribute data is being extracted to an AutoCAD table, as shown in the lower left corner of this dialog box. You can also choose to extract attributes to an external file. The following file extensions are available: CVS, MDB, XLS, and TXT. In this example, the attributes will be extracted to an AutoCAD table.

Figure 17–57

STEP 5

This next dialog box deals with the components of the table that the attributes will be extracted to. In the following image, you need to enter a title for the table. In this exercise, the title of this table will be ATTRIBUTE EXTRACTION RESULTS. You can also select a table style. In this exercise, a table style called Attributes was already created. This table also displays in the dialog box, as shown on the right in the following image.

Figure 17–58

STEP 6

This final dialog box gives you the option to save the changes made in this wizard out to a block template. The information saved in this template will match the attribute specifications you selected in Step 3. The purpose of saving a template is to use it again in the future. If you want create this type of template, click the Save Template button. This will display a Save As dialog box. Give the template a name and notice that the BLK extension will be used as the file extension. After clicking the Save button, you will return to the Attribute Extraction dialog box; click the Finish> button.

Figure 17–59

STEP 7

Clicking the Finish button in the last step will bring you back into the drawing editor. You will be prompted to insert the table; do so in a convenient location of the drawing, as shown in the following image. In this image, notice that the Quantity, COST, MANUFACTURER, Block Name, and PRODUCT_NAME were all extracted.

Figure 17–60

SYSTEM VARIABLES THAT CONTROL ATTRIBUTES

ATTREQ

Determines whether the INSERT command uses default attribute settings during insertion of blocks. The following settings can be used:

    0   No attribute values are requested; all attributes are set to their default values.

Turns on prompts or a dialog box for attribute values, as specified by ATTDIA.

Command: **ATTREQ**
Enter new value for ATTREQ <1>:

ATTMODE

Controls the display of attributes. The following settings can be used:

    0   Off: Makes all attributes invisible.

    1   Normal: Retains current visibility of each attribute; visible attributes are displayed, invisible attributes are not.

        On: Makes all attributes visible.

Command: **ATTMODE**
Enter new value for ATTMODE <1>:

ATTDIA

Controls whether the INSERT command uses a dialog box for attribute value entry. The following settings can be used:

    0   Issues prompts on the command line.

    1   Initiates a dialog box for attribute value entry.

Command: **ATTDIA**
Enter new value for ATTDIA <1>:

CHAPTER 18

Working with External References and Raster Image and DWF Files

This chapter begins the study of External References and how they differ from blocks. When changes are made to a drawing that has been externally referenced into another drawing, these changes are updated automatically. There is no need to redefine the external reference in the same way blocks are redefined. When sharing of data is important, the ETRANSMIT command is used to collect all support files that make up a drawing. This guarantees to the individual reviewing the drawing that all support files come with the drawing. Working with raster images is also discussed in this chapter. Raster images can be attached to a drawing file for presentation purposes. The brightness, contrast, and fade factor of the images can also be adjusted. To control the order in which images display in your drawing, the DRAWORDER command will be explained.

COMPARING EXTERNAL REFERENCES AND BLOCKS

You have already seen in chapter 16 how easy and popular it is to insert blocks and drawing files into other drawing files. One advantage in performing these insertions is a smaller file size, which results because all objects connected with each block are grouped into a single object. It is only when blocks are exploded that the drawing size once again increases because the objects that made up the block have been separated into their individual objects. Another advantage of blocks is their ability to be redefined in a drawing.

External references are similar to blocks in that they are considered one object. However, external references are attached to the drawing file, whereas blocks are inserted in the drawing. This attachment actually sets up a relationship between the current drawing file and the external referenced drawing. For example, take a floor plan file and externally reference it into a current drawing file. All objects associated with the floor plan are brought in, including their current layer and linetype qualities. With the floor plan acting as a guide, such items as electrical symbols, furniture, and dimensions can be added to the floor plan and saved to the current drawing file. Now a design change needs to be made to the floor plan. You open the floor plan, stretch a few doors and walls to new locations, and save the file. The next time the drawing holding the electrical symbols and furniture arrangement is opened, the changes to the floor plan are made automatically to the current drawing file. This is one of the primary advantages of external references over blocks. Also, even though blocks are known to reduce file size, using external references reduces the file size even more than using blocks. One disadvantage of an external reference is that items such as layers and other blocks that belong to the external reference can be viewed but have limited capabilities for manipulation.

CHOOSING EXTERNAL REFERENCE COMMANDS

The XREF (XR is the alias) command activates a palette that allows you to attach and control external references in a drawing. This command can be entered at the keyboard, selected from the Reference toolbar by clicking External Reference, or selected through either the Tools or Insert pull-down menus, as shown in the following image.

Figure 18–1

Whichever method you use, the External References palette displays, as shown in the following image.

Figure 18–2

Located in the upper corner of the External References palette are buttons that are used for attaching or reloading external references. Clicking on the button located on the left, as shown in the following image, will display three different attach modes. Attach DWG activates the Select Reference File dialog box, used for selecting an existing drawing file to merge into the current drawing. Attach Image activates the Select Image File dialog box, used for selecting a raster image for merging into the current drawing. Attach DWF activates the Select DWF File dialog box, used for selecting a DWF (Drawing Web Format) file for merging into the current drawing.

When changes to original drawing files that are referenced into the current drawing occur, click on Refresh to update the information located in the External References palette. You can also choose Reload All References as a means of updating your current drawing to any changes that have occurred to the external references in the drawing.

Figure 18–3

External references can also be attached and reloaded by right-clicking in the palette area and activating a shortcut menu as shown on the left in the following image. Once the External References palette is populated with various references, select one of the references and right-click to display the shortcut menu, shown on the right in the following image. With this menu you can elect to open the original drawing file being referenced or click Attach to merge another drawing file. The Unload, Reload, Detach, and Bind options are discussed in detail later in this chapter.

Figure 18–4

One of the more popular ways to access all commands that deal with External References is through the Reference toolbar previously shown. The following table outlines the main commands found in this toolbar.

Button	Tool	Function
	External References	Launches the External References Palette, used for managing external references in a drawing
	Attach Xref	Launches the External Reference dialog box, used for attaching an external reference into a drawing
	Clip Xref	Used to isolate a portion of an external reference by clipping away unnecessary objects
	Xbind	Used to convert named objects, such as blocks and layers that belong to an external reference, to usable items
	Xref Frame	Controls the frame used to clip an external reference

ATTACHING AN EXTERNAL REFERENCE

Use the ATTACH XREF command to merge or attach an external drawing file into the current drawing file. This can be compared with the INSERT command for merging blocks into drawing files. The Attach option sets up a path, which looks for the external reference every time the drawing containing it is loaded.

Note: To accomplish the Try It! exercises in this chapter, copy the chapter 18 files to a convenient location/folder on your hard drive. This way you will be able to modify and save the drawings as required.

Try It! – Start a new drawing file from scratch. Then, click on the Attach Xref button in the Reference toolbar to activate the Select Reference File dialog box, as shown in the following image. This dialog box is very similar to the one used for selecting drawing files to initially load into AutoCAD. Click on the file 18_Asesmp1 to attach and click the Open button.

Figure 18–5

Once the file 18_Asesmp1 is selected, the External Reference dialog box appears, as shown in the following image. Some of the information contained in this dialog box is similar to the information in the Insert dialog box, such as the insertion point, scale, and rotation angle of the external reference. Of importance, in the upper part of the dialog box, is the path information associated with the external reference. If, during file management, an externally referenced file is moved to a new location, the new path of the external reference must be reestablished; otherwise, it does not load into the drawing it was attached to. Remove the check from the Specify On-screen box under Insertion point. This will insert this external reference at absolute coordinate 0,0,0. Clicking the OK button returns you to the drawing editor and attaches the file to the current drawing.

Figure 18–6

Since you will not see the floor plan on the small drawing sheet, perform a ZOOM-All. Your results should appear similar to the illustration in the following image.

Figure 18–7

With the external reference attached to the drawing file, layers and blocks that belong to the external reference are displayed in a unique way. Illustrated in the following image are the current layers that are part of the drawing. However, notice how a number of layers begin with the same name (18_Asesmp1); also, what appears to be a vertical bar separates 18_Asesmp1 from the names of the layers. Actually, the vertical bar represents the Pipe symbol on the keyboard, and it designates that the layers belong to the external reference, namely 18_Asesmp1. These layers can be turned off, can be frozen, or can even have the color changed. However, you cannot make these layers current for drawing, because they belong to the external reference.

Figure 18–8

Just as layers are affected by the presence of an external reference, blocks display the same behavior. When blocks are listed, as in the following table, the name of the external reference is given first, with the Pipe symbol following, and finally the actual name of the block, as in 18_Asesmp1|DESK2. As with layers, these blocks cannot be used in the drawing because of the presence of the Pipe symbol, which is not supported in the name of the block. This does, however, provide a quick way of identifying the valid blocks and those belonging to an external reference. Study the following prompt sequence and table for identifying a partial list of blocks that belong to external references. To view the blocks defined in this drawing, follow the command sequence below to perform this operation.

Command: **-B** *(for -BLOCK. The "-" preceding the command provides a command line version of the block command)*
Enter block name or [?]: **?**
Enter block(s) to list <*>: *(Press* ENTER *to accept the default and display the list, as shown in the following image)*

```
Defined blocks.
    "18_Asesmp1"                    Xref: resolved
    "18_Asesmp1|CHAIR7"             Xdep: "18_Asesmp1"
    "18_Asesmp1|COPIER2"            Xdep: "18_Asesmp1"
    "18_Asesmp1|DESK2"              Xdep: "18_Asesmp1"
    "18_Asesmp1|DESK3"              Xdep: "18_Asesmp1"
    "18_Asesmp1|DESK4"              Xdep: "18_Asesmp1"
    "18_Asesmp1|TABLE1"             Xdep: "18_Asesmp1"
    "18_Asesmp1|TABLE2"             Xdep: "18_Asesmp1"
```

Figure 18–9

One of the real advantages of using external references is the way they are affected by drawing changes. To demonstrate this, first save your drawing file under the name 18_Facilities. Close this drawing and open the original floor plan called 18_Asesmp1. Use the Insert dialog box to insert a block called ROOM NUMBERS, as shown in the following image, into the 18_Asesmp1 file. Use an insertion point of 0,0,0 for placing this block and click OK in the Insert dialog box. Save this drawing file and open the drawing 18_Facilities.

Figure 18–10

In the following image, the drawing file 18_Asesmp1 is opened again. Notice how room tags have been added to the floor plan. Once the drawing holding the external reference is loaded, all room tags, which belong to the external reference, are automatically displayed.

Figure 18-11

OVERLAYING AN EXTERNAL REFERENCE

Suppose in the last example of the facilities floor plan that some design groups need to see the room number labels while other design groups do not. This is the purpose of overlaying an external reference instead of attaching it. All design groups will see the information if the external reference is attached. If information is overlaid and the entire drawing is externally referenced, the overlaid information does not display; it's as if it is invisible. This option is illustrated in the next Try It! exercise.

 Try It! – Open the drawing 18_Floor Plan. Then use the External Reference dialog box, shown in the following image, to attach the file 18_Floor Furniture (the steps for accomplishing this were demonstrated in the previous Try It! exercise). Use an insertion point of 0,0,0 and be sure the reference type is Attachment.

Figure 18-12

Your display should be similar to the illustration in the following image. You want all design groups to view this information.

Figure 18–13

Now we will overlay an external reference. In the External Reference palette click the Attach Dwg button. When the Select Reference File dialog box appears in the following image, pick the file 18_Floor Numbers and click the Open button.

Figure 18–14

When the External Reference dialog box appears, as shown in the following image, change the Reference Type by clicking on the radio button next to Overlay. Under the Insertion Point heading, be sure this external reference will be inserted at 0,0,0. When finished, click the OK button.

Figure 18–15

Your display should appear similar to the following image. Save this drawing file under its original name of 18_Floor Plan and then close the drawing file.

Figure 18–16

Now start a new drawing file from scratch. In the External Reference palette click the Attach Dwg button and attach the drawing file 18_Floor Plan. When the External Reference dialog box appears, as shown in the following image, be sure that the Reference Type is set to Attachment and that the drawing will be inserted at 0,0,0. When finished, click the OK button.

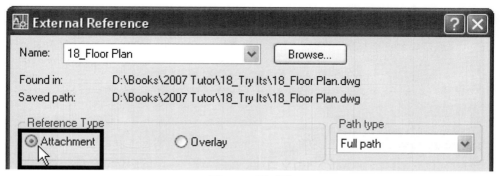

Figure 18–17

Your drawing will not be displayed because the default limits are too small, so perform a ZOOM-All operation and observe the results, shown in the following image. Notice that the room numbers do not display because they were originally overlaid in the file 18_Floor Plan. This completes this Try It! exercise.

Figure 18–18

THE XBIND COMMAND

Earlier, it was mentioned that blocks and layers belonging to external references cannot be used in the drawing they were externally referenced into. There is a way, however, to convert a block or layer into a usable object through the XBIND command.

Try It! – Open the drawing file 18_Facilities Plan. Then select the XBIND command from the Reference toolbar, as shown in the following image. This activates the Xbind dialog box, which lists the external reference 18_ASESMP1. Expand the listing of all named objects, such as blocks and layers associated with the external reference. Then expand the Block heading to list all individual blocks associated with the external reference, as shown in the following image.

Figure 18–19

Click on the block 18_Asesmp1|DESK2 in the listing on the left; then click on the Add-> button. This moves the block name over to the right under the listing of Definitions to Bind, as shown in the following image. Do the same for 18_Asesmp1|DESK3 and 18_Asesmp1|DESK4. When finished adding these items, click on the OK button to bind the blocks to the current drawing file.

Figure 18–20

Test to see that new blocks have in fact been bound to the current drawing file. Activate the Insert dialog box, click on the Name drop-down list box, and notice the display of the blocks, as shown in the following image. The three symbols just bound from the external reference still have the name of the external reference, namely, 18_Asesmp1. However, instead of the Pipe symbol separating the name of the external reference and block names, the characters 0 are now used. This is what makes the blocks valid in the drawing. Now these three blocks can be inserted in the drawing file even though they used to belong only to the external reference 18_Asesmp1.

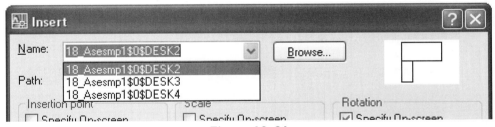

Figure 18–21

IN-PLACE REFERENCE EDITING

In-Place Reference Editing allows you to edit a reference drawing from the current drawing, which is externally referencing it. You then save the changes back to the original drawing file. This becomes an efficient way of making a change to a drawing file from an externally referenced file.

 Try It! – Open the drawing file 18_Pulley Assembly, shown in the following image. Both holes located in the Base Plate, Left Support, and Right Support need to be stretched 0.20 units to the left and right. This will center the holes along the Left and Right Supports. Since all images that make up the Pulley Assembly are external references, the In-Place Reference Editing feature will be illustrated.

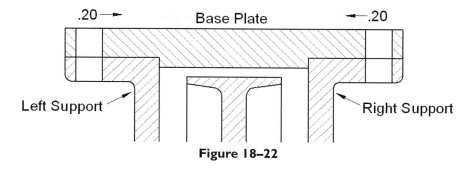

Figure 18–22

Begin by accessing the Edit Reference In-Place (REFEDIT) command from the Refedit toolbar, as shown in the following image.

Once in the REFEDIT command, you are prompted to select the reference you wish to modify. Pick the top line of the Base Plate, as shown in the following image. This displays the Reference Edit dialog box, as shown on the right in the following image. The reference to be edited, which is 18_Base Plate in our case, should be selected. Nested references may also be displayed; one of these could be selected for editing instead, if desired. Clicking on the OK button returns you to the screen.

 Note: Double-clicking on an external reference will also launch the Reference Edit dialog box.

Figure 18–23

If not already displayed, the Refedit toolbar will be automatically provided, as shown in the following image, and you are returned to the command prompt. To better identify the external reference being edited, the other external references in the drawing take on a slightly faded appearance. This faded effect returns to normal when exiting the Refedit process. You can now make modifications to the Base Plate by using the STRETCH command to stretch both holes a distance of 0.20 units to the inside.

Figure 18–24

When performing the STRETCH operation, notice that even if the crossing window were to extend across both parts, only the holes in the Base Plate will be modified, as shown in the following image. When you are satisfied with the changes, select the Save Reference Edits button on the toolbar. An AutoCAD alert box will ask you to confirm the saving of reference changes, as shown in the following image.

Figure 18–25

After you click OK, the results can be seen, as shown in the following image. Notice that only the objects that belong to the external reference (in this case, the holes that are part of the Base Plate) are affected. Also, the Refedit toolbar automatically closes.

Figure 18–26

Perform the same series of steps using In-Place Reference Editing separately for the Left and Right supports. Stretch the hole located on the Left Support a distance of 0.20 units to the right. Stretch the hole located on the Right Support a distance of 0.20 units to the left. The final results of this In-Place Reference Editing Try It! exercise are displayed in the following image.

Figure 18–27

BINDING AN EXTERNAL REFERENCE

Binding an external reference to a drawing makes the external reference a permanent part of the drawing and no longer an externally referenced file. In actuality, the external reference is converted into a block object. To bind the entire database of the Xref drawing, including all its xref-dependent named objects such as blocks, dimension styles, layers, linetypes, and text styles, use the Bind option of the External Reference palette. Since binding an external reference breaks the link with the original drawing file, this becomes a very popular technique at the end of a project when archiving final drawing files is important.

To bind an external reference, click on the External References button located in the Reference toolbar to activate the External References palette. Select the external reference that you want to bind to the current drawing, right-click, and pick Bind from the menu, as shown on the left in the following image. This option activates the Bind Xrefs dialog box, illustrated on the right in the following image. Two options are available inside this dialog box: Bind and Insert.

Figure 18–28

The Bind option binds to the current drawing file all blocks, layers, dimension styles, and so on that belonged to an external reference. After you perform this operation, layers can be made current and blocks inserted in the drawing. For example, a typical block definition belonging to an external reference is listed in the symbol table as XREFname|BLOCKname. Once the external reference is bound, all block definitions are converted to XREFname0BLOCKname. In the layer display in the following image, the referenced layers were converted to usable layers with the Bind option.

Referenced Layers	Converted Layers
18_Pulley|Hatch	18_Pulley0Hatch
18_Pulley|Object	18_Pulley0Object
18_Pulley|Text	18_Pulley0Text

 Note: The same naming convention is true for blocks, dimension styles, and other named items. The result of binding an external reference is similar to a drawing that was inserted into another drawing.

Figure 18–29

The Insert option of the Bind Xrefs dialog box is similar to the Bind option. However, instead of named items such as blocks and layers being converted to the format XREFname0BLOCKname, the name of the external reference is stripped, leaving just the name of the block, layer, or other named item (BLOCKname, LAYERname, etc.). In the following image, the following referenced layers were converted to usable layers with the Insert option:

Referenced Layers	Converted Layers
18_Pulley|Hatch	Hatch
18_Pulley|Object	Object
18_Pulley|Text	Text

Figure 18–30

 Note: It is considered an advantage to use the Bind option over the Insert option. This way, you will be able to identify the layer that was tied to the previously used external reference.

USING LIST VIEW AND TREE VIEW

The External References palette makes it easy to keep track of external references in a drawing. By default, external references are listed as shown in the following image. This is called List View. A wealth of information, including reference name, status, file size, type, date, and saved path, is listed when reviewing a certain file.

Figure 18–31

Clicking on the Tree View button lists external references in hierarchical form, similar to the form shown in the following image.

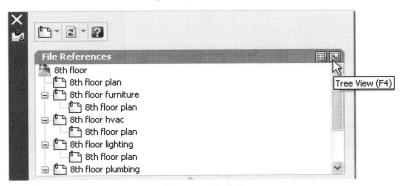

Figure 18–32

CLIPPING AN EXTERNAL REFERENCE

Since attaching an external reference displays the entire reference file, you have the option of displaying a portion of the file. This is accomplished by clipping the external reference with the XCLIP command. Choose this command from either the Reference toolbar or the Modify > Clip > Xref pull-down menu, as shown in the following image.

Figure 18–33

This operation is very useful when you want to emphasize a particular portion of your external reference file. Clipping boundaries include polylines in the form of rectangles, regular polygonal shapes, or even irregular polyline shapes. All polylines must form closed shapes.

 Try It! – Open the drawing file 18_Facilities Plan. Follow the illustration on the left in the following image and the command prompt sequence below for performing this operation.

 Command: **XCLIP**
Select objects: *(Pick the external reference)*
Select objects: *(Press* ENTER *to continue)*
Enter clipping option
[ON/OFF/Clipdepth/Delete/generate Polyline/New boundary] <New>: **N** *(For New)*
Specify clipping boundary:
[Select polyline/Polygonal/Rectangular] <Rectangular>: *(Press* ENTER*)*
Specify first corner: *(Pick a point at "A," as shown on the left in the following image)*
Specify opposite corner: *(Pick a point at "B")*

The results are displayed on the right in the following image. If you want to return the clipped image to the full external reference, use the XCLIP command and use the OFF clipping option. This temporarily turns off the clipping frame. To permanently remove the clipping frame, use the Delete clipping option.

Figure 18–34

OTHER OPTIONS OF THE EXTERNAL REFERENCES PALETTE

The following additional options of the External References palette will now be discussed, using the following image as a guide.

Figure 18–35

UNLOAD

Unload is similar to the Detach option, with the exception that the external reference is not permanently removed from the database of the drawing file. When an external reference is unloaded, it is still listed in the External References palette, as shown in the following image. Notice that an arrow (facing down) is displayed to signify the unloaded status. Since this option suppresses the external reference from any drawing regenerations, it is used as a productivity technique. Reload the external reference when you want it returned to the screen.

Figure 18–36

RELOAD

This option loads the most current version of an external reference. It is used when changes to the external reference are made while the external reference is currently being used in another drawing file. This option works well in a networked environment, where all files reside on a file server.

DETACH

Use this option to permanently detach or remove an external reference from the database of a drawing.

EXTERNAL REFERENCE NOTIFICATION TOOLS

A series of tools and icons are available to assist with managing external references. When a drawing consisting of external references opens, an icon appears in the extreme lower right corner of your screen, as shown on the left in the following image. Clicking on this icon will launch the External References palette. In the event the source file was changed or modified, the next time you return to the external reference drawing, a yellow caution icon appears, as shown on the right in the following image, informing you that the external reference was changed. This example refers to the Pulley Assembly, in which the Base Plate was modified and saved. When you return to the Pulley Assembly, the yellow caution icon signifies the change to the file 18_Base Plate.

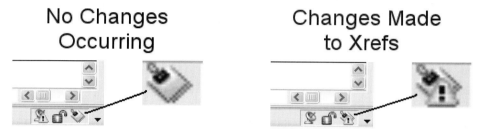

Figure 18–37

Clicking on the yellow caution icon will launch the External References palette. When you click on the Refresh button, you will notice the appearance of a yellow caution triangle, shown in the following image. The status of this file alerts you that 18_Base Plate needs to be reloaded in order for the change to be reflected in the Pulley Assembly drawing.

Figure 18–38

Right clicking on 18_Base Plate displays a menu, as shown in the following image. Clicking on Reload while the external reference is highlighted will reload the external reference.

Figure 18–39

Illustrated in the following image is another feature of external references. The left support of the pulley assembly was selected. When you right-click, the shortcut menu appears, as shown on the right in the following image. Use this menu to perform the following tasks:

Edit Xref In-place—This option activates the Reference Edit dialog box for the purpose of editing the external reference in-place.

Open Xref—This option will open the selected external reference in a separate window.

Clip Xref—This option launches the xclip command for the purpose of clipping a portion of the external reference

External References—This option launches the External References palette.

Figure 18–40

USING ETRANSMIT

This AutoCAD utility is helpful in reading the database of your drawing and listing all support files needed. Once these files are identified, you can have this utility gather all files into one zip file. You can then copy these files to a disk or CD, or transmit the files over the Internet. Clicking on eTransmit…, located in the File pull-down menu, as shown on the left in the following image, displays the Create Transmittal dialog box, as shown in the middle in the following image. Notice in the Files Tree tab a listing of all support files grouped by their specific category. For instance, an External References category exists. Clicking on the "+" sign will list all external references set to be transmitted.

Clicking on the Files Table tab, as shown on the right in the following image, displays a list of all files that will be included in the transmittal set.

When you are finished examining all the support files, clicking the OK button will take you to the Specify Zip File dialog box. It is here that you enter a file name. All support files listed under the Files Tree tab will be grouped into a single zip file for easy sharing with other individuals or companies.

Figure 18–41

WORKING WITH RASTER IMAGES

Raster images in the form of JPG, GIF, PCX, and so on can easily be merged with your vector-based AutoCAD drawings. The addition of raster images can give a new dimension to your drawings. Typical examples of raster images are digital photographs of an elevation of a house or an isometric view of a machine part. Whatever the application, working with raster images is very similar to what was just covered with external references. You attach the raster image to your drawing file. Once it is part of your drawing, additional tools are available to manipulate and fine-tune the image for better results. To attach a raster image to a drawing, you can use the External References palette (Attach Image button), the Insert pull-down menu (click Raster Image Reference, as shown on the left in the following image), or the Reference toolbar (click the Attach Image button). The Reference toolbar also contains a number of tools for manipulating raster images, as shown on the right in the following image.

Figure 18–42

The following table outlines the main raster image commands found in the Reference toolbar.

Button	Tool	Function
	Attach Image	Used for attaching a raster image to a drawing
	Clip Image	Used for cropping or clipping the raster image
	Adjust Image	Used for adjusting the brightness, contrast, and fade factor of a raster image
	Image Quality	Uses two settings, high or draft, to control the quality of raster images
	Image Transparency	Controls whether the background of a raster image is transparent or opaque
	Image Frame	Used for turning the rectangular frame on or off for a raster image

To attach a raster image utilizing the Reference toolbar, click on the Attach Image button. This will launch the Select Image File dialog box, as shown in the following image. Choose the desired raster image to attach from the list and then click on the Open button.

Figure 18–43

After choosing the correct raster image, the Image dialog box will appear as shown on the left in the following image. From this dialog box, change the insertion point, scale, or rotation parameters if desired. Clicking the OK button will return you to your drawing, where you pick an insertion point and change the scale factor of the image. A typical raster image is illustrated on the right in the following image.

Figure 18–44

To manage all images in a drawing, click on the External References button located in the Reference toolbar. The same dialog box used for managing external references is used for managing image files, as shown in the following image.

Figure 18–45

 Try It! – Begin this exercise on working with raster images by opening the drawing file 18_Linkage, illustrated in the following image. In this exercise a raster image will be attached and placed in the blank area to the right of the two-view drawing.

Figure 18–46

Click on the Attach Image button located in the Reference toolbar. When the Select Image File dialog box appears, click on the file Linkage1.jpg. Then click on the Open button.

Figure 18–47

When the Image dialog box appears, as shown in the following image, leave all default settings. You will be specifying the insertion point on the drawing screen. Click the OK button to continue.

Figure 18–48

Once back in the drawing file, pick a point anchoring the lower left corner of the graphic, as shown in the following image. Now move your cursor in an upward right direction to scale the image as necessary. Pick a second point at a convenient location to display the image.

Figure 18–49

As the image displays on your screen, you can adjust its size very easily. Click on the edge of the image (the image frame) and notice the grips appearing at the four corners of the image. Clicking on a grip and then moving your cursor will increase or decrease the image's size, as shown in the following image.

Figure 18–50

From the Reference toolbar, shown in the following image, identify the Adjust Image button and pick it now. Picking the edge of the raster image frame will display the Image Adjust dialog box, shown in the following image. Adjust the Brightness, Contrast, and Fade settings and notice that each of these affects the image previewed to the right of this dialog box. You can easily return to the original image by clicking the Reset button in the lower left corner of this dialog box.

Figure 18–51

Clicking the OK button in the Image Adjust dialog box will return you to the drawing. The final results are illustrated in the following image, with the vector drawing and raster image sharing the same layout.

Figure 18–52

USING THE DRAWORDER COMMAND

With the enhancements made to raster images and the ability to merge raster images with vector graphics, it is important to control the order in which these images are displayed. The DRAWORDER command is used to provide this level of control over raster images. The DRAWORDER command can be selected from the Modify II toolbar or from the Tools pull-down menu, as shown in the following image. A dedicated Draw Order toolbar is also available.

Bring to Front—The selected object is brought to the top of the drawing order.

Send to Back—The selected object is sent to the bottom of the drawing order.

Bring Above Object—The selected object is brought above a specified reference object.

Send Under Object—The selected object is sent below the specified reference object.

Figure 18-53

The four modes of the Draw Order toolbar are described as follows:

Button	Tool	Function
	Bring to Front	The selected object is brought to the top of the drawing order
	Send to Back	The selected object is sent to the bottom of the drawing order
	Bring Above Objects	The selected object is brought above a specified reference object
	Send Under Objects	The selected object is sent below the specified reference object

In the illustration on the left in the following image, the image of clouds has been merged with the Multiline text object CLOUD FORMATIONS. Unfortunately, because of the drawing order of operations, the image is masking part of the mtext object.

To correct the problem, select the Bring to Front tool. The mtext object is selected as the object to bring to the top of the drawing order, and the results are illustrated on the right in the following image. The mtext object is now at the top of the drawing order and can be read in full.

Figure 18-54

ATTACHING DWF FILES

DWF stands for Design Web Format and consists of a highly compressed file format created from a DWG file. DWF files can be easily shared and viewed on the Web. DWF files are created through the ePlot device found in the typical AutoCAD plotting dialog box. Once the DWF file is created, special viewers are available to perform real-time panning and zooming operations in addition to displaying layers and named views. Following are reasons for attaching a DWF file to a drawing:

> Security: When a DWF file is attached as an underlay, the information contained in the DWF file can be easily viewed but cannot be modified.
>
> Clarity: Since DWF files do not consist of individual pixels like raster images, the DWF file looks more presentable than image files.
>
> Object Snap Usage: You can snap to geometry in a DWF file the same way that you would use object snap in a DWG file.

DWF files are attached to a drawing utilizing the same methods we have seen for external reference drawings and raster images. They can be attached through the External Reference palette (attach DWF) or through the Insert pull-down menu (DWF Underlay). To attach a DWF file through the Insert pull-down menu, click on DWF Underlay, as shown on the left in the following image. This will launch the Select DWF File dialog box, as shown on the right in the following image, where you choose the DWF file to attach from a list. Once the DWF file is placed in a drawing, it is referenced through a path the same way that external referenced drawings and raster images are referenced. Once attached to a drawing file, the DWF files do not greatly increase the drawing file size.

Figure 18–55

 Note: Even though the DWF underlay file was created from a source AutoCAD drawing, the DWF file does not carry the same precision as the DWG format. Care needs to be taken when using a DWF underlay to design from.

When the Attach DWF Underlay dialog box appears, as shown in the following image, make changes to the insertion point, scale, and rotation angle if desired.

Figure 18–56

Once a DWF underlay has been attached, certain features of the file can be changed for special effects. Double-clicking on the frame of the DWF file will launch the Properties palette, as shown in the following image. An area called Underlay Adjust will allow you to change the Contrast, Fade, Monochrome, and Adjust Colors for Background. In the following image, the fade factor was changed from 25 to 75. The results show an image with a noticeable faded appearance. Changing the Monochrome setting from No to Yes would set the lines and background to black and white.

Figure 18–57

As with external references and raster images, the External References palette is used to manage DWF underlay files, as shown in the following image.

Figure 18–58

TUTORIAL EXERCISE: EXTERNAL REFERENCES

Figure 18–59

Purpose
This tutorial is designed to use the office floor plan to create an interior plan consisting of various interior symbols, such as desks, chairs, shelves, and plants. The office floor plan will be attached to another drawing file through the XREF command.

System Settings
Since these drawings are provided on the CD, all system settings have been made.

Layers
The creation of layers is not necessary because layers already exist for both drawing files you will be working on.

Suggested Commands
Begin this tutorial by opening the drawing 18_Office.Dwg, which is located on the CD, and view its layers and internal block definitions. Then open the drawing 18_Interiors.Dwg, also located on the CD, and view its layers and internal blocks. The file 18_Office.Dwg will then be attached to 18_Interiors.Dwg. Once this is accomplished, chairs, desks, shelves, and plants will be inserted in the office floor plan for laying out the office furniture. Once 18_Interiors.Dwg is saved, a design change needs to be made to the original office plan; open 18_Office.Dwg and stretch a few doors to new locations. Save this file and open 18_Interiors.Dwg; notice how the changes to the doors are automatically made. The Xbind dialog box will also be shown as a means for making a block that had previously belonged to an external reference usable in the file 18_Interiors.Dwg.

STEP 1

Open 18_Office.Dwg, which can be found on the CD, and observe a simple floor plan consisting of three rooms, as shown in the following image. Furniture will be laid out using the floor plan as a template.

Figure 18–60

STEP 2

While in 18_Office.Dwg, use the Layer Properties Manager dialog box and observe the layers that exist in the drawing for such items as doors, walls, and floor, as shown in the following image. These layers will appear differently once the office plan is attached to another drawing through the XREF command.

Figure 18–61

STEP 3

While in the office plan, activate the Insert dialog box through the INSERT command. At times, this dialog box is useful for displaying all valid blocks in a drawing. Clicking on the Name drop-down list displays the results shown in the following image. Two blocks are currently defined in this drawing; as with the layers, once the office plan is merged into another drawing through the XREF command, these block names will change. When finished viewing the defined blocks, close 18_Office.Dwg.

Figure 18–62

STEP 4

This next step involves opening 18_Interiors.Dwg, also found on the CD, and looking at the current layers found in this drawing. Once this drawing is open, use the Layer Properties Manager dialog box to observe that layers exist in this drawing for such items as floor and furniture, as shown in the following image.

Figure 18–63

STEP 5

As with the office plan, activate the Insert dialog box through the INSERT command to view the blocks internal to the drawing, as shown in the following image. The four blocks listed consist of various furniture items and will be used to lay out the interior plan.

Figure 18–64

STEP 6

Close 18_Office.Dwg and verify that 18_Interiors.Dwg is still open and active. The office floor plan will now be attached to the interior plan in order for the furniture to be properly laid out. Make the Floor layer current. Rather than insert the office plan as a block, use the External Reference palette to attach the drawing. This palette will activate when you enter the XREF command from the keyboard or choose External References from the Insert pull-down menu, as shown on the left in the following image. After the palette displays, click on the Attach DWG button (or the drop-down list, if the button is not displayed), as shown on the right in the following image.

Figure 18–65

STEP 7

Clicking on Attach DWG, shown in the previous image, displays the Select Reference File dialog box, as shown in the following image. Find the appropriate folder and click on the drawing file 18_Office.Dwg.

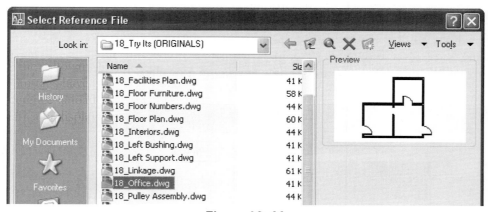

Figure 18–66

STEP 8

Selecting the file 18_Office.Dwg displays the External Reference dialog box, as shown on the left in the following image. Notice that 18_OFFICE is the name of the external reference file chosen for attachment in the current drawing. Verify that the Reference Type is an attachment. If selected, remove the check marks from the Specify On-screen boxes for the Insertion point, Scale, and Rotation. In the External Reference dialog box, click on the OK button to attach 18_Office.Dwg to 18_Interiors.Dwg.

The floor plan is now attached to the Interiors drawing at the insertion point 0,0, as shown on the right in the following image.

Figure 18–67

With the floor plan attached to the interiors drawing, notice how both drawings appear in the External References palette, as shown in the following image. When you have finished studying the information located in this palette, you can dismiss the palette from the screen by clicking on the "X" located in the upper corner.

Figure 18–68

STEP 9

Once again, activate the Layer Properties Manager dialog box, paying close attention to the display of the layers. Using the following image as a guide, you can see the familiar layers of Floor and Furniture. However, notice the group of layers beginning with 18_Office; the layers actually belonging to the external reference file have the designation of XREF|LAYER. For example, the layer 18_Office|DOORS represents a layer located in the file 18_Office.Dwg that holds all door symbols. The "|," or Pipe symbol, is used to separate the name of the external reference from the layer. The layers belonging to the external reference file may be turned on or off, locked, or even frozen. However, these layers cannot be made current for drawing on.

Figure 18–69

STEP 10

Make the Furniture layer current and begin inserting the desk, chair, shelf, and plant symbols in the drawing using the INSERT command or through the DesignCenter. The external reference file 18_OFFICE is to be used as a guide throughout this layout. It is not important that your drawing match exactly the illustration shown on the left in the following image. After positioning all symbols in the floor plan, save your drawing under its original name of 18_Interiors.Dwg but do not close this file.

Even with the interior drawing file still open, make the original drawing file, 18_Office.Dwg, current by opening this file, and make the following modifications: stretch all three doors as indicated to new locations and mirror one of the doors so it is positioned closer to the wall; stretch the wall opening over to the other end of the room, as shown on the right in the following image. Finally, save these changes under the original name of 18_Office.Dwg and close the file.

Figure 18–70

STEP 11

After closing the office plan, your display will return to the interiors plan. At this point, nothing in the drawing will appear to have changed. However, notice in the lower right corner of your display screen a tray of buttons. One of these buttons (Manage Xrefs) deals specifically with external reference files. Once the office plan is changed and saved, a Manage Xrefs button appears in this portion of your screen. It is also displayed on the left in the following image. Right-clicking on this button will display two options. One of the options allows you to reload all external reference files. Click this option and notice the effects on the floor plan, illustrated on the left, where changes to the office plan are now reflected in the interiors drawing.

In this case, observe how some of the furniture is now in the way of the doorways. If the office plan had been inserted in the interiors drawing as a block, these changes would not have occurred this easily.

Because of the changes in the door openings, edit the drawing by moving the office furniture to better locations. The illustration on the right in the following image can be used as a guide, although your drawing may appear slightly different.

Figure 18–71

STEP 12

A door needs to be added to an opening in one of the walls of the office plan. However, the door symbol belongs to the externally referenced file 18_Office.Dwg. The door block is defined as 18_Office|DOOR; the "|" character is not valid in the naming of the block and therefore cannot be used in the current drawing. The block must first be bound to the current drawing before it can be used. Use the XBIND command to make the external reference's Door block available in the 18_Interiors.Dwg. This command can be selected from the Reference toolbar, as shown in the following image, which will display the Xbind dialog box. While in this dialog box, click on the "+" symbol next to the file 18_Office and then click on the "+" symbol next to Block. This will display all blocks that belong to the external reference.

Figure 18–72

STEP 13

Clicking on 18_OFFICE|DOOR followed by the Add-> button in the previous image moves the block of the door to the Definitions to Bind area in the following image. Click OK to dismiss the dialog box, and the door symbol is now a valid block that can be inserted in the drawing.

Figure 18–73

STEP 14

Activate the Insert dialog box through the INSERT command, click on the Name drop-down list, and notice the name of the door, shown in the following image. It is now listed as 18_Office0DOOR; the "|" character was replaced by the "0," making the block valid in the current drawing. This is AutoCAD's standard way of converting blocks that belong to external references to blocks that can be used in the current drawing file. This same procedure works on layers belonging to external references as well.

Figure 18–74

STEP 15

Make the Floor layer current and insert the door symbol into the open gap, as illustrated in the following image, to complete the drawing.

Figure 18–75

CHAPTER 19

Creating Advanced Layouts

This chapter is a continuation of chapter 14, "Working with Drawing Layouts." First, multiple images of the same drawing file at different scales will be demonstrated. This will include a number of layering and dimensioning techniques to achieve success. This chapter continues by demonstrating how to arrange multiple details of different drawing files in the same layout. Also, a new layer technique of freezing layers in certain viewports will be utilized.

ARRANGING DIFFERENT VIEWS OF THE SAME DRAWING

This chapter begins immediately with a tutorial exercise designed to lay out two views of the same drawing in the same layout. Both images inside their respective viewports will be scaled at different values. Follow the next series of steps for creating this type of layout.

TUTORIAL EXERCISE: 19_ROOF PLAN.DWG

Figure 19–1

Purpose
This tutorial exercise is designed to lay out the two architectural views displayed in the previous image in the Layout mode (Paper Space environment). The two views consist of an overall and a detail view of a roof plan.

System Settings
Open an existing drawing called 19_Roof Plan. Keep all default settings for the units and limits of the drawing.

Layers
Layers have already been created for this drawing.

Suggested Commands
A new viewport will be created to house a detail view of the roof plan. Dimensions will also be added to this viewport. However, the Dimension scale setting of the drawing needs to be set to Paper Space scaling. This will enable the scale of the dimensions in the new viewport to match those dimensions in the main roof plan viewport.

STEP 1

Open the drawing file 19_Roof Plan. Create a new viewport, as shown in the following image. This viewport should be long and narrow to accommodate the detail. Notice that when you are creating the viewport, the entire image zooms to the extents of the viewport. This is normal and you will arrange the detail in the next series of steps.

Figure 19–2

STEP 2

Double-click inside the new viewport to make it active. Then pan the area of the roof plan so it appears similar to the following image. When applying a scale to this viewport, it is easier to see the results if the image you are detailing is approximately in the middle of the viewport. If it is not, it may be difficult to locate the area of the detail.

Figure 19–3

STEP 3

Activate the various scales from the Viewports toolbar and pick 1/2″ = 1′-0″. This should increase the size of the detail in this viewport. Use the PAN command to center the long rectangular portion of the roof in the viewport, as shown in the following image.

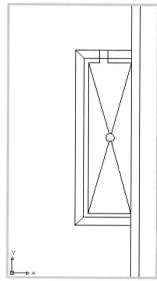

Figure 19-4

STEP 4

Make the Detail Dim layer current. This layer will be used to hold all dimensions that will be added to the detail view. Before adding dimensions, one more item needs to be taken care of. First activate the main floor plan viewport by clicking inside it to make it active. Then launch the Dimension Styles Manager dialog box, click on the Modify button, and then pick the Fit tab. Under the Scale for dimension features area, click the radio button next to Scale dimensions to layout as shown in the following image. When finished, click the OK button to dismiss this dialog box and close the main Dimension Style Manager dialog box. If the dimension scale does not automatically set for you (the dimensions are unreadable), you will have to update all dimensions in this viewport to reflect the changes to this dimension style. This can be accomplished by picking Update from the Dimension pull-down menu and selecting all the dimensions in the viewport.

 Note: You can verify that the scale is correct by performing the LIST command on a dimension and noting that the Dimscale system variable has been overridden and set to 96.

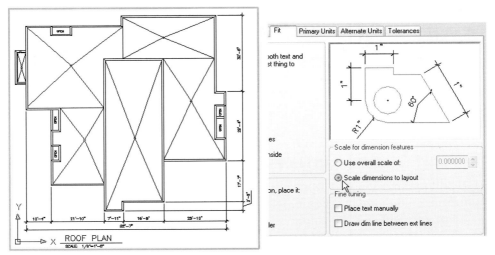

Figure 19–5

STEP 5

Activate the viewport containing the roof detail and place the linear dimensions as shown in the following image. Notice that the settings of these dimensions exactly match those in the main floor plan. The scale of the viewport, which we set earlier to 1/2″ = 1′-0″, sets the scale of the dimensions. This occurs automatically because the Scale dimensions to layout button was selected, as shown in the previous step. Notice that as these dimensions were placed in the detail, they also appear in the main floor plan viewport, as shown in the following image.

Figure 19–6

STEP 6

Activate the main floor plan viewport by clicking inside it. Then launch the Layer Properties Manager dialog box. Identify the Dim Detail layer and freeze it in the current viewport by clicking the button as shown in the following image. This action will freeze the Dim Detail layer in the main floor plan viewport while keeping the layer visible in the detail viewport.

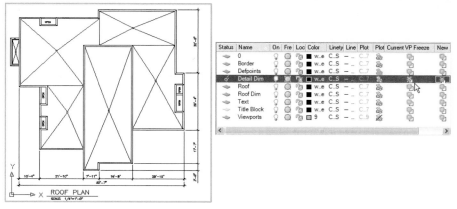

Figure 19–7

STEP 7

The completed layout is shown in the following image. When you are performing a plot preview, the viewports should turn invisible so as not to plot out.

Figure 19–8

CREATING A DETAIL PAGE IN LAYOUT MODE

This next discussion focuses on laying out on the same sheet a series of details composed of different drawings, which can be at different scales. As the viewports are laid out in Paper Space and images of the drawings are displayed in floating Model Space, all images will appear in all viewports. This is not a major problem, because Paper Space allows for layers to be frozen in certain viewports and not in others.

Try It! – Open the drawing file 19_Bearing Details. The following image shows three objects: a body, a bushing, and a bearing all arranged in Model Space. The body and bearing will be laid out at a scale of 1 = 1. The bushing will be laid out at a scale of 2 = 1 (enlarged to twice its normal size). The dimension scales have been set for all objects in order for all dimension text to be displayed at the same size. Also, layers have been created for each object.

Figure 19–9

The following items have already been added to this drawing:

- A Vports layer was created. This layer will hold all viewport information and when the drawing is completed can be frozen or turned off.
- In Layout mode, an ANSI D-size sheet was selected for Page Setup.
- A D-size title block was inserted onto the Title Block layer.

Click on the layout name ANSI-D. Except for the title block, the drawing sheet is empty. Using the Single Viewport tool located in the Viewports toolbar, create three viewports as shown in the following image. The exact size of these viewports is not important at this time since you will have to adjust them at the end of the exercise.

When creating these viewports, notice that all three objects (body, bearing, and bushing) appear in all three viewports. You will now freeze layers inside these viewports in order to show a different part in each viewport.

Figure 19–10

Begin the process of isolating one part per viewport by making the large viewport on the left active by double-clicking inside it. You will notice the viewport taking on the familiar thick border appearance, and the UCS icon is present in the lower left corner of the viewport, as shown on the left in the following image. Then activate the Layer Properties Manager dialog box. All layers that begin with Body need to be visible in this viewport. Pick all layers that begin with Bearing and Bushing and freeze these layers only in this viewport, as shown on the right in the following image.

 Note: To select multiple layers at one time, hold down the CTRL key as you select each layer name.

Figure 19–11

Next, click inside the viewport located in the upper right corner to make it active, as shown on the left in the following image. Activate the Layer Properties Manager dialog box. All layers that begin with Bearing need to be visible in this viewport. Pick all layers that begin with Body and Bushing and freeze these layers only in this viewport, as shown on the right in the following image.

Note: To select all the layers at one time, select the first layer (Body-Center) then hold down the SHIFT key as you select the last layer name (Bushing-Object).

Figure 19–12

Finally, click inside the viewport located in the lower right corner to make it active, as shown on the left in the following image. Activate the Layer Properties Manager dialog box. All layers that begin with Bushing need to be visible in this viewport. Pick all layers that begin with Body and Bearing and freeze these layers only in this viewport, as shown on the right in the following image.

Figure 19–13

When finished, each viewport should contain a different part file. Items dealing with the part body are visible in the large viewport on the left. The bearing views are visible in the upper right viewport and the bushing views are visible in the lower right viewport. One other step is needed to better organize your work. Click in each port and pan each image so it appears centered in each viewport, as shown in the following image. Return to Paper Space by double-clicking outside the viewports.

Figure 19–14

Each image will now be scaled to each viewport. Be sure the Viewports toolbar is visible somewhere on the display screen. Start with the large viewport that contains the body views. Click on the edge of the viewport to highlight it. Then select the scale 1:1 from the Viewports toolbar to scale the body views to this viewport, as shown in the following image.

Figure 19–15

Next, click on the edge of the viewport located in the upper right corner to highlight it. Then select the scale 1:1 from the Viewports toolbar to scale the bearing views to this viewport, as shown in the following image.

Figure 19–16

Next, click on the edge of the viewport located in the lower right corner to highlight it. Then select the scale 2:1 from the Viewports toolbar to scale the bushing views to this viewport, as shown in the following image. The scale of 2:1 will double the size of this view since it is smaller than the others in Model Space.

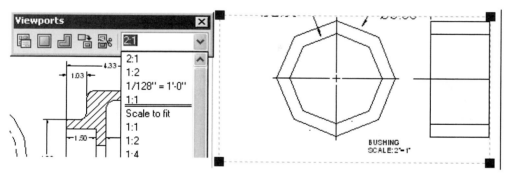

Figure 19–17

Adjust the viewports by clicking on the edges, picking corner grips, and stretching the viewports in order to see all the drawings and dimensions. The dimensions in the lower right corner viewport may appear larger than the others due to the larger scale. Activate this viewport by double-clicking inside it. Then launch the Dimension Style Manager dialog box, click on the Fit tab, and click on the button next to Scale dimensions to layout (paperspace). Then update all dimensions inside this viewport. You will also have to update the dimensions in other viewports in a similar fashion. The final layout, consisting of three different details, some at different scales, is shown in the following image.

Figure 19-18

A good test is to perform a Plot Preview on this drawing. Notice in the following image the absence of any viewports. This is due to the No Plot setting being applied to the VPORTS layer.

Figure 19-19

ADDITIONAL VIEWPORT CREATION METHODS

When constructing viewports, you are not limited to rectangular or square shapes. Although you have been using the Scale drop-down list box in the Viewports toolbar, as shown in the following image, to scale the image inside the viewport, other buttons are available and act on the shape of the viewport. You can clip an existing viewport to reflect a different shape, convert an existing closed object into a viewport, construct a multisided closed or polygonal viewport, or display the Viewports dialog box (you have already constructed a single viewport in a previous exercise).

Figure 19–20

Refer to the following table for a brief description of the extra commands available in the Viewports toolbar.

Button	Command	Description
	VPORTS	Displays the Viewports dialog box
	VPORTS Single	Creates a single viewport
	VPORTS Polygonal	Used for creating a polygonal viewport
	VPORTS Object	Converts existing object into a viewport
	VPCLIP	Used for clipping an existing viewport

Try It! – Open the drawing file 19_Floor Viewports. Notice that Vports is the current layer. You will convert the large rectangular viewport into a polygonal viewport by a clipping operation. First pick on the Clip Existing Viewport button. Pick the rectangular viewport and begin picking points to construct a polygonal viewport around the perimeter of the floor plan dimensions, as shown in the following image. You can turn ORTHO on to assist with this operation. It is not critical that all lines are orthogonal (horizontal or vertical). The following command sequence will also aid with this operation.

Command: **VPCLIP**
Select viewport to clip: *(Select the rectangular viewport)*
Select clipping object or [Polygonal] <Polygonal>: *(Press* ENTER*)*
Specify start point: *(Pick at "A")*
Specify next point or [Arc/Length/Undo]: *(Pick at "B")*
Specify next point or [Arc/Close/Length/Undo]: *(Pick at "C")*
Specify next point or [Arc/Close/Length/Undo]: *(Pick at "D")*
Specify next point or [Arc/Close/Length/Undo]: *(Pick at "E")*
Specify next point or [Arc/Close/Length/Undo]: *(Pick at "F")*
Specify next point or [Arc/Close/Length/Undo]: **C** *(To close the shape)*

Figure 19–21

When finished, move the viewport with the image of the floor plan to the right of the screen. Then construct a circle in the upper left corner of the title block. Pick the Convert Object to Viewport button and select the circle you just constructed. Notice that the circle converts to a viewport with the entire floor plan displayed inside its border. Double-click inside this new viewport to make it current and change the scale of the image to the 1/2″ = 1′-0″ scale using either the Viewports toolbar or the ZOOM command.

 Command: **-VPORTS**
Specify corner of viewport or
[ON/OFF/Fit/Shadeplot/Lock/Object/Polygonal/Restore/2/3/4] <Fit>: **O** *(For Object)*
Specify object to clip viewport: *(Pick the circle)*

Pan inside the circular viewport until the laundry and bathroom appear, as shown in the following image. Adjust the size of the circular viewport with grips if the image is too large or too small. When finished, double-click outside the edge of the viewport to switch to Paper Space.

Figure 19–22

Click on the Polygonal Viewport button and construct a multisided viewport similar to the one located in the following image. Close the shape and do not be concerned that a few lines may be inclined.

Command: **-VPORTS**
Specify corner of viewport or
[ON/OFF/Fit/Shadeplot/Lock/Object/Polygonal/Restore/2/3/4] <Fit>: **P** *(For Polygonal)*
Specify start point: *(Pick at "A")*
Specify next point or [Arc/Length/Undo]: *(Pick at "B")*
Specify next point or [Arc/Close/Length/Undo]: *(Pick at "C")*
Specify next point or [Arc/Close/Length/Undo]: *(Pick at "D")*
Specify next point or [Arc/Close/Length/Undo]: *(Pick at "E")*
Specify next point or [Arc/Close/Length/Undo]: *(Pick at "F")*
Specify next point or [Arc/Close/Length/Undo]: **C** *(To close the shape)*
Regenerating model.

As the image of the floor plan appears in this new viewport, double-click inside the new viewport to make it current. Scale the image inside the viewport to the scale 3/8″ = 1′0″. Pan until the kitchen area and master bathroom are visible. Your display should appear similar to the following image.

Figure 19–23

Double-click outside this viewport to return to Paper Space. Make any final adjustments to viewports using grips. When finished, turn off the Vports layer. Your display should appear similar to the following image.

Figure 19–24

MATCHING THE PROPERTIES OF VIEWPORTS

In chapter 7, the MATCHPROP (Match Properties) command was introduced as a means of transferring all or selected properties from a source object to a series of destination objects. In addition to transferring layer, color, and linetype information, dimension styles, hatch properties, and text styles can also be transferred from one object to another. You can also transfer viewport information from a source viewport to other viewports. Information such as viewport layer, whether the viewport is locked or unlocked, and the viewport scale are a few of the properties to transfer to other viewports. When you enter the MATCHPROP command and select the Settings option, the dialog box in the following image will appear. When transferring viewport properties, be sure the Viewport option of this dialog box is checked.

Figure 19–25

Try It! – Open the drawing file 19_Matchprop Viewports, as shown in the following image. This drawing consists of four viewports holding different object types. The object in the first viewport (labeled "A" in the following image) is already scaled to 1 = 1. This viewport belongs to the Viewport layers; this viewport is also locked to prevent accidental zooming while in floating Model Space. All other viewports do not belong to the Viewports layer. They are all scaled differently. Also, none of these viewports are locked. These three viewports need to have the same properties as the first; namely, all viewports need to belong to the Viewports layer; all viewports need to be locked; and all images inside all viewports need to be scaled to 1 = 1. Rather than perform these operations on each individual viewport, the MATCHPROP command will be used to accomplish this task.

Command: **MA** *(For MATCHPROP)*
Select source object: *(Pick the edge of Viewport "A")*
Current active settings: Color Layer Ltype Ltscale Lineweight Thickness PlotStyle Text Dim Hatch Polyline Viewport Table
Select destination object(s) or [Settings]: *(Pick the edge of Viewport "B")*
Select destination object(s) or [Settings]: *(Pick the edge of Viewport "C")*
Select destination object(s) or [Settings]: *(Pick the edge of Viewport "D")*
Select destination object(s) or [Settings]: *(Press* ENTER *to exit this command)*

Figure 19–26

The results are illustrated in the following image. All viewports share the same layer, are scaled to 1 = 1, and are locked after using the MATCHPROP command.

Figure 19–27

NOTES TO THE STUDENT AND INSTRUCTOR

The Try It! and tutorial exercises in this chapter have been designed around the topic of creating multiple viewports in Paper Space. As with all tutorials, you will tend to follow the steps very closely, taking care not to make a mistake. This is to be expected. However, most individuals rush through the tutorials to get the correct solution only to forget the steps used to complete the tutorial. This is also to be expected.

Of all tutorial exercises in this book, this Paper Space tutorial will probably pose the greatest challenge. Certain operations have to be performed in Paper Space while other operations require floating Model Space. It is recommended to both student and instructor that all tutorial exercises, and especially those tutorials that deal with Paper Space, be performed two or even three times. Completing the tutorial the first time will give you the confidence that it can be done; however, you may not understand all the steps involved. Completing the tutorial a second time will allow you to focus on where certain operations are performed and why things behave the way they do. This still may not be enough. Completing the tutorial exercise a third time will allow you to anticipate each step and have a better idea of which operation to perform in each step.

TUTORIAL EXERCISE: 19_ARCHITECTURAL DETAILS.DWG

Figure 19–28

Purpose

This tutorial exercise is designed to lay out three architectural details, displayed in the previous image, in the Layout mode (Paper Space environment). The three details consist of a floor plan, cornice detail, and foundation detail.

System Settings

Open an existing drawing called 19_Architectural Details. Keep all default settings for the units and limits of the drawing.

Layers

Layers have already been created for this drawing.

Suggested Commands

An existing drawing will be used to create a new layout. This layout will hold three viewports displaying three different architectural details at different scales. The Layout wizard will be used to create the initial layout in paper space. Additional viewports will be added to hold the remaining details. The Viewports toolbar will be used to scale the details relative to their individual viewports. Grips will be used to adjust the viewports. A plot preview will be performed to observe the images floating on the drawing sheet.

STEP 1

Open the drawing 19_Architectural Details.Dwg. Notice the floor plan and the two small details. The floor plan will need to be laid out at a scale of 1/4″ = 1′-0″.

Figure 19–29

STEP 2

Before continuing, zoom in to each of the smaller details. Both are illustrated in the following image. On the left, a Cornice (Wall Detail) is to be laid out at a scale of 3 = 1′-0″. The Foundation detail on the right is to be laid out at a scale of 1-1/2″ = 1′-0″.

Figure 19–30

STEP 3

Activate the Layout wizard by picking in the pull-down menu, as shown on the left in the following image. When the Create Layout dialog box appears, change the name of the layout to Detail Sheet, as shown on the right in the following image. Continue inside this wizard by making the following changes in the additional dialog boxes that will appear:

Figure 19–31

Printer—Select DWF6 ePlot.pc3 device from the list of available printers.

Paper Size—Select ARCH expanded D (36.00 × 24.00 Inches).

Orientation—Verify that Landscape is the current setting in this dialog box.

Title Block—From the available list, pick Architectural Title Block.dwg.

Define Viewports—Verify that Single is set for the Viewport Setup. Then pick 1/4″ = 1′-0″ in the Viewport scale drop-down list box.

Pick Location—Click on the Select Location button. This will return you momentarily to the drawing editor. Pick two points at "A" and "B" as the corners of the viewport, as shown in the following image.

Finish—Click on the Finish button to exit the wizard and return to the drawing editor.

Creating Advanced Layouts 997

The viewport scale is already set; however, you should pan the image as necessary to better center it in the viewport. It may also be necessary to adjust the viewport in Paper Space to crop the details from the view.

Figure 19–32

STEP 4

Create two additional viewports using the Single Viewport button, as shown in the following image. Construct the first viewport from "A" to "B" and the second viewport from "C" to "D." These locations are approximate. Notice that the images of all details appear in both viewports. This actually occurred in the first viewport created by the wizard, although the smaller details may have been cropped out of view.

Figure 19–33

STEP 5

Click on the edge of all three viewports and notice in the Layer Control box the name of the layer. Clicking the down arrow and selecting the Viewports layer will change all three viewports to this layer, as shown in the following image.

Figure 19–34

STEP 6

Double-click inside the upper viewport to make it current. Pan the image of the Cornice approximately to the middle of the top viewport. It is helpful to perform this step on small details that are about to be scaled. Sometimes images get lost in the viewport if this step is not performed. In the Viewports toolbar, change the scale of this image to 3″ = 1′-0″, as shown in the following image.

Figure 19–35

STEP 7

The results of this operation are illustrated in the following image. You may have to pan the image in order to center it better inside of the viewport.

Figure 19–36

STEP 8

An alternate step is to activate the Layer Properties Manager dialog box, select all layers beginning with Floor and Foundation, and freeze these layers in the current viewport. This will ensure that these layers never display in this viewport. However, the images controlled by these layers were cropped out and this step does not need to be performed.

Figure 19–37

STEP 9

Perform the same operations on the bottom viewport in order to scale the foundation detail inside the viewport. Click inside the bottom viewport to make it current. Pan the image of the Foundation detail approximately to the middle of the bottom viewport. In the Viewports toolbar, change the scale of this image to 1-1/2″ = 1′-0″, as shown in the following image.

Figure 19–38

STEP 10

The results of this operation are illustrated in the following image. You may have to pan the image in order to center it better inside the viewport.

Figure 19–39

STEP 11

Return to Layout mode and stretch the corners of the viewpoints as shown in the following image.

Figure 19–40

STEP 12

Perform a plot preview. Your display should appear similar to the following image.

Figure 19–41

CHAPTER 20

Solid Modeling Fundamentals

Solid models are mathematical models of actual objects that can be analyzed through the calculation of such items as mass properties, center of gravity, surface area, moments of inertia, and much more. Before solid models are built, various options of the UCS command need to be understood. The use and creation of visual styles are also discussed. The solid model starts the true design process by defining objects as a series of primitives. Boxes, cones, cylinders, spheres, and wedges are all examples of primitives. These building blocks are then joined together or subtracted from each other through modify commands. Intermediate solid modeling commands such as EXTRUDE, REVOLVE, SWEEP, and LOFT are discussed, as well as the ability to create a spiral or helix object. Fillets and chamfers can be created to give the solid model a more realistic appearance and functionality. Two-dimensional views can be extracted from the solid model along with a cross section of the model.

3D MODEL TYPES

A 3D model can be broken down into various types. The more popular types of 3D models include wireframe, surface, and solid. Each of these is described in greater detail below.

WIREFRAME MODELS

A 3D wireframe model is shown on the left in the following image. This is the simplest type of 3D model and, unlike the 2D isometric drawing, it can be viewed from any direction by simply changing our viewpoint. A wireframe model, however, as lines intersect from the front and back of the object, is sometimes difficult to visualize and interpret.

In 2D drawings the Z value of an XYZ coordinate remains set to zero. In 3D modeling, height is shown by providing values for Z. Values for height can be entered at the command prompt, selected using object snaps, or selected through the use of aids, such as XYZ point filters. In this method, values are temporarily saved for later use inside a command. The values can be retrieved and a coordinate value entered to complete the command. Manipulation of a user-defined coordinate system (UCS) is another especially useful method for providing height when creating a 3D object. A new coordinate system can be defined along any plane on which to place objects.

SURFACE MODELS

Surfacing picks up where the wireframe model leaves off. Because of the complexity of the wireframe and the number of intersecting lines, surfaces are applied to the wireframe. A major advantage of a surfaced model over a wireframe model is the capability of performing a hidden line removal operation. A HIDE is performed to view the model without interference of other faces. The surfaces created on the right in the following image are in the form of opaque objects, called 3D faces. Three-dimensional faces are created by selecting either three or four points in space, as required, to create the surface needed. Surfacing an existing wireframe can be easily accomplished by utilizing OSNAP options to assist in point selection. The 3DFORBIT (3D Free Orbit) command may be used to view the model in different 3D positions to make sure all sides of the model have been surfaced.

Figure 20–1

SOLID MODELS

The creation of a solid model remains the most complete way to represent a model in 3D, as shown in the following image. It is also the most versatile 3D representation of an object. Wireframe models and surfaced models can be analyzed by taking distance readings and the identification of key points of the model. Key orthographic views such as Front, Top, and Right Side views can be extracted from wireframe models. Surfaced models can be imported into shading packages for increased visualization. Solid models do all these operations and more. This is because the solid model, as it is called, is a solid representation of the actual object. From cylinders to spheres, wedges to boxes, all objects that go into the creation of a solid model have volume. This allows a model to be constructed of what are referred to as primitives. These primitives are then merged into one using addition, subtraction, and intersection operations. What remains is the most versatile of 3D drawings. This method of creating models will be the main focus of this chapter.

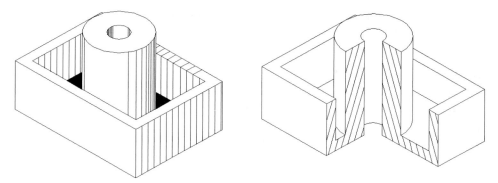

Figure 20–2

SOLID MODELING BASICS

All objects, no matter how simple or complex, are composed of simple geometric shapes, or primitives. The shapes include boxes, wedges, cylinders, cones, spheres, and tori. Solid modeling allows for the creation of these primitives. Once created, the shapes are either merged or subtracted to form the final object. Follow the next series of steps to form the object shown on the left in the following image.

Begin the process of solid modeling by constructing a solid slab that will represent the base of the object. You can do this by using the BOX command. Supply the length, width, and height of the box, and the result is a solid slab, as shown in the middle of the following image.

Next, construct a cylinder using the CYLINDER command. This cylinder will eventually form the curved end of the object. One of the advantages of constructing a solid model is the ability to merge primitives together to form composite solids. Using the UNION command, combine the cylinder and box to form the complete base of the object, as shown on the right in the following image.

Figure 20–3

As the object progresses, create another solid box and move it into position on top of the base. There, use the UNION command to join this new block with the base. As you add new shapes during this process, they all become part of the same solid, as shown on the left in the following image.

Yet another box is created and moved into position. However, instead of being combined with the block, this new box is removed from the solid. Use the SUBTRACT command to perform this subtraction operation and create the step, as shown on the right in the following image.

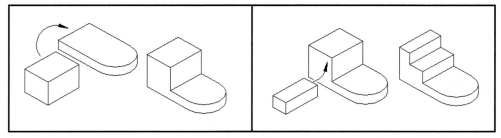

Figure 20–4

You can form holes in a similar fashion. First, use the CYLINDER command to create a cylinder with the diameter and depth of the desired hole. Next, move the cylinder into the solid and then subtract it using the SUBTRACT command. Again, the object shown on the left in the following image represents a composite solid object.

Using existing AutoCAD tools such as User Coordinate Systems (discussed in detail later in this chapter), create another cylinder with the CYLINDER command. It too is moved into position, where it is subtracted through the SUBTRACT command, as shown on the right in the following image.

Figure 20–5

The complete solid model of the object is shown on the left in the following image. Advantages of constructing a solid model out of an object come in many forms.

Two-dimensional profiles of different surfaces of the solid model can be taken to create orthographic views. Section views of solids can automatically be formed and crosshatched, as shown on the right in the following image.

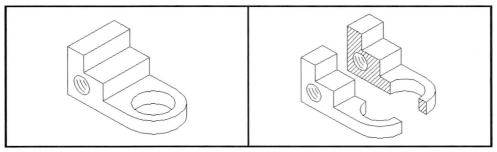

Figure 20–6

A very important analysis tool is the MASSPROP command, which is short for "mass property." Information such as the mass and volume of the solid object is automatically calculated, along with centroids and moments of inertia, components that are used in computer-aided engineering (CAE). See the following image.

```
Mass:     223.2 gm
Volume: 28.39 cu cm (Err: 3.397)
Bounding box:        X:–1.501—6.001 cm
       Y: –0.0009944—2.75 cm
       Z: –3.251—3.251 cm
Centroid:    X: 2.36 cm
       Y: 0.7067 cm
       Z: 0.07384 cm
Moments of inertia:   X: 612.2 gm sq cm (Err: 76.16)
       Y: 2667 gm sq cm (Err: 442.6)
Z: 2444 gm sq cm (Err: 383.2)
```

Figure 20–7

CONSTRUCTING IN THE 3D MODELING WORKSPACE

As a means of allowing you to work in a dedicated custom, task-orientated environment, predefined workspaces are already created in AutoCAD. These workspaces consist of menus, toolbars, and palettes that are organized around a specific task. When you select a workspace, only those menus, toolbars, and palettes that relate directly to the task are displayed.

The workspaces can be found by picking Workspaces from the Tools pull-down menu, as shown in the following image. Of these workspaces, you will find it productive to use 3D Modeling throughout this chapter in the book.

Figure 20–8

THE 3D MODELING WORKSPACE

Clicking on the 3D Modeling workspace in the previous image changes your AutoCAD screen to appear similar to the following image. In this image, your screen contains only 3D-related toolbars, menus, and palettes. One other addition is the dashboard, which is displayed on the right in the following image.

 Note: In addition to the dashboard, the tool palette also displays when you activate the 3D Modeling workspace. The tool palette has been turned off in the following image.

When you make changes to your drawing display (such as moving, hiding, or displaying a toolbar or a tool palette group) and you want to preserve the display settings for future use, you can save the current settings to a workspace.

Figure 20–9

USING THE DASHBOARD

A dashboard is a special palette that consists of buttons and controls used primarily for 3D modeling, 3D navigation, controlling lights, controlling visual styles, creating and applying materials and producing renderings as shown on the left in the following image. The use of a dashboard eliminates the need to display numerous toolbars, which tend to clutter up your screen. This enables you to have more screen real estate for constructing your 3D models. When you activate the 3D Modeling workspace, the dashboard automatically displays. The dashboard can also be displayed in any workspace by clicking the Tools pull-down menu, followed by Palettes and then Dashboard.

Right-clicking on a panel inside the dashboard displays a menu, as shown on the right in the following image. Use this menu to hide a specific panel or turn on panels from a list provided. In addition to the 3D controls, you can also display a panel specifically designed for 2D drawing and modification commands.

Figure 20–10

Notice the large icons displayed at the left side of the dashboard; these are called control panel icons. If you click this icon or on the double arrows, as shown on the left in the following image, a slide-out panel opens that contains additional tools and controls, as shown on the right in the following image. Only one slide-out panel can be displayed at a time.

Figure 20–11

THE UCS COMMAND

Two-dimensional computer-aided design still remains the most popular form of representing drawings for most applications. However, in applications such as architecture and manufacturing, 3D models are becoming increasingly popular for creating rapid prototype models or for creating tool paths from the 3D model. To assist in this creation process, User Coordinate Systems are used to create construction planes where features such as holes and slots are located. In the illustration on the left in the following image, a model of a box is displayed along with the User Coordinate System icon. The appearance of the UCS icon can change depending on the current visual style the model is displayed in. Visual styles are discussed later in this chapter. For now, the UCS icons illustrated on the right can take on a 2D or 3D appearance. Both UCS icon examples show the directions of the three User Coordinate System axes. The icon arrows indicate the current positive coordinate direction.

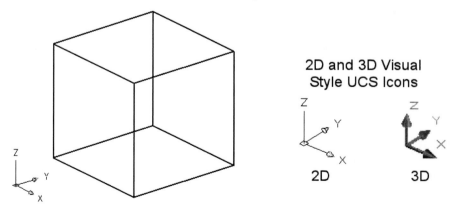

Figure 20–12

Phase 1 of creating a User Coordinate system begins with an understanding of how the UCS command operates. The command line sequence follows, along with all options; the command and options can also be selected from the UCS toolbar, illustrated in the following image.

Command: **UCS**
Current UCS name: *WORLD*
Specify origin of UCS or [Face/NAmed/OBject/Previous/View/World/X/Y/Z/ZAxis] <World>: **N** *(For New)*
Specify origin of new UCS or [ZAxis/3point/OBject/Face/View/X/Y/Z] <0,0,0>:

Figure 20–13

The following table gives a brief description of each User Coordinate System mode.

Button	Tool	Function
	UCS	Activates the UCS command located in the Command Prompt area
	World	Switches to the World Coordinate System from any previously defined User Coordinate System
	UCS Previous	Sets the UCS icon to the previously defined User Coordinate System
	Face UCS	Creates a UCS based on the selected face of a solid object
	Object	Creates a UCS based on an object selected
	View	Creates a UCS parallel to the current screen display
	Origin	Used to specify a new origin point for the current UCS
	Z Axis Vector	Creates a new UCS based on two points that define the Z axis
	3 Point	Creates a new UCS by picking three points
	X	Used for rotating the current UCS along the X axis
	Y	Used for rotating the current UCS along the Y axis
	Z	Used for rotating the current UCS along the Z axis
	Apply	Sets the current User Coordinate System setting to a specific viewport(s)

The following additional options are available when you use the UCS command:

Move—Identifies a new User Coordinate System by shifting origin in the XY plane or changing the Z coordinate.

orthoGraphic—Identifies new User Coordinate System from six preset systems (Top/Bottom/Front/Back/Left/Right).

Restore—Restores a previously saved User Coordinate System.

Save—Saves the position of a User Coordinate System under a unique name given by the CAD operator.

Del—Deletes a User Coordinate System from the database of the current drawing.

?—Lists all previously saved User Coordinate Systems.

THE UCS–SPECIFY ORIGIN OF UCS AND ORIGIN OPTIONS

The default sequence of the UCS command defines a new User Coordinate System by first defining a new origin (translating – picking a new 0,0,0 position) and second, if needed, changing the direction of the X-axis (rotating the coordinate system).

The illustration on the left in the following image is a sample model with the current coordinate system being the World Coordinate System. Follow the illustrations in the middle and on the right in the following image to define a new User Coordinate System using the default command sequence.

Try It! – Open the drawing file 20_Ucs Origin. Activate the UCS command. Identify a new origin point for 0,0,0 at "A," as shown in the middle in the following image. This should move the User Coordinate System icon to the point that you specify. If the icon remains in its previous location and does not move, use the UCSICON command with the Origin option to display the icon at its new origin point. To prove that the corner of the box is now 0,0,0, construct a circle at the bottom of the 5″ cube, as shown on the right in the following image.

 Command: **UCS**
Current UCS name: *WORLD*
Specify origin of UCS or [Face/NAmed/OBject/Previous/View/World/X/Y/Z/ZAxis] <World>: **End**
of (Select the endpoint of the line at "A")
Specify point on X-axis or <Accept>: (Press ENTER to accept since you do not want to rotate the coordinate system)

 Command: **C** (For CIRCLE)
Specify center point for circle or [3P/2P/Ttr (tan tan radius)]: **2.5,2.5,0**
Specify radius of circle or [Diameter]: **2**

Notice a small "plus" at the corner of the User Coordinate System icon (see the illustration on the right in the following image). This indicates that the UCS icon is displayed at the origin (0,0,0). A small box, as shown on the left in the following image, indicates that the icon is displaced at the World Coordinate System (WCS).

It is also important to note that the circle shown on the right in the following image is a 2D object and can only be drawn in or parallel to the XY plane (for example, the bottom or top of the box). To draw a circle in the side of the box, we will need to not only move (translate) our coordinate system but rotate it as well.

Figure 20-14

Another popular method of changing the UCS origin is provided through the Origin option. This option is available through the UCS toolbar and Tools pull-down menu. It allows you select a new 0,0,0 position but does not provide any prompts for rotating the coordinate system. The Origin option is not listed in the command prompts but can be entered as shown below.

 Command: **UCS**
Current UCS name: *WORLD*
Specify origin of UCS or [Face/NAmed/OBject/Previous/View/World/X/Y/Z/ZAxis] <World>: **O** *(For Origin)*
Specify new origin point <0,0,0>: *(Pick desired point)*

THE UCS-3POINT OPTION

Use the 3point option of the ucs command to specify a new User Coordinate System by identifying an origin and new directions of its positive X and Y axes (translate and rotate). This option, like the Origin option, is not listed in the command sequence but can be entered anyway. The option is displayed in the UCS toolbar and Tools pull-down menu (New UCS).

 Try It! – Open the drawing file 20_Ucs 3p. The illustration on the left in the following image shows a 3D cube in the World Coordinate System. To construct objects on the front panel, first define a new User Coordinate System parallel to the front. Use the following command sequence to accomplish this task.

 Command: **UCS**
Current UCS name: *WORLD*
Specify origin of UCS or [Face/NAmed/OBject/Previous/View/World/X/Y/Z/ZAxis] <World>: **3** *(For 3point)*
Specify new origin point <0,0,0>: **End**
of *(Select the endpoint of the model at "A" as shown in the middle of the following image)*
Specify point on positive portion of X-axis <>: **End**
of *(Select the endpoint of the model at "B")*
Specify point on positive-Y portion of the UCS XY plane <>: **End**
of *(Select the endpoint of the model at "C")*

With the Y axis in the vertical position and the X axis in the horizontal position, any type of object can be constructed along this plane, such as the lines shown on the right in the following image.

Figure 20–15

 Try It! – Open the drawing file 20_Incline. The 3point method of defining a new User Coordinate System is quite useful in the example shown in the following image, where a UCS needs to be aligned with the inclined plane. Use the intersection at "A" as the origin of the new UCS, the intersection at "B" as the direction of the positive X axis, and the intersection at "C" as the direction of the positive Y axis.

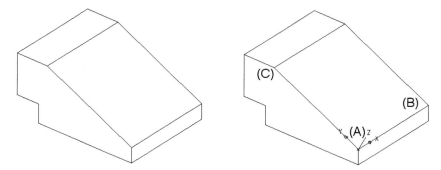

Figure 20–16

THE UCS-X/Y/Z ROTATION OPTIONS

Using the X/Y/Z rotation options will rotate the current user coordinate around the specific axis. Select the X, Y, or Z option to establish the axis that will act as the pivot; a prompt appears asking for the rotation angle about the pivot axis. The right-hand rule is used to determine the positive direction of rotation around an axis. Think of the right hand gripping the pivot axis with the thumb pointing in the positive X, Y, or Z direction. The curling of the fingers on the right hand determines the positive direction of rotation. The following image illustrates positive rotation about each axis. By viewing down the selected axis, the positive rotation is seen to be counterclockwise.

Axis Rotations

Figure 20–17

 Try It! – Open the drawing file 20_Ucs Rotate. Given the cube shown on the left in the following image in the World Coordinate System, the X option of the UCS command will be used to stand the icon straight up by entering a 90° rotation value, as in the following prompt sequence.

Command: **UCS**
Current UCS name: *NO NAME*
Specify origin of UCS or [Face/NAmed/OBject/Previous/View/World/X/Y/Z/ZAxis]
   <World>: **X**
Specify rotation angle about X-axis <90>: *(Press ENTER to accept 90° of rotation)*

The X axis is used as the pivot of rotation; entering a value of 90° rotates the icon the desired degrees in the counterclockwise direction, as shown on the right in the following image.

Figure 20–18

THE UCS-OBJECT OPTION

Another option for defining a new User Coordinate System is to select an object and have the User Coordinate System align to that object (translate and rotate).

 Try It! – Open the drawing file 20_Ucs Object. Given the 3D cube shown on the left in the following image, use the following command sequence and the illustration on the right to accomplish this.

 Command: **UCS**
Current UCS name: *WORLD*
Specify origin of UCS or [Face/NAmed/OBject/Previous/View/World/X/Y/Z/ZAxis] <World>: **OB** *(For Object)*
Select object to align UCS: *(Select the circle shown on the right in the following image)*

The type of object selected determines the alignment (translation and rotation) of the User Coordinate System. In the case of the circle, the center of the circle becomes the origin of the User Coordinate System. Where the circle was selected becomes the point through which the positive X axis aligns. Other types of objects that can be selected include arcs, dimensions, lines, points, plines, solids, traces, 3dfaces, text, and blocks.

Figure 20–19

 THE UCS-FACE OPTION

The Face option of the UCS command allows you to establish a new coordinate system aligned to the selected face of a solid object.

 Try It! – Open the drawing file 20_Ucs Face. Given the current User Coordinate System, as shown on the left in the following image, and a solid box, follow the command sequence below to align the User Coordinate System with the Face option.

Command: **UCS**
Current UCS name: *WORLD*
Specify origin of UCS or [Face/NAmed/OBject/Previous/View/World/X/Y/Z/ZAxis] <World>: **F** *(For Face)*
Select face of solid object: *(Select the edge of the model at point "A," as shown on the right in the following image—select the edge near the corner that you want as the origin)*
Enter an option [Next/Xflip/Yflip] <accept>: *(Press* ENTER *to accept the UCS position)*

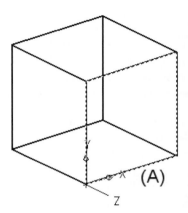

Figure 20–20

THE UCS–VIEW OPTION

The View option of the UCS command allows you to establish a new coordinate system where the XY plane is perpendicular to the current screen viewing direction; in other words, it is parallel to the display screen.

 Try It! – Open the drawing file 20_Ucs View. Given the current User Coordinate System, as shown on the left in the following image, follow the prompts below along with the illustration on the right to align the User Coordinate System with the View option.

 Command: **UCS**
Current UCS name: *WORLD*
Specify origin of UCS or [Face/NAmed/OBject/Previous/View/World/X/Y/Z/ZAxis] <World>: **V** *(For View)*

The results are displayed on the right in the following image, with the User Coordinate System aligned parallel to the display screen.

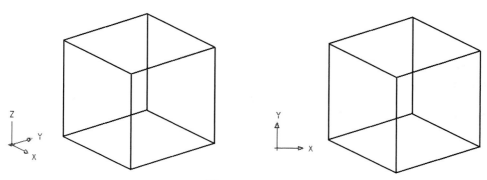

Figure 20–21

USING DYNAMIC UCS MODE

While inside the UCS command, you can automatically switch the plane of the UCS by simply hovering your cursor over the face of a 3D solid object. This special function is available when the DUCS (Dynamic UCS) button is turned on in the status bar, as shown in the following image. The next Try It! exercise illustrates how this method of manipulating the UCS dynamically is accomplished.

Figure 20–22

 Try It! – Open the drawing file 20_Dynamic Ucs. Given the current User Coordinate System, as shown on the left in the following image, follow the prompts below, along with the illustrations, to dynamically align the User Coordinate System to a certain face and location.

In this first example of dynamically setting the UCS, hover your cursor along the front face of the object illustrated on the left in the following image and pick the endpoint at "A" to locate the UCS, as shown on the right in the following image.

Command: **UCS**
Current ucs name: *WORLD*
Specify origin of UCS or [Face/NAmed/OBject/Previous/View/World/X/Y/Z/ZAxis]
 <World>: *(Move the Dynamic UCS icon over the front face, as shown on the left in the following image. Then pick the endpoint at "A")*
Specify point on X-axis or <Accept>: *(Press* ENTER *to accept)*

With the new UCS defined, it is considered good practice to save the position of the UCS under a unique name. These named User Coordinate Systems can then be easily retrieved for later use.

Command: **UCS**
Current ucs name: *NO NAME*
Specify origin of UCS or [Face/NAmed/OBject/Previous/View/World/X/Y/Z/ZAxis]
  <World>: **NA** *(For NAmed)*
Enter an option [Restore/Save/Delete/?]: **S** *(For Save)*
Enter name to save current UCS or [?]: **Front**

Figure 20–23

This next example requires you to pick the endpoint to better define the X-axis while dynamically locating the UCS. Hover your cursor along the top face of the object illustrated on the left in the following image and pick the endpoint at "A" to locate the origin of the UCS. Continue by picking the endpoint at "B" as the X-axis. Save this UCS as "Top."

Command: **UCS**
Current ucs name: Front
Specify origin of UCS or [Face/NAmed/OBject/Previous/View/World/X/Y/Z/ZAxis]
  <World>: *(Move the Dynamic UCS icon over the top face as shown on the left in the following image. Then pick the endpoint at "A")*
Specify point on the X-axis or <Accept>: *(Pick the endpoint at "B" to align the X axis)*
Specify point on the XY plane or <Accept>: *(Press* ENTER *to accept)*

Command: **UCS**
Current ucs name: *NO NAME*
Specify origin of UCS or [Face/NAmed/OBject/Previous/View/World/X/Y/Z/ZAxis]
  <World>: **NA** *(For NAmed)*
Enter an option [Restore/Save/Delete/?]: **S** *(For Save)*
Enter name to save current UCS or [?]: **Top**

Figure 20–24

Next, hover your cursor along the side face of the object illustrated on the left in the following image and pick the endpoint at "A" to locate the origin of the UCS. Continue by picking the endpoint at "B" as the X axis and the endpoint at "C" to define the XY plane. Save this UCS as "Side."

Command: **UCS**
Current ucs name: Top
Specify origin of UCS or [Face/NAmed/OBject/Previous/View/World/X/Y/Z/ZAxis]
    <World>: *(Move the Dynamic UCS icon over the side face as shown on the left in the following image. Then pick the endpoint at "A")*
Specify point on X-axis or <Accept>: *(Pick the endpoint at "B" to align the X axis)*
Specify point on the XY plane or <Accept>: *(Pick the endpoint at "C" to align the XY plane)*

Command: **UCS**
Current ucs name: *NO NAME*
Specify origin of UCS or [Face/NAmed/OBject/Previous/View/World/X/Y/Z/ZAxis]
<World>: **NA** *(For NAmed)*
Enter an option [Restore/Save/Delete/?]: **S** *(For Save)*
Enter name to save current UCS or [?]: **Side**

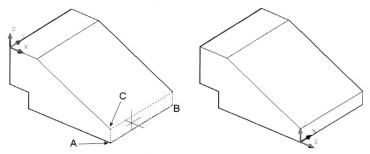

Figure 20–25

Finally, hover your cursor along the inclined face of the object illustrated on the left in the following image and pick the endpoint at "A" to locate the origin of the UCS. Continue by picking the endpoint at "B" as the X axis and the endpoint at "C" to define the XY plane. Save this UCS as "Auxiliary."

Command: **UCS**
Current ucs name: Side
Specify origin of UCS or [Face/NAmed/OBject/Previous/View/World/X/Y/Z/ZAxis]
    <World>: *(Move the Dynamic UCS icon over the inclined face as shown on the left in the following image. Then pick the endpoint at "A")*
Specify point on X-axis or <Accept>: *(Pick the endpoint at "B" to align the X axis)*
Specify point on the XY plane or <Accept>: *(Pick the endpoint at "C" to align the XY plane)*

Command: **UCS**
Current ucs name: *NO NAME*
Specify origin of UCS or [Face/NAmed/OBject/Previous/View/World/X/Y/Z/ZAxis]
    <World>: **NA** *(For NAmed)*
Enter an option [Restore/Save/Delete/?]: **S** *(For Save)*
Enter name to save current UCS or [?]: **Auxiliary**

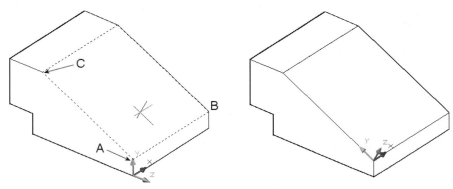

Figure 20–26

USING THE UCS DIALOG BOX

As stated earlier, considerable drawing time can be saved by assigning a name to a User Coordinate System once it has been created. Once numerous User Coordinate Systems have been defined in a drawing, using their names instead of re-creating each coordinate system easily restores them. You can accomplish this on the command line by using the Save and Restore options of the ucs command. A method of retrieving previously saved User Coordinate Systems is to choose Named UCS from the Tools pull-down menu illustrated on the left in the following image (the ucsman command). This displays the UCS dialog box, as shown on the right in the following image, with the Named UCSs tab selected. All User Coordinate Systems previously defined in the drawing are listed here. To make one of these coordinate systems current, highlight the desired UCS name and select the Set Current button, as shown on the right in the following image. A named UCS can also be made current by simply double-clicking it. To define (save) a coordinate system you must have first translated and rotated the UCS into a desired new position; then use the dialog box to select the "Unnamed" UCS and rename it. The UCS dialog box provides a quick method of restoring previously defined coordinate systems without entering them at the keyboard.

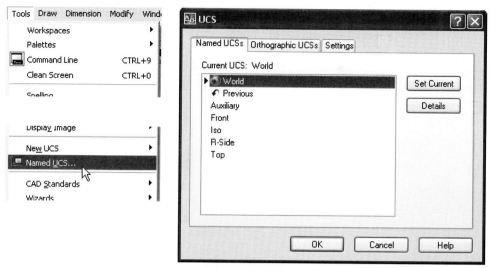

Figure 20–27

Clicking on the Orthographic UCSs tab of the UCS dialog box will allow you to rotate the User Coordinate System so that the XY plane is parallel and oriented to one of the six orthographic views: Front, Top, Back, Right Side, Left Side, and Bottom.

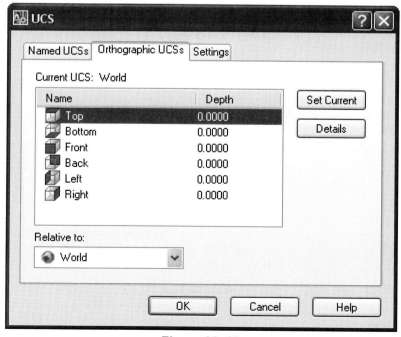

Figure 20–28

THE UCS ICON DIALOG BOX

A dialog box is provided to control the appearance of the User Coordinate System icon. Choosing UCS Icon from the Display cascading menu on the View menu followed by Properties, as shown on the left in the following image, displays the UCS Icon dialog box shown on the right (UCSICON command, Properties option). You can control the icon style, size, and color in this dialog box. Two different controls are provided to change the color of the icon in model space and in layout mode.

Figure 20–29

THE APPEARANCE OF THE 3D COORDINATE SYSTEM ICON

The 3D Coordinate System icon can take on different forms during the drawing process. Viewing the icon in plan view displays the following image at "A." The box at the intersection of the X and Y axes indicates the World Coordinate System (WCS). In the following image at "B," a "+" inside the box indicates that the WCS icon is displayed at 0,0,0. The icon must be totally displayed on the screen; if the icon cannot be totally shown at its origin, the icon locates itself in the lower left corner of the display screen and no "+" is shown.

 Note: The UCSICON command can be used to turn the coordinate system icon on or off, shown in 3D or 2D style, and determine whether it will be displayed at the origin or always in the lower left corner (the Noorigin option).

The icon displayed at "C" represents the Paper Space environment or layout mode. This environment is strictly two-dimensional and is normally used as a layout tool for arranging views and displaying title block information.

A User Coordinate System (UCS) icon (notice that the box is missing) is shown in the following image at "D." This coordinate system has been translated and/or rotated from the World Coordinate System. In the following image at "E," the "+" indicates that the UCS icon is displayed at 0,0,0.

Fig. 20–30

THE PLAN COMMAND

When you design in 3D space, it is sometimes helpful after you change the User Coordinate System to view your model in plan view or parallel to the XY plane. This is the purpose of the PLAN command. For example, illustrated on the left in the following image is a 3D model in which the X and Y axes are positioned along the front face of the object. Activating the PLAN command and accepting the default value of <Current> will display the model as shown on the right in the following image. This gives you a 2D view of the solid model, which can be used to better see how certain features such as holes and slots are located along a 2D plane. To switch back to the 3D view, perform a ZOOM-Previous operation.

Figure 20–31

 Note: A system variable called UCSFOLLOW is available to automate the PLAN command. Here is how it works. When UCSFOLLOW is set to a value of 1 (or turned on), a plan view is automatically generated whenever you change to a new User Coordinate System. Setting UCSFOLLOW to 0 (zero) will turn this mode off.

VIEWING 3D MODELS

Various methods can be used to view a model in 3D. One of the more efficient ways is through the 3DFORBIT (3D Free Orbit) command, which can be selected from the View menu, as shown on the left in the following image. Command options as well as other viewing tools are located in the 3D Navigation toolbar, shown on the right in the following image.

Figure 20–32

The following table gives a brief description of each command found in the 3D Navigation toolbar.

Button	Tool	Function
	3D Pan	Used to pan a 3D model around the display screen
	3D Zoom	Used to perform real-time zooming operations in 3D
	Constrained Orbit	Constrains the 3D orbit along the XY plane or Z axis
	Free Orbit	Used to dynamically rotate a 3D model around the display screen
	Continuous Orbit	Allows you to view an object in a continuous orbit motion
	Swivel	Allows you to view an object with a motion that is similar to looking through a camera viewfinder
	Adjust Distance	Allows you to view an object closer or farther away
	Walk	Changes the view of a 3D display so that you appear to be walking through the model
	Fly	Allows you to fly through a 3D model
	Walk and Fly Settings	Allows you to change settings used for producing an animation of a 3D Walk or 3D Fly

VIEWING WITH FREE ORBIT

Choosing the Free Orbit button (3DFORBIT) from the 3D Navigation toolbar displays your model, similar to the following image. Use the large circle to guide your model through a series of dynamic rotation maneuvers. For instance, moving your cursor outside the large circle at "A" allows you to dynamically rotate (drag) your model only in a circular direction. Moving your cursor to either of the circle quadrant identifiers at "B" and "C" allows you to rotate your model horizontally. Moving your cursor to either of the circle quadrant identifiers at "D" or "E" allows you to rotate your model vertically. Moving your cursor inside the large circle at "F" allows you to dynamically rotate your model to any viewing position.

Figure 20-33

VIEWING WITH CONSTRAINED ORBIT

Another way to rotate a 3D model is through the Constrained Orbit button (3DORBIT), which can be selected from the 3D Navigation toolbar. However unlike the FREE ORBIT command, performing a constrained orbit prevents you from rolling the 3D model completely over, adding to confusion in interpreting the model. It is easy to orbit around the geometry; however, as you begin to attempt to orbit above or below the 3D model, the orbiting stops when you reach the top or bottom.

Note: A quick and efficient way of activating the Constrained Orbit tool is to press and hold down the SHIFT key while pressing on the middle button or wheel of the mouse.

VIEWING WITH CONTINUOUS ORBIT

Performing a continuous orbit (the 3DCORBIT command) rotates your 3D model continuously. After entering the command, press and drag your cursor in the direction you want the continuous orbit to move. Then, when you release the mouse button, the 3D model continues to rotate in that direction.

ADDITIONAL OPTIONS OF THE ORBIT COMMANDS

Right-click while in the one of the Orbit commands to gain access to the numerous options, such as Pan, Zoom, and Shading Modes, provided by this command, as shown in the following image.

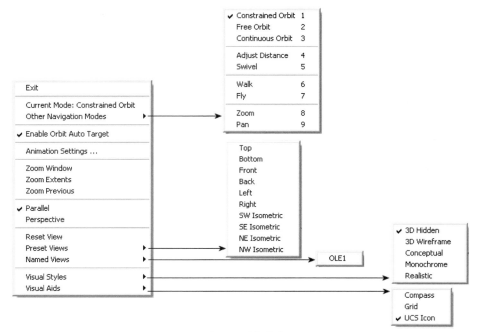

Figure 20–34

THE VIEW TOOLBAR

A View toolbar is available, as shown in the following image. The View toolbar has the extra advantage of displaying icons that guide you in picking the desired viewpoint. You can also access most 3D viewing modes through the View pull-down menu by clicking on 3D Views, also shown in the following image.

Solid Modeling Fundamentals 1029

Figure 20–35

The following table gives a brief description of each View mode.

Button	Tool	Function
	Named Views	Launches the View Manager dialog box used for creating named views
	Top View	Orientates a 3D model to display the top view
	Bottom View	Orientates a 3D model to display the bottom view
	Left View	Orientates a 3D model to display the left view
	Right View	Orientates a 3D model to display the right view
	Front View	Orientates a 3D model to display the front view
	Back View	Orientates a 3D model to display the back view
	SW Isometric	Orientates a 3D model to display the southwest isometric view
	SE Isometric	Orientates a 3D model to display the southeast isometric view
	NE Isometric	Orientates a 3D model to display the northeast isometric view
	NW Isometric	Orientates a 3D model to display the northwest isometric view
	Create Camera	Used to set up a point from which to view a 3D model and the point that you are viewing

VISUALIZING SOLID MODELS

Various shading modes are available to help you better visualize the solid model you are constructing. Access the five shading modes by choosing Visual Styles from the View pull-down menu, as shown on the left in the following image, or by clicking a button on the Visual Styles toolbar. An area is also available in the dashboard for working with visual styles, as shown on the right in the following image. Each of these modes is explained in the next series of paragraphs.

Figure 20–36

The following table gives a brief description of each visual style mode.

Options of the SHADEMODE Command

Button	Visual Style	Description
	2D Wireframe	Displays the 3D model as a series of lines and arcs that represent boundaries
	3D Wireframe	Displays the 3D model that is similar in appearance to the 2D option; the UCS icon appears as a color shaded image
	Hidden	Displays the 3D model with hidden edges removed
	Realistic	Shades the objects and smooths the edges between polygon faces; if materials are attached to the model, they will display when this visual style is chosen
	Conceptual	This mode also shades the objects and smooths the edges between polygon faces; however, the shading transitions between cool and warm colors
	Manage Visual Styles	Launches the Visual Styles Manager palette, used for creating new visual styles and applying existing visual styles to a 3D model

2D WIREFRAME

The 2D wireframe visual style mode, illustrated in the following image, displays a wireframe image of a solid model. Note that the User Coordinate System icon is displayed in the normal 3D or 2D style as determined by the property settings of the UCSICON command.

Figure 20–37

3D WIREFRAME VISUAL STYLE

The 3D wireframe mode is identical to the 2D mode regarding the wireframe mode. Notice, however, that the User Coordinate System icon in the following image has switched to a better-defined, color-coded icon that illustrates the directions of the X, Y, and Z axes.

Figure 20–38

HIDDEN VISUAL STYLE

The Hidden visual style mode performs a hidden line removal, which is illustrated by the model in the following image. Only those surfaces in front of your viewing plane are visible.

Figure 20–39

REALISTIC VISUAL STYLE

The Realistic visual style shades the objects and smooths the edges between polygon faces, as shown in the following image. If materials are attached to the model, they will display when this visual style is chosen.

Figure 20–40

CONCEPTUAL VISUAL STYLE

The Conceptual mode also shades the objects and smooths the edges between polygon faces, as does the Realistic visual style mode. However, the shading is transitioned between cool and warm colors, as shown in the following image.

Figure 20-41

Try It! – Open the drawing file 20_Visual Styles. Experiment with the series of shading modes. Your results should be similar to the images provided in the previous illustrations.

Note: When you perform such operations as Free, Constrained, and Continuous Orbit, the current visual style mode remains persistent. This means that if you are in a Realistic visual style and you rotate your model using one of the previously mentioned operations, the model remains shaded throughout the rotation operation.

CREATING A VISUAL STYLE

Custom visual styles can be created to better define how a 3D model will appear when shaded. Click on the Manage Visual Styles button located in the Visual Styles toolbar, as shown on the left in the following image, to launch the Visual Styles Manager palette, as shown in the middle of the following illustration. The five visual styles that are located in the Visual Styles toolbar can also be found in this palette. Use the palette for creating new visual styles and applying them to the current viewport. A visual style area is also located in the dashboard, as shown on the right in the following image. The dashboard consists of various buttons that can be turned on or off that affect the visual style. Slider bars are also available to experiment with changing the values of certain settings that will automatically be reflected in the 3D solid model.

Figure 20–42

The current visual style will always be indicated by a yellow border, as shown on the left in the following image. Also, notice the presence of the AutoCAD product icon, located in the lower right corner of each visual style box. This indicates a visual style that shipped with the software and cannot be deleted. Also, various settings associated with this visual style, such as Face Settings and Materials and Color, are displayed in the panel settings area below. You can also access all visual styles from the dashboard, as shown on the right in the following image.

Figure 20–43

Four buttons are displayed in the tool strip of the Visual Style Manager palette. The following is a brief description of each button, which is designed to perform the most frequently used option for dealing with visual styles.

Button	Tool	Description
	Create	Creates a new visual style
	Apply	Applies a selected visual style to the current viewport
	Export	Exports a selected visual style to a tool palette
	Delete	Deletes a visual style

Right-clicking on an existing visual style will display the menu shown in the following image. This is another way of performing operations that deal with visual styles. In addition to creating, applying, exporting, and deleting operations, you can also control the size of the icons displayed in the palette.

Figure 20–44

VISUAL STYLE SETTINGS

The heart of a visual style is in the various settings that can be made to achieve a desired effect. A few groups of these settings are displayed in the following image. Depending on the current visual style, only certain settings appear. For instance, if the current visual style is 3D Wireframe, as shown on the left in the following image, only five visual style settings are available. However, if the current visual style is 3D

Hidden, seven visual style settings are available. The following table gives a brief description of each visual style setting. Also, a tutorial on creating a new visual style and manipulating a few settings is available at the end of the chapter.

Visual Style Setting	Description
Face Settings	Controls the appearance of faces in a viewport
Materials and Color	Controls the display of materials and color on faces
Environment Settings	Controls the appearance of shadows and the ability to display a background
Edge Settings	Controls how edges are displayed
Edge Modifiers	Controls settings that apply to all edge modes
Fast Silhouette Edges	Controls settings that apply to silhouette edges; silhouette edges are not displayed on wireframe or transparent objects
Obscured Edges	Controls settings that apply to obscured edges when edge mode is set to Facet Edges
Intersecting edges	Controls settings that apply to intersection edges when edge mode is set to Facet Edges

Figure 20–45

CONTROLLING VISUAL STYLE FROM THE DASHBOARD

The dashboard holds a number of buttons that allow you to experiment with visual styles. The next series of images showcase what effects the Edge Overhang, Edge Jitter, Silhouette Edges, and Obscured Edges setting have on a 3D model. The area of the dashboard that holds these setting is illustrated on the right in the following image.

Figure 20–46

Edge Overhang

The Edge Overhand setting makes lines extend beyond their intersection points, as shown on the left in the following image. This gives your model a hand-drawn effect. Clicking on the button located in the dashboard, as shown on the right in the following image, turns this setting on or off. Also, you can move the slider to the right or left and observe the overhang increasing or decreasing in the 3D model.

Figure 20–47

Edge Jitter

The Edge Jitter setting makes all edges of a 3D model appear as though they were hand sketched, as shown on the left in the following image. Clicking on the button located in the dashboard, as shown on the right in the following image, turns this setting on or off. Also, you can move the slider to the right or left and observe the amount of jitter applied to the 3D model.

Figure 20–48

Silhouette Edges

The Silhouette Edges setting displays silhouette edges, as shown on the left in the following image. This mode is not available when displaying wireframe or transparent. Clicking on the button located in the dashboard, as shown on the right in the following image, turns this setting on or off. Also, you can move the slider to the right or left and observe the silhouette edges getting thicker or thinner on the 3D model.

Figure 20–49

Obscured Edges

The Obscured Edges setting displays edges that are normally hidden whenever shading a 3D model, as shown on the left in the following image. Clicking on the button located in the dashboard, as shown on the right in the following image, turns this setting on or off.

Figure 20–50

SOLID MODELING COMMANDS

The following image illustrates the pull-down menu and the toolbar for accessing solid modeling commands. Choosing Modeling from the Draw pull-down menu displays six groupings of commands. The first grouping displays POLYSOLID, BOX, SPHERE, CYLINDER, CONE, WEDGE, TORUS, and PYRAMID, which are considered the building blocks of the solid model and are used to construct basic "primitives." The second grouping displays the PLANESURF command, designed to create a planar surface between two closed shapes or through two diagonal points. The third grouping displays the "sweep" commands—EXTRUDE, REVOLVE, SWEEP, and LOFT—which provide an additional way to construct solid models. Polyline outlines or circles can be extruded (swept) to a thickness that you designate. You can also revolve (sweep) other polyline outlines about an axis. The fourth grouping of solid modeling commands enables you to create a section plane in order to look at the interior of a solid model. The fifth grouping displays a set of commands that create or manipulate mesh surfaces. Commands include SOLID, 3DFACE, 3DMESH, EDGE, REVSURF, TABSURF, RULESURF, and EDGESURF. The last grouping, Setup, displays three commands designed to extract orthographic views from a solid model. The three commands are SOLDRAW, SOLVIEW, and SOLPROF. Many of these same solid modeling commands can be accessed from the Modeling toolbar and Dashboard, as shown in the following image.

Figure 20–51

The following table gives a brief description of each command located in the Modeling toolbar.

Button	Tool	Shortcut	Function
	Polysolid	POLYSOLID	Creates a solid shape based on a direction, width and height of the solid
	Box	BOX	Creates a solid box
	Wedge	WEDGE	Creates a solid wedge
	Cone	CONE	Creates a solid cone
	Sphere	SPHERE	Creates a solid sphere
	Cylinder	CYLINDER	Creates a solid cylinder
	Torus	TOR	Creates a solid torus
	Pyramid	PYRAMID	Creates a solid pyramid
	Helix	HELIX	Creates a 2D or 3D helix
	Planar Surface	PLANESURF	Creates a planar surface
	Extrude	EXT	Creates a solid by extruding a 2D profile
	Presspull	PRESSPULL	Presses or pulls closed areas resulting in a solid shape or a void in a solid
	Sweep	SWEEP	Creates a solid based on a profile and a path
	Revolve	REV	Creates a solid by revolving a 2D profile about an axis of rotation
	Loft	LOFT	Creates a lofted solid based on a series of cross-section shapes
	Union	UNI	Joins two or more solids together
	Subtraction	SU	Removes one or more solids from a source solid shape
	Intersect	INT	Extracts the common volume shared by two or more solid shapes
	3D Move	3DMOVE	Moves objects a specified distance in a specified direction based on the position of a move grip tool
	3D Rotate	3DROTATE	Revolves objects around a base point based on the rotate grip tool
	3D Align	3DALIGN	Aligns objects with other objects in 2D and 3D

CREATING A BOX

Use the BOX command to create a 3D solid box. Choose this command in one of the following ways:

- From the Modeling Toolbar
- From the pull-down menu (Draw > Modeling > Box)
- From the keyboard (BOX)

Methods for creating this type of solid include selecting 3D diagonal corners, selecting diagonal corners in the XY plane and then specifying the height, or simply entering values for its length, width, and height. A Cube option simplifies construction if all three dimensions are the same.

Try It! – Open the drawing file 20_Box. Use the following image and command sequences to show examples of creating solid boxes.

Figure 20–52

 Command: **BOX**

Specify first corner or [Center] <0,0,0>: **4.00,7.00** *(At "A")*
Specify other corner or [Cube/Length]: **@2.00,2.00,1.00** *(At "B")*

Command: **BOX**

Specify first corner or [Center] <0,0,0>: **-1.50,6.50** *(At "C")*
Specify other corner or [Cube/Length]: **@3.50,2.50** *(At "D")*
Specify height: **1**

Note: Turn off Dynamic Input (DYN button on the Status Bar) to ensure your image looks like the one shown above. If it is on, you will have to position the cursor to determine the direction of the box.

 Command: **BOX**

Specify first corner or [Center] <0,0,0>: **2.00,2.00** *(At "E")*
Specify other corner or [Cube/Length]: **L** *(To define the length of the box)*
Specify length: **5** *(In X direction)*
Specify width: **2** *(In Y direction)*
Specify height: **1** *(In Z direction)*

CREATING A WEDGE

The WEDGE command is very similar to the BOX command. Choose this command in one of the following ways:

- From the Modeling Toolbar
- From the pull-down menu (Draw > Modeling > Wedge)
- From the keyboard (WEDGE or WE)

This primitive is simply a box that has been diagonally cut. Use the WEDGE command to create a wedge with the base parallel to the current User Coordinate System and the slope of the wedge constructed along the X axis.

 Try It! – Open the drawing file 20_Wedge. Use the following prompts and image as examples of how to use this command.

 Command: **WE** *(For WEDGE)*

Specify first corner or [Center]: **5.00,5.00** *(At "A")*
Specify other corner or [Cube/Length]: **@6.5,2.75** *(At "B")*
Specify height or [2Point] <4.0000>: **3**

Note: Turn off Dynamic Input (DYN button on the Status Bar) to ensure your image looks like the one shown above. If it is on, you will have to position the cursor to determine the direction of the wedge.

Command: **WE** (For WEDGE)
Specify first corner or [Center]: **8.50,-1.50** (At "C")
Specify other corner or [Cube/Length]: **L** (For Length)
Specify length: **5** (In X direction)
Specify width: **1** (In Y direction)
Specify height: **3** (In Z direction)

Figure 20–53

CREATING A CYLINDER

The CYLINDER command is similar to extruding a circle or ellipse without a taper. The central axis of the cylinder is along the Z axis of the current User Coordinate System. Choose this command in one of the following ways:

- From the Modeling Toolbar
- From the pull-down menu (Draw > Modeling > Cylinder)
- From the keyboard (CYLINDER)

 Try It! – Open the drawing file 20_Cylinder. Use this command and the following image to construct a cylinder with either a circular or an elliptical base.

Figure 20-54

 Command: **CYLINDER**
Specify center point of base or [3P/2P/Ttr/Elliptical]: **7.00,7.00** *(At "A")*
Specify base radius or [Diameter]: **D** *(For Diameter)*
Specify diameter: **2**
Specify height or [2Point/Axis endpoint] <3.0000>: **4**

 Command: **CYLINDER**
Specify center point of base or [3P/2P/Ttr/Elliptical]: **E** *(For Elliptical)*
Specify endpoint of first axis or [Center]: **C** *(For Center)*
Specify center point: **8.00,0.50** *(At "B")*
Specify distance to first axis <1.0000>: **@2<0**
Specify endpoint of second axis: **@1<90**
Specify height or [2Point/Axis endpoint] <4.0000>: **4**

CREATING A CONE

A solid cone can be created by first defining a circular or elliptical base. This base tapers to a point perpendicular to the base. Choose this command in one of the following ways:

- From the Modeling Toolbar
- From the pull-down menu (Draw > Modeling > Cone)
- From the keyboard (CONE)

 Try It! – Open the drawing file 20_Cone. Use the CONE command and the following image to construct a cone of specified height with the base of the cone either circular or elliptical.

Figure 20–55

 Command: **CONE**
Specify center point of base or [3P/2P/Ttr/Elliptical]: **5.00,5.00** *(At "A")*
Specify base radius or [Diameter]: **1**
Specify height or [2Point/Axis endpoint/Top radius]: **4**

 Command: **CONE**
Specify center point of base or [3P/2P/Ttr/Elliptical]: **E** *(For Elliptical)*
Specify endpoint of first axis or [Center]: **C** *(For Center)*
Specify center point: **6.5,-.50**
Specify distance to first axis: **@2<0**
Specify endpoint of second axis: **@1<90**
Specify height or [2Point/Axis endpoint/Top radius]: **4**

CREATING A SPHERE

Use the SPHERE command to construct a sphere by defining the center of the sphere along with a radius or diameter. As with the cylinder, the central axis of a sphere is along the Z axis of the current User Coordinate System. Choose this command in one of the following ways:

- From the Modeling Toolbar
- From the pull-down menu (Draw > Modeling > Sphere)
- From the keyboard (SPHERE)

In the illustration on the left in the following image, the sphere is constructed in wireframe mode and does not look like much of a sphere. Perform a hidden line removal using the HIDE command to get a better view of the sphere.

An application of using a sphere is illustrated on the right in the following image, in which the sphere is used to construct a 45° elbow pipe fitting.

Figure 20–56

 Try It! – Open the drawing file 20_Sphere. Use the following command sequence for constructing a sphere.

 Command: **SPHERE**
Specify center point or [3P/2P/Ttr]: **7.00,7.00**
Specify radius or [Diameter] <1.0000>: D *(For Diameter)*
Specify diameter <2.0000>: **4**

CREATING A TORUS

A torus is formed when a circle is revolved about a line in the same plane as the circle. In other words, a torus is similar to a 3D donut. Choose this command in one of the following ways:

- From the Modeling Toolbar
- From the pull-down menu (Draw > Modeling > Torus)
- From the keyboard (TORUS or TOR)

The torus can be constructed by either the radius or diameter method. When you use the radius method, two radius values must be used to define the torus; one for the radius of the tube and the other for the distance from the center of the torus to the center of the tube. Use two diameter values when specifying a torus by the diameter method. Once you construct the torus, it lies parallel to the current User Coordinate System.

Try It! – Open the drawing file 20_Torus. Use the prompts below to construct a torus similar to the following image.

Figure 20–57

 Command: **TOR** *(For TORUS)*
Specify center point or [3P/2P/Ttr]: **10.00,10.00**
Specify radius or [Diameter] <2.0000>: **5** *(For the radius of the torus at "A")*
Specify tube radius or [2Point/Diameter]: **1** *(For the radius of the tube at "B")*

 Command: **TOR** *(For TORUS)*
Specify center point or [3P/2P/Ttr]: **20.00,12.00** *(Identify the center of the torus through coordinate entry or by picking)*
Specify radius or [Diameter] <5.0000>: **D** *(For the diameter of the torus at "C")*
Specify diameter of torus <10.0000>: **4**
Specify tube radius or [2Point/Diameter] <1.0000>: **D** *(For the diameter of the tube at "D")*
Specify tube diameter <2.0000>: **2**

 Try It! – Open the drawing file 20_O Ring. A half section is illustrated on the left in the following image. Using this figure and the following command prompt sequence, create two O-rings, illustrated on the right using the TORUS command.

Command: **TOR** *(For TORUS)*
Specify center point or [3P/2P/Ttr]: **0,0,0**
Specify radius or [Diameter] <2.0000>: **D** *(For Diameter)*
Specify diameter of torus <4.0000>: **3**
Specify tube radius or [2Point/Diameter] <1.0000>: **D** *(For Diameter)*
Specify tube diameter <2.0000>: **0.125**

 Command: **CP** *(For COPY)*
Select objects: *(Pick the O-ring you just constructed)*
Select objects: *(Press ENTER to continue)*
Specify base point or [Displacement] <Displacement>: **End**
of *(Pick the endpoint at "A")*
Specify second point or <use first point as displacement>: **End**
of *(Pick the endpoint at "B")*

Specify second point or [Exit/Undo] <Exit>: *(Press ENTER to exit)*

Figure 20–58

CREATING A PYRAMID

A solid pyramid can be created through the PYRAMID command by defining the number of sides, specifying a center point for the base, specifying the base radius, and specifying the height, as shown on the left in the following image.

 Try It! – Open the drawing file 20_Pyramid. Use the following prompts and image as examples for how to use this command.

Command: PYRAMID
4 sides Circumscribed
Specify center point of base or [Edge/Sides]: **10.00,10.00**
Specify base radius or [Inscribed] <9.1404>: **4**
Specify height or [2Point/Axis endpoint/Top radius] <5.0000>: **5**

In addition to a base radius, you can also specify a top radius for the pyramid. This is commonly referred to as a pyramid frustrum and is illustrated on the right in the following image. Follow the prompts below for creating this solid.

 Command: PYRAMID
4 sides Circumscribed
Specify center point of base or [Edge/Sides]: **0,0,0**
Specify base radius or [Inscribed]: **4**
Specify height or [2Point/Axis endpoint/Top radius]: **T** *(For Top radius)*
Specify top radius <0.0000>: **1**
Specify height or [2Point/Axis endpoint]: **5**

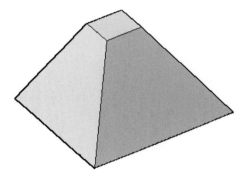

Figure 20–59

USING BOOLEAN OPERATIONS ON SOLID PRIMITIVES

To combine one or more primitives to form a composite solid, a Boolean operation is performed. Boolean operations must act on at least a pair of primitives, regions, or solids. These operations in the form of commands are located in the Modify menu under Solids Editing. They can also be obtained through the Modeling or Solids Editing toolbars. Boolean operations allow you to add two or more objects together, subtract a single object or group of objects from another, or find the overlapping volume—in other words, form the solid common to both primitives. Displayed in the following image are the UNION, SUBTRACT, and INTERSECT commands that you use to perform these Boolean operations.

Figure 20–60

In the following image, a cylinder has been constructed along with a box. Depending on which Boolean operation you use, the results could be quite different. In the following image at "Union," both the box and cylinder are considered one solid object. This is the purpose of the UNION command: to join or unite two solid primitives into one. The image at "Subtract" goes on to show the result of removing or subtracting the cylinder from the box—a hole is formed inside the box as a result of using the SUBTRACT command. The image at "Intersect" illustrates the intersection of the two solid primitives or the area that both solids have in common. This solid is obtained through the INTERSECT command. All Boolean operation commands can work on numerous solid primitives; that is, if you want to subtract numerous cylinders from a box, you can subtract all of the cylinders at one time.

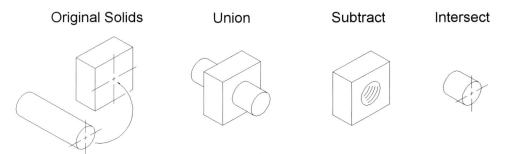

Figure 20–61

CREATING SOLID UNIONS

This Boolean operation, the UNION command, joins two or more selected solid objects together into a single solid object. Choose the UNION command in one of the following ways:

- From the Solids Editing or Modeling toolbars
- From the pull-down menu (Modify > Solids Editing > Union)
- From the keyboard (UNI or UNION)

 Try It! – Open the drawing file 20_Union. Use the following command sequence and image for performing this task.

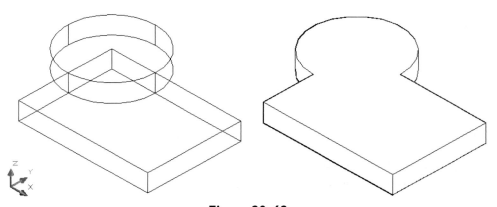

Figure 20–62

Command: **UNI** *(For UNION)*
Select objects: *(Pick the box and cylinder)*
Select objects: *(Press ENTER to perform the union operation)*

SUBTRACTING SOLIDS

Use the SUBTRACT command to subtract one or more solid objects from a source object, as shown in the following image. Choose this command in one of the following ways:

- From the Solids Editing or Modeling toolbars
- From the pull-down menu (Modify > Solids Editing > Subtract)
- From the keyboard (SU or SUBTRACT)

Try It! – Open the drawing file 20_Subtract. Use the following command sequence and image for performing this task.

Figure 20–63

 Command: **SU** *(For SUBTRACT)*
Select solids and regions to subtract from...
Select objects: *(Pick the box)*
Select objects: *(Press ENTER to continue with this command)*
Select solids and regions to subtract...
Select objects: *(Pick the cylinder)*
Select objects: *(Press ENTER to perform the subtraction operation)*

CREATING INTERSECTIONS

Use the INTERSECT command to find the solid common to a group of selected solid objects, as shown in the following image. Choose this command in one of the following ways:

- From the Solids Editing or Modeling toolbars
- From the pull-down menu (Modify > Solids Editing > Intersect)
- From the keyboard (IN or INTERSECT)

 Try It! – Open the drawing file 20_Intersection. Use the following command sequence and image for performing this task.

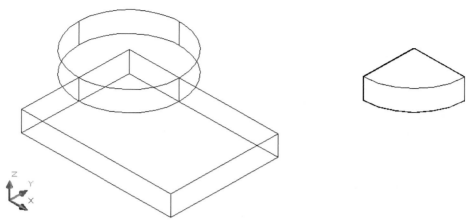

Figure 20-64

Command: **IN** (For INTERSECT)
Select objects: *(Pick the box and cylinder)*
Select objects: *(Press ENTER to perform the intersection operation)*

3D APPLICATIONS OF UNIONING SOLIDS

The following image shows an object consisting of one horizontal solid box, two vertical solid boxes, and two extruded semicircular shapes. All primitives have been positioned with the MOVE command. To join all solid primitives into one solid object, use the UNION command. The order of selection of these solids for this command is not important.

 Try It! – Open the drawing file 20_3D App Union. Use the following prompts and image for performing these tasks.

Figure 20-65

Command: **UNI** *(For UNION)*
Select objects: *(Select the solid extrusion at "A")*
Select objects: *(Select the vertical solid box at "B")*
Select objects: *(Select the horizontal solid box at "C")*
Select objects: *(Select the vertical solid box at "D")*
Select objects: *(Select the solid extrusion at "E")*
Select objects: *(Press ENTER to perform the union operation)*

3D APPLICATIONS OF MOVING SOLIDS

Using the same problem from the previous example, let us now add a hole in the center of the base. The cylinder will be created through the CYLINDER command. It will then be moved to the exact center of the base. You can use the MOVE command along with the OSNAP-Tracking mode to accomplish this. Tracking mode will automatically activate the ORTHO mode when it is in use. See the following image.

 Command: **CYLINDER**
Specify center point of base or [3P/2P/Ttr/Elliptical]: **3.00,3.00**
Specify base radius or [Diameter]: **0.75**
Specify height or [2Point/Axis endpoint]: **0.25**

Turn off dynamic input (DYN) on the status bar to ensure that Tracking will function properly in the next step.

 Command: **M** *(For MOVE)*
Select objects: *(Select the cylinder at "A")*
Select objects: *(Press ENTER to continue with this command)*
Specify base point or [Displacement] <Displacement>: **Cen**
of *(Select the bottom of the cylinder at "A")*
Specify second point or <use first point as displacement>: **TK** *(To activate tracking mode)*
First tracking point: **Mid**
of *(Select the midpoint of the bottom of the base at "B")*
Next point (Press ENTER to end tracking): **Mid**
of *(Select the midpoint of the bottom of the base at "C")*
Next point (Press ENTER to end tracking): *(Press ENTER to end tracking and perform the move operation)*

Figure 20–66

3D APPLICATIONS OF SUBTRACTING SOLIDS

Now that the solid cylinder is in position, use the SUBTRACT command to remove the cylinder from the base of the main solid and create a hole in the base, as shown in the following image.

Figure 20–67

Command: **SU** *(For SUBTRACT)*
Select solids and regions to subtract from...
Select objects: *(Select the main solid as source at "A")*
Select objects: *(Press ENTER to continue with this command)*
Select solids and regions to subtract...
Select objects: *(Select the cylinder at "B")*
Select objects: *(Press ENTER to perform the subtraction operation)*

Command: **HIDE** *(For hidden line removal view)*

Command: **RE** *(For REGEN to return to wireframe view)*

UNDERSTANDING INTERSECTIONS

Try It! – Open the drawing file 20_Int1. One of the most misunderstood yet powerful Boolean operations is the Intersection. In the following image, the object at "C" represents the finished model. Look at the sequence beginning at "A" to see how to prepare the solid primitives for an Intersection operation. Two separate 3D Solid objects are created at "A." One object represents a block that has been filleted along with the placement of a hole drilled through. The other object represents the U-shaped extrusion. With both objects modeled, they are moved on top of each other at "B." The OSNAP-Midpoint was used to accomplish this. Finally the INTERSECT command is used to find the common volume shared by both objects; namely the illustration at "C."

Figure 20–68

Try It! – Open the drawing file 20_Int2. The object in the following image is another example of how the INTERSECT command may be applied to a solid model. For the results at "B" to be obtained of a cylinder that has numerous cuts, the cylinder is first created as a separate model. Then the cuts are made in another model at "A." Again, both models are moved together (use the Quadrant and Midpoint OSNAP modes), and then the INTERSECT command is used to achieve the results at "B." Before undertaking any solid model, first analyze how the model is to be constructed. Using intersections can create dramatic results, which would normally require numerous union and subtraction operations.

Figure 20–69

Try It! – Open the drawing file 20_Int3. In the following image, use the MOVE command to move object "A" from the midpoint at "A" to the quadrant at "B." Run the INTERSECT command on both objects and observe the results in the following image.

Figure 20–70

 Try It! – Open the drawing file 20_Int4. In the following image, use the MOVE command to move object "A" from the midpoint at "A" to the midpoint at "B." Run the INTERSECT command on both objects and observe the results in the following image.

Figure 20–71

 Try It! – Open the drawing file 20_Int5. In the following image, use the MOVE command to move object "A" from the midpoint at "A" to the midpoint at "B." Run the INTERSECT command on both objects and observe the results in the following image.

Figure 20–72

CREATING SOLID EXTRUSIONS

The EXTRUDE command creates a solid by extrusion. Choose this command in one of the following ways:

- From the Modeling Toolbar
- From the pull-down menu (Draw > Modeling > Extrude)
- From the keyboard (EXT or EXTRUDE)

Only regions or closed, single-entity objects such as circles and closed polylines can be extruded. Other options of the EXTRUDE command include the following: Extrude by Direction, in which you specify two points that determine the length and direction of the extrusion; Extrude by Path, in which the extrusion is created based on a path that consists of a predefined object; and Extrude by Taper Angle, in which a tapered extrusion is created based on an angle value between −90 and +90.

Use the following prompts to construct a solid extrusion of the closed polyline object in the following image. For the height of the extrusion, you can enter a numeric value or you can specify the distance by picking two points on the display screen.

 Try It! – Open the drawing file 20_Extrude1. Use the following command sequence and image for performing this task.

Command: **EXT** *(For EXTRUDE)*
Current wire frame density: ISOLINES=4
Select objects to extrude: *(Select the polyline object at "A")*
Select objects to extrude: *(Press ENTER to continue with this command)*
Specify height of extrusion or [Direction/Path/Taper angle]: **1.00**

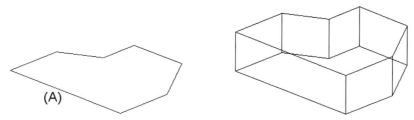

Figure 20–73

You can create an optional taper along with the extrusion by utilizing the Taper angle option provided.

 Try It! – Open the drawing file 20_Extrude2. Use the following command sequence and image for performing this task.

Command: **EXT** *(For EXTRUDE)*
Current wire frame density: ISOLINES=4
Select objects to extrude: *(Select the polyline object at "B" in the following image)*
Select objects to extrude: *(Press ENTER to continue with this command)*
Specify height of extrusion or [Direction/Path/Taper angle]: **T** *(For Taper angle)*
Specify angle of taper for extrusion <0>: **15**
Specify height of extrusion or [Direction/Path/Taper angle]: **1.00**

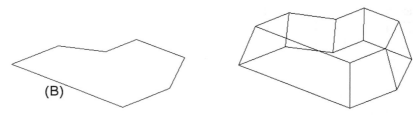

Figure 20–74

You can also create a solid extrusion by selecting a path to be followed by the object being extruded. Typical paths include regular and elliptical arcs, 2D and 3D polylines, or splines. The extruded pipe in the following image was created using the following steps: First the polyline path was created. Then, a new User Coordinate System was established through the UCS command along with the Z-axis option; the new UCS was positioned at the end of the polyline with the Z axis extending along the polyline. A circle was constructed with its center point at the end of the polyline. Finally, the circle was extruded along the polyline path.

Try It! – Open the drawing file 20_Extrude Pipe. Use the following command sequence and image for performing this task.

Figure 20–75

 Command: **EXT** (For EXTRUDE)
Current wire frame density: ISOLINES=4
Select objects to extrude: (Select the small circle as the object to extrude)
Select objects to extrude: (Press ENTER to continue)
Specify height of extrusion or [Direction/Path/Taper angle]: **P** (For Path)
Select extrusion path or [Taper angle]: (Select the polyline object representing the path)

EXTRUDING AN EXISTING FACE

Once a solid object is created, existing faces of the model can be used as profiles to further extrude shapes. Illustrated on the left in the following image is a wedge-shaped 3D model. After entering the EXTRUDE command, press and hold down the CTRL key and select the inclined face. You can either enter a value or drag your cursor to define the height of the extrusion, as shown in the middle in the following image. The results are shown on the right in the following image. The new extruded shape created from the inclined face is considered a separate solid object. If both shapes need to be considered one, use the UNION command and select both extruded boxes to join them as one solid.

 Try It! – Open the drawing file 20_Extrude Face. Use the following command sequence and image for performing an extrusion on an existing face.

 Command: **EXT** (For EXTRUDE)
Current wire frame density: ISOLINES=4
Select objects to extrude: (Press and hold down the CTRL key and select the inclined face
   as shown on the left in the following image)
Select objects to extrude: (Press ENTER to continue)
Specify height of extrusion or [Direction/Path/Taper angle]: **5.00**

Figure 20–76

CREATING REVOLVED SOLIDS

The REVOLVE command creates a solid by revolving an object about an axis of revolution. Choose this command in one of the following ways:

- From the Modeling Toolbar
- From the pull-down menu (Draw > Modeling > Revolve)
- From the keyboard (REV or REVOLVE)

Only regions or closed, single-entity objects such as polylines, polygons, circles, ellipses, and 3D polylines can be revolved. If a group of objects is not in the form of a single entity, group them together using the PEDIT command or create a closed polyline/region with the BOUNDARY command. The following image represents a revolved 3D Solid object.

 Try It! – Open the drawing file 20_Revolve1. Use the following command sequence and image for performing this task.

 Command: **REV** *(For REVOLVE)*
Current wire frame density: ISOLINES=4
Select objects to revolve: *(Select profile "A" as the object to revolve)*
Select objects to revolve: *(Press ENTER to continue with this command)*
Specify axis start point or define axis by [Object/X/Y/Z] <Object>:**O** *(For Object)*
Select an object: *(Select line "B")*
Specify angle of revolution or [STart angle] <360>: *(Press ENTER to accept the default and perform the revolving operation)*

Try It! – Open the drawing file 20_Revolve2. Many practical applications require creating a composite solid from primitives and extrusions. The CYLINDER command was used to construct the primitive and the REVOLVE command to create the extrusion. Existing OSNAP options are fully supported in solid modeling. Use the Center option of OSNAP along with the MOVE command to position the revolved solid inside the cylinder. See the following image.

Command: **M** (For MOVE)
Select objects: *(Select the revolved solid in the following image)*
Select objects: *(Press ENTER to continue with this command)*
Specify base point or [Displacement] <Displacement>: **Cen**
of *(Select the center of the revolved solid at "C")*
Specify second point or <use first point as displacement>: **Cen**
of *(Select the center of the cylinder at "D")*

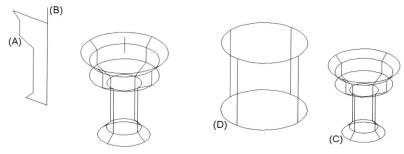

Figure 20–77

Once the revolved solid is positioned inside the cylinder, use the SUBTRACT command to subtract the revolved solid from the cylinder, as shown in the following image. Use the HIDE command to perform a hidden line removal at "B" to check that the solid is correct (this would be difficult to interpret in wireframe mode).

 Command: **SU** (For SUBTRACT)
Select solids and regions to subtract from...
Select objects: *(Select the cylinder as source)*
Select objects: *(Press ENTER to continue with this command)*
Select solids and regions to subtract...
Select objects: *(Select the revolved solid)*
Select objects: *(Press ENTER to perform the subtraction operation)*
Command: **HI** (For HIDE)

Figure 20–78

Note: As with the EXTRUDE command, you can also revolve an existing face of a 3D model to create a revolved feature of a 3D model.

CREATING A SOLID BY SWEEPING

The SWEEP command creates a solid by sweeping a profile along an open or closed 2D or 3D path. The result is a solid in the shape of the specified profile along the specified path. If the sweep profile is closed, a solid is created. If the sweep profile is open, a swept surface is created. Choose this command in one of the following ways:

- From the Modeling toolbar
- From the pull-down menu (Draw > Modeling > Sweep)
- From the keyboard (SWEEP)

 Try It! – Open the drawing file 20_Sweep. Use the following command sequence and image for performing this task.

Illustrated on the left in the following image is an example of the geometry required to create a swept solid. Circles "A" and "B" represent profiles, while arc "C" represents the path of the sweep. Notice in this illustration that the circles do not have to be connected to the path; however, both circles must be constructed in the same plane in order for both to be included in the sweep operation. Use the following command sequence for creating a swept solid. The results of this operation are illustrated on the right in the following image.

 Command: **SWEEP**
Current wire frame density: ISOLINES=4
Select objects to sweep: *(Pick circles "A" and "B")*
Select objects to sweep: *(Press ENTER to continue)*
Select sweep path or [Alignment/Base point/Scale/Twist]: *(Pick arc "C")*

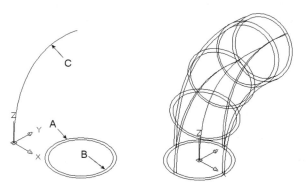

Figure 20–79

Illustrated on the left is the shaded solution to sweeping two circles along a path consisting of an arc. Notice, however, that an opening is not created in the shape; instead, the inner swept shape is surrounded by the outer swept shape. Both swept shapes are considered individual objects. To create the opening, subtract the inner shape from the outer shape using the SUBTRACT command. The results are illustrated on the right in the following image.

 Command: **SU** *(For SUBTRACT)*
Select solids and regions to subtract from:
Select objects: *(Select the outer sweep shape)*
Select objects: *(Press ENTER)*
Select solids and regions to subtract ...
Select objects: *(Select the inner sweep shape)*
Select objects: *(Press ENTER to perform the subtraction)*

Figure 20–80

CREATING A SOLID BY LOFTING

The LOFT command creates a solid based on a series of cross sections. These cross sections define the shape of the solid. If the cross sections are open, a surface loft is created. If the cross sections are closed, a solid loft is created. When you are performing lofting operations, at least two cross sections must be created. Choose this command in one of the following ways:

- From the Modeling toolbar
- From the pull-down menu (Draw > Modeling > Loft)
- From the keyboard (LOFT)

The following image illustrates a lofting operation based on open spline shaped objects. In the illustration on the left, the cross sections of the plastic bottle are selected individually and in order starting with the left of the bottle and ending with the profile on the right of the bottle. The results are illustrated on the right in the following image, with a surface that is generated from the open profiles.

 Command: **LOFT**
Select cross-sections in lofting order: *(Select all cross sections from the rear to the front)*
Select cross-sections in lofting order: *(Press* ENTER *to continue)*
Enter an option [Guides/Path/Cross-sections only] <Cross-sections only>: **G** *(For Guides)*
Select guide curves: *(Select both guide curves)*
Select guide curves: *(Press* ENTER *to create the loft)*

Figure 20–81

In the following image, instead of open profiles, all of the cross sections consist of closed profile shapes. The same rules apply when creating lofts; all profiles must be selected in the proper order. The results of lofting closed profiles are illustrated on the right in the following image with the creation of a solid shape.

 Command: **LOFT**
Select cross-sections in lofting order: *(Select all cross sections from the rear to the front)*
Select cross-sections in lofting order: *(Press* ENTER *to continue)*
Enter an option [Guides/Path/Cross-sections only] <Cross-sections only>: **G** *(For Guides)*
Select guide curves: *(Select both guide curves)*
Select guide curves: *(Press* ENTER *to create the loft)*

Figure 20–82

Try It! – Open the drawing file 20_Bowling Pin. You will first create a number of cross-section profiles in the form of circles that represent the diameters at different stations of the bowling pin. The LOFT command is then used to create the solid shape.

Begin by turning Dynamic Input (DYN) off. Then, create all the circles that make up the cross sections of the bowling pin. Use the table shown on the left to construct the circles shown on the right in the following image.

Circle Coordinates	Circle Diameters
0,0,9.5	Ø0.75
0,0,8.5	Ø1.50
0,0,6.5	Ø1.00
0,0,3.5	Ø3.00
0,0,0.5	Ø2.00
0,0,0	Ø1.75

Figure 20–83

With all profiles created, activate the LOFT command and pick the cross sections of the bowling pin beginning with the bottom circle and working your way up to the top circle. When the Loft Settings dialog box appears, as shown on the left in the following image, verify that Smooth Fit is selected and click the OK button to produce the loft that is illustrated in wireframe mode in the middle of the following image.

 Command: **LOFT**

Select cross-sections in lofting order: *(Select the six cross-sections of the bowling pin in order)*
Select cross-sections in lofting order: *(Press ENTER to continue)*
Enter an option [Guides/Path/Cross-sections only] <Cross-sections only>: *(Press ENTER to accept this default value)*
Enter an option [Guides/Path/Cross-sections only] <Cross-sections only>: *(Press ENTER to perform the loft)*

To complete the bowling pin, use the FILLET command to round off the topmost circle of the bowling pin. Then view the results by clicking on the Realistic or Conceptual visual style. Your display should appear similar to the illustration as shown on the right in the following image.

Command: **F** *(For FILLET)*
Current settings: Mode = TRIM, Radius = 0.00
Select first object or [Undo/Polyline/Radius/Trim/Multiple]: **R** *(For Radius)*
Specify fillet radius <0.00>: **0.50**
Select first object or [Undo/Polyline/Radius/Trim/Multiple]: *(Pick the edge of the upper circle)*
Enter fillet radius <0.50>: *(Press ENTER to accept this value)*
Select an edge or [Chain/Radius]: *(Press ENTER to perform the fillet operation)*
1 edge(s) selected for fillet.

Figure 20–84

CREATING A HELIX

The HELIX command creates a 2D or 3D spiral object. Choose this command in one of the following ways:

- From the Modeling toolbar
- From the pull-down menu (Draw > Modeling > Helix)
- From the keyboard (HELIX)

 Try It! – Open the drawing file 20_Helix. Use the following command sequences and images for creating a 2D, 3D, and spiral helix.

Using the table below, experiment with the following command prompts for constructing a 2D helix, 3D helix, and 3D spiral, as shown on the left in the following images.

Helix Type	Helix Command Prompt
	Use the following command prompt sequence to create a 2D Helix: Command: **HELIX** Number of turns = 3.0000 Twist=CCW Specify center point of base: **0,0,0** Specify base radius or [Diameter] <1.0000>: Specify top radius or [Diameter] <1.0000>: Specify helix height or [Axis endpoint/Turns/turn Height/tWist] <1.0000>: **T** Enter number of turns <3.0000>: **6** Specify helix height or [Axis endpoint/Turns/turn Height/tWist] <1.0000>: **0**
	Use the following command prompt sequence to create a 3D Helix: Command: **HELIX** Number of turns = 3.0000 Twist=CCW Specify center point of base: **0,0,0** Specify base radius or [Diameter] <1.0000>: **3** Specify top radius or [Diameter] <3.0000>: Specify helix height or [Axis endpoint/Turns/turn Height/tWist] <1.0000>: **T** Enter number of turns <3.0000>: **6** Specify helix height or [Axis endpoint/Turns/turn Height/tWist] <1.0000>: **2**
	Use the following command prompt sequence to create a 3D Spiral: Command: **HELIX** Number of turns = 3.0000 Twist=CCW Specify center point of base: **0,0,0** Specify base radius or [Diameter] <1.0000>: **5** Specify top radius or [Diameter] <5.0000>: **2** Specify helix height or [Axis endpoint/Turns/turn Height/tWist] <1.0000>: **T** Enter number of turns <3.0000>: 6 Specify helix height or [Axis endpoint/Turns/turn Height/tWist] <1.0000>: **10**

Figure 20–85

HELIX APPLICATIONS

Typically, a wireframe model of a helix does not fully define how an object like a spring should look. It would be beneficial to show the spring as a thin wire wrapping around a cylinder to form the helical shape. To produce this type of object, use the wireframe of a circle as the object to sweep around the helix to produce the spring.

 Try It! – Open the drawing file 20_Helix Spring. A helix is already created, along with small circular profile. Using the helix as a path and the circle as the object to sweep, use the following command prompt and images to create a spring.

 Command: **SWEEP**

Current wire frame density: ISOLINES=4
Select objects to sweep: *(Select the small circle)*
Select objects to sweep: *(Press* ENTER *to continue)*
Select sweep path or [Alignment/Base point/Scale/Twist]: *(Select the helix)*

Figure 20–86

CREATING POLYSOLIDS

A polysolid is created in a fashion similar to the one used to create a polyline. The POLYSOLID command is used to create this type of object. You pick points and can use the direct distance mode of entry to designate the distances of the polysolid, as shown on the left in the following image, where a width has been entered. The main difference between polysolids and polylines is that you can designate the height of a polysolid. In this way, polysolids are ideal for creating such items as walls in your model, as shown on the right in the following image.

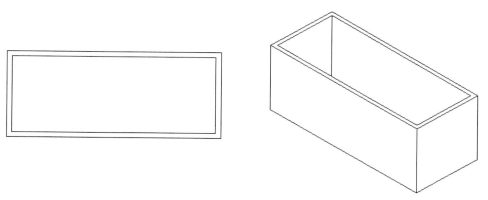

Figure 20–87

Polysolids can also be created from existing 2D geometry such as lines, polylines, arcs, and even circles. Illustrated on the left in the following image is a 2D polyline that has had fillets applied to a number of corners. Activating the POLYSOLID command and selecting the polyline will change the appearance of the object to match the illustration, as shown in the middle in the following image. Here a width has been automatically applied to the polysolid. When the polysolid is viewed in 3D using the SE Isometric, the polysolid is displayed with a height, as shown on the right in the following image.

 Command: **POLYSOLID**
Specify start point or [Object/Height/Width/Justify] <Object>: *(Press* ENTER *to accept Object)*
Select object: *(Pick the polyline object as shown on the left in the following image)*

Figure 20–88

Try It! – Open the drawing file 20_Polysolid Walls. Use the following command sequence and image to construct the 3D walls using the POLYSOLID command.

 Command: **POLYSOLID**
Specify start point or [Object/Height/Width/Justify] <Object>: **H** *(For Height)*
Specify height <0'-4">: **8'**
Specify start point or [Object/Height/Width/Justify] <Object>: **W** *(For Width)*
Specify width <0'-0 1/4">: **4**
Specify start point or [Object/Height/Width/Justify] <Object>: *(Pick a point in the lower left corner of the display screen)*
Specify next point or [Arc/Undo]: *(Move your cursor to the right and enter 30')*
Specify next point or [Arc/Undo]: *(Move your cursor up and enter 10')*
Specify next point or [Arc/Close/Undo]: *(Move your cursor to the left and enter 5')*
Specify next point or [Arc/Close/Undo]: *(Move your cursor up and enter 5')*
Specify next point or [Arc/Close/Undo]: *(Move your cursor to the left and enter 25')*
Specify next point or [Arc/Close/Undo]: **C** *(For Close)*

When finished, rotate your model using the SE Isometric view or the 3DFORBIT (3D Free Orbit) command to view the results.

Figure 20–89

FILLETING SOLID MODELS

Filleting of simple or complex objects is easily handled with the FILLET command. This is the same command as the one used to create a 2D fillet.

 Try It! – Open the drawing file 20_Tee Fillet. Use the following prompt sequence and image for performing this task.

Figure 20–90

 Command: **F** *(For FILLET)*
Current settings: Mode = TRIM, Radius = 0.5000
Select first object or [Undo/Polyline/Radius/Trim/Multiple]: *(Select the edge at "A," which represents the intersection of both cylinders)*
Enter fillet radius <0.5000>: **0.25**
Select an edge or [Chain/Radius]: *(Press* ENTER *to perform the fillet operation)*
1 edge(s) selected for fillet.

 Try It! – Open the drawing file 20_Slab Fillet. A group of objects with a series of edges can be filleted with the Chain option of the FILLET command. Use the following prompt sequence and image for performing this task.

Figure 20–91

Command: **F** *(For FILLET)*
Current settings: Mode = TRIM, Radius = 0.5000
Select first object or [Undo/Polyline/Radius/Trim/Multiple]: *(Select the edge at "A")*
Enter fillet radius <0.5000>: *(Press ENTER to accept the default value)*
Select an edge or [Chain/Radius]: **C** *(For chain mode)*
Select an edge chain or [Edge/Radius]: *(Select edge "A" again; notice how the selection is chained until it reaches an abrupt corner)*
Select an edge chain or [Edge/Radius]: *(Select the edge at "B")*
Select an edge chain or [Edge/Radius]: *(Press ENTER when finished selecting all edges to perform the fillet operation)*
10 edge(s) selected for fillet.

CHAMFERING SOLID MODELS

Just as the FILLET command uses the Chain mode to group a series of edges together, the CHAMFER command uses the Loop option to perform the same type of operation.

 Try It! – Open the drawing file 20_Slab Chamfer. Use the following prompt sequence and image for performing this task.

Figure 20–92

 Command: **CHA** *(For CHAMFER)*
(TRIM mode) Current chamfer Dist1 = 0.5000, Dist2 = 0.5000
Select first line or [Undo/Polyline/Distance/Angle/Trim/mEthod/Multiple]: *(Pick the edge at "A")*
Base surface selection...
Enter surface selection option [Next/OK (current)] <OK>: **N** *(This option selects the other surface shared by edge "A"; the top surface should highlight. If not, use this option again)*
Enter surface selection option [Next/OK (current)] <OK>: *(Press ENTER to accept the top surface)*
Specify base surface chamfer distance <0.5000>: *(Press ENTER to accept the default)*
Specify other surface chamfer distance <0.5000>: *(Press ENTER to accept the default)*
Select an edge or [Loop]: **L** *(To loop all edges together into one)*
Select an edge loop or [Edge]: *(Pick any top edge; notice that the loop option does not stop at abrupt corners)*
Select an edge loop or [Edge]: *(Press ENTER to perform the chamfering operation)*

MODELING TECHNIQUES USING VIEWPORTS

Using viewports can be very important and beneficial when laying out a solid model. Use the VPORTS command to lay out the display screen in a number of individually different display areas. Select this command by choosing Viewports from the View menu, as shown on the left in the following image. A cascading menu appears with the different types of viewports available. Choosing New Viewports displays the Viewports dialog box, as shown on the right in the following image. A number of viewport configurations can be selected that will convert the display screen to the viewport configuration desired. Selecting 3D in the Setup area automatically sets up 3D viewpoints for the viewports.

Figure 20–93

The following image shows the results of the dialog box settings and the way viewports can be used in the construction of a 3D solid model. The image in the right viewport represents the solid model viewed from a southeast isometric viewpoint displaying all three dimensions. The image on the left shows the model as seen from the top. The images are viewed based on the World Coordinate System. This gives you two viewports to help visualize the construction of the solid model.

Figure 20–94

Try It! – Open the drawing file 20_Tee Vports. Use the previous images and text to set up a series of 3D viewports of the intersecting tee pipe object.

 Tip: The VPORTS command can be used to create either tiled viewports in model space or floating viewports while in layout mode.

OBTAINING MASS PROPERTIES OF A SOLID MODEL

The MASSPROP command calculates the mass properties of a solid model. Choose this command in one of the following ways:

- From the Inquiry toolbar
- From the pull-down menu (Tools > Inquiry > Region/Mass Properties)
- From the keyboard (MASSPROP)

 Try It! – Open the drawing file 20_Tee Massprop. Use the MASSPROP command to calculate the mass properties of a selected solid, as shown in the following image. All calculations are based on the current position of the User Coordinate System. You will be given the option of writing this information to a file if desired.

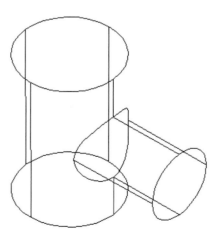

Figure 20–95

Command: **MASSPROP**
Select objects: *(Select the model)*
Select objects: *(Press ENTER to continue with this command)*

```
---------------- SOLIDS ----------------
Mass:                   50.1407
Volume:                 50.1407
Bounding box:           X: 5.0974 -- 10.6984
                        Y: 1.3497 -- 4.5516
                        Z: 0.0000 -- 5.0000
Centroid:               X: 7.2394
                        Y: 2.9507
                        Z: 2.5000
Moments of inertia:     X: 865.8985
                        Y: 3118.9425
                        Z: 3184.2145
Products of inertia:    XY: 1071.0602
                        YZ: 369.8728
                        ZX: 907.4699
Radii of gyration:      X: 4.1556
                        Y: 7.8869
                        Z: 7.9690
Principal moments and X-Y-Z directions about centroid:
                        I: 115.9688 along [1.0000 0.0000 0.0000]
                        J: 177.7542 along [0.0000 1.0000 0.0000]
                        K: 119.8558 along [0.0000 0.0000 1.0000]
```

Figure 20–96

SYSTEM VARIABLES USED FOR SOLID MODELS

System variables are available to control the appearance of your solid model whenever you perform shading or hidden line removal operations. Three in particular are useful: ISOLINES, FACETRES, and DISPSILH. The following text describes each system variable in detail.

THE ISOLINES SYSTEM VARIABLE

Tessellation refers to the lines that are displayed on any curved surface, such as those shown in the following image, to help you visualize the surface. Tessellation lines are automatically formed when you construct solid primitives such as cylinders and cones. These lines are also calculated when you perform solid modeling operations such as SUBTRACT and UNION.

Figure 20–97

The number of tessellation lines per curved object is controlled by the system variable called ISOLINES. By default, this variable is set to a value of 4. The following image shows the results of setting this variable to other values, such as 9 and 20. After the isolines have been changed, regenerate the screen to view the results. The more lines used to describe a curved surface, the more accurate the surface will look in wireframe mode; however, it will take longer to process screen regenerations.

 Try It! – Open the drawing file 20_Tee Isolines. Experiment by changing the ISOLINES system variable to numerous values. Perform a drawing regeneration using the REGEN command each time that you change the number of isolines.

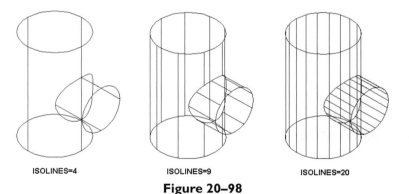

Figure 20–98

THE FACETRES SYSTEM VARIABLE

When you perform hidden line removals on solid objects, the results will be similar to those displayed in the following image. The curved surfaces are now displayed as flat triangular surfaces (faces), and the display of these faces is controlled by the FACETRES system variable. The cylinder with FACETRES set to 0.50 processes much more quickly than the cylinder with FACETRES set to 2, because there are fewer surfaces to process in such operations as hidden line removals. However, the image with FACETRES set to a large value (as high as 10) shows a more defined circle. The default value for FACETRES is 0.50, which seems adequate for most applications.

Try It! – Open the drawing file 20_Tee Facetres. Experiment by changing the FACETRES system variable to numerous values (valid values are .01 to 10). Perform a HIDE each time you change the number of facet lines to view the results.

Figure 20–99

THE DISPSILH SYSTEM VARIABLE

To have the edges of your solid model take on the appearance of an isometric drawing when displayed as a wireframe, use the DISPSILH system variable. This system variable means "display silhouette" and is used to control the display of silhouette curves of solid objects while in either the wireframe or hide mode. Silhouette edges are turned either on or off, with the results displayed in the following image; by default they are turned off. When a HIDE is performed, this system variable controls whether faces are drawn or suppressed on a solid model.

 Try It! – Open the drawing file 20_Tee Dispsilh. Experiment by turning the display of silhouette edges on and off. Regenerate your display each time you change this mode to view the results. Also use the HIDE command to see how the hidden line removal image is changed.

Figure 20–100

TUTORIAL EXERCISE: 20_BRACKET01.DWG

Figure 20–101

Purpose
This tutorial exercise is designed to show you how to create a custom visual style.

System Settings
No special settings need to be made for this tutorial.

Layers
Special layers do not have to be created for this tutorial exercise.

Suggested Commands
Activate the Visual Styles palette and create a new visual style. You will change a few of the settings under the Edge Modifiers area of the palette. When finished, your 3D model should resemble a hand-sketched isometric view due to the overhang value and jitter setting.

STEP 1

Open the existing file 20_Bracket01.Dwg. Activate the Visual Styles palette, as shown on the left in the following image, and pick the button labeled Create New Visual Style. This will launch the Create New Visual Style dialog box, as shown on the right in the following image. Enter the name of the new visual style as Sketched Appearance and add the following description: Displaying a 3D Model as a Sketch. When finished, click the OK button to create the visual style.

Figure 20–102

STEP 2

With the new visual style created, notice the various settings located in the bottom half of the palette. Identify the Edge Modifiers and notice two buttons, as shown in the following image. Clicking on these buttons turns on the Overhang and Jitter settings. For the purposes of this tutorial, these settings will keep their default values.

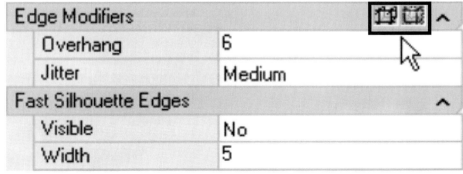

Figure 20–103

STEP 3

With the overhang and jitter values set to the current visual style, click on the button labeled Apply Selected Visual Style to Current Viewport. Notice also in the following image, a yellow rectangle is present identifying the current visual style.

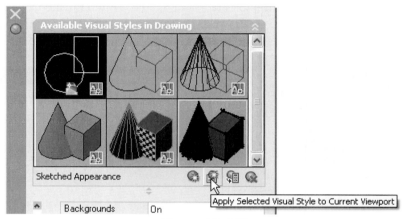

Figure 20–104

STEP 4

Applying the selected visual style to the current viewport displays the model, as shown on the right in the following image. Notice how all corners have an overhang. The jitter value of Medium displays the 3D model as if it was hand sketched.

Figure 20–105

STEP 5

Activate the dashboard and in the Visual Styles area, expand all available visual styles and pick the Conceptual style, as shown on the right in the following image. Your model will change to a different visual style and display similar to the illustration shown on the left in the following image.

Figure 20–106

TUTORIAL EXERCISE: COLLAR.DWG

Figure 20–107

Purpose
This tutorial is designed to construct a solid model of the Collar using the dimensions in the previous image.

System Settings
Use the current limits set to 0,0 for the lower left corner and (12,9) for the upper right corner. Change the number of decimal places from four to two using the Drawing Units dialog box.

Layers
Create the following layer:

Name	Color	Linetype
Model	Cyan	Continuous

Suggested Commands
Begin this tutorial by laying out the Collar in plan view and drawing the basic shape outlined in the Top view. Convert the objects to a polyline and extrude the objects to form a solid. Draw a cylinder and combine this object with the base. Add another cylinder and then subtract it to form the large hole through the model. Add two small cylinders and subtract them from the base to form the smaller holes. Construct a solid box and subtract it to form the slot across the large cylinder.

STEP 1

Begin the Collar by setting the Model layer current. Then draw the three circles shown on the left in the following image using the CIRCLE command. Perform a ZOOM-All after all three circles have been constructed.

STEP 2

Draw lines tangent to the three arcs using the LINE command and the OSNAP-Tangent mode, as shown on the right in the following image. Notice that the User Coordinate System icon has been turned off in this image.

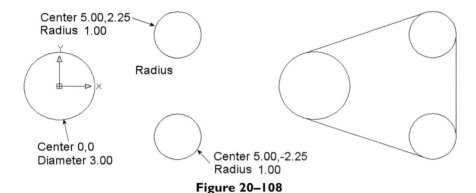

Figure 20–108

STEP 3

Use the TRIM command to trim the circles. When prompted to select the cutting edge object, press ENTER; this will make cutting edges out of all objects in the drawing. Perform this trimming operation so your display will appear similar to the illustration on the left in the following image.

STEP 4

Prepare to construct the base by viewing the object in 3D. Select a viewpoint by choosing 3D Views from the View pull-down menu and then SE Isometric. Your display should appear similar to the illustration on the right in the following image.

Figure 20–109

STEP 5

Convert all objects to a polyline using the Join option of the PEDIT command, as shown on the left in the following image.

STEP 6

Use the EXTRUDE command to extrude the base to a thickness of 0.75 units, as shown on the right in the following image.

Figure 20–110

STEP 7

Create a cylinder using the CYLINDER command. Begin the center point of the cylinder at 0,0,0, with a diameter of 3.00 units and a height of 2.75 units. You may have to perform a ZOOM-All to display the entire model, as shown on the left in the following image.

STEP 8

Merge the cylinder just created with the extruded base using the UNION command, as shown on the right in the following image.

Figure 20–111

STEP 9

Use the CYLINDER command to create a 2.00-unit-diameter cylinder representing a through hole, as shown on the left in the following image. The height of the cylinder is 2.75 units, with the center point at 0,0,0.

STEP 10

To cut the hole through the outer cylinder, use the SUBTRACT command. Select the base as the source object; select the inner cylinder as the object to subtract. Use the HIDE command to view the results, as shown on the right in the following image.

Figure 20–112

STEP 11

Use the REGEN command to regenerate your screen and return to Wireframe mode. Begin placing the two small drill holes (1.00 diameter and .75 high) in the base using the CYLINDER command. Use the OSNAP-Center mode to place each cylinder at the center of arcs "A" and "B," as shown on the left in the following image.

STEP 12

Subtract both 1.00-diameter cylinders from the base of the model using the SUBTRACT command. Use the HIDE command to view the results, as shown on the right in the following image.

Figure 20–113

STEP 13

Begin constructing the rectangular slot that will pass through the two cylinders. Use the BOX command and Center option to accomplish this. Locate the center of the box at 0,0,2.75 and make the box 4 units long, 1 unit wide, and -.75 units high as shown in the following image.

Figure 20–114

STEP 14

Use the SUBTRACT command to subtract the rectangular box from the solid model, as shown on the left in the following image.

STEP 15

Change the facet resolution to a higher value using the FACETRES system variable and a value of 5. Then perform a hidden line removal to see the appearance of the completed model, as shown on the right in the following image.

Figure 20–115

TUTORIAL EXERCISE: VACUUM ATTACHMEN.DWG

Figure 20–116

Purpose
This tutorial is designed to construct a solid model of the vacuum cleaner attachment using the Loft tool.

System Settings
Use the current limits, which are set to 0,0 for the lower left corner and (12,9) for the upper right corner. Change the number of decimal places from four to two using the Drawing Units dialog box.

Layers
Create the following layer:

Name	Color	Linetype
Model	Cyan	Continuous

Suggested Commands
Begin this tutorial by laying out a slot shape and circle. Next move the midpoint of the bottom of the slot to 0,0,0 and then copy it to 0,0,2. Move the bottom quadrant of the circle to 0,0,3.5 and then copy it to 0,0,5. These steps form the four cross sections used for creating the loft.

STEP 1

First create a new drawing file. Then, begin the construction of this 3D model by constructing 2D geometry that will define the final shape of the vacuum cleaner attachment. Use the PLINE command to construct the slot shape as a closed polyline using the dimensions shown on the left in the following image. Then construct the circle using the diameter dimension shown on the right in the following image.

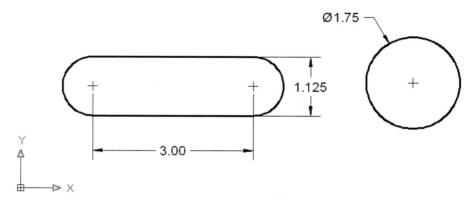

Figure 20–117

STEP 2

Switch to SE Isometric viewing, as shown in the following image. This viewing position can be found by selecting View from the pull-down menu, then selecting 3D Views. Next, move the slot shape from the midpoint of the bottom line to point 0,0,0.

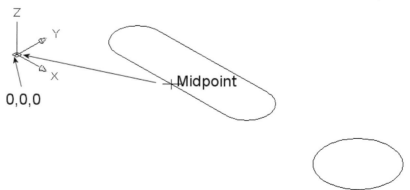

Figure 20–118

STEP 3

Before continuing, be sure that Dynamic Input is turned off for this segment. With the slot moved to the correct position, use the COPY command to create a duplicate shape of the slot using a base point of 0,0,0 and a second point at 0,0,2. When finished, your display should appear similar to the following image.

Figure 20–119

STEP 4

You will next move the circle. Enter the MOVE command and pick the base point of the move using the bottom quadrant of the circle, as shown on the right in the following image. For the second point of displacement, enter the coordinate value of 0,0,3.5 from the keyboard.

Figure 20–120

STEP 5

With the circle moved to the correct location, use the COPY command to duplicate this circle. Copy this circle from the bottom quadrant to a second point located at 0,0,5, as shown on the left in the following image. When finished, your display should appear similar to the illustration on the right in the following image. These form the four cross-sectional shapes that will make up the vacuum cleaner attachment.

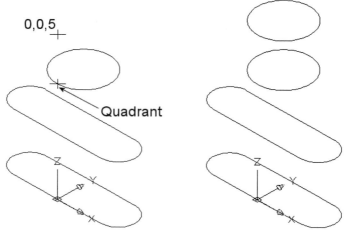

Figure 20–121

STEP 6

Issue the PLAN command through the keyboard and observe that the bottom midpoints of both slots and the bottom quadrants of both circles are all aligned to 0,0,0 as shown in the following image. When finished, perform a ZOOM-Previous operation to return to the SE Isometric view.

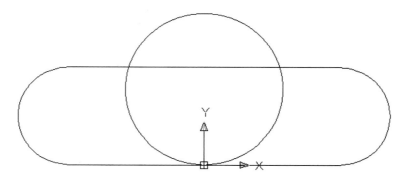

Figure 20–122

STEP 7

Use the 3DFORBIT (Free Orbit) command and rotate your model so it appears similar to the following image.

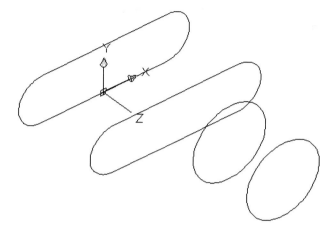

Figure 20–123

STEP 8

Activate the LOFT command and pick the four cross sections in the order labeled 1 through 4, as shown on the left in the following image. When the Loft Settings dialog box appears, keep the default Surface control at cross sections setting at Smooth Fit and click the OK button, as shown on the right in the following image.

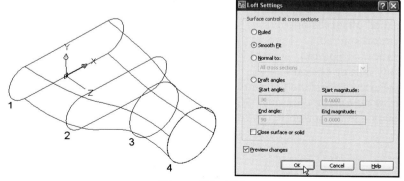

Figure 20–124

STEP 9

The results are illustrated in the following image.

Figure 20–125

STEP 10

An alternate step would be to activate the SOLIDEDIT command and use the Shell option to produce a thin wall of .10 units on the inside of the vacuum attachment. This command and option will be discussed in greater detail in the next chapter. Use the following command prompt sequence and images to perform this operation.

Command: **SOLIDEDIT**
Solids editing automatic checking: SOLIDCHECK=1
Enter a solids editing option [Face/Edge/Body/Undo/eXit] <eXit>: **B** *(For Body)*
Enter a body editing option
[Imprint/seParate solids/Shell/cLean/Check/Undo/eXit] <eXit>: **S** *(For Shell)*
Select a 3D solid: *(Select the vacuum attachment)*
Remove faces or [Undo/Add/ALL]: *(Pick the face at "A")*
Remove faces or [Undo/Add/ALL]: *(Pick the face at "B")*
Remove faces or [Undo/Add/ALL]: *(Press ENTER to continue)*
Enter the shell offset distance: **.10**
Solid validation started.
Solid validation completed.
Enter a body editing option
[Imprint/seParate solids/Shell/cLean/Check/Undo/eXit] <eXit>: *(Press ENTER)*
Solids editing automatic checking: SOLIDCHECK=1
Enter a solids editing option [Face/Edge/Body/Undo/eXit] <eXit>: *(Press ENTER)*

Figure 20–126

STEP 11

When finished completing the Shell option of the SOLIDEDIT command, your model should appear similar to the following image.

Figure 20–127

PROBLEMS FOR CHAPTER 20

Directions for Problems 20–1 through 20–20

1. Create a 3D solid model of each object on Layer "Model."
2. When completed, calculate the volume of the solid model using the *MASSPROP* command.

PROBLEM 20–1

PROBLEM 20–2

PROBLEM 20-3

PROBLEM 20-4

PROBLEM 20-5

PROBLEM 20-6

PROBLEM 20-7

PROBLEM 20-8

PROBLEM 20-9

PROBLEM 20-10

PROBLEM 20-11

PROBLEM 20-12

PROBLEM 20-13

PROBLEM 20-14

PROBLEM 20-15

PROBLEM 20-16

PROBLEM 20-17

PROBLEM 20-18

ALL UNMARKED RADII = R.13

PROBLEM 20-19

PROBLEM 20-20

CHAPTER 21

Concept Modeling and Editing Solids

This chapter begins with the study of creating concept models using a press-and-pull drag method. The use of subshapes for editing solid models is discussed, as is the use of grips for editing solids. Once a solid model is created, various methods are available to edit the model. The methods discussed in this chapter include the following 3D operation commands: (3DMOVE, 3DROTATE, 3DALIGN, MIRROR3D, and ARRAY3D). Also included in this chapter is a segment on using the many options of the SOLIDEDIT command. Various 3D models will be opened and used to illustrate these commands and options.

CONCEPT MODELING

Chapter 20 dealt with the basics of creating solid primitives and how to join, subtract, or intersect these shapes to form complex solid models. Sometimes you do not need the regimented procedures outlined in the previous chapter to get your point across about a specific product you have in mind. You can simply create a concept solid model by dragging shapes together to form your idea. Various concept-modeling techniques will be explained in the next segment of this chapter.

DRAGGING BASIC SHAPES

Constructing a 3D model with a concept in mind is easier than ever. Where exact distances are not important, the following image illustrates the construction of a solid block using the BOX command. First view your model in one of the many 3D viewing positions, such as SE (Southeast) Isometric. Next enter the BOX command and pick first and second corner points for the box on the screen, as shown on the left in the following image. When prompted for the height of the box, move your cursor up and notice the box increasing in height, as shown on the right in the following image. Click to locate the height, and exit the BOX command. All solid primitives can be constructed using this technique.

Figure 21–1

 Try It! – Open the drawing file 21_Concept Box. This drawing file does not contain any objects. Also, it is already set up to be viewed in the Southeast Isometric position (SE Iso). Activate the BOX command, pick two points to define the first and second corner points of the rectangular base, and move your cursor up and pick to define the height of the box. You could also experiment using this technique for creating cylinders, pyramids, spheres, cones, wedges, and a torus. When finished, exit this drawing without saving any changes.

USING DYNAMIC UCS TO CONSTRUCT ON FACES

Once a basic shape is created, it is very easy to create a second shape on an existing face with Dynamic UCS (DUCS) mode. Here is how it works. In the following image, a cylinder will be constructed on one of the faces. Turn on DUCS (located in the status bar), activate the CYLINDER command, and when prompted for the base or center point, hover your cursor over the face, as shown on the left in the following image. The face will highlight (the edge will appear dashed) to indicate that it has been acquired. The Dynamic UCS cursor adjusts itself to this face by aligning the XY plane parallel to the face. When you click a point on this face for the start of the cylinder, the UCS icon changes to reflect this change and the base of the cylinder can be seen, as shown in the middle in the following image. Pick a point to specify the radius. To specify the height of the cylinder, simply drag your cursor away from the face and you will notice the cylinder taking shape, as shown on the right in the following image. Clicking a point will define the height and exit the CYLINDER command.

Figure 21–2

 Try It! – Open the drawing file 21_Concept Cylinder. Issue the CYLINDER command and, for the base or center point, hover your cursor over the front face in the previous image and pick a point. The UCS will change to reflect this new position. Move your cursor until you see the circle forming on the face and pick to define the radius of the cylinder. When prompted for the height, move your cursor forward and notice the cylinder taking shape. Pick to define the cylinder height and to exit the command.

When the cylinder is created, it is considered an individual object separate from the solid block, as shown on the left in the following image. After the primitives are constructed, the UNION command is used to join all primitives together as a single solid object, as shown on the right in the following image.

Figure 21–3

USING GRIPS TO MODIFY SOLID MODELS

Whenever a solid primitive is selected when no command is active, grips are displayed, as with all types of objects that make up a drawing. The grips that appear on solid primitives, as shown in the following image, range in shape from squares to arrows. You can perform an edit operation by selecting either the square or arrow shapes. The type of editing that occurs will depend on the type of grip selected.

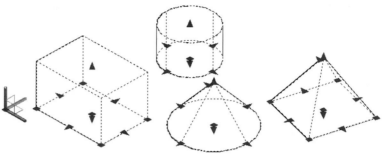

Figure 21–4

KEY GRIP LOCATIONS

The difference between the square and arrow grips is illustrated in the following image of a solid box primitive. The square grip located in the center of a primitive allows you to change the location of the solid. Square grips displayed at the corner (vertex) locations of a primitive will allow you to resize the base shape. The arrows located along the edges of the rectangular base allow each individual side to be modified. Arrow grips that point vertically also appear in the middle of the top and bottom faces of the box primitive. These grips allow you to change the height of the primitive.

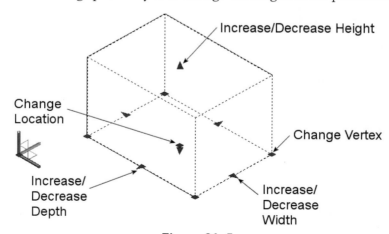

Figure 21–5

GRIP EDITING A CONE

The previous image outlined the various types of grips that display on a box primitive. The grips that appear on cylinders, pyramids, cones, and spheres have similar editing capabilities. The following Try It! exercise illustrates the affects of editing certain arrow grips on a cone.

 Try It! – Open the drawing file 21_GripEdit Cone. Click on the cone and the grips will appear, as shown on the left in the following image. Click the arrow grip at "A" and stretch the base of the cone in to match the object illustrated on the right in the following image. Next, click the arrow grip at "B" (not the grip that points up) and stretch this grip away from the cone to create a top surface similar to the object illustrated on the right in the following image. Experiment further with grips on the cone by clicking on the top grip, which points up, and stretch the cone (now referred to as a frustrum) up.

Figure 21–6

EDITING WITH GRIPS AND DYNAMIC INPUT

You have already seen how easy it is to select grips that belong to a solid primitive and stretch the grips to change the shape of the primitive. This next Try It! exercise deals with grip editing of primitives that make up a solid model in an accurate manner. To accomplish this, Dynamic Input must be turned on.

 Try It! – Open the drawing file 21_GripEdit Shape01. This 3D model consists of three separate primitives, namely, two boxes and one cylinder. The height of the cylinder needs to be lowered. To accomplish this, click on the cylinder and observe the positions of the grips and numeric values, as shown on the left in the following image. The total height of 6.0000 needs to be changed to 3.50 units. Click on the arrow grip at the top of the cylinder that is pointing up, press the TAB key to highlight the overall height value of 6.0000, and change this value to 3.50. Pressing ENTER to accept this new value will lower the cylinder, as shown on the right in the following image. Feel free to experiment with both boxes by either increasing or decreasing their heights using this grip-editing method.

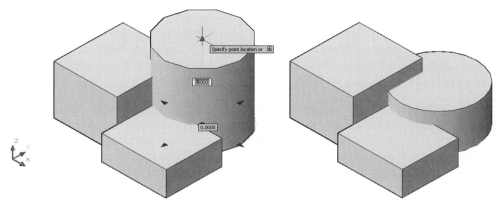

Figure 21–7

MANIPULATING SUBOBJECTS

A subobject is any part of a solid. It could be a face, an edge, or a vertex (corner). The following image illustrates a 3D box and pyramid. Pressing the CTRL key will allow you to select a subobject. Notice in this illustration that each subobject selected has a grip associated with it. If you accidentally select a subobject, press CTRL + SHIFT and pick it again to deselect it.

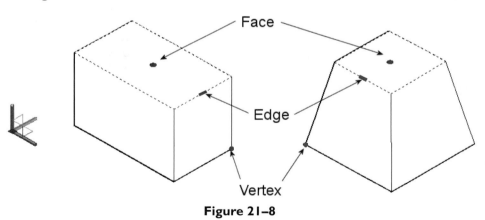

Figure 21–8

Once a subobject is selected, click on the grip to activate the grip Stretch, Move, Rotate, Scale, and Mirror modes. You can drag your cursor to a new location or enter a direct distance value from the keyboard. In the following image, the edge of each solid object was selected as the subobject and dragged to a new location.

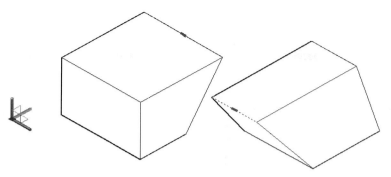

Figure 21–9

EDITING SUBOBJECTS OF A SOLID MODEL

You have seen how easy it is to isolate a subobject of a solid model by pressing the CTRL key while selecting the subobject. The same technique can be used to isolate a primitive that is already consumed or made part of a solid model. This would allow you to edit a specific primitive while leaving other primitives of the solid model unselected. The next Try It! exercise illustrates this technique.

Try It! – Open the drawing file 21_GripEdit Shape02. This drawing represents three separate primitives joined into one solid model using the UNION command. The height of the front block needs to be increased. First, verify that ORTHO is turned on to assist in this operation. Then, press and hold down the CTRL key while clicking on the front block. Notice that only this block highlights and displays various grips, as shown on the left in the following image. The height of this block needs to be increased by 3.00 units. Click on the arrow grip at the top of the block that is pointing up, and drag the geometry up until the tool tip reads approximately 3.00 units, as shown in the middle in the following image. Type a value of 3.00 and press ENTER to increase the block, as shown on the right in the following image. Feel free to experiment with the other block and cylinder by either increasing or decreasing their heights using this grip-editing method.

Figure 21–10

ADDING EDGES AND FACES TO A SOLID MODEL

When building solid models, use the UNION, SUBTRACT, and INTERSECT commands as the primary means of joining, removing, or creating a solid that is common to two intersecting primitives. You can also imprint regular objects such as lines, circles, or polylines—for example, the line segment shown on the left in the following image—directly onto the face of a solid model. This line, once imprinted, becomes part of the solid and in our case divides the top face into two faces. Then, using the subobject technique, we can change the shape of the solid model. The command used to perform this operation is IMPRINT, which can be found in the dashboard, as shown on the right in the following image.

Command: **IMPRINT**

Select a 3D solid: *(Select the 3D solid model)*
Select an object to imprint: *(Pick the line segment constructed across the top surface of the solid model)*
Delete the source object [Yes/No] <N>: **Y** *(For Yes)*
Select an object to imprint: *(Press ENTER to perform this operation and exit the command)*

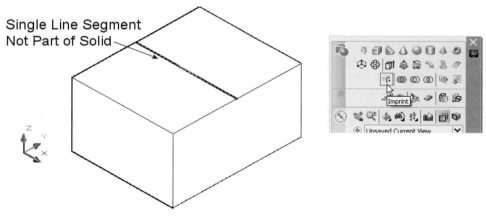

Figure 21–11

After the imprint operation is performed on the line segment, this object becomes part of the solid model. When the imprinted line is selected as a subobject, the grip located on the line can be selected and dragged up or down, as shown in the following image. The results can dramatically change the shape of the solid model.

Figure 21–12

 Try It! – Open the drawing file 21_Imprint Roof. First, verify that ORTHO is turned on to assist in this operation. Issue the IMPRINT command, select the solid block, as shown on the left in the following image, and pick the single line segment as the object to imprint. Next, press and hold down the CTRL key while clicking on the imprinted line. Notice that only this line highlights and displays a grip at its midpoint, as shown in the middle of the following image. Pick this grip, slowly move your cursor up, and notice the creation of the roof peak. You could also move your cursor down to create a V-shaped object.

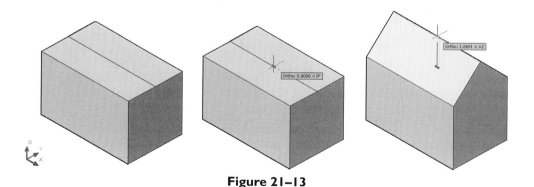

Figure 21–13

PRESSING AND PULLING BOUNDING AREAS

An additional technique used for constructing solid models is available to speed up the construction and modification processes. The technique is called pressing and pulling, which can be accessed by picking the PRESSPULL command from the dashboard, as shown in the following image. Any closed area that can be hatched can be manipulated using the PRESSPULL command.

Figure 21–14

 Try It! – Open the drawing file 21_PressPull Shapes. Activate the PRESSPULL command from the Dashboard, pick inside of the circular shape, as shown on the left in the following image, move your cursor up, and enter a value of 6 units. Activate the PRESSPULL command again, pick inside of the closed block shape, as shown in the middle in the following image, move your cursor up, and enter a value of 4 units. Use the PRESSPULL command one more time on the remaining closed block shape, as shown on the right in the following image, move your cursor up, and enter a value of 2 units. Performing this task results in the creation of three separate solid shapes. Use the UNION command to join all shapes into one 3D solid model.

Figure 21–15

Another technique of activating the press-and-pull feature is to press and hold the CTRL + ALT keys, and then picking the area. You will be able to perform the operation the same way as using the PRESSPULL command from the dashboard. The next Try It! exercise illustrates this technique.

 Try It! – Open the drawing file 21_PressPull Hole. Press and hold down the CTRL+ALT keys and pick inside of the circular shape, as shown on the left in the following image. Moving your cursor up will create the cylinder, as shown on the right in the following image.

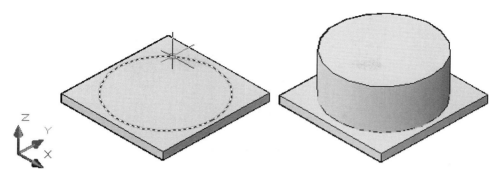

Figure 21–16

Undo the previous operation, activate press and pull by pressing and holding down the CTRL + ALT keys, and pick inside of the circular shape, as shown on the left in the following image. However, instead of moving your cursor up to form a cylinder, drag your cursor down into the thin block to perform a subtraction operation and create a hole in the block.

Figure 21–17

 Try It! – Open the drawing file 21_PressPull Plan. Press and hold down the CTRL + ALT keys while picking inside of the bounding area created by the inner and outer lines, as shown on the left in the following image. Move your cursor up and enter a value of 8' to construct a solid model of the walls, as shown on the right in the following image.

Figure 21-18

 Try It! – Open the drawing file 21_PressPull Openings. Press and hold down the CTRL + ALT keys while picking inside of one of the rectangles that signify a door or window opening. Move your cursor into the wall and pick inside of the model to create the opening. Use this technique for the other openings as well.

Figure 21-19

USING PRESS AND PULL ON BLOCKS

Pressing and pulling to create solid shapes is not limited to just closed shapes such as rectangles, circles, or polylines. The press-and-pull feature can also be used on block objects. In some cases, depending on how you use the press-and-pull feature, it is possible to convert the 2D block into a 3D solid model. The next Try It! exercise illustrates this.

 Try It! – Open the drawing file 21_PressPull Bed. All objects that describe this 2D bed are made up of a single block. You will use the PRESSPULL command to highlight certain closed boundary areas and pull the area to a new height. In this way, PRESSPULL is used to convert a 2D object into a 3D model.

Activate the PRESSPULL command, move your cursor inside of the area, as shown on the left in the following image, and press and hold down the left mouse button as you move your cursor up. When you let go of your cursor, type a value of 12 to extrude this area a distance of 12 units up, as shown on the right in the following image.

Command: **PRESSPULL**
Click inside bounded areas to press or pull. *(Move your cursor over the area, as shown on the left in the following image. Press and hold down the mouse button and pull in an upward direction. Let go of the mouse and enter a value of 12 in the designated edit box and press ENTER when finished)*
1 loop extracted.
1 Region created.
12 (Value entered)

Figure 21–20

Activate the PRESSPULL command, move your cursor inside of the small triangular area, as shown on the left in the following image, and press and hold down the left mouse button as you move your cursor up. When you let go of your cursor, type a value of 12.5 to extrude this area a distance of 12.5 units up, as shown on the right in the following image.

Command: **PRESSPULL**
Click inside bounded areas to press or pull. *(Move your cursor over the triangular, area, as shown on the left in the following image. Press and hold down the mouse button and pull in an upward direction. Let go of the mouse and enter a value of 12.5 in the designated edit box and press ENTER when finished)*
1 loop extracted.
1 Region created.
12.5 (Value entered)

Figure 21–21

Activate the PRESSPULL command, move your cursor inside of the area, as shown on the left in the following image, and press and hold down the left mouse button as you move your cursor up. When you let go of your cursor, type a value of 12 to extrude this area a distance of 12 units up, as shown on the right in the following image.

Command: **PRESSPULL**
Click inside bounded areas to press or pull. *(Move your cursor over the back area of the bed, as shown on the left in the following image. Press and hold down the mouse button and pull in an upward direction. Let go of the mouse, enter a value of 12 in the designated edit box, and press* ENTER *when finished)*
1 loop extracted.
1 Region created.
12 (Value entered)

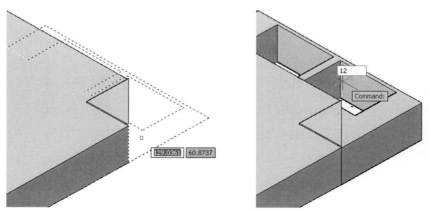

Figure 21–22

Activate the PRESSPULL command, move your cursor inside of the rectangular area represented as a pillow, as shown on the left in the following image, and press and hold down the left mouse button as you move your cursor up. When you let go of your cursor, type a value of 15 to extrude this area a distance of 15 units up, as shown on the right in the following image.

Command: **PRESSPULL**
Click inside bounded areas to press or pull. *(Move your cursor over the area representing the pillow, as shown on the left in the following image. Press and hold down the mouse button and pull in an upward direction. Let go of the mouse button, enter a value of 15 in the designated edit box, and press ENTER when finished)*
1 loop extracted.
1 Region created.
15 (Value entered)

Figure 21–23

Perform the same press-and-pull operation on the second pillow. Extrude this shape a value of 15 units, as shown on the left in the following image. Since the 3D objects created by the press-and-pull operations are all considered individual primitives, use the UNION command to join all primitives into a single 3D solid model. An alternate step would be to use the FILLET command to round off all corners and edges of the bed and pillows, as shown on the right in the following image.

Figure 21–24

ADDITIONAL METHODS FOR EDITING SOLID MODELS

Additional methods are available to edit the solid model. These methods include the 3D Operation commands (3DMOVE, 3DALIGN, 3DROTATE, MIRROR3D, and 3DARRAY) and the SOLIDEDIT command. Various 3D drawings will be opened and used to illustrate these commands.

Figure 21–25

USING THE 3DMOVE COMMAND

To assist in the positioning of objects, the 3DMOVE tool is available. This tool displays the move grip tool, which displays an axis for the purpose of moving objects a specified direction and distance. Select this command either from the dashboard, as shown on the left in the following image, or from the Modify pull-down menu, as shown on the right in the following image.

Figure 21–26

When using this command, you can select either objects or subobjects to move. To select subobjects, press and hold down the CTRL key as you select. After you press ENTER to signify that you are done with the selection process, the move grip tool will display attached to your cursor, as shown in the following image.

Figure 21–27

Pick any convenient point to utilize as the base point for the move. Object snap modes can be used to accomplish this task if desired. You then move your cursor over one of the three axis handles to define the direction of the move. As you hover over one of the handles, it will turn yellow and a direction vector will display. Click on this axis handle to lock in the direction vector. Then enter a value to move the solid or move your cursor along the direction vector, as shown in the following image, and pick a location to move.

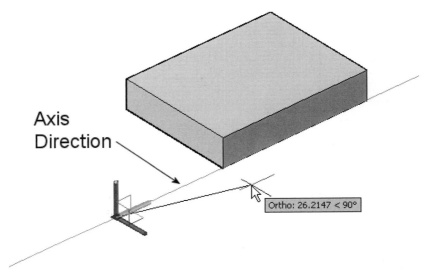

Figure 21–28

THE 3DALIGN COMMAND

Use the 3DALIGN command to specify up to three points to define the source plane of one 3D solid model, followed by up to three points to define the destination plane where the first solid model will be moved or aligned to.

When you specify points, the first source point is referred to as the base point. This point is always moved to the first destination point. Selecting second and third source or destination points will result in the 3D solid model being rotated into position.

 Try It! – Open the drawing file 21_3DAlign. The objects on the right in the following image need to be positioned or aligned to form the assembled object shown in the small isometric view in this image. At this point, it is unclear at what angle the objects are currently rotated. When you use the 3DALIGN command, it is not necessary to know this information. Rather, you line up source points with destination points. When the three sets of points are identified, the object moves and rotates into position. The first source point acts as a base point for a move operation. The first destination point acts as a base point for rotation operations. The second and third sets of source and destination points establish the direction and amount of rotation required to align the objects.

Figure 21–29

Follow the prompt sequence below and the illustration in the following image for aligning the hole plate with the bottom base. The first destination point acts as a base point to which the cylinder locates.

Command: **3DALIGN**
Select objects: *(Select the object with the hole)*
Select objects: *(Press ENTER to continue)*
Specify source plane and orientation ...
Specify base point or [Copy]: *(Select the endpoint at "A")*
Specify second point or [Continue] <C>: *(Select the endpoint at "B")*
Specify third point or [Continue] <C>: *(Select the endpoint at "C")*
 Specify destination plane and orientation ...
Specify first destination point: *(Select the endpoint at "D")*
Specify second destination point or [eXit] <X>: *(Select the endpoint at "E")*
Specify third destination point or [eXit] <X>: *(Select the endpoint at "F")*

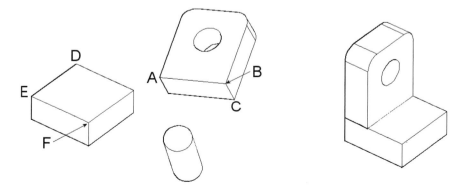

Figure 21–30

The results of the previous step are illustrated on the left in the following image. Next, align the cylinder with the hole. Circular shapes often need only two sets of source and destination points for the shapes to be properly aligned.

Command: **3DALIGN**
Select objects: 1 found
Select objects:
Specify source plane and orientation ...
Specify base point or [Copy]: *(Select the center of circle "A")*
Specify second point or [Continue] <C>: *(Select the center of circle "B")*
Specify third point or [Continue] <C>: *(Press ENTER to continue)*
Specify destination plane and orientation ...
Specify first destination point: *(Select the center of circle "C")*
Specify second destination point or [eXit] <X>: *(Select the center of circle "D")*
Specify third destination point or [eXit] <X>: *(Press ENTER to exit)*

The completed 3D model is illustrated on the right in the following image. The ALIGN command provides an easy means of putting solid objects together to form assembly models.

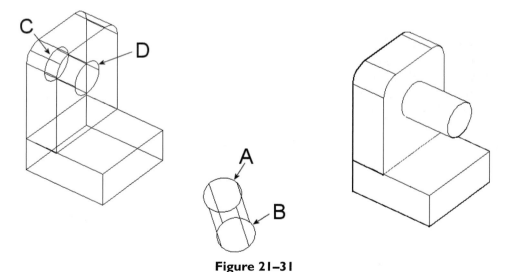

Figure 21–31

THE 3DROTATE COMMAND

The 3DROTATE command uses a special rotate grip tool to rotate objects around a base point. Select this command from either the Modify pull-down menu, as shown on the left in the following image, or from the dashboard, as shown on the right in the following image.

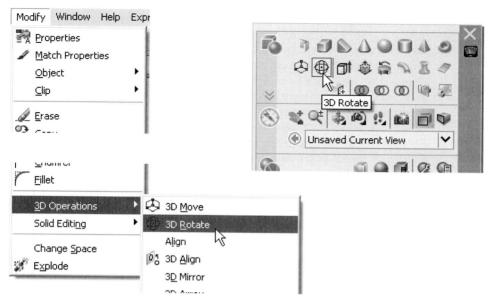

Figure 21–32

After you select the object or objects to rotate, you are prompted to pick a base point, which will act as the pivot point of the rotation. After you pick this base point, the rotate grip tool will appear, as shown in the following image. You then hover your cursor over an axis handle until it turns yellow and an axis vector appears. If this axis is correct, click it with the mouse to establish the axis of rotation. Then enter the start and end angles to perform the rotation.

Figure 21–33

Try It! – Open the drawing 21_3DRot. In the following image, a base containing a slot needs to be joined with the two rectangular boxes to form a back and side. First select box "A" as the object to rotate in 3D. You will be prompted to define the base point of rotation. Next you will be prompted to choose a rotation axis by picking the appropriate axis handle on the grip tool. This axis will serve as a pivot point where the rotation occurs. Entering a negative angle of 90° will rotate the box in the counterclockwise direction (as you would look down the Y axis). Negative angles rotate in the clockwise direction. An easy way to remember the rotation direction is to use the right-hand rule. Point your thumb in the axis direction and your fingers will curl in the positive rotation direction.

Command: **3DROTATE**
Current positive angle in UCS: ANGDIR=counterclockwise ANGBASE=0
Select objects: *(Select box "A")*
Select objects: *(Press ENTER to continue)*
Specify base point: *(Pick the endpoint at "A")*
Pick a rotation axis: *(When the rotate grip tool appears, hover on the axis handle until the proper axis appears, as shown in the following image; then pick with the mouse)*
Specify angle start point: *(Pick the endpoint at "A")*
Specify angle end point: **-90**

Figure 21–34

Next, the second box in the following image is rotated 90° after the proper rotation axis is selected, as shown in the following image.

Command: **3DROTATE**
Current positive angle in UCS: ANGDIR=counterclockwise ANGBASE=0
Select objects: *(Select box "A")*
Select objects: *(Press ENTER to continue)*
Specify base point: *(Pick the endpoint at "A")*
Pick a rotation axis: *(When the rotate grip tool appears, click on the axis handle until the proper axis appears, as shown in the following image; then pick this axis with the mouse)*
Specify angle start point: *(Pick the endpoint at "A")*
Specify angle end point: **90**

Figure 21–35

The results of these operations are illustrated on the left in the following image. Once the boxes are rotated to the correct angles, they are moved into position using the MOVE command and the appropriate Object Snap modes. Box "A" is moved from the endpoint of the corner at "A" to the endpoint of the corner at "C." Box "B" is moved from the endpoint of the corner at "B" to the endpoint of the corner at "C." Once moved, they are then joined to the model through the UNION command, as shown on the right in the following image.

Figure 21–36

THE MIRROR3D COMMAND

The MIRROR3D command is a 3D version of the MIRROR command. In this command, however, instead of flipping over an axis, you mirror over a plane. A thorough understanding of the User Coordinate System is a must in order to properly operate this command.

 Try It! – Open the drawing 21_Mirror3D. As illustrated on the left in the following image, only half of the object is created. The symmetrical object on the right in the following image is needed and can easily be created by using the MIRROR3D and UNION commands.

Command: **MIRROR3D**
Select objects: *(Select the part)*
Select objects: *(Press ENTER to continue)*
Specify first point of mirror plane (3 points) or
[Object/Last/Zaxis/View/XY/YZ/ZX/3points] <3points>: **YZ** *(For YZ plane)*
Specify point on YZ plane <0,0,0>: *(Select the endpoint at "A")*
Delete source objects? [Yes/No] <N>: **N** *(Keep both objects)*

Command: **UNI** *(For UNION)*
Select objects: *(Select both solid objects)*
Select objects: *(Press ENTER to perform the union operation)*
Command: **HI** *(For HIDE; the solid should appear as illustrated on the right in the following image)*

Figure 21–37

THE 3DARRAY COMMAND

The 3DARRAY command, like the ARRAY command, allows you to create both polar and rectangular arrays. For the 3D polar array, however, you select any 3D axis to rotate about, and in the 3D version of the rectangular array, you have not only rows and columns but levels as well.

Try It! – Open the drawing 21_3DArray. As illustrated on the left in the following image, six arms need to be arrayed around the hub in the center. To accomplish this we will perform a polar array about an axis running through the hub. We will rotate throughout a full 360°.

Note: If you perform a polar array at an angle other than 360°, take note of the direction in which you select the axis because this will affect the direction of rotation (right-hand rule).

Command: **3A** *(For 3DARRAY)*
Select objects: *(Select the Arm "A")*
Select objects: *(Press ENTER to continue)*
Enter the type of array [Rectangular/Polar] <R>: **P** *(For Polar)*
Enter the number of items in the array: **6**
Specify the angle to fill (+=ccw, -=cw) <360>: *(Press ENTER to accept the default)*
Rotate arrayed objects? [Yes/No] <Y>: **Y**
Specify center point of array: **Cen**
of *(Specify the center of the back hub circle at "B")*
Specify second point on axis of rotation: **Cen**
of *(Specify the center of the front hub circle at "C")*

 Command: **UNI** *(For UNION)*
Select objects: *(Select the six arms and the hub)*
Select objects: *(Press ENTER to perform the union operation)*
Command: **HI** *(For HIDE, the solid should appear as illustrated on the right in the following image)*
Command: **RE** *(For REGEN; this will convert the image back to wireframe mode)*

Figure 21–38

DETECTING INTERFERENCES OF SOLID MODELS

The INTERFERE command will identify any interference and highlight the solid models that overlap. Choose this command in one of the following ways:

- From the 3D Make control panel located in the dashboard
- From the pull-down menu (Modify > 3D Operations > Interference Checking)
- From the keyboard (INTERFERE)

Try It! – Open the drawing file 21_Pipe Interference. Use this command to find any interference shared by a series of solid pipe objects, as shown on the left in the following image. Click on the Interference checking button, located in the dashboard, as shown on the right in the following image, to begin this command.

Figure 21–39

After launching this command, you will be prompted to select first and second sets of solids. In each case, pick both pipes separately. Pressing ENTER at the end of the command sequence will change the solid objects to a wireframe display to expose the areas of the objects considered to be interfering with one another, as shown on the left in the following image. An Interference Checking dialog box also appears, as shown on the right in the following image. The information in the dialog box allows you to verify the first and second sets of interfering objects. If they exist, you can also cycle through additional interferences using the Previous and Next buttons. In addition, a check box is provided, which will allow you to create the interference solid, if desired.

 Command: INTERFERE
Select first set of objects or [Nested selection/Settings]: *(Pick one of the pipe objects)*
Select first set of objects or [Nested selection/Settings]: *(Press* ENTER *to continue)*
Select second set of objects or [Nested selection/checK first set] <checK>: *(Pick the second pipe object)*
Select second set of objects or [Nested selection/checK first set] <checK>: *(Press* ENTER *to perform the interference check)*

Figure 21–40

SECTIONING A SOLID

The SECTIONPLANE command is used to create a section plane that passes through a solid model. When prompted with the following command sequence, you construct a line that signifies the section line, as shown on the left in the following image. You can also select multiple points to create a jogged section plane line or identify the section plane based on an orthographic view, such as the Front, Top, or Side view.

Command: SECTIONPLANE
Select face or any point to locate section line or [Draw section/Orthographic]: *(Pick a first point to define the section line)*
Specify through point: *(Pick a second point to define the section line)*

Once the section plane line is created, picking on this line and right-clicking will display the menu, as shown in the middle in the following image. Clicking on Activate live sectioning will create the section cut and reveal the inner details of the 3D model, as shown on the right in the following image.

Figure 21–41

Clicking again on the section plane line and right-clicking to display the menu will expose other tools used for controlling the appearance of the section cut. Picking Show cut-away geometry from the menu, as shown in the middle in the following image, will display all geometry cut away by the section, as shown on the right in the following image. Notice the red default color that signifies the geometry that is cut away.

Figure 21–42

Clicking on Live section settings from the menu in the previous image will display a dialog box that allows you to control the appearance of a number of items. Clicking on 2D section / elevation block creation settings will display the settings used when generating a 2D section from the solid model as shown on the left in the following image. You can control such items as the hatch pattern and hatch scale in addition to the background, cut-away, and curve tangency lines.

Clicking on 3D section block creation settings, as shown in the middle in the following image, will display a similar number of settings used for controlling the display of a 3D section created from the 3D model.

Clicking on Live Section settings, as shown on the right in the following image, will display a menu with settings that deal with the initial section cut of the solid model. The settings in this area deal with Intersection Boundary, Intersection Fill, and Cut-away Geometry.

Figure 21–43

One of the advantages of cutting a solid model with a section plane is to generate either a 2D or 3D section. The purpose of the 2D section would be to further document a model by creating a section view, as shown on the left in the following image. To generate this type of section, click on the section plane line, as shown in the middle in the following image, right-click, and pick Generate 2D/3D section from the menu, as shown in the following image.

Figure 21–44

Clicking on Generate 2D/3D section in the previous menu will launch the Generate Section/Elevation dialog box, as shown in the following image. First choose whether you want to create a 2D or 3D section. Then determine the source geometry by either accepting the ability to include all objects in the section or deciding to manually select the objects to include in the section. Whether you are creating a 2D or 3D section, the geometry will be grouped into a block ready for insertion. Or, if changes need to be made to an existing block, you can elect to replace this block with the latest information. You can even export the file into a DWG format, which can then be inserted into a drawing at a later time. The Section Setting button launches the dialog box that controls various effects when the 2D and 3D sections are being created. When you are satisfied with all the settings in this dialog box, click the Create button to make the section. The arrow pointing up that is displayed in the lower left corner of the dialog box collapses the dialog box into a smaller version, similar to the illustration on the right in the following image.

Figure 21–45

Illustrated on the left in the following image is an example of a 3D section being created in the same manner that a 2D section was created.

Figure 21–46

 SLICING SOLID MODELS

Yet another way of cutting sections in 3D solid models is through the SLICE command. This command will create new solids from the existing ones that are sliced. You can also retain one or both halves of the sliced solid. Slicing a solid requires some type of cutting plane. The default method of creating this plane is by picking two points. You can also define the cutting plane or surface by specifying three points, by picking a surface, by using another object, or by basing the cutting plane line on the current positions of the XY, YZ, or ZX planes.

Choose the SLICE command in one of the following ways:

- From the 3D Make control panel located in the dashboard
- From the pull-down menu (Modify > 3D Operations > Slice)
- From the keyboard (SL or SLICE)

 Try It! – Open the drawing file 21_Tee Slice. This command is similar to the SECTION command except that the solid model is actually cut or sliced at a plane that you define. In the example on the left in the following image, this plane is defined by the User Coordinate System. Before the slice is made, you also have the option of keeping either one or both halves of the object. The MOVE command is used to separate both halves, as shown on the right in the following image.

Command: **SL** *(For SLICE)*
Select objects: *(Select the solid object)*
Select objects: *(Press ENTER to continue with this command)*
Specify first point on slicing plane by [Object/Zaxis/View/XY/YZ/ZX/3points] <3points>: **XY**
Specify a point on XY-plane <0,0,0>: *(Press ENTER to accept this default value)*
Specify a point on desired side of the plane or [keep Both sides]: **B** *(To keep both sides)*

Figure 21–47

SLICING A SOLID WITH A SURFACE

A solid object can also be sliced by a surface. A surface is created by performing a 3D operation such as extrusion or revolution on an open object. Once the surface is created, it is positioned inside of the 3D solid model, where a slicing operation is performed. The next Try It! exercise illustrates the use of this technique.

 Try It! – Open the drawing file 21_Surface Flow. You will first extrude a spline to create a surface. The surface will then be used to slice a solid block. You will keep the bottom portion of the solid.

Before slicing the solid block, first extrude the spline object, as shown on the left in the following image, a distance equal to the depth of the block (from "A" to "B"). Since the spline represents an open shape, the result of performing this operation will be the creation of a surface instead of a solid, as shown on the right in the following image.

Command: **EXT** *(For EXTRUDE)*
Current wire frame density: ISOLINES=4
Select objects to extrude: *(Pick the spline object)*
Select objects to extrude: *(Press ENTER to continue)*
Specify height of extrusion or [Direction/Path/Taper angle]: **D** *(For Direction)*
Specify start point of direction: *(Pick the endpoint at "A")*
Specify end point of direction: *(Pick the endpoint at "B")*

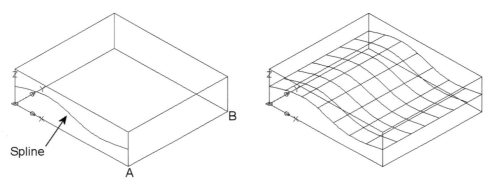

Figure 21–48

With the newly created surface positioned inside of the solid block, issue the SLICE command. Pick the solid block as the object to slice and select the surface as the slicing plane, as shown on the left in the following image. You will also be prompted to select the portion of the solid to keep. Here is where you pick the bottom of the solid, as shown on the left in the following image.

 Command: **SL** *(For SLICE)*
Select objects to slice: *(Pick the solid block)*
Select objects to slice: *(Press ENTER to continue)*
Specify start point of slicing plane or [planar Object/Surface/Zaxis/View/XY/YZ/ZX/3points] <3points>: **S** *(For Surface)*
Select a surface: *(Pick the surface, as shown on the left in the following image)*
Select solid to keep or [keep Both sides] <Both>: *(Pick the bottom of the solid)*

The results are displayed on the right in the following image, with the solid block being cut by the surface.

Figure 21-49

TERRAIN MODELING

A unique type of solid model can be created when it is sliced by a surface created using a lofting operation. In the following image, four different splines have been applied to the edge faces of a solid block. Using the LOFT command, two splines are selected as cross sections and the other two splines as guides or rails. Once this specialized surface is created, you slice the solid block using the surface and keep the lower portion of the model.

 Try It! – Open the drawing file 21_Terrain. You will create a lofted surface by selecting the two cross sections and the two guide curves, as shown on the left in the following image. The results are displayed on the right in the following image, with a complex surface being created from the loft operation.

Command: LOFT
Select cross-sections in lofting order: *(Select cross section #1)*
Select cross-sections in lofting order: *(Select cross section #2)*
Select cross-sections in lofting order: *(Press ENTER to continue)*
Enter an option [Guides/Path/Cross-sections only] <Cross-sections only>: **G** *(For Guides)*
Select guide curves: *(Select guide curve #1)*
Select guide curves: *(Select guide curve #2)*
Select guide curves: *(Press ENTER to create the surface)*

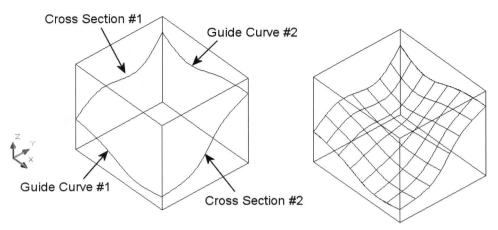

Figure 21–50

With the surface created, activate the SLICE command, pick the solid block as the object to slice, select the surface as the slicing plane, and, finally, pick the lower portion of the solid as the portion to keep, as shown on the left in the following image. The results are illustrated on the right in the following image.

Command: SLICE
Select objects to slice: *(Select the solid block)*
Select objects to slice: *(Press* ENTER *to continue)*
Specify start point of slicing plane or [planar Object/Surface/Zaxis/View/XY/YZ/ZX/3points] <3points>: **S** *(For Surface)*
Select a surface: *(Select the surface)*
Select solid to keep or [keep Both sides] <Both>: *(Pick the lower portion of the solid block)*

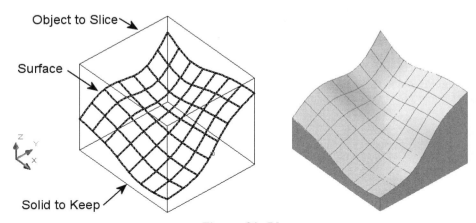

Figure 21–51

THE SOLIDEDIT COMMAND

Once features such as holes, slots, and extrusions are constructed in a solid model, the time may come to make changes to these features. This is the function of the SOLIDEDIT command. This command contains numerous options, which can be selected from the Solids Editing toolbar, as shown in the following image. You could also choose Solids Editing from the Modify pull-down menu. The menus are arranged in three groupings; namely, Face, Edge, and Body editing. These groupings are discussed in the pages that follow. Also, it is recommended that instead of entering the SOLIDEDIT command at the command prompt, you use the Solids Editing toolbar or pull-down menu to perform editing operations. This will eliminate a number of steps and make it easier to locate the appropriate option under the correct grouping.

Figure 21–52

The following table gives a brief description of each mode for performing solid editing operations.

Button	Tool	Function
	Extrude Faces	Used for lengthening or shortening faces on a solid.
	Move Faces	Used for moving a solid shape to a new location.
	Offset Faces	Used for offsetting faces at a specified distance on a solid.
	Delete Faces	Used for removing faces and fillets from a solid.
	Rotate Faces	Used for rotating faces on a selected solid.
	Taper Faces	Used for tapering selected faces of a solid at a draft angle along a vector direction.
	Copy Faces	Used for copying selected faces of a solid. These copied faces can take the form of regions or bodies.
	Color Faces	Used for assigning unique colors to individual faces.
	Copy Edges	Used to copy edges from a solid. These new edges are often used to create new solids.
	Color Edges	Used for assigning unique colors to individual solid edges.
	Imprint	Used for adding construction geometry to a solid model.
	Clean	Used to remove imprints from a solid.
	Separate	Used to separate a solid into multiple parts as long as those parts do not intersect at any point. Solids sometimes act as a single entity even though they appear to be separate solids (unioning solids together that do not touch or removing part of a solid so that the remaining pieces do not touch).
	Shell	Used to create a thin wall in a solid model.
	Check	Used to prove a solid is valid.

EXTRUDING (FACE EDITING)

Faces may be lengthened or shortened through the Extrude option of the SOLIDEDIT command. A positive distance extrudes the face in the direction of its normal. A negative distance extrudes the face in the opposite direction.

Try It! – Open the drawing file 21_Extrude. In the following image, the highlighted face at "A" needs to be decreased in height.

Tip: You will achieve better results when selecting a face for the SOLIDEDIT command if you pick on the inside of the face rather than on the edge of the face.

Command: **SOLIDEDIT**

Solids editing automatic checking: SOLIDCHECK=1
Enter a solids editing option [Face/Edge/Body/Undo/eXit] <eXit>: **F** *(For Face)*
Enter a face editing option
[Extrude/Move/Rotate/Offset/Taper/Delete/Copy/coLor/mAterial/Undo/eXit] <eXit>:
   E *(For Extrude)*
Select faces or [Undo/Remove]: *(Select the face inside the area represented by "A")*
Select faces or [Undo/Remove/ALL]: *(Press ENTER to continue)*
Specify height of extrusion or [Path]: **-10.00**
Specify angle of taper for extrusion <0>: *(Press ENTER)*
Solid validation started.
Solid validation completed.
Enter a face editing option
[Extrude/Move/Rotate/Offset/Taper/Delete/Copy/coLor/mAterial/Undo/eXit] <eXit>:
   (Press ENTER)
Solids editing automatic checking: SOLIDCHECK=1
Enter a solids editing option [Face/Edge/Body/Undo/eXit] <eXit>: *(Press ENTER)*

The result is illustrated in the following image at "B."

Figure 21–53

MOVING (FACE EDITING)

 Try It! – Open the drawing 21_Move. The object in the following image illustrates two intersecting cylinders. The two horizontal cylinders need to be moved 1 unit up from their current location. The cylinders are first selected at "A" and "B" through the SOLIDEDIT command along with the Move option.

Command: **SOLIDEDIT**
Solids editing automatic checking: SOLIDCHECK=1
Enter a solids editing option [Face/Edge/Body/Undo/eXit] <eXit>: **F** *(For Face)*
Enter a face editing option
[Extrude/Move/Rotate/Offset/Taper/Delete/Copy/coLor/mAterial/Undo/eXit] <eXit>:
  M *(For Move)*
Select faces or [Undo/Remove]: *(Select both highlighted faces at "A" and "B")*
Select faces or [Undo/Remove/ALL]: *(Press* ENTER *to continue)*
Specify a base point or displacement: *(Pick any point on the screen)*
Specify a second point of displacement: **@0,0,1**
Solid validation started.
Solid validation completed.
Enter a face editing option
[Extrude/Move/Rotate/Offset/Taper/Delete/Copy/coLor/mAterial/Undo/eXit] <eXit>:
  (Press ENTER*)*
Solids editing automatic checking: SOLIDCHECK=1
Enter a solids editing option [Face/Edge/Body/Undo/eXit] <eXit>: *(Press* ENTER*)*

The results are illustrated in the following image at "C."

Figure 21-54

ROTATING (FACE EDITING)

 Try It! – Open the drawing file 21_Rotate. In the following image, the triangular cutout needs to be rotated 45° in the clockwise direction. Use the Rotate Face option of the SOLIDEDIT command to accomplish this. You must select all faces of the triangular cutout at "A," "B," and "C." During the selection process you will have to remove the face that makes up the top of the rectangular base at "D" before proceeding.

 Command: **SOLIDEDIT**
Solids editing automatic checking: SOLIDCHECK=1
Enter a solids editing option [Face/Edge/Body/Undo/eXit] <eXit>: **F** *(For Face)*
Enter a face editing option
[Extrude/Move/Rotate/Offset/Taper/Delete/Copy/coLor/mAterial/Undo/eXit] <eXit>:
  R *(For Rotate)*
Select faces or [Undo/Remove]: *(Select all faces that make up the triangular extrusion. Pick inside areas at "A," "B," and "C"; select "C" twice to highlight it)*
Select faces or [Undo/Remove/ALL]: **R** *(For Remove)*
Remove faces or [Undo/Add/ALL]: *(Select the face at "D" to remove)*
Remove faces or [Undo/Add/ALL]: *(Press ENTER to continue)*
Specify an axis point or [Axis by object/View/Xaxis/Yaxis/Zaxis] <2points>: **Z** *(For Zaxis)*
Specify the origin of the rotation <0,0,0>: *(Select the endpoint at "E")*
Specify a rotation angle or [Reference]: **-45** *(To rotate the triangular extrusion 45° in the clockwise direction)*
Solid validation started.
Solid validation completed.
Enter a face editing option
[Extrude/Move/Rotate/Offset/Taper/Delete/Copy/coLor/mAterial/Undo/eXit] <eXit>:
  (Press ENTER)
Solids editing automatic checking: SOLIDCHECK=1
Enter a solids editing option [Face/Edge/Body/Undo/eXit] <eXit>: *(Press ENTER)*

The results are illustrated in the following image at "F."

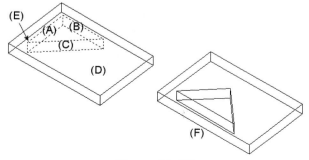

Figure 21–55

OFFSETTING (FACE EDITING)

 Try It! – Open the drawing file 21_Offset. In the following image, the holes need to be resized. Use the Offset Face option of the SOLIDEDIT command to increase or decrease the size of selected faces. Using positive values actually increases the volume of the solid. Therefore, the feature being offset gets smaller, similar to the illustration at "B." Entering negative values reduces the volume of the solid; this means that the feature being offset gets larger, as in the figures at "C." Study the prompt sequence and the following image for the mechanics of this command option.

Command: **SOLIDEDIT**
Solids editing automatic checking: SOLIDCHECK=1
Enter a solids editing option [Face/Edge/Body/Undo/eXit] <eXit>: **F** *(For Face)*
Enter a face editing option
[Extrude/Move/Rotate/Offset/Taper/Delete/Copy/coLor/mAterial/Undo/eXit] <eXit>:
  O *(For Offset)*
Select faces or [Undo/Remove]: *(Select inside the edges of the two holes at "D" and "E"; you can select the edges of the holes twice, but if you do, you will need to use the Remove option to deselect the bottom and top edges at "F" and "G")*
Select faces or [Undo/Remove/ALL]: *(Press ENTER to continue)*
Specify the offset distance: **0.50**
Solid validation started.
Solid validation completed.
Enter a face editing option
[Extrude/Move/Rotate/Offset/Taper/Delete/Copy/coLor/mAterial/Undo/eXit] <eXit>:
  (Press ENTER)
Solids editing automatic checking: SOLIDCHECK=1
Enter a solids editing option [Face/Edge/Body/Undo/eXit] <eXit>: *(Press ENTER)*

(A)　　　　　　　　　(B)　　　　　　　　　(C)

Figure 21–56

TAPERING (FACE EDITING)

 Try It! – Open the drawing 21_Taper. The object at "A" in the following image represents a solid box that needs to have tapers applied to its sides. Using the Taper Face option of the SOLIDEDIT command allows you to accomplish this task. Entering a positive angle moves the location of the second point into the part, as shown at "F." Entering a negative angle moves the location of the second point away from the part, as shown at "G."

Command: **SOLIDEDIT**
Solids editing automatic checking: SOLIDCHECK=1
Enter a solids editing option [Face/Edge/Body/Undo/eXit] <eXit>: **F** *(For Face)*
Enter a face editing option
[Extrude/Move/Rotate/Offset/Taper/Delete/Copy/coLor/mAterial/Undo/eXit] <eXit>:
   T *(For Taper)*
Select faces or [Undo/Remove]: *(Select faces "A" through "D")*
Select faces or [Undo/Remove/ALL]: *(Press ENTER to continue)*
Specify the base point: *(Select the endpoint at "E")*
Specify another point along the axis of tapering: *(Select the endpoint at "B")*
Specify the taper angle: **10** *(For the angle of the taper)*
Solid validation started.
Solid validation completed.
Enter a face editing option
[Extrude/Move/Rotate/Offset/Taper/Delete/Copy/coLor/mAterial/Undo/eXit] <eXit>:
   (Press ENTER)
Solids editing automatic checking: SOLIDCHECK=1
Enter a solids editing option [Face/Edge/Body/Undo/eXit] <eXit>: *(Press ENTER)*

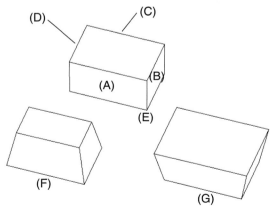

Figure 21–57

DELETING (FACE EDITING)

Faces can be erased through the Delete Face option of the SOLIDEDIT command.

 Try It! – Open the drawing file 21_Delete. In the following image, select the hole at "A" as the face to erase.

Command: **SOLIDEDIT**
Solids editing automatic checking: SOLIDCHECK=1
Enter a solids editing option [Face/Edge/Body/Undo/eXit] <eXit>: **F** *(For Face)*
Enter a face editing option
[Extrude/Move/Rotate/Offset/Taper/Delete/Copy/coLor/mAterial/Undo/eXit] <eXit>:
   D *(For Delete)*
Select faces or [Undo/Remove]: *(Select inside the hole at "A")*
Select faces or [Undo/Remove/ALL]: *(Press* ENTER *to continue)*
Solid validation started.
Solid validation completed.
Enter a face editing option
[Extrude/Move/Rotate/Offset/Taper/Delete/Copy/coLor/mAterial/Undo/eXit] <eXit>:
   (Press ENTER*)*
Solids editing automatic checking: SOLIDCHECK=1
Enter a solids editing option [Face/Edge/Body/Undo/eXit] <eXit>: *(Press* ENTER*)*

The results are illustrated in the following image at "B."

Figure 21–58

COPYING (FACE EDITING)

You can copy a face for use in the creation of another solid model using the Copy Face option of the SOLIDEDIT command.

 Try It! – Open the drawing file 21_Copy. In the following image, the solid model at "A" will be used to create a region from the face at "B." While in the command, select the face by picking in the area at "B." Notice that all objects making up the face, such as the rectangle and circles, are highlighted. Picking a base point and second point copies the face at "C." The resulting object at "C" is actually a region. The region could be exploded back into individual lines and circles, which could then be used to create a new object. A region can also be extruded, using the EXTRUDE command, to create another solid model such as the one illustrated at "D."

 Command: **SOLIDEDIT**
Solids editing automatic checking: SOLIDCHECK=1
Enter a solids editing option [Face/Edge/Body/Undo/eXit] <eXit>: **F** *(For Face)*
Enter a face editing option
[Extrude/Move/Rotate/Offset/Taper/Delete/Copy/coLor/mAterial/Undo/eXit] <eXit>:
  C *(For Copy)*
Select faces or [Undo/Remove]: *(Select the top face in area "B")*
Select faces or [Undo/Remove/ALL]: *(Press* ENTER *to continue)*
Specify a base point or displacement: *(Pick a point to copy from)*
Specify a second point of displacement: *(Pick a point to copy to)*
Enter a face editing option
[Extrude/Move/Rotate/Offset/Taper/Delete/Copy/coLor/mAterial/Undo/eXit] <eXit>:
  (Press ENTER*)*
Solids editing automatic checking: SOLIDCHECK=1
Enter a solids editing option [Face/Edge/Body/Undo/eXit] <eXit>: *(Press* ENTER*)*

Figure 21–59

IMPRINTING (BODY EDITING)

An interesting and powerful method of adding construction geometry to a solid model is though the process of imprinting.

 Try It! – Open the drawing file 21_Imprint. In the following image at "A," a box along with slot is already modeled in 3D. A line was constructed from the midpoints of the top surface of the solid model. The SOLIDEDIT command will be used to imprint this line to the model, which results in dividing the top surface into two faces.

 Command: **SOLIDEDIT**

Solids editing automatic checking: SOLIDCHECK=1
Enter a solids editing option [Face/Edge/Body/Undo/eXit] <eXit>: **B** *(For Body)*
Enter a body editing option
[Imprint/seParate solids/Shell/cLean/Check/Undo/eXit] <eXit>: **I** *(For Imprint)*
Select a 3D solid: *(Select the solid model)*
Select an object to imprint: *(Select line "B")*
Delete the source object [Yes/No] <N>: **Y** *(For Yes; this erases the line)*
Select an object to imprint: *(Press ENTER to complete the imprint operation)*
Enter a body editing option
[Imprint/seParate solids/Shell/cLean/Check/Undo/eXit] <eXit>: *(Press ENTER)*
Solids editing automatic checking: SOLIDCHECK=1
Enter a solids editing option [Face/Edge/Body/Undo/eXit] <eXit>: *(Press ENTER)*

The segments of the lines that come in contact with the 3D solid remain on the part's surface. However, these are no longer line segments; rather, these lines now belong to the part. The lines actually separate the top surface into two faces.

Use the Extrude Face option of the SOLIDEDIT command on one of the newly created faces at "C" and increase the face in height by an extra 1.50 units. The results are illustrated in the following image at "D."

Figure 21-60

SEPARATING SOLIDS (BODY EDITING)

Sometimes when performing union and subtraction operations on solid models, you can end up with models that do not actually touch (intersect) but act as a single object. The Separating Solids option of the SOLIDEDIT command is used to correct this condition.

 Try It! – Open the drawing file 21_Separate. In the following image, the model at "A" is about to be sliced in half with a thin box created at the center of the circle and spanning the depth of the rectangular shelf.

Use the SUBTRACT command and subtract the rectangular box from the solid object. After you subtract the box, pick the solid at "B" and notice that both halves of the object highlight even though they appear separate. To convert the single solid model into two separate models, use the SOLIDEDIT command followed by the Body and Separate options.

Command: **SOLIDEDIT**
Solids editing automatic checking: SOLIDCHECK=1
Enter a solids editing option [Face/Edge/Body/Undo/eXit] <eXit>: **B** *(For Body)*
Enter a body editing option
[Imprint/seParate solids/Shell/cLean/Check/Undo/eXit] <eXit>: **P** *(For Separate)*
Select a 3D solid: *(Pick the solid model at "B")*
Enter a body editing option
[Imprint/seParate solids/Shell/cLean/Check/Undo/eXit] <eXit>: *(Press ENTER)*
Solids editing automatic checking: SOLIDCHECK=1
Enter a solids editing option [Face/Edge/Body/Undo/eXit] <eXit>: *(Press ENTER)*

This action separates the single model into two. When you select the model in the following image at "C," only one half highlights.

Figure 21–61

SHELLING (BODY EDITING)

Shelling is the process of constructing a thin wall inside or outside of a solid model. Positive thickness will produce the thin wall inside; negative values for thickness will produce the thin wall outside. This wall thickness remains constant throughout the entire model. Faces may be removed during the shelling operation to create an opening.

 Try It! – Open the drawing file 21_Shell. For clarity, it is important to first rotate your model or your viewpoint such that any faces to be removed are visible. Use the 3DFORBIT (Free Orbit) command and the following image to change the model view from the one shown at "A" to the one shown at "B." Use the 3D Hidden Visual Style option in the 3DFORBIT command to ensure that the bottom surface at "C" is visible. Now the Shell option of the SOLIDEDIT command can be used to "hollow out" the part and remove the bottom face. An additional note: only one shell is permitted in a model.

 Command: **SOLIDEDIT**

Solids editing automatic checking: SOLIDCHECK=1
Enter a solids editing option [Face/Edge/Body/Undo/eXit] <eXit>: **B** *(For Body)*
Enter a body editing option
[Imprint/seParate solids/Shell/cLean/Check/Undo/eXit] <eXit>: **S** *(For Shell)*
Select a 3D solid: *(Select the solid model at "B")*
Remove faces or [Undo/Add/ALL]: *(Pick a point at "C"; because the model is already highlighted, it is not obvious that the face is selected but "1 face found, 1 removed" will be indicated)*
Remove faces or [Undo/Add/ALL]: *(Press* ENTER *to continue)*
Enter the shell offset distance: **0.20**
Solid validation started.
Solid validation completed.
Enter a body editing option
[Imprint/seParate solids/Shell/cLean/Check/Undo/eXit] <eXit>: *(Press* ENTER*)*
Solids editing automatic checking: SOLIDCHECK=1
Enter a solids editing option [Face/Edge/Body/Undo/eXit] <eXit>: *(Press* ENTER*)*

The results are illustrated in the following image at "D."

Figure 21–62

CLEANING (BODY EDITING)

When imprinted lines that form faces are not used, they can be deleted from a model by the Clean (Body Editing) option of the SOLIDEDIT command.

 Try It! – Open the drawing file 21_Clean. The object at "A" in the following image illustrates lines originally constructed on the top of the solid model. These lines were then imprinted at "B." Since these lines now belong to the model, the Clean option is used to remove them. The results are illustrated at "C."

 Command: **SOLIDEDIT**
Solids editing automatic checking: SOLIDCHECK=1
Enter a solids editing option [Face/Edge/Body/Undo/eXit] <eXit>: **B** *(For Body)*
Enter a body editing option
[Imprint/sePparate solids/Shell/cLean/Check/Undo/eXit] <eXit>: **L** *(For Clean)*
Select a 3D solid: *(Select the solid model at "B")*
Enter a body editing option
[Imprint/sePparate solids/Shell/cLean/Check/Undo/eXit] <eXit>: *(Press ENTER)*
Solids editing automatic checking: SOLIDCHECK=1
Enter a solids editing option [Face/Edge/Body/Undo/eXit] <eXit>: *(Press ENTER)*

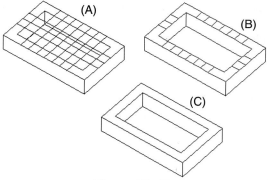

Figure 21–63

Additional options of the SOLIDEDIT command include the ability to apply a color or material to a selected face. You can also apply color to an edge or copy edges from a model so that they can be used for construction purposes on other models.

TUTORIAL EXERCISE: 21_3DROTATE.DWG

Figure 21–64

Purpose
This tutorial exercise is designed to use the 3DROTATE command to rotate the objects into position before creating unions or performing subtractions, as shown in the following image.

System Settings
The drawing units, limits, grid, and snap values are already set for this drawing. Check to see that the following Object Snap modes are already set: Endpoint, Center, Intersection, and Extension. Also check to see that OSNAP is turned on. These are located in the bottom status bar.

Layers
Make sure the current layer is Model.

Suggested Commands
Open the drawing file 21_3DRotate.dwg.

Use the 3DROTATE command to rotate the sides to the proper angle before they are moved into place and joined. The cylinders are also rotated and moved into place before being subtracted from the base and sides to create holes.

STEP 1

Use the MOVE command to position the center of the large cylinder "A" in the middle of the base, between "B" and "C," as identified in the following image. Use the OSNAP-m2p option (midpoint between two points) to accomplish this.

Figure 21–65

STEP 2

Begin rotating the shapes into place using the 3DROTATE command. Select solid "A" as the object to rotate. Pick the endpoint at "B" as the base point of the rotation. When the rotate grip tool appears, click on the axis handle to display the axis, as shown in the following image. Enter a value of -90 degrees to perform the rotate operation.

 Command: **3DROTATE**
Current positive angle in UCS: ANGDIR=counterclockwise ANGBASE=0
Select objects: *(Select solid "A")*
Select objects: *(Press ENTER to continue)*
Specify base point: *(Pick the endpoint at "B")*
Pick a rotation axis: *(When the rotate grip tool appears, hover on the axis handle until the proper axis appears, as shown in the following image; then pick with the mouse)*
Specify angle start point: *(Pick the endpoint at "B")*
Specify angle end point: **-90**

Figure 21–66

STEP 3

Position the endpoint of the side panel at "A" that you just rotated to the endpoint of the top of the base at "B" using the MOVE command, as shown in the following image.

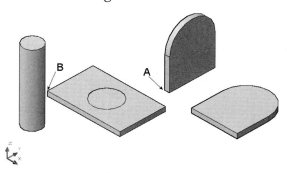

Figure 21–67

STEP 4

Now rotate the second side panel into position. This will take two steps to accomplish.

First, rotate the panel 90° using the rotation axis, as shown in the following image.

 Command: **3DROTATE**
Current positive angle in UCS: ANGDIR=counterclockwise ANGBASE=0
Select objects: *(Select the other side panel)*
Select objects: *(Press* ENTER *to continue)*
Specify base point: *(Pick the endpoint at "A")*
Pick a rotation axis: *(When the rotate grip tool appears, hover on the axis handle until the proper axis appears, as shown in the following image; then pick with the mouse)*
Specify angle start point: *(Pick the endpoint at "A")*
Specify angle end point: **90**

Figure 21–68

STEP 5

Now rotate the same panel -90° using the rotation axis, as shown in the following image.

Command: **3DROTATE**
Current positive angle in UCS: ANGDIR=counterclockwise ANGBASE=0
Select objects: *(Select the same side panel)*
Select objects: *(Press ENTER to continue)*
Specify base point: *(Pick the endpoint at "A")*
Pick a rotation axis: *(When the rotate grip tool appears, hover on the axis handle until the proper axis appears, as shown in the following image; then pick with the mouse)*
Specify angle start point: *(Pick the endpoint at "A")*
Specify angle end point: **-90**

Figure 21-69

STEP 6

Now position the endpoint of the panel at "A" along the top of the endpoint of the base at "B" using the MOVE command, as shown on the left in the following image.

When finished, the display should appear similar to the illustration on the right in the following image.

Figure 21-70

STEP 7

Next, rotate the long cylinder -90° along the rotation axis, as shown on the left in the following image. Then, move the cylinder from its center at "B" to the center of the arc on the side panel at "C," as shown on the right in the following image.

 Command: **3DROTATE**

Current positive angle in UCS: ANGDIR=counterclockwise ANGBASE=0
Select objects: *(Select the tall cylinder)*
Select objects: *(Press ENTER to continue)*
Specify base point: *(Pick the center of the cylinder at "A")*
Pick a rotation axis: *(When the rotate grip tool appears, click on the axis handle until the proper axis appears, as shown in the following image; then pick this axis with the mouse)*
Specify angle start point: *(Pick the center of the cylinder at "A")*
Specify angle end point: **-90**

Figure 21–71

STEP 8

Finally, use the SUBTRACT command to subtract the two cylinders from the 3 plates using the following prompt sequence and illustration on the left in the following image. The side panels will automatically be unioned to the base when you complete this command. The completed object is displayed on the right in the following image. The ALIGN command could also have been used to rotate and move the extruded shapes into their proper positions.

Command: **SU** *(For SUBTRACT) (can be selected from the pull-down menu [modify>solids editing>subtract])*
Select solids and regions to subtract from
Select objects: *(Select the two side panels and the base)*
Select objects: *(Press ENTER to continue)*
Select solids and regions to subtract.
Select objects: *(Select the two cylinders to subtract)*
Select objects: *(Press ENTER to perform the union/subtraction operations)*

Figure 21–72

TUTORIAL EXERCISE: 21_VALVE HEAD SECTION.DWG

Figure 21–73

Purpose

This tutorial exercise is designed to create a cutting plane through a 3D model and generate both a flat 2D section view along with a 3D section using the SECTIONPLANE command.

System Settings

The drawing units, limits, grid, and snap values are already set for this drawing.

Layers

Layers have already been created for this tutorial exercise.

Suggested Commands

Open the drawing file 21_Valve Head Section.dwg.

Use the SECTIONPLANE command to create a cutting plane through the 3D model, generating a 2D section and a 3D section.

STEP 1

Begin this tutorial by clicking on Section Plane, located in the dashboard, as shown in the middle in the following image. When prompted to "Select face or any point to locate section line," click on the center of the small hole, as shown on the right in the following image. When prompted for the "Through point," turn Ortho mode on and construct the line to the other side of the valve head, as shown on the right in the following image. This line forms the section plane.

Figure 21–74

STEP 2

Next, click on the line just created. Notice the appearance of grips that perform different functions. Then right-click and pick Activate live sectioning from the menu, as shown in the following image.

Figure 21–75

STEP 3

The results are illustrated on the left in the following image. Here, active sectioning lining is now active and shows only the portion of the 3D model cut by the section plane in the direction identified by the Flip arrow. Since we need to look in the opposite direction at the cut, click on this arrow to view the section cut, as shown on the right in the following image. When finished, press the ESC key to deselect all items.

Figure 21–76

STEP 4

Select the section plane line again, right-click, and pick Show cut-away geometry from the menu, as shown in the following image. The geometry cut by the section plane is displayed in the default color of red.

Figure 21–77

STEP 5

A jog needs to be created in the section plane so that more details of the valve head can be viewed. Begin this process by picking on the section plane line, right-clicking, and selecting Add jog to section, as shown in the following image.

Figure 21-78

STEP 6

When prompted to "Specify a point on the section line to add jog," click on the approximate location on the section plane line, as shown on the left in the following image. The Osnap-Nearest mode automatically activates when you select this point. The results are displayed on the right in the following image. A predefined jog distance repositions the section plane beginning at the selected point. Unfortunately, the jog distance does not cut through the required features of the model. This will be corrected in the next step.

Figure 21-79

STEP 7

Activate the PLAN command and accept the default value of Current UCS to view the valve head as a flat image. Then pick the arrow grip, as shown on the left in the following image, and stretch the section plane jog down until it is located approximately in the middle of the hole feature, as shown on the right in the following image.

Figure 21–80

STEP 8

Since it is no longer necessary to display the cut-away geometry, pick the section plane line, right-click, and select Show cut-away geometry to turn this feature off. Perform a ZOOM-Previous to return to the Isometric view and display the 3D model with the section plane jog, as shown in the following image.

Figure 21–81

STEP 9

With all section plane parameters made, a 2D and 3D section will now be created. Click on the section plane line, right-click, and pick Generate 2D/3D section from the menu, as shown in the following image. This will launch the Generate Section/Elevation dialog box, as shown on the right in the following image. A number of settings need to be changed before the sections are created. Click on the Section Settings button, located in the lower portion of the dialog box.

Figure 21–82

STEP 10

A number of categories are available to fine-tune the creation of the section. The first category to make changes to is Intersection Fill. By default, the hatch pattern used in the section creation is Solid. Click in the Select hatch pattern type next to Face Hatch, as shown in the following image. This will launch the Hatch Pattern Type dialog box, where the default pattern is changed to ANSI31, as shown on the right in the following image. Also, the Hatch Scale was changed from 1.00 to 25.40 to enlarge the spacing between the hatch lines, since this is a metric 3D model.

Figure 21–83

STEP 11

Continue making section settings changes in the three areas illustrated in the following image. In the Background Lines category, click in the area next to Hidden Line and change the setting from Yes to No, as shown on the left in the following image. Background hidden lines will not be displayed when the section is created. In the Cut-away Geometry category, click in the area next to Hidden Line and change the setting from Yes to No, as shown in the middle of the following image. Finally, identify the Curve Tangency Lines category, click in the area next to Show, and change the setting from Yes to No, as shown on the right in the following image.

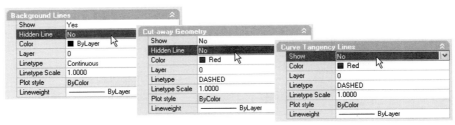

Figure 21–84

STEP 12

When finished making changes to the Section Settings, click the OK button to return to the Generate Section/Elevation dialog box. Click the Create button, as shown on the left in the following image, to begin generating the section. When prompted for an insertion point, pick a point approximately at "A," as shown in the middle in the following image. Keep the default X and Y scale factors but change the rotation angle to 180°. Your section view should appear similar to the illustration shown in the middle in the following image.

Figure 21–85

STEP 13

Perform one other task on this section view. Use the MIRROR command and flip the section view without making a copy, in order for it to appear similar to the illustration shown on the left in the following image.

Figure 21–86

STEP 14

Before continuing, make the 3D Section layer current. Then, click on the section plane line, right-click, and pick Generate 2D/3D section from the menu to launch the Generate Section/Elevation dialog box, as shown in the following image. The next section will be 3D, so click on the 3D Section box at the top of this dialog box and click the Create button. Place the section, as shown on the right in the following image, in a convenient location on your screen.

Figure 21–87

STEP 15

After locating the 3D section, click once more on the section plane line and delete it. This will return the 3D model of the valve cover back to its original display, as shown on the right in the following image.

Figure 21–88

STEP 16

Since the 2D and 3D sections were created with anonymous block names, use the RENAME command to launch the Rename dialog box, as shown on the left in the following image, and rename these block names to something that can be more easily recognized, such as 2D Section and 3D Section, as shown on the right in the following image.

Figure 21–89

PROBLEMS FOR CHAPTER 21

PROBLEM 21-1

Utilize the skills that are laid out in this chapter on problems that are found in Chapters 9, 10, and 20. Many of the problems in these chapters will apply the editing skills found in this chapter.

One option is to take a drawing that was completed as a two-dimensional drawing and re-create it as a 3D solid.

CHAPTER 22

Creating 2D Multiview Drawings from a Solid Model

One of the advantages of creating a 3D image in the form of a solid model is the ability to use the data of the solid model numerous times for other purposes. The purpose of this chapter is to generate 2D multiview drawings from the solid model. Two commands are used together to perform this operation: SOLVIEW and SOLDRAW. The SOLVIEW command is used to create a layout for the 2D views. This command automatically creates layers used to organize visible lines, hidden lines, and dimensions. The SOLDRAW command draws the requested views on the specific layers that were created by the SOLVIEW command (this includes the drawing of hidden lines to show hidden features, and even hatching if a section is requested).

Included in the layout of 2D multiview projections will be the creation of orthographic, auxiliary, section, and isometric views for a drawing. Also explained will be the FLATSHOT command, used to create 2D views that are saved as blocks. A tutorial is available at the end of this chapter to demonstrate how to dimension orthographic views laid out with the SOLVIEW/SOLDRAW command.

THE SOLVIEW AND SOLDRAW COMMANDS

Once you create the solid model, you can lay out and draw the necessary 2D views using a number of Modeling Setup commands. Choosing Modeling from the Draw pull-down menu and then Setup, as shown in the following image, exposes the Drawing (SOLDRAW), View (SOLVIEW), and Profile (SOLPROF) commands. Only SOLDRAW and SOLVIEW will be discussed in this chapter.

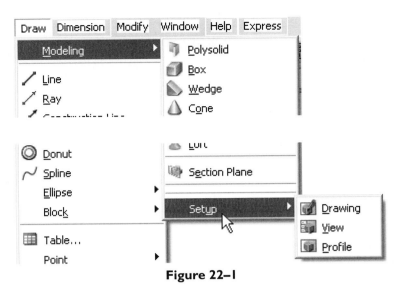

Figure 22–1

Once the SOLVIEW command is entered, the display screen automatically switches to the first layout or Paper Space environment. Using SOLVIEW will lay out a view based on responses to a series of prompts, depending on the type of view you want to create. Usually, the initial view that you lay out serves as the starting point for other orthogonal views and is based on the current User Coordinate System. This needs to be determined before you begin this command. Once an initial view is created, it is very easy to create Orthographic, Auxiliary, Section, and even Isometric views.

As SOLVIEW is used as a layout tool, the images of the views created are still simply plan views of the original solid model. In other words, after you lay out a view, it does not contain any 2D features, such as hidden lines. As shown in the previous image, clicking on Drawing activates the SOLDRAW command, which is used to actually create the 2D profiles once it has been laid out through the SOLVIEW command.

 Try It! – Open the drawing file 22_Solview. Before using the SOLVIEW command, study the illustration of this solid model, as shown at "A" in the following image. In particular, pay close attention to the position of the User Coordinate System icon. The current position of the User Coordinate System will start the creation process of the base view of the 2D drawing.

 Tip: Before you start using the SOLVIEW command, remember to load the Hidden linetype. This will automatically assign this linetype to any new layer that requires hidden lines for the drawing mode. If the linetype is not loaded at this point, it must be manually assigned later to each layer that contains hidden lines through the Layer Properties Manager dialog box.

Activating SOLVIEW automatically switches the display to the layout or Paper Space environment. Since this is the first view to be laid out, the UCS option will be used to create the view based on the current User Coordinate System. The view produced is similar to looking down the Z axis of the UCS icon. A scale value may be entered for the view. For the View Center, click anywhere on the screen and notice the view being constructed. You can pick numerous times on the screen until the view is in a desired location. The placement of this first view is very important because other views will most likely be positioned from this one. When this step is completed, press ENTER to place the view. Next, you are prompted to construct a viewport around the view. Remember to make this viewport large enough for dimensions to fit inside. Once the view is given a name, it is laid out similar to the illustration at "B" in the following image.

 Command: **SOLVIEW**
Enter an option [Ucs/Ortho/Auxiliary/Section]: **U** *(For Ucs)*
Enter an option [Named/World/?/Current] <Current>: *(Press* ENTER*)*
Enter view scale <1.0000>: *(Press* ENTER*)*
Specify view center: *(Pick a point near the center of the screen to display the view; keep picking until the view is in the desired location)*
Specify view center <specify viewport>: *(Press* ENTER *to place the view)*
Specify first corner of viewport: *(Pick a point at "D")*
Specify opposite corner of viewport: *(Pick a point at "E")*
Enter view name: **FRONT**
Enter an option [Ucs/Ortho/Auxiliary/Section]: *(Press* ENTER *to exit this command)*

Once the view has been laid out through the SOLVIEW command, use SOLDRAW to actually draw the view in two dimensions. The Hidden linetype was loaded for you in this drawing and since it was loaded prior to using the SOLVIEW command, hidden lines will automatically be assigned to layers that contain hidden line information. The result of using the SOLDRAW command is shown at "C" in the following image. You are no longer looking at a 3D solid model but at a 2D drawing created by this command.

 Command: **SOLDRAW**
(If in Model Space, you are switched to Paper Space)
Select viewports to draw.
Select objects: *(Pick anywhere on the viewport—at "C" in the following image)*
Select objects: *(Press* ENTER *to perform the* SOLDRAW *operation)*
One solid selected.

Figure 22–2

The use of layers in 2D-view layout is so important that when you run the SOLVIEW command, the layers shown in the following image are created automatically. With the exception of Model and 0, the layers that begin with "FRONT" and the VPORTS layer were all created by the SOLVIEW command. The FRONT-DIM layer is designed to hold dimension information for the Front view. FRONT-HID holds all hidden lines information for the Front view; FRONT-VIS holds all visible line information for the Front view. All Paper Space viewports are placed on the VPORTS layer. The Model layer has automatically been frozen in the current viewport to hide the 3D model and show the 2D visible and hidden lines.

Status	Name	On	Fre	Loc	Color	Linetype	Lineweight	Plot Styl	Plot	Cur	Ne	Description
	0				white	CONTINUOUS	Default	Color_7				
	FRONT-DIM				white	CONTINUOUS	Default	Color_7				
	FRONT-HID				white	HIDDEN	Default	Color_7				
	FRONT-VIS				white	CONTINUOUS	Default	Color_7				
✓	MODEL				white	CONTINUOUS	Default	Color_7				
	VPORTS				white	CONTINUOUS	Default	Color_7				

Figure 22–3

In order for the view shown on the left in the following image to be dimensioned in model space, three operations must be performed. First, double-click inside the Front view to be sure it is the current floating Model Space viewport. Next, make FRONT-DIM the current layer. Finally, if it is not already positioned correctly, set the User Coordinate System to the current view using the View option of the UCS command. The UCS icon should be similar to the illustration on the left in the following image (you should be looking straight down the Z axis). Now add all dimensions to the view using conventional dimensioning commands with the aid of Object snap modes. When you work on adding dimensions to another view, the same three operations must be made in the new view: make the viewport active by double-clicking inside it, make the appropriate dimension layer current, and update the UCS, if necessary, to the current view with the View option.

When you draw the views using the SOLDRAW command and then add the dimensions, switching back to the solid model by clicking on the Model tab displays the illustration shown on the right in the following image. In addition to the solid model of the object, the constructed 2D view and dimensions are also displayed. All drawn views from Paper Space will display with the model. To view just the solid model you would have to use the Layer Properties Manager dialog box along with the Freeze option and freeze all drawing-related layers.

 Tip: Any changes made to the solid model will not update the drawing views. If changes are made, you must erase the previous views and run SOLVIEW and SOLDRAW again to generate a new set of views.

Figure 22–4

CREATING ORTHOGRAPHIC VIEWS

Once the first view is created, orthographic views can easily be created with the Ortho option of the SOLVIEW command.

 Try It! – Open the drawing file 22_Ortho. Notice that you are in a layout and a Front view is already created. Follow the command prompt sequence below to create two orthographic views. When finished, your drawing should appear similar to the following image.

Command: **SOLVIEW**
Enter an option [Ucs/Ortho/Auxiliary/Section]: **O** *(For Ortho)*
Specify side of viewport to project: *(Select the top of the viewport at "A"—a midpoint Osnap will be automatically provided)*
Specify view center: *(Pick a point above the front view to locate the top view)*
Specify view center <specify viewport>: *(Press* ENTER *to place the view)*
Specify first corner of viewport: *(Pick a point at "B")*
Specify opposite corner of viewport: *(Pick a point at "C")*
Enter view name: **TOP**
Enter an option [Ucs/Ortho/Auxiliary/Section]: **O** *(For Ortho)*
Specify side of viewport to project: *(Select the right side of the viewport at "D")*
Specify view center: *(Pick a point to the right of the front view to locate the right side view)*
Specify view center <specify viewport>: *(Press* ENTER *to place the view)*
Specify first corner of viewport: *(Pick a point at "E")*
Specify opposite corner of viewport: *(Pick a point at "F")*
Enter view name: **R_SIDE**
Enter an option [Ucs/Ortho/Auxiliary/Section]: *(Press* ENTER *to exit this command)*

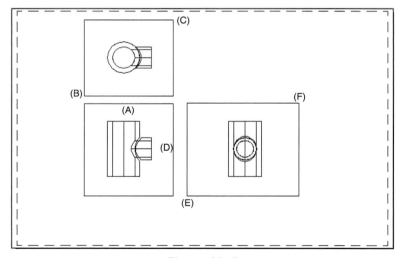

Figure 22–5

Running the SOLDRAW command on the three views displays the illustration shown in the following image. Notice the appearance of the hidden lines in all views. The VPORTS layer is turned off to display only the three views.

Command: **SOLDRAW**
Select viewports to draw.
Select objects: *(Select the three viewports that contain the Front, Top, and Right Side view information)*
Select objects: *(Press* ENTER *to perform the* SOLDRAW *operation)*

Figure 22–6

CREATING AN AUXILIARY VIEW

In the illustration on the left in the following image, the true size and shape of the inclined surface containing the large counterbore hole cannot be shown with the standard orthographic view. An auxiliary view must be used to properly show these features.

 Try It! – Open the drawing file 22_Auxiliary. From the 3D model in the following image use the SOLVIEW command to create a Front view based on the current User Coordinate System. The results are shown on the right in the following image.

 Command: **SOLVIEW**
Enter an option [Ucs/Ortho/Auxiliary/Section]: **U** *(For Ucs)*
Enter an option [Named/World/?/Current] <Current>: *(Press* ENTER*)*
Enter view scale <1.0000>: *(Press* ENTER*)*
Specify view center: *(Pick a point to locate the view, as shown on the right in the following image)*
Specify view center <specify viewport>: *(Press* ENTER *to place the view)*
Specify first corner of viewport: *(Pick a point at "A")*
Specify opposite corner of viewport: *(Pick a point at "B")*
Enter view name: **FRONT**
Enter an option [Ucs/Ortho/Auxiliary/Section]: *(Press* ENTER *to exit this command)*

Tip: Normally you do not end the SOLVIEW command after each view is laid out. Once you finish creating a view you simply enter the appropriate option (Ucs, Ortho, Auxiliary, or Section) and create the next one. This process can continue until all necessary views are provided.

Figure 22–7

Now begin the process of constructing an auxiliary view, as shown on the left in the following image. After selecting the Auxiliary option of the SOLVIEW command, click on the endpoints at "A" and "B" to establish the edge of the surface to view. Pick a point at "C" to indicate the side from which to view the auxiliary view. Notice how the Paper Space icon tilts perpendicular to the edge of the auxiliary view. Pick a location for the auxiliary view and establish a viewport. The result is illustrated in the middle in the following image.

 Command: **SOLVIEW**
Enter an option [Ucs/Ortho/Auxiliary/Section]: **A** *(For Auxiliary)*
Specify first point of inclined plane: **End**
of *(Pick the endpoint at "A")*
Specify second point of inclined plane: **End**
of *(Pick the endpoint at "B")*
Specify side to view from: *(Pick a point inside of the viewport at "C")*
Specify view center: *(Pick a point to locate the view, as shown in the middle in the following image)*
Specify view center <specify viewport>: *(Press ENTER to place the view)*
Specify first corner of viewport: *(Pick a point at "D")*
Specify opposite corner of viewport: *(Pick a point at "E")*
Enter view name: **AUXILIARY**
Enter an option [Ucs/Ortho/Auxiliary/Section]: *(Press ENTER to exit this command)*

Run the SOLDRAW command and turn off the VPORTS layer. The finished result is illustrated on the right in the following image. Hidden lines display only because this linetype was previously loaded.

Figure 22–8

CREATING A SECTION VIEW

The SOLVIEW and SOLDRAW commands can also be used to create a full section view of an object. This process will automatically create section lines and place them on a layer (*-HAT) for you.

 Try It! – Open the drawing file 22_Section. From the model illustrated on the left in the following image, create a Top view based on the current User Coordinate System, as shown on the right in the following image.

 Command: **SOLVIEW**
Regenerating layout.
Enter an option [Ucs/Ortho/Auxiliary/Section]: **U** *(For Ucs)*
Enter an option [Named/World/?/Current] <Current>: *(Press ENTER)*
Enter view scale <1.0000>: *(Press ENTER)*
Specify view center: *(Pick a point to locate the view, as shown on the right in the following image)*
Specify view center <specify viewport>: *(Press ENTER to place the view)*
Specify first corner of viewport: *(Pick a point at "A")*
Specify opposite corner of viewport: *(Pick a point at "B")*
Enter view name: **TOP**
Enter an option [Ucs/Ortho/Auxiliary/Section]: *(Press ENTER to exit this command)*

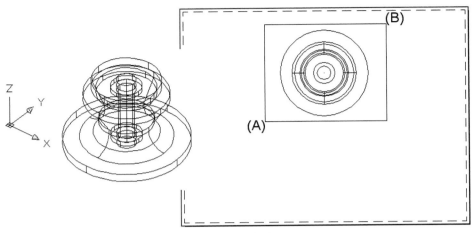

Figure 22–9

Begin the process of creating the section. You must first establish the cutting plane line in the Top view, as shown on the left in the following image. After the cutting plane line is drawn, you select the side from which to view the section. You then locate the section view. This is similar to the process of placing an auxiliary view.

Command: **SOLVIEW**
Enter an option [Ucs/Ortho/Auxiliary/Section]: **S** *(For Section)*
Specify first point of cutting plane: **Qua**
of *(Pick a point at "A")*
Specify second point of cutting plane: *(Turn Ortho on, pick a point at "B")*
Specify side to view from: *(Pick a point inside of the viewport at "C")*
Enter view scale <1.0000>: *(Press ENTER)*
Specify view center: *(Pick a point below the top view to locate the view, as shown on the right in the following image)*
Specify view center <specify viewport>: *(Press ENTER to place the view)*
Specify first corner of viewport: *(Pick a point at "D")*
Specify opposite corner of viewport: *(Pick a point at "E")*
Enter view name: **FRONT_SECTION**
Enter an option [Ucs/Ortho/Auxiliary/Section]: *(Press ENTER to continue)*

Figure 22–10

Running the SOLDRAW command on the viewports results in the illustration on the left in the following image. You can also activate the viewport displaying the section view and use the HATCHEDIT command to edit the hatch pattern. In the illustration on the right in the following image, the hatch pattern scale was increased to a value of 2.00 and the viewports turned off.

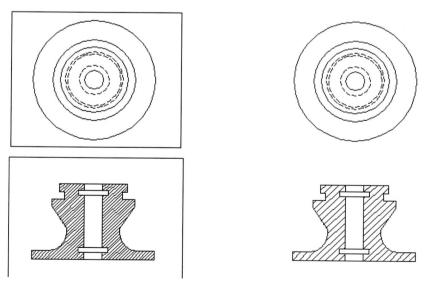

Figure 22–11

CREATING AN ISOMETRIC VIEW

Once orthographic, section, and auxiliary views are projected, you also have an opportunity to project an isometric view of the 3D model. This type of projection is accomplished using the UCS option of the SOLVIEW command and relies entirely on the viewpoint and User Coordinate System setting for your model.

 Try It! – Open the drawing file 22_Iso. This 3D model should appear similar to the illustration on the left in the following image. To prepare this image to be projected as an isometric view, first define a new User Coordinate System based on the current view. See the prompt sequence below to accomplish this task. Your image and UCS icon should appear similar to the illustration on the right in the following image.

 Command: **UCS**
Current UCS name: *WORLD*
Specify origin of new UCS or [Face/NAmed/OBject/Previous/View/World/X/Y/Z/ ZAxis] <World>: **V** *(For View)*

Figure 22–12

Next, run the SOLVIEW command based on the current UCS. Locate the view and construct a viewport around the isometric, as shown in the sample layout in the following image. Since dimensions are normally placed in the orthographic view drawings and not on an isometric, you can tighten up on the size of the viewport.

Command: **SOLVIEW**
Regenerating layout.
Enter an option [Ucs/Ortho/Auxiliary/Section]: **U** *(For Ucs)*
Enter an option [Named/World/?/Current] <Current>: *(Press ENTER)*
Enter view scale <1.0000>: *(Press ENTER)*
Specify view center: *(Pick a point to locate the view in the following image)*
Specify view center <specify viewport>: *(Press ENTER to place the view)*
Specify first corner of viewport: *(Pick a point at "A")*
Specify opposite corner of viewport: *(Pick a point at "B")*
Enter view name: **ISO**
Enter an option [Ucs/Ortho/Auxiliary/Section]: *(Press ENTER to exit this command)*

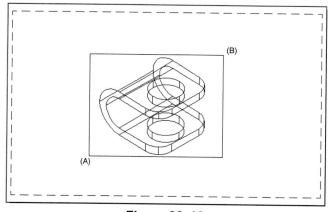

Figure 22–13

Running the SOLDRAW command on the isometric results in visible lines as well as hidden lines being displayed, as shown on the left in the following image. Generally, hidden lines are not displayed in an isometric view. The layer called ISO-HID, which was created by SOLVIEW, contains the hidden lines for the isometric drawing. Use the Layer Properties Manager dialog box to turn off this layer. The results of this operation are illustrated on the right in the following image.

Figure 22–14

EXTRACTING 2D VIEWS WITH FLATSHOT

The FLATSHOT command is used to create a quick 2D view of a 3D solid model based on the current view. You first align your view of the 3D solid and Flatshot projects an object onto the XY plane. The object created is in the form of a block and can be inserted and modified if necessary, since the block consists of 2D geometry. Choose Flatshot from the 3D Make control panel of the dashboard, as shown in the following image.

Figure 22–15

Begin the process of creating a flattened 2D view from a 3D solid model by first aligning the screen for the view that you want captured. Illustrated on the left in the following image is a 3D solid model that is currently being viewed in the Front direction. Notice also the alignment of the XY plane; Flatshot will project the geometry to this plane.

When you activate the FLATSHOT command, the dialog box illustrated on the right in the following image displays. The Destination area is used for inserting a new block or replacing an existing block. You can even export the geometry to a file with the familiar DWG extension that can be read directly by AutoCAD.

The Foreground lines area allows you to change the color and linetype of the lines considered visible.

In the Obscured lines area, you have the option of showing or not showing these lines. Obscured lines are considered invisible to the view and should be assigned the HIDDEN linetype if showing this geometry.

Figure 22–16

When clicking on the Create button, you will be prompted to insert the block based on the view. In the following image, the object on the left is the original 3D solid model, and the object on the right is the 2D block generated by the Flatshot operation. An isometric view is used in this image to illustrate the results performed by Flatshot.

Figure 22–17

It was pointed out that Flatshot creates a block. However, during the creation process, you are never asked to input the name of the block. This is because Flatshot creates a block with a randomly generated name (sometimes referred to as anonymous). This name is illustrated in the following image, where the Rename dialog box (RENAME command) is used to change the name of the block to something more meaningful, such as Front View. In fact, it is considered good practice to immediately rename the block generated by Flatshot to something more recognizable.

Figure 22–18

By default, whenever creating a block with Flatshot, you must immediately insert this block in Model space. It is also considered good practice to insert all blocks in a layout. In this way, Model space will hold the 3D solid model information and the layout will hold the 2D geometry, as shown in the following image.

Figure 22–19

 Note: While FLATSHOT can be used with layouts, it also provides a quick way to generate a 2D view of a 3D model while in Model Space.

TUTORIAL EXERCISE: 22_SOLID DIMENSION.DWG

Figure 22–20

Purpose
This tutorial exercise is designed to add dimensions to a solid model that has had its views extracted using the SOLVIEW and SOLDRAW commands, as shown in the previous image.

System Settings
Drawing and dimension settings have already been changed for this drawing.

Layers
Layers have already been created for this tutorial exercise.

Suggested Commands
In this tutorial you will activate the front viewport of the 22_Solid Dimension drawing and make the Front-DIM layer current. Then add dimensions to the Front view. As these dimensions are placed in the Front view, the dimensions do not appear in the other views. This is because the SOLVIEW command automatically creates layers and then freezes those layers in the appropriate viewports. Next, activate the top viewport, make the Top-DIM layer current, and add dimensions to the Top view. Finally, activate the right side viewport, make the Right Side-DIM layer current, and add the remaining dimensions to the Right Side views.

STEP 1

Begin this tutorial by opening the drawing 22_Solid Dimension. Your display should appear similar to the following image. Viewports have already been created and locked in this drawing. A locked viewport means that as you activate the viewport and accidentally zoom, the image inside of the viewport will not zoom; rather, the entire drawing will be affected by the zoom operation. Centerline layers have also been created and correspond to the three viewports. Centerlines have already been placed in their respective views.

Be sure Osnap is turned on with Endpoint being the active mode.

Figure 22–21

STEP 2

Activate the viewport that contains the Front view by double-clicking inside of it. You know you have accomplished this if the floating model space icon appears. Then make the Front-DIM layer current, as shown in the following image.

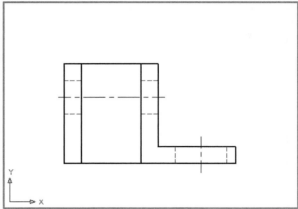

Figure 22–22

STEP 3

Most of the layers displayed in the Layer Control box were created by the SOLVIEW command. Notice how they correspond to a particular viewport. For example, study the following table regarding the layer names dealing with the front viewport:

Layer Name	Purpose
Front-CENTER	Center lines
Front-DIM	Dimension lines
Front-HID	Hidden lines
Front-VIS	Visual (Object) lines

Notice the Layer Front-DIM in the following image. Because the viewport holding the Front view is current, this layer is thawed, meaning the dimensions placed in this viewport will be visible. Notice that the other dimension layers (Right Side-DIM and Top-DIM) are frozen. The SOLVIEW command automatically set up the dimension layers to be visible in the current viewport and frozen in the other viewports.

Figure 22–23

STEP 4

Add the two dimensions to the Front view, as shown in the following image. Linear and baseline dimension commands could be used to create these dimensions. Grips can be used to stretch the text of the 0.50 dimension so that it is located in the middle of the extension lines.

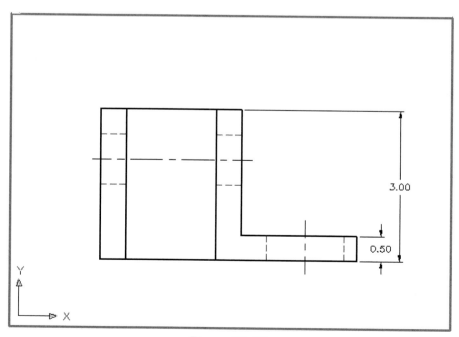

Figure 22–24

STEP 5

Activate the viewport that contains the Top view by clicking inside of it. Then make the Top-DIM layer current, as shown in the following image. This layer is designed to show dimensions visible in the top viewport and make dimensions in the front and right side viewports frozen (or invisible).

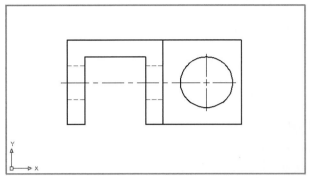

Figure 22–25

STEP 6

Add dimensions to the Top view using the following image as a guide.

 Tip: Because the viewports are locked, use the zoom and pan operations freely while dimensioning in floating model space.

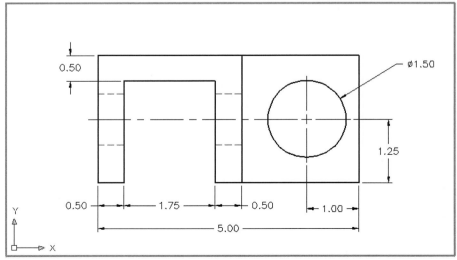

Figure 22–26

STEP 7

Activate the viewport that contains the Right Side view by clicking inside of the viewport. Then make the Right Side-DIM layer current, as shown in the following image. This layer is designed to make dimensions visible in the right side viewport and make dimensions in the front and top viewports frozen (or invisible).

Figure 22–27

STEP 8

Add dimensions to the Right Side view using the following image as a guide. Edit the diameter dimension text to reflect two holes. Use the DDEDIT command to accomplish this. When the Text Formatting dialog box appears, type "2X" from the keyboard and click the OK button.

Figure 22–28

STEP 9

The completed dimensioned solid model is illustrated in the following image.

Figure 22–29

PROBLEMS FOR CHAPTER 22

1. Create an engineering drawing of Problems 20–1 through 20–20 consisting of Front, Top, Right Side, and Isometric views.
2. Add proper dimensions to each drawing.

CHAPTER 23

Producing Renderings and Motion Studies

This chapter introduces you to renderings in AutoCAD and how to produce realistic renderings of 3D models. The heart of the rendering operation is the controls found in the dashboard. Adding lights will be the next topic of discussion. With this important feature you produce unlimited special effects depending on the location and intensity of the lights. The Point, Distant, and Spotlights features are discussed, as is the ability to simulate sunlight. You will be able to attach materials to your models to provide additional realism. Materials can be selected from a library available in AutoCAD, or you can even make your own materials. You will be shown how to walk and fly through a 3D model. You will also be given instruction on how to make a motion path animation of a 3D model.

AN INTRODUCTION TO RENDERINGS

Engineering and architectural drawings are able to pack a vast amount of information into a 2D outline drawing supplemented with dimensions, some symbols, and a few terse notes. However, training, experience, and sometimes imagination are required to interpret them, and many people would rather see a realistic picture of the object. Actually, realistic pictures of a 3D model are more than just a visual aid for the untrained. They can help everyone visualize and appreciate a design, and can sometimes even reveal design flaws and errors.

Shaded, realistic pictures of 3D models are called renderings. Until recently, they were made with colored pencils and pens or with paintbrushes and airbrushes. Now they are often made with computers, and AutoCAD comes with a rendering program that is automatically installed by the typical AutoCAD installation and is ready for your use. The following image shows, for comparison, the solid model of a bracket in its wireframe form, as it looks when the HIDE command has been invoked, and when it is rendered.

This chapter is designed to give you an overview on how to create pleasing, photorealistic renderings of your 3D models.

Figure 23–1

The following image displays numerous ways of accessing rendering commands. Clicking on View in the pull-down menu, followed by Render, will display a number of tools used for rendering, as shown on the left in the following image. The Render toolbar, shown in the middle in the following image, is another convenient way to access rendering commands. The dashboard, shown on the right in the following image, can also be used to access light, material, and rendering commands.

Figure 23–2

The following table gives a brief description of each rendering tool.

Button	Tool	Function
	Hide	Performs a hidden line removal on a 3D model
	Render	Switches to the Render window, where a true rendering of the 3D model is performed
	Lights	Contains six additional buttons used for controlling lights in a 3D model
	Light List	Displays the Lights in Model palette used for managing lights that already exist in a 3D model
	Materials	Displays the Materials palette, used for creating and applying materials to a 3D model
	Planar Mapping	Contains four additional buttons, used for mapping materials to planar, box, cylindrical, and spherical surfaces
	Render Environment	Displays the Rendering Environment dialog box that is used mainly for controlling the amount of fog applied to a 3D model
	Advanced Render Settings	Displays the Advanced Rendering Settings palette, used for making changes to various rendering settings

AN OVERVIEW OF PRODUCING RENDERINGS

The object illustrated in the following image consists of a 3D model that has a Realistic visual style applied. The color of the model comes from the color set through the Layer Properties Manager. After a series of lights are placed in a 3D model, and when materials have been applied, the next step in the rendering process is to decide how accurate a rendering to make.

Figure 23–3

As shown on the left in the following image of the dashboard, the Render control panel displays 5 different rendering modes, or presets, as they are called. These presets range from Draft to Presentation and control the quality of the final rendered image. For example, when you perform a rendering in draft mode, the processing speed of the rendering will be very fast; however, the quality of the rendering will be very poor. This render preset is used, for example, to perform a quick rendering when you are unsure about the positioning of lights. When you are pleased with the lighting in the 3D model, you can switch to a higher render preset, such as High or Presentation. These modes will process the rendering very slowly; however, the quality will be considered photo-realistic. To perform a rendering, click on the Render button, as shown on the right in the following image.

Figure 23–4

Clicking on the Render button, as shown in the previous image, switches your screen to the render window, as shown in the following image. Notice the shadows that are cast to the base of the 3D model. Shadow effects are one of many special rendering tools used to make a 3D model appear more realistic. Illustrated on the right of the rendering window is an area used for viewing information regarding the rendered image. Also, when you produce a number of renderings, they are saved in a list at the bottom of the rendering window.

Figure 23–5

As mentioned earlier, the following image displays the results of performing a Draft versus Presentation rendering. In the draft image, notice that the edges of the model do not look as sharp as they do in the presentation model. Also, the draft image does not apply shadows when being rendered. All these factors speed up the rendering of the draft image; however, the quality of the image suffers.

Figure 23–6

When you have produced a quality rendered image, you can save this image under one of the many file formats illustrated in the following image. Supported raster image formats include BMP, PCX, TGA, TIF, JPEG, and PNG.

Figure 23–7

 Try It! – This exercise consists of an overview of the rendering process. Open the drawing file 23_Render Presets. You will notice a container and two glasses resting on top of a flat platform, as shown in the following image. Also shown in this image are two circular shapes with lines crossing through their centers. These shapes represent lights in the model. A third light representing a spotlight is also present in this model although it is not visible in the following image.

Figure 23–8

Activate the dashboard, locate the Render control panel, and click the Render button, as shown on the right in the following image. The results are illustrated on the following image, with materials, lights, and a background being part of the rendered scheme. Notice in the dashboard the presence of Medium. This represents one of the many render presets used to control the quality of the rendered image.

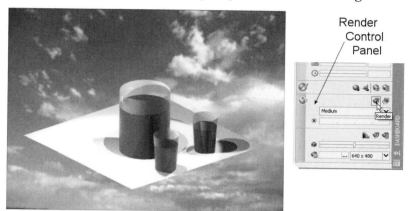

Figure 23–9

To see how the Render Preset value affects the rendered image, change the render preset in the dashboard from Medium to Draft, as shown on the right in the following image. Notice that the quality of the rendered image looks choppy; also, shadows are lost. However, the Draft Render Preset is always useful when testing out the lighting of the rendered scene. The processing time of this preset is very fast compared to other render presets, and the quality of the Draft Render Preset does not look very appealing.

Figure 23–10

In the previous image, the lighting looks too bright and overpowering in the rendered image. To edit the intensity of existing lights, click on the Lights List button, located in the Light control panel of the dashboard, as shown on the right in the following image. This will launch the Lights in Model palette, as shown on the left in the following image. Double-clicking on Pointlight1 will launch the Properties palette for that light, as shown in the middle of the following image. Locate the Intensity factor under the General category and change the default intensity value of 1.00 to a new value of 0.50. This will reduce the intensity of this light by half. Perform this same operation on Pointlight2 and Spotlight1 by changing their intensities from 1.00 to 0.50.

Figure 23–11

After changing the intensity of each light in the model, change the render preset from Draft to High in the dashboard and then click the Render button, as shown on the right in the following image. The results are displayed in the following image, with the lights being less intense and the shadows being more pronounced.

Figure 23–12

Next, click on the Materials button, located in the Materials control panel of the dashboard, as shown on the right in the following image. The Materials palette displays all materials created in the model. Clicking on the material will display the properties of this material, as shown on the left in the following image. We will experiment more with materials later in this chapter. This concludes this exercise.

Figure 23–13

CREATING AND PLACING LIGHTS FOR RENDERING

The ability to produce realistic renderings is dependent on what type of lighting is used and how these lights are placed in the 3D model. Once the lights are placed, their properties can be edited through the Properties palette. Before placing the light, you need to decide on the type of light you wish to use: Point Light, Spotlight, or Distant Light. The following image displays three areas for obtaining light commands: the View pull-down menu, on the left, the Lights toolbar, in the middle, and the Light control panel, located in the dashboard on the right.

Figure 23–14

You can place numerous lights in a single 3D model and then adjust these lights depending on the desired affect. Each of the standard light types also allows for shadows to be cast, giving depth to your 3D model. The following table displays each light button along with the name of the light and its function.

Button	Tool	Function
	New Point Light	Creates a new point light, similar to a lightbulb
	New Spotlight	Creates a new spotlight given a source and target
	New Distant Light	Creates a new distant light, similar to the sun
	Light List	Displays a list of all lights defined in a model
	Geographic Location	Displays a dialog box allowing you to select a geographic location from which to calculate the sun angle
	Sun Properties	Displays the Sun Properties palette, allowing you to make changes to the properties of the sun

The Geographic Location button is unique in that it will allow you to perform sunlight studies. The Geographic Location dialog box will allow you to control the position of the sun depending on a location, as shown in the following image. Clicking the down arrow in the Region area will display regional maps of the world, which allow you to produce sun studies at these locations.

Figure 23–15

 Try It! – Open the file 23_Valve Head Lights. This drawing file, shown in the following image, contains the solid model of a machine component that we will use for a rendering. You will place a number of point lights and one spotlight to illuminate this model.

Figure 23–16

Switch to Plan view, as shown in the following image. It will be easier to place the lights while viewing the model from this position. Activate the dashboard, and in the Light control panel, click on the Create a point light button, as shown on the right in the following image.

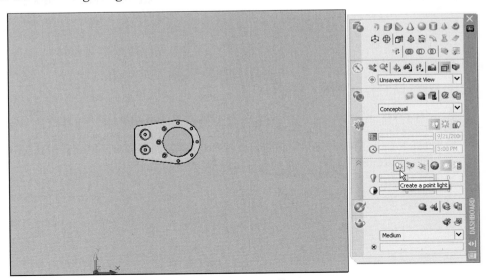

Figure 23–17

Place three point lights using the following prompts. You will place the lights at the approximate locations indicated in the following image. You will change the intensity and the default names of each light through the following command prompts.

For the first light, enter the following information:

Command: _**pointlight** (Pick from the Light control panel of the dashboard)
Specify source location <0,0,0>: (Pick at "A")
Enter an option to change [Name/Intensity/Status/shadoW/Attenuation/Color/eXit] <eXit>: **I** (For Intensity)
Enter intensity (0.00 - max float) <1.0000>: **.25**
Enter an option to change [Name/Intensity/Status/shadoW/Attenuation/Color/eXit] <eXit>: **N** (For Name)
Enter light name <Pointlight1>: **Overhead Light**
Enter an option to change [Name/Intensity/Status/shadoW/Attenuation/Color/eXit] <eXit>: (Press ENTER to create the light)

For the second light, enter in the following information:

Command: _**pointlight** (Pick from the Light control panel of the dashboard)
Specify source location <0,0,0>: (Pick at "B")
Enter an option to change [Name/Intensity/Status/shadoW/Attenuation/Color/eXit] <eXit>: **I** (For Intensity)
Enter intensity (0.00 - max float) <1.0000>: **.25**
Enter an option to change [Name/Intensity/Status/shadoW/Attenuation/Color/eXit] <eXit>: **N** (For Name)
Enter light name <Pointlight2>: **Lower Left Light**
Enter an option to change [Name/Intensity/Status/shadoW/Attenuation/Color/eXit] <eXit>: (Press ENTER to create the light)

For the third light, enter in the following information:

Command: _**pointlight** (Pick from the Light control panel of the dashboard)
Specify source location <0,0,0>: (Pick at "C")
Enter an option to change [Name/Intensity/Status/shadoW/Attenuation/Color/eXit] <eXit>: **I** (For Intensity)
Enter intensity (0.00 - max float) <1.0000>: **.25**
Enter an option to change [Name/Intensity/Status/shadoW/Attenuation/Color/eXit] <eXit>: **N** (For Name)
Enter light name <Pointlight3>: **Upper Left Light**
Enter an option to change [Name/Intensity/Status/shadoW/Attenuation/Color/eXit] <eXit>: (Press ENTER to create the light)

Figure 23–18

When you have finished placing all three point lights, activate the View Manager dialog box (the VIEW command), select the SE Zoomed view, and click the Set Current button, as shown in the following image. This will change your model to a zoomed-in version of the Southeast Isometric view.

Figure 23–19

When all three point lights were placed, unfortunately all of these lights were located on the top of the base plate, as shown in the following image. The lights need to be assigned an elevation, or Z coordinate. To perform this task, activate the dashboard and click on the Light List button to display the Lights in Model palette, as shown on the right in the following image.

Figure 23–20

Double-click on the Lower Left Light to display the Properties palette on this light. Locate the Position Z coordinate, located under the Geometry category of this palette, as shown in the following image, and change the value from 0 to 200. This will elevate the Lower Left Light a distance of 200 mm, as shown in the following image.

Figure 23–21

Continue changing the elevations of the remaining point lights. Change the Position Z coordinate value of the Overhead Light from 0 mm to 300 mm and the Position Z coordinate value of the Upper Left Light from 0 mm to 200 mm. Your display should appear similar to the following image.

Figure 23–22

Next, place a spotlight into the 3D model by clicking on the Create a spotlight button, as shown on the right in the following image. Place the source for the spotlight at the approximate location at "A" and the spotlight target at the center of the bottom of the valve head at "B," as shown in the following image.

Command: **spotlight** *(Pick from the Light control panel of the dashboard)*
Specify source location <0,0,0>: *(Pick the approximate location for the spotlight at "A")*
Specify target location <0,0,-10>: *(Pick the bottom center of the valve head at "B")*
Enter an option to change
[Name/Intensity/Status/Hotspot/Falloff/shadoW/Attenuation/Color/eXit] <eXit>: **I**
　(For Intensity)
Enter intensity (0.00 - max float) <1.0000>: **.50**
Enter an option to change
[Name/Intensity/Status/Hotspot/Falloff/shadoW/Attenuation/Color/eXit] <eXit>: **N**
　(For Name)
Enter light name <Spotlight5>: **Spotlight**
Enter an option to change
[Name/Intensity/Status/Hotspot/Falloff/shadoW/Attenuation/Color/eXit] <eXit>:
　(Press ENTER *to create the light)*

Figure 23–23

As with the point lights, the source of the spotlight is located at an elevation of 0 and needs to be changed to a different height. Double-click on the spotlight icon to display the Properties palette, and change the Position Z coordinate value to 200 mm, as shown on the right in the following image.

Figure 23–24

Finally, check to see that the render preset value is set to Medium and click the Render button to display the model, as shown in the following image. Shadows are automatically applied to the model from the lights. This concludes this exercise.

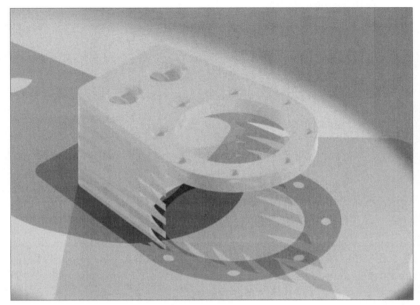

Figure 23–25

AN INTRODUCTION TO MATERIALS

Another way to make models more realistic and lifelike is to apply a material to the 3D model. A library of materials is available through a number of tabs located in the Tool palette, as shown in the following image. The materials can be dragged and dropped onto the model.

Figure 23–26

Materials can also be created through the Materials palette, as shown on the right in the following image. This palette can be activated from the dashboard, as shown on the left, or from the View pull-down menu, as shown in the middle.

Figure 23–27

Located in the Materials palette are a number of buttons used for creating and applying materials to the 3D model. These buttons and their functions are explained in the following table.

Button	Tool	Function
	Create New Material	Displays the Create New Material dialog box, where you enter a name and description of a new material
	Purge from Drawing	Used to eliminate a material from the database of a drawing
	Indicate Materials in Use	Materials currently used in a drawing are identified by a drawing icon located in the lower right corner of the material swatch
	Apply Material to Objects	Attaches a material to a single object or group of objects
	Remove Materials from Selected Objects	Removes a material from a selected object(s) but keeps the material loaded in the Materials palette

Before a material can be applied to a model, it must first be loaded. One method of accomplishing this is to drag the material from the Tool palette, as shown on the right in the following image, and drop it anywhere in the drawing. The material will then appear in the Materials palette, as shown on the left in the following image.

Figure 23–28

The Create New Material button in the Materials palette allows you to create your own custom materials. For creating materials, a number of templates are available, as shown on the left in the following image. These templates already have properties set depending on the purpose of the material. For instance, for the Mirror template, a number of settings have already been made to ensure that the material reflects correctly. As you create materials, you can further modify these settings, as shown on the right in the following image. Clicking in the Diffuse area will allow you to change the color of the material. You can also control such properties as shininess, translucency, and opacity in this palette.

Figure 23–29

WORKING WITH MATERIALS

Once a material or group of materials has been loaded into a 3D model drawing, the next step is to attach these materials to drawing shapes and components. This can be accomplished in a number of ways, as outlined in the following Try It! exercise.

 Try It! – Open the drawing file 23_Connecting Rod. A single point light source has already been created and placed in this model. A realistic visual style is currently being applied to the model, as shown in the following image. You will attach a material from the Tool palette and observe the rendering results. After removing this material, you will create a new material, change a few settings, and observe these rendering results.

Figure 23–30

First, activate the Tool palette and locate the Metal tab. Then, locate the Galvanized material in this tab, as shown on the right in the following image. Click once on this material and, at the Select Objects prompt, pick the edge of the 3D model, as shown on the left in the following image.

Figure 23–31

Since you are in a realistic visual style, materials and textures are automatically turned on through the dashboard, as shown on the right in the following image. The results of applying the galvanized steel material to the model are shown on the left in the following image. To display shadows, perform a rendering with the rendering style set to Medium.

Figure 23–32

The results of performing the rendering are shown in the following image. In addition to the galvanized material being applied, shadows are cast along a flat surface from the existing light source.

Figure 23–33

Before continuing to the next material, click the Remove Materials from Selected Objects button in the Materials palette and pick the edge of the model to remove this material from the model but not from the Materials palette.

Figure 23–34

To create a new material, click on the Create New Material button, as shown on the left in the following image. When the Create New Material dialog box appears, enter the name Red Metal, as shown on the right in the following image. Click OK.

Figure 23–35

Keep the default template set to Realistic and click in the Diffuse box, as shown on the left in the following image, to change the color of the material. When the Select Color dialog box appears, change to the Index Color tab and change the color to the one shown on the right in the following image. Click the OK button to accept the material and return to the Materials palette.

Figure 23–36

When you return to the Materials palette, click on the Apply Material to Objects button, as shown in the following image. Select the 3D model, and it should update to display this new material.

Figure 23–37

Producing a rendering of the 3D model with the new material should result in an image similar to the following. When finished, exit the rendering mode and return back to the 3D model.

Figure 23–38

While still using the red material, expand the Material Editor to expose the Opacity heading, and change the value to 20 by sliding the bar, as shown on the left in the following image. The material will update to reflect this change in opacity, which will change the material to include a degree of transparency.

Figure 23–39

Producing a new rendering will display the model, as shown in the following image, complete with transparent material and shadows. This concludes this exercise.

Figure 23–40

USING MATERIAL TEMPLATES

Using a materials template is one way of to automate the creation process for new materials. The next Try It! exercise illustrates the use of materials templates.

 Try It! – Open the file 23_Piston Mirror, as shown in the following image. A number of lights have already been placed in this model. Also, the model is being viewed through the Realistic visual style. In this exercise, you will create two new materials. One of the materials will contain mirror properties and be applied to the piston. When you perform a render operation, the reflection of one of the piston rings will be visible in the top of the piston.

Figure 23–41

Begin by launching the Materials palette and clicking the Create New Material button, as shown on the left in the following image. Enter a new material name called Mirror in the Create New Material dialog box, as shown on the right in the following image. Click OK.

Figure 23–42

A number of templates are available to assist with the creation of special material types such as a mirror property. Click the down arrow in the Template area and pick the Mirror template from this list, as shown on the left in the following image. A few other items need to be fine-tuned to produce the best mirror property. First, click in the Diffuse area; this will launch the Color dialog box. Click on the Index Color tab and choose one of the shades of yellow. Then, in the Shininess area, adjust the slider bar to read approximately 75%. Adjust the Self-Illumination slider bar to read approximately 50%. The Materials palette should appear similar to the illustration shown on the right in the following image.

Figure 23–43

Next, click on the Apply Material to Objects button and pick the edge of the piston to apply the material to this 3D model, as shown in the following image.

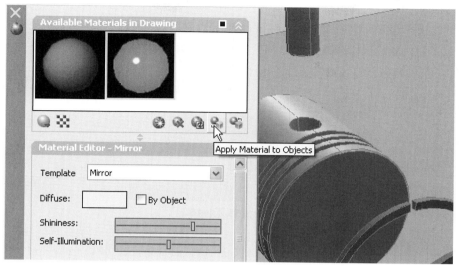

Figure 23–44

Test the mirror property by performing a render with the render style set to Medium. Notice in the following image that the piston ring and shadows are visible in the piston due to the mirror material.

Figure 23–45

Create another material called Piston Support Parts, as shown in the following image. This material will be applied to the remainder of the parts that form the piston assembly.

Figure 23–46

Keep the default material template set to Realistic. Click in the Diffuse area and change the color of this material to a different shade of green. When finished, click on the Apply Material to Objects button and pick the remainder of the piston parts, as shown in the following image. (Leave the main piston set to the mirror material.)

Figure 23–47

Perform another rendering test. Your image should appear similar to the following image. This concludes this exercise.

Figure 23-48

ASSIGNING MATERIALS BY LAYER

One way to render detailed reflections is to use ray tracing, which gives the most realistic reflections. We can create our own material with attributes that will take advantage of this type of rendering.

 Try It! – Open the file 23_Interior Materials, as shown in the following image. You will drag and drop existing materials from the Tool palette into the drawing. You will then assign these materials to specific layers and perform the rendering. Lights have already been created for this exercise.

Figure 23–49

Activate the Materials palette and notice that two materials have already been created in this drawing. The first material, shown on the left in the following image, is a fabric color designed to be applied to the chair. The second material, shown on the right in the following image, is a paint color to be applied to the walls of the 3D model.

Figure 23–50

All other materials will be obtained from the Tool palette. You will be dragging and dropping a number of materials into a blank area of your drawing. This drag-and-drop action will load the materials into the drawing for your use. You will be selecting materials from three tabs located in this palette. The first set of materials will be loaded from the Doors and Windows Materials Library, as shown on the left in the following image. You will have to move your cursor over a material and leave it stationary in order for the whole material name to be displayed. Locate the first material,

Doors – Windows.Door Hardware.Chrome.Satin, as shown on the right in the following image. Press and hold down your mouse button on this material, drag the material icon into your drawing, and drop it to load it.

Figure 23–51

Continue loading the following materials located in the appropriate tabs using the table below as a guide:

Tool Palette Tab	Material
Doors and Windows	Doors – Windows.Door Hardware.Chrome.Satin
Doors and Windows	Doors – Windows.Glazing.Glass.Mirrored
Doors and Windows	Doors – Windows.Wood Doors.Ash
Finishes	Finishes.Flooring.Wood.Parquet.1
Woods and Plastics	Wood – Plastics.Finish Carpentry.Plastic Laminates.Light Beige
Woods and Plastics	Wood – Plastics.Finish Carpentry.Wood Cherry
Woods and Plastics	Wood – Plastics.Finish Carpentry.Wood Mahogany
Woods and Plastics	Wood – Plastics.Finish Carpentry.Wood Pine

When you have finished loading all materials, you can check the status of the load by clicking on the Materials button, located in the dashboard, as shown on the right in the following image. This will launch the Materials palette and display all materials that can be applied to 3D models in the drawing.

Figure 23–52

The next step is to assign a material to a layer. From the dashboard, click on the Attach By Layer button, as shown on the left in the following image. This will launch the Material Attachment Options dialog box, as shown on the right in the following image. Now you will drag a material located in the left column of the dialog box and drop it onto a layer located in the right column of the dialog box. In this example, the Chrome.Satin material has been dropped onto the Door Knob layer.

Figure 23–53

Using the following image as a guide, continue assigning materials to layers by dragging and dropping the materials onto the appropriate layers.

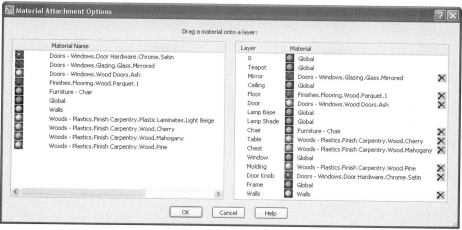

Figure 23–54

When all material assignments have been made to the layers, click OK. Render out the design and verify that the materials are properly assigned to the correct 3D objects, as shown in the following image.

Figure 23–55

One final step needs to be performed on the 3D model. The light beige plastic laminate material was mistakenly loaded and not used. Activate the PURGE command, expand the Materials item, as shown in the following image, and purge this material from the drawing. This concludes this exercise.

Figure 23–56

APPLYING A BACKGROUND

Images can be placed behind a 3D model for the purpose of creating a background effect. Backgrounds can further enhance a rendering. For example, if you have designed a 3D house, you could place a landscape image behind the rendering. To place a background image, the image must be in a raster format such as BMP, TGA, or TIF. AutoCAD comes supplied with many background images for you to choose from. The process of assigning a background image begins with the View Manager dialog box, as shown in the following image. You create a new view and associate a background with the view. In the following image, the New button is picked, which will launch the New View dialog box. Enter a new name for the view and place a check in the box next to Override default background. This will launch the Background dialog box, where you can pick a background from three different types.

Figure 23–57

These background types are explained as follows.

Solid—A solid background means that AutoCAD replaces the default white (or black) background of the drawing screen with another color. You choose the color from the Colors section of the dialog box.

Gradient—A gradient means that the color changes from one end of the screen to the other, such as from red at the bottom to light blue at the top (to simulate a sunset). AutoCAD gives you three ways to control a linear gradient—look carefully: they are tucked into the lower right corner. Horizon specifies where the lower color ends; a value closer to 0 moves the lower color lower down. When Height is set to 0, you get a two-color gradient; any other value gives you a three-color gradient. Rotation rotates the gradient.

Image—You select a raster image for the background. The image can be in BMP (Bitmap), GIF, PNG, TGA (Targa), TIF (Tagged Image File Format), JPG (JPEG File Interchange Format), or PCX (PC Paintbrush) format.

Try It! – Open the drawing file 23_Piston Background. The Realistic visual style is applied to this model, as shown in the following image. Also applied are four point lights. A special mirror material is attached to the sphere.

Figure 23–58

Begin by performing a render using the Medium rendering style. Your image should appear similar to the following image, in which the reflections of various piston parts appear in the sphere due to its mirror property. All that is missing is a background that will be applied to the 3D models.

Figure 23–59

All visual style and rendering backgrounds are controlled in the View Manager dialog box, as shown in the following image. Begin the selection of a background by clicking on the New button.

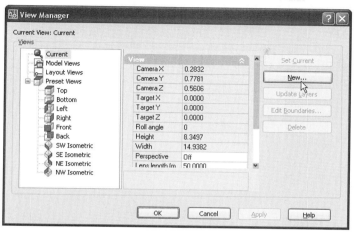

Figure 23–60

You must first create a new view and then assign a background to this view. When the New View dialog box appears, as shown on the left in the following image, enter a name for the view, such as Sky Background. You also must place a check in the box next to Override default background in order to be taken to other dialog boxes to select the background. After you check this box, the Background dialog box appears, as shown on the right in the following image. Choose Image from the Type list.

Figure 23–61

Once you click the Browse button in the Background dialog box, the Select File dialog box will appear, as shown in the following image. A folder called Textures is automatically created and stocked with various image files when AutoCAD is loaded. From this list, find sky.tga, select it, and click the Open button.

Figure 23–62

This will take you to the Background dialog box again. The sky graphic will appear small but centered on the sheet, as shown on the left in the following image. To control the display of this file in the final rendering, click on the Adjust Image button to launch the Adjust Background Image dialog box and change the Image position to Stretch. This should make the sky graphic fill the entire screen, as shown on the right in the following image.

Figure 23–63

Click the OK buttons in the Adjust Background Image and Background dialog boxes to return to the View Manager dialog box, where the Sky Background is now part of the Views list. Click on the Set Current button to make this view current in the drawing and click the OK button to dismiss this dialog box.

Figure 23–64

Perform a render using the Medium rendering style and notice the results, as shown in the following image. With the image applied, it appears that the flat base sheet is floating in air. Also, since the sphere still has the mirror material property, the sky is reflected here in addition to the piston parts. This concludes this exercise.

Figure 23–65

WALKING AND FLYING THROUGH A MODEL

To further aid with visualization of a 3D model, walking and flying actions can be simulated through the 3DWALK and 3DFLY commands. Both can be selected from the View pull-down menu, as shown on the left in the following image. When walking through a 3D model, you travel along the XY plane. When flying through the model, you move the mouse to look over the top of the model.

Additional controls for walking and flying through a model can be found under the 3D Navigation control panel of the dashboard, as shown on the right in the following image.

Figure 23–66

When you first activate the 3DWALK command, a warning dialog box appears stating that you must be in Perspective mode to walk or fly through your model. Click the OK button to enter Perspective mode. Clicking OK will display a second dialog box, as shown in the following image. This dialog box acts as a reminder of the controls used for walking through a model. The controls are all from the keyboard and consist of the directional arrow keys found between the main body of the keyboard and the numeric keypad. When dismissing this dialog box, you can easily make it reappear by pressing the Tab key.

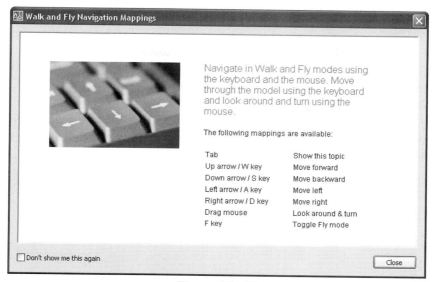

Figure 23–67

When you first enter the walk or fly mode, a Position Locator palette appears, as shown on the left in the following image. It gives you an overall view of the position of the camera and target in relation to the 3D model. You can drag on the camera location inside the preview pane of the Position Locator to change its position. You can also change the target as you adjust the viewing points of the 3D model. Right-clicking will display the shortcut menu, as shown in the following image. Use this menu for changing to various modes that assist in the rotating of the model.

Figure 23–68

ANIMATING THE PATH OF A CAMERA

The ability to walk or fly through a model has just been discussed. This last segment will concentrate on creating a motion path animation by which a camera can follow a predefined polyline path to view the contents of a 3D model. Clicking on Motion Path Animations, found under the View pull-down menu, as shown on the left in the following image, will launch the Motion Path Animation dialog box, as shown on the right in the following image. You select the path for the camera and target in addition to changing the number of frames per second and the number of frames that will make up the animation.

Figure 23–69

After making changes to the Motion Path Animation dialog box, you have the opportunity to preview the animation before actually creating it. A sample animation preview is shown in the following image. You can even see the relative position the camera is in as it passes through the 3D model along the polyline path. After the preview is finished, clicking the OK button in the main Motion Path Animation dialog box will create the animation and write the results out to a dedicated file format. Supported formats include AVI, MOV, MPG, and WMV. Depending on the resolution and number of frames, this process could take a long time. However it gives you the capability of creating an animation using any kind of 3D model.

Figure 23-70

 Try It! – Open the drawing file 3D_House Motion. A polyline path has already been created at an elevation of 4' to simulate an individual walking through this house. Once you have created the motion path animation, this polyline path will not be visible when the animation is played back.

Begin by clicking on Motion Path Animations, which is found under the View pull-down menu. When the Motion Path Animation dialog box appears, as shown in the following image, make the following changes:

Pick the Select Path button in the Camera area and pick the polyline displayed in the floor plan. If necessary, change the path name to Path1. If an AutoCAD Alert box appears, click the Yes button to override the existing path name.

Pick the Select Path button in the Target area and pick the polyline displayed in the floor plan. If necessary, change the path name to Path2. If an AutoCAD Alert box appears, click the Yes button to override the existing path name. Both the Camera and Target will share the same polyline path.

Change the Frame rate (frames per second) from 30 to 60.

Change the Number of frames from 30 to 600. This will update the Duration from 1 to 10 seconds.

Change the Visual style to Realistic.

Change the Format to AVI.

Keep the resolution set to 320 × 240.

Figure 23–71

When you have finished making these changes, click on the Preview button to preview the results of the motion animation. Play the animation preview as many times as you like. Once you close the preview, you will return to the Motion Path Animation dialog box. Click the OK button and enter the name of the AVI file as House Motion Study. Clicking the OK button in this dialog box will begin processing the individual frames that will make up the animation. The total processing time to produce the animation should be between 5 and 10 minutes.

When finished, launch one of the many Windows Media Player applications and play the AVI file. This concludes this exercise.

TUTORIAL EXERCISE: SUNLIGHT STUDY

Figure 23-72

Purpose

This tutorial is designed to simulate sunlight on a specific day and time, and to observe the shadows that are cast by the house.

System Settings

Since this drawing is provided on the CD-ROM, all system settings have already been made.

Layers

The creation of layers is not necessary.

Suggested Commands

Begin this tutorial by opening up the drawing 23_House Plan Rendering, as shown in the previous image. You will be performing a study based on the current location of the house and the position of the sun on a certain date, time, and geographic location. Shadow casting will be utilized to create a more realistic study.

STEP 1

Activate the dashboard and click on the Sun Status button, as shown on the right in the following image, to turn on the sun. A dialog box will appear informing you that you cannot display sunlight if the default lighting mode is turned on and asking you whether you want to turn off default lighting. Click the Yes button in the dialog box to turn off the default and activate the lighting of the sun.

Figure 23–73

STEP 2

Next, click on the Edit the Sun button, located in the dashboard, as shown on the right in the following image, to display the Sun Properties palette, as shown on the left in the following image. In this palette, you can change various properties that deal with the sun, such as shadows, date, time, azimuth, altitude, and source vector of the sun.

Figure 23–74

STEP 3

Click on the launch Geographic Location button, as shown on the left in the following image, to change the location of the 3D model. When the map of North America displays in the Geographic Location dialog box, click on the coast of South Carolina and check to see that Charleston, SC appears, as shown on the right in the following image. You could also use the Nearest City drop-down list to select the desired location. Other maps from throughout the world also are available. When you are satisfied with the location, click the OK button to leave the Geographic Location dialog box.

Figure 23–75

STEP 4

A dialog box will appear informing you that the time zone has been automatically updated with the change in the geographic location. Click the OK button to accept this.

Figure 23–76

STEP 5

Next you will change the date and time to perform a sun study when the sun is positioned on a fall day. Click on the Date area in the Sun Properties palette, click on the three dots (ellipses), and change the date to October 18, 2006, as shown in the following image.

Figure 23–77

STEP 6

Then, click in the Time area. When an arrow appears, click on it to display a number of times of day in 15-minute increments, and click on 11:00 AM, as shown on the left in the following image. When you perform this change, the date and time information is also updated in the dashboard, as shown on the right in the following image.

Figure 23–78

STEP 7

Clicking on the Render button in the dashboard, as shown on the right in the following image, will render the house, as shown on the left in the following image. The shadows cast by the house reflect the time of 11:00 AM in mid-October in Charleston, South Carolina, in the United States.

Figure 23–79

STEP 8

To see the effects of the sun on the same date but at a different time, activate the Sun Properties palette and change the time to 4:30 PM, as shown on the left in the following image. The results are displayed on the right, with the shadows being cast based on the different position of the sun.

Figure 23–80

STEP 9

Experiment with other locations and times. The following image illustrates the Geographic Location dialog box and the other world locations available when you click the down arrow found in the Region category.

Figure 23–81

INDEX

Note: Terms in SMALL CAPS signify AutoCAD commands and system variables.

Symbols

-ATTEDIT command, 882
2D multiview drawings
  auxiliary views, 1177–1178
  FLATSHOT command and, 1184–1186
  from solid model, 1171
  isometric views, 1182
  orthographic views, 1175
  section views, 1179–1181
  SOLDRAW command, 1171–1175
  SOLVIEW command, 1171–1175
2D shading mode, 1030
2D wireframes, solid modeling and, 1031
2 Points mode
  CIRCLE command, 268
3DALIGN command, 1124, 1126
3DARRAY command
  solid model editing, 1131
3D Coordinate System icon, 1024–1025
3DCORBIT command, 1028
3D drawings
  moving solids, 1055
  subtracting solids, 1056–1057
  unioning solids, 1054
3DFLY command, 1238
3DFORBIT command, 1004, 1026, 1028, 1072
3D Modeling workspace, 1, 1008
3D models
  3DFORBIT command, 1026, 1028
  3DORBIT button, 1027
  Constrained Orbit button, 1027
  faces, selection, 1004
  solid, creating, 1004
  surface, 1004
  View toolbar and, 1028–1029
  wireframe, 1003, 1031

3DMOVE command, 1123–1124
3D Navigation toolbar, 1026
3DORBIT button, 1027
3D Orbit toolbar, 1026
3DROTATE command, 1126–1127, 1129
3D shading mode, 1030
3D solid modeling
  BOX command, 1042, 1046
  tutorial, 1080
3DWALK command, 1238
3point option, UCS command, 1014
3 Points mode
  arcs, 263
  CIRCLE command, 268
  DRAWORDER command, 959
  VPORTS command, 721
  XCLIP command, 947

A

absolute coordinates, 48–49
absolute Polar Angles, 41
Acad.Dwt template, 131
Acadiso.Dwt template, 131
Add-A-Plotter Wizard, 766
Add Plotter-Begin dialog box, 767
Advanced wizard, 94
Aligned Dimensioning mode, 563
  rotated linear dimensions, 564
Aligned Dimensioning System, 556
aligned sections, 480
alignment
  dimension text, 625
  text, justified, 333
alignment parameters, dynamic block creation, 839–841

Alternate Units tab, Modify Dimension Style: Mechanical dialog box, 634–635
angle offset parameters, dynamic block creation, 842, 844–845
angle of hatch pattern, 495
angles
  chamfers and, 200
  dimensioning and, 577
  DIST command, 684
Angular dimension units, 632
animation
  motion path, 1240–1242
ANSI31 pattern, 475
ARC command, 262
  3 Points method, 263
  center point, 264
  Continue mode, 264
  end point, 264
  start point, 264
arc dimensions, 573
architectural details tutorial, 994–999
architectural dimensions, 558
architectural drawings
  dimensions tutorial, 659–660, 662–666
  Paper Space arrangement, 740–741
architectural sections, 484
architectural units, 88
architectural values, drawing lines, 88
arcs, 262
  3 Points method, 263
  Center (Cen) mode Object Snap, 25
  center mark settings, 620
  center point, 264
  Continue mode, 264
  end point, 264

1249

arcs *(continued)*
    Endpoint (End) Object Snap mode, 25
    Extension (Ext) Object Snap mode, 25
    grips and, 389
    lines tangent to, 276
    Midpoint (Mid) Object Snap mode, 28–29
    Quadrant (Qua) Object Snap mode, 31
    start point, 264
    tangent to one line and an arc (CIRCLE-TRR command), 270
    tangent to two arcs (CIRCLE-TRR command), 272–274
    tangent to two lines (CIRCLE-TRR command), 269
arc symbol, dimension lines, 620
area
    circles, 675–677
    fields in calculation, 680
    polylines, enclosed, 675–677
    shapes, 674
    subtraction and, 678
AREA command, 674–678
Area tool, Inquiry toolbar, 674
arranging views, 973
ARRAY command, 437
Array dialog boxes
    rectangular arrays, 218–220
    tutorial, 245–247
ARRAY modify command, 218–220
    tutorial, 239, 241, 243–244
arrays
    creating, 218–220
    dynamic blocks, 850–852
    polar, 222, 1131
    rectangular, 218–220, 1131
arrowhead terminators, arrowhead settings, 618–619
arrows, grips, 389
assembly sections, 479–480
associative dimensions, 749–750
associative hatch modifications, 497–499
associative layouts, 749–750
associativity of dimensions
    DIMASSOC variable, 636
    reassociating, 637, 639

At symbol (@), 20, 49
attaching a raster image to a drawing, 952
attaching external references, 932–934
Attach option, External Reference dialog box, 932–934
ATTACH XREF command, 931
ATTDEF command, 882
ATTDIA command, 925
ATTDISP command, 882, 893
    modes, 893
ATTEDIT command, 882
ATTEXT command, 882, 903
ATTMODE command, 925
ATTREDEF command, 907
ATTREQ command, 925
Attribute Definition dialog box, 882
    attribute creation, 883–884
    attribute modes, 884
    attribute prompt, 884
    attribute tags, 883
    attribute values, 884
Attribute Extraction dialog box, 882, 903
Attribute Extraction wizard, 883, 913, 919
    tutorial, 920–924
attribute modes, Attribute Definition dialog box, 884
attribute prompt, Attribute Definition dialog box, 884
attributes
    commands, 882–883
    creating, 883–884
    displaying, 893
    editing, tutorial, 900–902
    extracting, 903
    fields, 892
    introduction, 881
    modes, 884
    properties, 894–899
    redefining, 905–912
    system variables, 925
    tags, 881
    tutorial, 885–890
attribute tags, Attribute Definition dialog box, 883

attribute values, Attribute Definition dialog box, 884
auto-hide, Properties Palette, 400
AutoCAD Classic workspace, 1
AutoCAD Default workspace, 2
Autodesk DWF Viewer, 788
AutoStack Properties dialog box, fractional text, 332
auxiliary views, 1177–1178
    constructing, 529–531
    overview, 525–528
    partial views, 528
    tutorial, 532–534, 536, 538–541

B
background
    exercise, 1234–1237
    gradient, 1233
    image as, 1233
    solid, 1233
Background dialog box
    Gradient option, 1233
    Image option, 1233
    Solid option, 1233
background masking, text objects, 338
backgrounds, 1232
Back view, 1029
    glass box, 435
    shape description and, 433
baseline
    dimension lines, 615
baseline dimensions, 555, 566
Baseline mode, dimensioning, 569
BATTMAN command, 883, 915
BHATCH command, 473, 485–487
    angle of pattern, 495
    scale value, 494–495
binding external references, 943–944
Block Attribute Manager dialog box, 883, 913, 915
    tutorial, 917–918
BLOCK command, 795–797
Block Definition dialog box, 795
Block Editor, 807
    dynamic block creation, 828
    toolbar, 829
blocks
    commands, selecting, 795

converting, XBIND command, 938–940
creating, 795–798, 812
description, 793
dividing, 808–809
dynamic, 827–842, 844–860
editing, 806–808
exploding, 804, 905
extending to, 803
external references and, 927–928
grid, 813
grips, 830
grips, enabling, 387
inserting, 799
inserting, applications, 801–802
inserting, DesignCenter, 817–818
introduction, 793
MEASURE command, 809–810
naming, 794
pressing and pulling, 1118–1121
redefining, 806–808
redefining attributes, 907
renaming, 810
tables and, 811
Tool Palette, 819–823
trimming, 802
tutorials, 861–863, 865–876
unused data, purging, 804–806
BMP (Bitmap) images, background, 1233
Body editing, 1141
cleaning, 1153
imprinting, 1150
separating solids, 1151
shelling, 1152
bold text, 331
Boolean operations, 1050
INTERSECT command, 1053, 1055–1056
intersections, 1057–1058
SUBTRACT command, 1052–1053
UNION command, 1052
border lines, object lines and, 437
borders, raster images, 958–959
Bottom view, 1029
glass box, 435
shape description and, 433

boundaries
edges, 211, 213
layers and, 265
polyline, 265–266
views, 159
BOUNDARY command, 262, 265–266
Boundary Creation dialog box, 265
Boundary Hatch dialog box, 473
bounding areas
pressing and pulling, 1115–1117
BOX command, 1005, 1042, 1046
breaking objects, 215–217
circles, 216
first break point, 216
points, 217
Break line, 101
BREAK modify command, 215–217
broken sections, 482
bulleted text, 335

C

calculations
drawing size limits, 91
QuickCalc, 407–409
tables, 379–383
camera, motion path animation, 1240–1242
capped multilines, 286
Cartesian coordinate system, 46
cells (tables), merging, 356
Center (Cen) Object Snap mode, 25
centering objects, dynamic blocks, 845–847
Center line, 101
centerlines, 437
dimensions and, 554
Center mark settings, 620
center point
arc drawing, 264
ZOOM-Center, 148
chain dimensions, 555
CHAMFER command, 1074–1075
chamfering, solid models, 1074–1075
CHAMFER modify command, 198
chamfers, 198
angles, 200

beam project, 202
equal distances, 199
polylines, 200
trimming and, 201
character mapping, dimension symbols, 590
CIRCLE-TRR command, 269–270, 272–274
CIRCLE command, 52, 262, 268
2 Points mode, 268
3 Points mode, 268
Diameter mode, 53
Dynamic Input mode, 54
OSNAP-Quadrant mode, 278
OSNAP-Tangent mode, 277–278
Radius mode, 53
circles
area, 675–677
breaking, 216
Cen mode (Object Snap) mode, 25
center mark settings, 620
CIRCLE command, 52
Diameter mode, 53
Dynamic Input mode, 54
ellipses, 281–282
filled-in, 279, 280
From (Fro) Object Snap mode, 26
lines tangent to, 276
Quadrant (Qua) Object Snap mode, 31
Radius mode, 53
tangent points, 277–278
torus and, 1047
cleaning, Body editing, 1153
clipboard, 826–827
clipping external references, 946–947
CLOSE command, 60–61
Collar, tutorial, 1084–1088
color
extension lines, 615
grips, 387
layers, 107–108
materials, 1216
Properties Palette, 400
Properties toolbar, 114
Color-Dependent plot style table, 777–784

Color Books tab, Select Color dialog box, 108
Color layer state, 103
command line, 3, 7
   prompt, 7
command prompt
   grips, 388
commands
   -ATTEDIT, 882
   3DALIGN, 1124–1126
   3DARRAY, 1131
   3DCORBIT, 1028
   3DFLY, 1238
   3DFORBIT, 1026, 1028, 1072
   3DFORBIT, 1004
   3DMOVE, 1123–1124
   3DROTATE, 1126–1129
   3DWALK, 1238
   aliases, 14
   ARC, 262–264
   AREA, 674–678
   ARRAY, 437
   ATTACH XREF, 931
   ATTDEF, 882
   ATTDIA, 925
   ATTDISP, 882, 893
   ATTEDIT, 882
   ATTEXT, 882, 903
   ATTMODE, 925
   ATTREDEF, 907
   ATTREQ, 925
   attributes, 882–883
   BATTMAN, 883, 915
   BHATCH, 473, 485–487
   BLOCK, 795–797
   BOUNDARY, 262, 265–266
   BOX, 1005, 1042, 1046
   CHAMFER, 1074–1075
   CIRCLE, 52–54, 262, 268
   CICLE-TRR, 269–270, 272–274
   CLOSE, 60–61
   CONE, 1045
   CONVERTPSTYLES, 784
   COPYBASE, 826
   COPYCLIP, 826
   COPYTOLAYER, 125
   CUTCLIP, 826
   CYLINDER, 1005, 1044
   DIMALIGNED, 563–564
   DIMARC, 573
   DIMBASELINE, 566
   DIMCONTINUE, 565
   DIMEDIT, 583–586
   dimensioning and, 549
   dimensions, selecting, 559
   DIMJOGGED, 573
   DIMLINEAR, 561
   DIST, 683
   DIVIDE, 262, 291, 808–809
   DONUT, 262, 279–280
   DRAWING commands, 16
   EATTEDIT, 883, 914
   EATTEXT, 883
   ELLIPSE, 262, 281–282
   ERASE, 57, 58
   EXPLODE, 804
   EXTEND, 803
   external reference, 928
   FILLET, 1072
   FLATSHOT, 1184–1186
   geometric, tutorial, 308–314
   GRID, 94
   HATCHEDIT, 500
   HELIX, 1068–1070
   ID, 684
   INSERT, 799
   INTERFERE, 1132
   INTERSECT, 1053, 1055–1058
   keyboard shortcuts, 14
   LAYCUR, 123
   LAYDEL, 130
   LAYERP, 115
   LAYFRZ, 127
   LAYISO, 124
   LAYLCK, 128
   LAYMCH, 122
   LAYMRG, 129
   LAYULK, 128
   LAYWALK, 126
   LEADER, 574
   LIMITS, 90–91
   LINE, 17–19
   LIST, 686
   LOFT, 1065–1066, 1068
   LTSCALE, 116–117, 438
   MASSPROP, 1007, 1076
   MATCHPROP, 415–418, 420, 422–423, 991, 993
   MATCHPROPR, 437
   MEASURE, 262, 293, 809, 810
   MIRROR3D, 1130
   MLEDIT, 287–288
   MLINE, 262, 284
   modify, 183–191, 193, 198, 203, 206, 208–212, 214–220, 224, 226–227, 229–238
   MTEXT, 747
   OPEN, 15, 825
   PAN, 153–154
   PASTEBLOCK, 827
   PASTECLIP, 827
   PLAN, 1025
   PLANSURF, 1040
   PLINE, 54, 56
   POINT, 262, 290
   POLYGON, 262, 294
   POLYSOLID, 1070–1072
   PRESSPULL, 1115–1121
   PURGE, 804–806
   PYRAMID, 1049
   QDIM, 567
   QLEADER, 575–576
   QSAVE, 58–59
   QSELECT, 168–170
   RAY, 262, 296
   REASSOCIATE, 637, 639
   RECTANG, 262, 296–298
   REFEDIT, 941
   RENAME, 810
   REVCLOUD, 262, 299, 301
   REVOLVE, 1062
   RMAT, 1217
   SAVE, 58–59
   SAVEAS, 58–59
   SECTIONPLANE, 1133–1136
   selecting, solid modeling, 1040–1041
   selection method, 8–9, 12, 14
   SHADEMODE, 1030
   SLICE, 1136–1137, 1140
   SNAP, 95–96
   SOLDRAW, 1171–1175
   SOLIDEDIT, 1141, 1143–1153
   SOLVIEW, 1171–1175
   SPACETRANS, 746
   SPHERE, 1046
   SPLINE, 262, 302
   SUBTRACT, 1052–1053
   SWEEP, 1064
   text, 329–332, 339–340, 342, 344, 347, 352–354, 358
   TORUS, 1047
   TRIM, 802
   UCS, 1010–1012

ucsicon, 1024, 1031
union, 1005, 1052
vpmax, 731
vports, 1075
wblock, 798
wedge, 1043
wipeout, 262, 303
xbind, 938–940
xline, 262, 304, 305
xr, 928
zoom, 143–150, 720–721, 731, 946–947, 958–959
concept modeling
    DUCS mode, 1108–1109
    faces, DUCS, 1108–1109
    shapes, dragging, 1107–1108
Conceptual visual style, 1033
cone command, 1045
cones
    creating, 1045
    grips and, 1111
configuring plotter, 765–769
Confirm Purge dialog box, 805
Constant mode, attributes, 884
Constrained Orbit button, 1027
continue dimensions, 555, 565
Continue mode, arcs, 264
Continue option, line command, 19
continuous mode, dimensining, 567
Convert distance between spaces tool, Text toolbar, 746
converting drawings to different plot, 784–786
convertpstyles command, 784
Coordinate Display mode, status bar, 4–5
coordinates
    absolute, 48–49
    axes, 47
    mode combining, line drawing and, 51
    polar, 20
    polar coordinate mode, 50
    quadrants, 47
    relative, 20, 49
coordinate systems, 46
    X values, 47
    Y values, 47
copybase command, 826
copyclip command, 826

copying
    face editing, 1149
    grips and, 391
    mirroring and, 224
    objects/groups, 187–188, 826
    offsetting, 203
    press and drag technique, 188
copy modify command, 187–188
copytolayer command, 125
corners
    chamfers, 198
    fillets, 193, 195
    rounds, 193
Create Camera view, 1029
Create Layout wizard, 739–740
Create New Dimension Style dialog box, 612
Create New Drawing dialog box, 15
Create New Material dialog box, 1220
Create Web Page dialog box, 788
crosshatching, hatching islands, 493
crossing box, trimming, 207
crossing polygons, object selection, 166
Crossing Window option, object selection, 164
crossing windows, stretching and, 228
cross sections
    lofting and, 1065
cursor menus, 13
curves
    fit curves, 231
    generating, 230
    ogee curves, 306
    reverse curves, 306
    splines, 302
cutclip command, 826
cutting edges, trimming, 206, 209
    all objects as, 207
    extending, 208
cutting objects, 826
Cutting Plane line, 100
cycling through objects. *See* Object Selection Cycling
cylinder
    creating, 1044–1045
cylinder command, 1005, 1044

cylinders
    dimensions, 556

D

dashboard, 11
    accessing visual styles, 1034
    control panels, 11
    in 3D Modeling workspace, 1008–1009
    visual styles and, 1037
ddedit command, 342, 344
Decimal Degrees, 88
Decimal units, 88
dee variable, QuickCalc, 409
Degrees/Minutes/Seconds, 88
DesignCenter, 793, 813
    blocks, 817–818
    Tree View, 816
Detach option, External References palette, 948
detail pages, Layout Mode, 981–986
diagonal points, rectangle creation, 296
dialog boxes, 11
    Add Plotter-Begin, 767
    Attribute Definition, 882
    Attribute Extraction, 882, 903
    Block Attribute Manager, 883, 913, 915
    Block Definition, 795
    Bondary Creation, 265
    Boundary Hatch, 473
    Confirm Purge, 805
    Create Layout wizard, 740
    Create New Dimension Style, 612
    Create New Drawing, 15
    Create New Material, 1220
    Create Web Page, 788
    Dimension Style Manager, 611–613
    Drafting Settings, 96
    Drawing Units, 87–88
    Enable i-drop, 790
    Enhanced Attribute Editor, 883, 913–914
    External Reference, 931
    Field, 680
    Generate Images, 791
    Hatch and Gradient, 487, 751
    Hatch Edit, 473

dialog boxes *(continued)*
  Image, 955
  Image Adjust, 957
  Insert, 799
  Insert a Block in a Table Cell, 811
  Insert Table, 354
  Layer Filter Properties, 118
  Layer Properties Manager, 102–104, 743
  Layer States Manager, 120
  Layer Walk, 126
  Leader Settings, 575
  Linetype Manager, 111–112
  Lineweight Settings, 111, 774
  Multiline Edit Tools, 287
  Multiline Styles, 284–286
  Page Setup Manager, 723–725
  Plot, 770
  Plot Style Table Editor, 780–781
  Preview and Post, 791
  Purge, 804–806
  Quick Select, 169, 402–405
  Reference Edit, 941
  Render, 1197
  Scale List, 728
  Select DWF File, 929
  Select Image File, 929, 952
  Select Image Type, 788
  Select Reference File, 929
  Select Template, 789
  sizing, 12
  Table Style, 358
  Text Style, 349–350
  UCS, 1022–1023
  UCS Icon, 1024
  Viewports, 721, 987, 1075
  Visibility States, 854
  Visual Styles, 1033
  Write Block, 798
dialogue boxes
  Layer Properties Manager, 127
diameter, dimensioning, 570
  QDIM and, 572
diameter method, torus creation, 1047
Diameter mode
  circles, 53
DIMARC command, 573
DIMASSOC variable, 636
DIMEDIT command, 583–586
  Oblique option, 585

dimensioning
  Aligned Dimensioning mode, 563–564
  angles, 577
  arrows, flipping, 586
  Baseline mode, 569
  commands for, 549
  Continuous mode, 567
  diameter, 570, 572
  GDT, 587–589
  Linear Dimensioning mode, 561, 564
  Ordinate mode, 570
  overview, 549
  radius, 570, 572
  slots, 578
  Staggered mode, 568
  symbols, 589–590
dimensioning systems
  Aligned Dimensioning System, 556
  architectural dimensions, 558
  ordinate dimensions, 557
  Unidirectional Dimensioning System, 556
Dimension Input, 98
Dimension line, 101
dimension lines, 549, 614
  arrowhead settings, 618–619
  baseline, 615
dimension properties, matching, 417
Dimension pull-down menu, 559
dimensions
  Angular dimension units, 632
  arc dimensions, 573
  architectural dimensions, 558
  arc symbol, 620
  associative, 749–750
  associativity, 636–637, 639
  baseline, 555
  baseline dimensions, 566
  centerlines and, 554
  chain, 555
  commands, 559, 561, 563–567
  continue, 555, 565
  cylinders, 556
  editing, 20, 583–586
  extensions and, 551
  grips and, 591–592
  grouping, 553
  hidden items, 554
  jogged, 573

  layers, 101
  leader lines, 574
  Linear dimension units, 632
  Measurement scale, 633
  measurement units, 634
  ordinate, 557, 579–581
  placement, 551–552
  prefixes, 633
  Quick Dimensions, 567
  radius dimension jog angle, 621
  rectangles, 298
  repetitive, 558
  rounding off values, 632
  scale, 630
  spacing, 550
  stretching and, 228
  suffixes, 633
  text, 622–625
  tutorial, 595, 599–602, 604, 646–649, 651–657, 659, 660, 662–666
  visibility, 549
  visible features and, 554
Dimension Style Manager dialog box, 611–613
  dimension types, 640
dimension styles
  dimension types and, 639–641
  modifying, 644–645
  overriding, 642–643
Dimension toolbar, 559
dimension types, dimension styles and, 639–641
DIMJOGGED command, 573
DIMLALIGNED command, 563–564
DIMLBASELINE command, 566
DIMLCONTINUE command, 565
DIMLINEAR command, 561
Direct Distance mode
  Angle Block tutorial, 64, 66
  Angle Plate tutorial, 66
  Gage Block tutorial, 62, 64
  line drawing, 20
  tutorials, one-view drawing, 73–74
directional arrows, grips, 389
display
  enlarging appearance, 7
  PAN command, 153–154
  views, named, 156–160

wheel mouse for zooming/
  panning, 155
ZOOM command, 143–150
displaying attributes, 893
  modes, 893
DISPSILH system variable, solid
  modeling, 1079
distance, offsetting and, 203
Distance tool, Inquiry toolbar,
  674
DIST command, 683
  angles and, 684
divide command, 262, 291
  ogee curves, 306
  reverse curves, 306
DIVIDE command, 808–809
dividing blocks, 808–809
dividing objects, 291
docking
  toolbars, 10
  Tool Palette, 822
DONUT command, 262, 279–280
Drafting Settings dialog box, 96,
  98–99
dragging shapes, 1107–1108
Draw commands tutorial, 316,
  318–320
drawing
  lines, 17–20, 22, 24
  three-view, 443
  two-view, 440, 442
drawing area size, 90
  calculating, 91
  wizards, 93–94
drawing commands, 16
drawing layout tutorial, 754–756,
  759, 762–764
drawing lines, architectural
  values, 88
drawings
  architectural, arranging in
    Paper Space, 740–741
  converting to different plot
    style, 784–786
  importing, text, 337
  managing images in, 954
  Multiple Design Environment,
    825
  Multiple Design Environment,
    opening multiple, 825
  Multiple Design Environment,
    working between
    multiple, 826–827

new, 15
one-view, 438
opening, 15
drawing sessions, exiting, 60–61
drawing units, 87–88
  Architectural, 88
  Decimal, 88
  Engineering, 88
  Fractional, 88
  measuring angles, 88
  Scientific, 88
  wizards, 93–94
Drawing Units dialog box, 87–88
DRAWORDER command, 958
Draw Order toolbar, 958–959
Draw pull-down menu, 261
  Hatch option, 485–487
Draw toolbar, 3, 16
  BLOCK commands, 795
Draw toolbars, 261
Draw tool palette, 261
DTEXT command, 330, 339–340
DUCS (Dynamic UCS),
  1108–1109
DUCS button
  status bar, 5
DWF files, 960
  attaching to a drawing, 960
  ePlot device for, 960
  External Reference palette
    and, 960
  External References palette
    and, 962
Dynamic Block Editor, 832
dynamic blocks
  array applications, 850–852
  creating, 827–829
  grips, 830
  lookup parameters, 857–860
  object centering, 845–847
  parameters, 831–842, 844–845
  stretching, 847–850
  tables, 857–860
  visibility states, 853–856
Dynamic Input, 8, 98–99
  arrow keys and, 20
  circles, 54
  command options, 20
  command prompts, 20
  coordinates and, 20
  Dimension Input, 98
  editing dimensions, 20

line drawing and, 19
lofting and, 1067
plines, 56
Pointer Input, 98
status bar, 4–5
dynamic input
  grips and, 1111
Dynamic UCS, 1019
Dynamic User Coordinate
  System
  status bar, 4–5
DYN button, 8
DYN toggle, status bar, 98–99

E

EATTEDIT command, 883, 914
EATTEXT command, 883
Edge editing, 1141
Edge Jitter setting, 1038
Edge Overhang setting, 1037
edges
  solid modeling, 1114–1115
editing
  attribute properties, 894–899
  attributes tutorial, 900–902
  blocks, 806–808
  dimensions, 583–586
  grips and, 1111
  In-Place Reference Editing,
    940–942
  objects, subobjects, 1113
  polylines, 229, 233–234
  revision clouds and, 299, 301
  solid models, 1123–1131,
    1141–1142
  text, 342, 344
ELLIPSE command, 262, 281–282
Enable i-drop dialog box, 790
Endpoint (End) Object Snap
  mode, 25
end point, arc drawing, 264
Endpoint mode, LINE command,
  22
Engineering units, 88
Enhanced Attribute Editor
  dialog box, 883,
  913–914
ePlot device
  creating DWF files with, 960
equal distances, chamfers and,
  199
ERASE command, 57–58

Erase option, OFFSET command, 205
eTransmit, 951
Excel spreadsheets, tables and, 357–358
exit drawing session, 60–61
Exit option, File menu, 60–61
EXPLODE command, 804
EXPLODE modify command, 235
exploding blocks, 804
   redefining attributes, 905
EXTEND command, 803
   toggle to/from TRIM command, 214
Extended Intersection (Int), Object Snap, 28
extending objects, 211–212, 214
   boundary edges, 211, 213
   multiple, 212
extending to blocks, 803
EXTEND modify command, 211–212, 214
Extension (Ext) Object Snap mode, 25
Extension line, 101
extension lines
   controls, 615
   fixed-length, 617
extension linetypes, 617
extensions
   dimensions and, 551
   placement, 552–553
extents, ZOOM-Extents, 149
External Reference dialog box, 931
   Attach option, 932–934
   Overlay option, 936
External Reference palette
   binding external references, 943
   for attaching DWF files, 960
External References
   attaching and reloading, 929, 931
External references
   attaching and reloading, 930
   shortcut menu, 930
external references
   attaching, 932–934
   binding, 943–944
   blocks and, 927–928
   clipping, 946–947
   command, 928
   In-Place Reference Editing, 940–942
   notification tools, 949–950
   overlaying, 935–937
   tutorial, 963–971
   XBIND command, 938–940
External References palette, 928
   attaching a raster image, 952
   binding external references, 943
   Detach option, 948
   DWF files and, 962
   List view option, 945
   refreshing, 929
   Reload option, 948
   Tree view option, 945
   Unload option, 948
extracting attributes, 903
EXTRUDE command, 1059, 1061
extrusions, 1059
   existing faces, 1061
   face editing, 1143

F

Face editing, 1141
   copying, 1149
   deleting, 1148
   extruding, 1143
   moving, 1144
   offsetting, 1146
   rotating, 1145
   tapering, 1147
Face option, UCS command, 1017
faces
   extrusions, 1061
   solid modeling, 1114–1115
faces, 3D models, 1004
FACETRES system variable, solid modeling, 1078
fences, object selection, 164
FIELD command, 352–353
Field dialog box, 680
fields
   area calculation, 680
   attributes and, 892
   multiline text, 352–353
   tables, 357
   text, 351–353
   tutorial, 699–700
File menu
   Exit option, 60–61
files
   Quick Save, 59
   Save As, 59
   saving, 58–59
   template files, creating, 903–905
   templates, creating, 131–133
filled-in circles, 279–280
FILLET command, 1072
filleting
   by radius, 194
   solid models, 1072
FILLET modify command, 193, 198
fillets, 193, 198
   cornering and, 195
   multiple, 195
   parallel lines, 198
   polylines, 196
   trimming, 197
fill patterns, trimming, 496
filters, layers, 118
fit curves, generating, 231
Fit tab, Modify Dimension Style: Mechanical dialog box, 626–631
fixed-length extension lines, 617
Flanged visibility state, blocks, 855
Flat + Edges shading mode, 1030
Flat Shade mode, 1030
FLATSHOT command, 1184–1186
flipping, mirroring and, 225
floating Model Space, activating, 726
flying through models, 1238
   motion path animation, 1240–1242
formatting text, fractional, 332
Form View tab, Plot Style Table Editor dialog box, 781
fractional text, formatting, 332
Fractional units, 88
Freeze layer state, 103
From (Fro) Object Snap mode, 26
Front view
   glass box, 435
   shape description and, 433
   tutorial, 448–450
Front View mode, 1029
Full Sections, 477

function keys
    status bar tools, 6

G

Galvanized material, 1217
GDT (Geometric Dimensioning and Tolerancing), 587–589
Generate Images dialog box, 791
geometric commands tutorial, 308–314
geometric tolerancing symbols, 587
GIF raster images, 951, 955–957
    background, 1233
glass box
    Back view and, 435
    Bottom view and, 435
    Front view and, 435
    Left Side view and, 435
    Right Side view and, 435
    shape description and, 434
    Top view and, 435
globally editing attributes, 896–899
Gouraud + Edges shading mode, 1030
Gouraud Shade mode, 1030
Gradient background, 1233
gradient hatch patterns, 491
Grads, 88
graphics cursor, 3
GRID command, 94
Grid mode, status bar, 4–5
grids, 94
    Adaptive Grid mode, 97
    Allow Subdivision Below Grid Spacing mode, 97
    behavior tools, 97
    blocks and, 813
    Display Grid Beyond Limits, 98
    Follow Dynamic UCS mode, 98
    menu, 96
    snap values, 95–96
GRID toggle, status bar
    Drafting Settings dialog box, 96
grips, 385
    arcs and, 389
    arrows, 389
    color, 387

copying, 391
dimensions and, 591–592
editing, command mode, 388
enabling, 387
Grip Size setting, 387
Hover grip color setting, 387
Lug.Dwg project, 398
modes, 387–394
offset snap, moving and, 396
offset snap rotation, 395–396
Options dialog box settings, 386
prompts, 388
Selected grip color setting, 387
selection limit, 387
shortcut menu, 389–396, 398
solid primitives and, 1110–1111
    cones, 1111
    dynamic input and, 1111
    editing and, 1111
    grip location, 1110
tables and, 356
tips, enabling, 387
Unselected grip color setting, 387
groups
    copying, 187–188
    deleting, TRIM commands, 206, 208–210
    dimensions, 553
    layers, 118–119
    moving, 185–186

H

Half Sections, 478
Hatch and Gradient dialog box, 487, 751
    advanced features, 500, 502
    associative hatches, 497–499
    Gradient tab, 491
    Hatch tab, 487–490
    Inherit Properties button, 503
HATCHEDIT command, 500
Hatch Edit dialog box, 473
hatching
    advanced techniques, 500–501
    associative hatches, 497–499
    gap tolerance, 502
    properties, 503, 511, 513,–516

hatch objects, Object Snaps and, 504
Hatch Pattern Palette, 489–490
hatch patterns, 487–489
    angle, 495
    editing, 499–500
    gradient, 491
    hatching islands, 493
    placement, 502
    scaling, 494–495, 751
    solid fill, 489–490
    symbols, 493
    trimming, 496
hatch properties, matching, 419
height of text, globally modifying, 344–345
HELIX command, 1068–1070
hidden items, dimensions and, 554
Hidden line, 100
Hidden shade mode, 1031
HIDE command, 3D models, 1004
Hide shading mode, 1030
hiding/showing
    objects, WIPEOUT command, 303
    Tool Palette, 822
holes
    CYLINDER command, 1006
    SUBTRACT command, 1006
horizontal text placement, dimension text, 624–625
Hover grip color setting, 387

I

ID command, 684
Ignore Hatch Objects option, 504
Image Adjust dialog box, 957
Image dialog box, 955
images. *See also* raster images
    background, 1233
    raster images, 951, 955–957
importing text to drawings, 337
imprinting, Body editing, 1150
In-Place Reference Editing, 940–942
indented text
    multiline, 334–335
    tabs, 335
Index Color tab, Select Color dialog box, 107

inheritance
  hatching properties, 503
  hatch patternstutorial, 511, 513–516
inquiry commands
  DIST, 683
  ID, 684
  LIST, 686
  tutorials, 688, 690–694, 698
Inquiry toolbar, 673–674
Insert (Ins) Object Snap mode, 27
Insert a Block in a Table Cell dialog box, 811
INSERT command, blocks, 799
Insert dialog box, 799
Insert Field button, Text Formatting toolbar, 680
Insert Table dialog box, 354
Insert toolbar, 799
INTERFERE command, 1132
interferences, solid models, 1132
INTERSECT command, 1053, 1055, 1057–1058
intersections, 1057–1058. *See also* runouts
  Intersection (Int) Object Snap mode, 27–28
Invisible mode, attributes, 884
ISOLINES system variable, solid modeling, 1077–1078
isometric sections, 484
isometric views, 1182
italic text, 331

J

jogged dimensions, 573
joining objects, 236
JOIN modify command, 236
JPG (JPEG File Interchange Format) images
  background, 1233
JPG raster images, 951, 955–957
justified text
  globally modifying, 346
  multiline text, 333

K

keyboard, modify commands, 184
keyboard shortcuts, commands, 14

L

LAYCUR command, 123
LAYDEL command, 130
Layer Control area, Layers toolbar
  properties, 113
Layer Control box, 127, 410–412
Layer Filter Properties dialog box, 118
Layer option, OFFSET command, 205
Layer Properties Manager dialog box, 102–104, 127, 743
  applying changes, 106
  layer creation, 104–105
  layer creation tutorial, 134–135, 137–138
  layer lineweight, 110
  layers, current, 114
  plot styles, 785
layers, 101–104
  boundary creation and, 265
  changes, applying, 106
  color, 107–108
  converting, XBIND command, 938–940
  creating, 104–105, 134–135, 137–138
  current, 114
  filters, 118
  groups, 118–119
  LAYERP command, 115
  linetypes, assigning, 108–109
  lineweight, 110
  Lineweight Settings dialog box, 111
  materials and, 1227–1231
  moving objects between, 410–411
  properties, 113
  Properties Palette, 400
  right-clicking, 116
  Select Color dialog box, 107–108
  settings, named views, 160
  states, 120
  Status property, 115
  Tree View pane, 118–119
layer states
  Color, 103
  Freeze, 103
  Linetype, 103
  Lineweight, 103
  Lock, 103
  On/Off, 103
  Plot, 104
  Plot Style, 104
  Thaw, 103
Layer States Manager dialog box, 120
Layers toolbar, 3, 113
Layer Tools pull-down menu, 121
Layer Walk dialog box, 126
LAYFRZ command, 127
LAYISO command, 124
LAYLCK command, 128
LAYMCH command, 122
LAYMRG command, 129
Layout mode, detail pages, 981–986
Layouts, 3
layouts
  associative, 749–750
  Create Layout wizard, 739–740
  introduction, 719
  Model Space, 719
  multiple, 742–745
  Paper Space, 733–739, 742–745
  tutorial, 754–756, 759, 762–764
LAYULK command, 128
LAYWALK command, 126
LEADER command, 574
leader lines, dimensions, 574
  QLEADER command, 575–576
leaders, arrowhead settings, 618–619
Leader Settings dialog box, 575
Left Side view
  glass box, 435
  shape description and, 433
Left View mode, 1029
LENGTHEN modify command, 235
  tutorial, 239, 241, 243–244
Level II modify commands, 218
  ARRAY, 218–220
  EXPLODE, 235
  JOIN, 236
  LENGTHEN, 235

MIRROR, 224, 226
PEDIT, 229–234
REDO, 238
STRETCH, 227
UNDO, 237
Level I modify commands, 184
  BREAK, 215–217
  CHAMFER, 198
  COPY, 187–188
  EXTEND, 211–212, 214
  FILLET, 193, 198
  MOVE, 185–186
  OFFSET, 203
  ROTATE, 191
  SCALE, 189–190
  TRIM, 206, 208–210
libraries
  materials, 1214
lighting
  rendering, 1204, 1206–1214
  sunlight tutorial, 1243, 1245, 1248
  tutorial, sunlight, 1243, 1245, 1248
Light List button, 1211
Lights in Model palette, 1211
LIMITS command, 90–91
  grids and, 94
limits of drawing size, 90
  calculating, 91
  wizards, 93–94
Linear Dimensioning mode, 561
  rotated linear dimensions, 564
Linear dimension units, 632
linear distance
  DIST command, 683
LINE command, 17–19
  Continue option, 19
  Endpoint mode, 22
  OSNAP-Tangent option, 276
  Undo option, 18
LINE command, 19
Line Generation option, PEDIT command, 232
lines
  architectural values, 88
  Break line, 101
  Center line, 101
  centerlines, dimensions and, 554
  coordinates, combining modes, 51
  Cutting Plane line, 100
  dimension lines, 101, 549, 614–615
  drawing, 17–20, 22
  Endpoint (End) Object Snap mode, 25
  Extension (Ext) Object Snap mode, 25
  extension lines, 101, 615
  Hidden line, 100
  Midpoint (Mid) Object Snap mode, 28
  Object line, 100
  Parallel (Par) Object Snap mode, 30
  parallel, filleting and, 198
  Phantom line, 101
  polar coordinate mode, 50
  polylines, 54, 56, 196
  relative coordinates, 49
  scale, 438
  section lines, errors, 476
  Tangent (Tan) Object Snap mode, 32
  tangent to two arcs/circles, 276
  xlines, 304–305
Lines tab, Modify Dimension Style: Mechanical dialog box, 614–617
Linetype layer state, 103
Linetype Manager dialog box, 111–112
linetypes
  extension linetypes, 617
  generating, 232
  layer assignment, 108–109
  multilines, 283
  Properties Palette, 400
  Properties toolbar, 114
  scale, 116–117
  text embedded, 109
Linetype Scale command, 438
Lineweight layer state, 103
Lineweight mode, status bar, 4–5
lineweights
  layers, 110
  LWT button, 774
  Paper Space, 775
  plots and, 772–777
  Properties toolbar, 114
  scale, 774
Lineweight Settings dialog box, 111, 774
LIST command, 686
list of scales, 728–729
lists, numbered, 335
List tool, Inquiry toolbar, 674
List view option, External References palette, 945
Locate tool, Inquiry toolbar, 674
locking viewports, 730
Lock layer state, 103
LOFT command, 1065–1066, 1068
lookup parameters, dynamic blocks, 857–860
LTSCALE command, 116–117, 438
Lug.Dwg
  grips and, 398
LWT button, 774

M

masking, WIPEOUT command, 303
MASSPROP command, 1007, 1076
mass properties, solid models, 1076, 1132
Mass Properties tool, Inquiry toolbar, 674
matching properties, 415–416, 418, 420, 422–423
  dimension properties, 417
  hatch properties, 419
  polyline properties, 420
  tables, 422
  text properties, 418
  viewport properties, 420, 422
  viewports, 991, 993
MATCHPROP command, 415–418, 420, 422–423, 437, 991, 993
Material Editor, 1222
materials
  applying, 1217
  color, 1216
  creating, 1215–1222, 1227
  Galvanized, 1217
  layers and, 1227–1231
  library, 1214
  loading, 1215
  opacity, 1222
  Remove Materials option, 1219
  rendering and, 1205
  shadows and, 1218
  templates, 1223–1226

Materials palette, 1215
  buttons, 1215
mathematical calculations
  QuickCalc, 407–409
MEASURE command, 262, 293, 809, 810
Measurement scale, dimensions, 633
measuring angles
  Decimal Degrees, 88
  Degrees/Minutes/Seconds, 88
  Grads, 88
  Radians, 88
  Surveyorís Units, 88
mee variable, QuickCalc, 409
menus, 11
  cursor menus, 13
  Dimension, 559
  pull-down, 8
  shortcut, 13, 389–396, 398
merging table cells, 356
Microsoft Excel spreadsheets, tables and, 357–358
Midpoint (Mid) Object Snap mode, 28
MIRROR3D command, solid model editing, 1130
mirroring
  copying and, 224
  flipping and, 225
  grips and, 392
  text, 226
MIRROR modify command, 224, 226
MLEDIT command, 287–288
MLINE command, 262, 284
Model/Paper mode, status bar, 4
modeling
  flying through models, 1238
    motion path animation, 1240–1242
  walking through models, 1238
    motion path animation, 1240–1242
Modeling toolbar
  Boolean operations, 1050
Model mode, status bar, 5
Model Space, 719
  dimensions, associative, 749–750
  floating, 726
  layouts, associative, 749–750
  locking, 730

Paper Space comparison, 722
  scaling viewports, 727
  text, 746–748
  viewports, tiled, 720
modes
  Aligned Dimensioning, 563–564
  grips, 387–394
  Linear Dimensioning, 561, 564
  UCS, 1012
  View modes, 1029
modify commands, 183
  Level I, 184–191, 193, 198, 203, 206, 208–212, 214–217
  Level II, 218–220, 224, 226–227, 229–238
  selecting, 183–184
Modify menu, Solids Editing, 1050
Modify pull-down menu, 183
Modify toolbar, 3
Modify toolbars, 183
Modify tool palette, 183
motion path animation, 1240–1242
mouse
  click, 13–14
  wheel mouse, zooming/panning, 155
MOVE command, tables, 356
MOVE mode, grips, 391
MOVE modify command, 185–186
moving
  face editing, 1144
  grips and, offset snap, 396
  objects/groups, 185–186
  press and drag technique, 186
  solids, 3D applications, 1055
  tables, 356
MTEXT command, 330–332
  plotted text, 747
  width, 336
Multiline Edit Tools dialog box, 287
multiline objects, 283–284
Multiline Styles dialog box, 284
  styles, 285–286
multiline text
  adding, 331–332
  bulleted, 335
  capped, 286

  editing, 287–288
  fields, 352–353
  justification, 333
  numbered, 335
  styles, 285–286
Multiline Text Editor, 331
  editing text, 342
  fractional text, 332
  importing text, 337
  justification, 333
multiline text symbols, 338
Multiple Design Environment, 825
  opening multiple drawings, 825
  working between drawings, 826–827
multiple ROTATE mode, grips, 394

N

named blocks, 794
  renaming, 810
Named plot style, 784
named views, 156–160
  boundaries, changing, 159
  layers, settings, 160
Named Views mode, 1029
Nearest (Nea) Object Snap mode, 30
nee variable, QuickCalc, 409
NE Isometric view, 1029
NEW command, 15
New Dimension Style Mechanical dialog box, 612
  Alternate Units tab, 634
  Fit tab, 626–631
  Lines tab, 614–617
  Primary Units tab, 631–633
  Symbols & Arrows tab, 618–620
  Text tab, 621–625
  Tolerances tab, 635
New option, UCS command, 1013
Node (Nod) Object Snap mode, 30
notification tools, external references, 949–950
numbered lists, 335
Numeric Pad tab, QuickCalc, 408
NW Isometric view, 1029

O

object grips, 385. *See also* grips
object line, 100
   border lines and, 437
   thickness, 549
Object option, UCS command, 1017
objects
   breaking, 215–217
   centering, dynamic block creation, 845–847
   construction line, rays, 296
   copying, 187–188, 826
   Crossing Window option, 164
   cutting, 826
   delete from selection set, 165
   deleting, TRIM command, 206, 208–210
   dimension style modification, 644–645
   dividing, 291
   erasing, 57–58
   extending, 211–214
   fences, 164
   joining, 230, 236
   lengthening, 235
   mirroring, 224–225
   moving, 185–186
   multiline, 283–284
   offsetting, 203
   pickbox, 162
   point objects, 290
   polyline, offsetting, 232
   pre-highlighting, 161
   preselecting, 167
   rotating, 191
   scaling, 189–190
   selecting, tutorial, 144, 170–171, 173, 175, 177–181
   subobjects, 1112
      editing, 1113
   Window selection mode, 163
Object Selection Cycling, 167
object selection sets, 161
   crossing polygons, 166
   Crossing Window option, 164
   fences, 164
   object deletion, 165
   options, 161
   pickbox, 162
   pre-highlighting, 161
   QSELECT command, 168–170
   select previous, 166
   Window selection mode, 163
Object Snap
   hatch objects, excluding, 504
   modes, 23, 25–32
   precision, 22, 24
   Running OSnap, 33
   status bar, 4–5
   Tracking mode, 35
Object Snap toolbar, 23
Object Snap Tracking mode, 42
   status bar, 5
Object Tracking mode, status bar, 4
obliquing angles, dimensioning and, 585
Obscured Edges setting, 1039
OFFSET command
   Erase option, 205
   Layer option, 205
OFFSET modify command, 203
   tutorial, 250–253
Offset Sections, 481
offset snap
   moving, grips and, 396
   rotations, grips and, 395–396
offsetting, 203
   distance and, 203
   face editing, 1146
   multiple, 205
   objects, polyline, 232
   rectangular arrays, 218–220
   through points and, 203
ogee curves, 306
On/Off layer state, 103
one-view drawing
   Direct Distance mode tutorial, 73–74
   Polar Tracking tutorial, 69–72
   relative coordinates tutorial, 73–74
one-view drawings, 438
opacity
   materials, 1222
OPEN command, 15
OPEN command, 825
opening drawings, 15
Options dialog box, grips settings, 386
ordinate dimensions, 557, 579–581
Ordinate mode, dimensioning, 570
orthographic views, 1175
Ortho mode, status bar, 4–5
OSNAP-Center mode, ID command and, 684
OSNAP-Quadrant mode, CIRCLE command, 278
OSNAP-Tangent mode, CIRCLE command, 277–278
OSNAP-Tangent option, LINE command, 276
Osnap button, status bar, 5
Otrack button, status bar, 5
overhead lighting values, 1212
overlaying external references, 935–937

P

Page Setup dialog box
   Paper size area, 724
   Plot style table area, 724
   Printer/plotter area, 724
Page Setup Manager, 723–725
palettes
   Hatch Pattern Palette, 489–490
   Lights in Model, 1211
   Materials, 1215
   Properties, 399–402
   Tool Palette, 819–823
   transparency, 824
PAN command, 153–154
panning with wheel mouse, 155
Paper mode, status bar, 5
Paper size area, Page Setup dialog box, 724
Paper Space
   architectural drawings, arranging, 740–741
   layouts, 733–740, 742–745
   lineweights, 775
   Model Space comparison, 722
   plotting from, 770–772
   reasons for using, 722–723
   scaling, hatch patterns, 751
Parallel (Par) Object Snap mode, 30
parallel lines, filleting, 198

parameters
  dynamic block creation, 831–835
    alignment, 839–841
    angle offset, 842, 844–845
    base applications, 837–838
    scale, 842, 844–845
    value set properties, 835–837
  dynamic blocks, lookup parameters, 857–860
PART_NAME attribute tag, 881
partial views, 528
PASTEBLOCK command, 827
PASTECLIP command, 827
patterns, ANSI31, 475
PCX raster images, 951, 955–957
  background, 1233
PEDIT modify command, 229–234
  Linetype Generation option, 232
Perpendicular (Per) Object Snap mode, 31
perspective
  walking and flying and, 1238
Phantom line, 101, 437
Phi variable, QuickCalc, 409
PICKADD button, Properties Palette, 407
pickbox, object selection, 162, 167
PLAN command, 1025
planes, mirroring over, 1130
PLANESURF command, 1040
PLINE command, 54, 56
Plot dialog box, 770, 782
Plot layer state, 104
plots
  lineweights, 772–777
  Publish to the Web utility, 787–792
Plot Settings tab, Plot dialog box, 782
Plot Style layer state, 104
plot styles, 779
  converting drawing to, 784–786
  Layer Properties Manager dialog box, 785
  Named, 784

Plot style table area, Page Setup dialog box, 724
Plot Style Table Editor dialog box, 780–781
Plot style tables
  creating, 779
  default, 725
plotted text size, 746–748
Plotter Manager, 765
plotting
  Color-Dependent plot style table, 777–784
  from Paper Space, 770–772
  plotter configuration, 765–769
PNG images, background, 1233
POINT command, 262, 290
Pointer Input, 98
point objects, 290
points
  area, enclosed shape, 674
  breaking at, 217
  diagonal, rectangle creation, 296
  Node (Nod) Object Snap mode, 30
  splines, 302
  Temporary Tracking Point tool, 44, 46
  user-defined, xlines, 304–305
Point Style dialog box, 290
Polar Angles
  Absolute, 41
  Relative to last segment, 41
polar arrays, 222, 1131
polar coordinate mode, 50
Polar mode, status bar, 5
Polar Snap mode
  status bar, 4
  values, 39
Polar Tracking, 3, 36
  Angle Block tutorial, 64, 66
  Angle Plate tutorial, 66
  Gage Block tutorial, 62, 64
  tooltips, 36
  tracking path, 36
  tutorials, 69–72, 77–79
POLYGON command, 262, 294
polygons
  creating, 294
  Polar Tracking, tutorial, 77–79
polylines, 54
  boundaries, 265–266

chamfers and, 200
curve generation, 230
Dynamic Input and, 56
editing, 229, 233–234
enclosed, area, 675–677
exploding, 235
filleting, 196
joining objects into single, 230
offsetting objects, 232
properties, matching, 420
spline generation, 230
POLYSOLID command, 1070–1072
polysolids, 1070–1072
pre-highlighting objects, 161
pre-selecting objects, 167
precision
  Object Snap, 22, 24
prefixes, dimensions, 633
Preset mode, attributes, 884
press and drag
  copying and, 188
  moving and, 186
pressing and pulling, 1115–1117
  blocks and, 1118–1121
PRESSPULL command, 1115–1121
Preview and Post dialog box, 791
Primary Units tab, Modify Dimension Style: Mechanical dialog box, 631–633
primitives
  solid
    Boolean operations, 1050
    grips and, 1110–1111
      cones and, 1111
      dynamic input and, 1111
      editing and, 1111
      grip location, 1110
primitives, solid modeling, 1005
Printer/plotter area, Page Setup dialog box, 724
prompt, command line, 7
properties
  attributes, 894–899
  hatch patterns, 503, 511, 513–516
  Layer Control box, 410–412
  layers, 113, 115
  MASSPROP command, 1007
  mass properties, solid models, 1076, 1132
  matching, 415–420, 422–423

text fields and, 353
tutorial, 424–432
viewports, matching, 991, 993
Properties Palette, 399
 color, 400
 hiding/showing, 400
 layers, 400
 linetypes, 400
 opening, 402
 PICKADD button, 407
 redefining attributes, 906–907
Properties palette
 DWF files and, 961
Properties toolbar, 3
 color, 114
 linetypes, 114
 lineweight, 114
Publish to the Web utility, plots, 787–792
pull-down menus, 8
 Dimension, 559
 Draw, 261
 Modify, 183
PURGE command, 804–806
Purge dialog box, 804–806
purging unused block data, 804–806
pyramid
 creating, 1049–1050
PYRAMID command, 1049

Q

QDIM command, 567
QLEADER command, 575–576
QSAVE command, 58–59
QSELECT command, 168–170
Quadrant (Qua) Object Snap mode, 31
quadrants in coordinate systems, 47
QuickCalc, 407–409
Quick Dimensions, 567
 Baseline mode, 569
 Continuous mode, 567
 diameter dimensions, 572
 Ordinate mode, 570
 radius dimensions, 572
 Staggered mode, 568
Quick Select dialog box, 169, 402–405
Quick wizard, 93–94

R

Radians, 88
radius
 dimensioning, 570, 572
 filleting by, 194
radius dimension jog angle, 621
radius method, torus creation, 1047
Radius mode, circles, 53
rad variable, QuickCalc, 409
raster images, 951, 955–957. *See also* images
 attaching to a drawing, 952
 borders, 958–959
 Reference toolbar and, 952
RAY command, 262, 296
rays, 296
Realistic visual style, 1032
Realtime PAN, 147
Realtime ZOOM, 147
realtime zooming, 146
REASSOCIATE command, 637, 639
reassociating dimensions, 637, 639
RECTANG command, 262, 296–298
rectangles, 296–298
 diagonal points and, 296
 dimensions, 298
rectangular arrays, 1131
 negative offset, 220
 positive offset, 218–220
redefining
 attributes, 905–912
 blocks, 806–808
REDO modify command, 238
REFEDIT command, 941
Refedit toolbar, 941
Reference Edit dialog box, 941
Reference option
 ROTATE command, 192
 SCALE command, 190
Reference toolbar, 928, 930, 952
 raster images and, 952
relationships, views, 436
relative coordinates, 49
 Angle Block tutorial, 64, 66
 tutorials, one-view drawing, 73–74
relative mode, Angle Plate tutorial, 66

Relative to last segment Polar Angle, 41
Reload option, External References palette, 948
removed sections, 483
RENAME command, 810
renaming blocks, 810
Render button, 1200
Render control panel, 1200
Render dialog box, 1197
rendering, 1197, 1199
 exercise, 1203–1205
 introduction, 1197
 lighting, 1206–1212, 1214
 lighting and, 1204
 materials and, 1205
 modes, 1200
 presets, 1203
 raster images, 1202
 Render Preset value, 1203
Render Preset value, 1203
Render toolbar
 tools, 1199
repetitive dimensions, 558
RESISTANCE attribute tag, 881
REVCLOUD command, 262, 299
reverse curves, 306
revision clouds, 299
REVOLVE command, 1062
revolved sections, 483
revolved solids, 1062
ribs, sectioning, 482
right-clicking
 command prompt area, 14
 layers, 116
 shortcut menus, 13
Right Side view
 glass box, 435
 shape description and, 433
 tutorial, 448–450
Right View mode, 1029
RMAT command, 1217
ROTATE3D command
 tutorial, 1154, 1159–1160
rotated linear dimensions, 564
ROTATE mode
 grips, 393–394
ROTATE modify command, 191
rotating
 face editing, 1145
 grips and, 393, 395–396

rotating *(continued)*
  objects, 191
  tutorial, 1154, 1159–1160
rounding off dimension values, 632
rounds, 193, 198
Running Osnap, 33
runouts, 447

S

SAVEAS command, 58–59
SAVE command, 58–59
saving
  files, 58–59
scale
  blocks, 813
  dimension features, 630
  linetype, 116–117, 438
  lineweights, 774
  paramaeters, dynamic block creation, 842, 844–845
scale factor, limit calculation, 91
Scale List dialog box, 728
SCALE mode, grips, 390
SCALE modify command, 189–190
scaling
  grips and, 390
  hatch patterns, 494–495, 499–500, 751
  images, ZOOM command, 729
  list, 728–729
  objects, 189–190
  referencing, 190
  viewport images, 727
  ZOOM-Scale option, 152
Scientific tab, QuickCalc, 408
Scientific units, 88
sectioning solids, 1133–1136
section lines
  BHATCH command, 485–487
  errors, 476
SECTIONPLANE command, 1133–1136
sections
  tutorial, 505–509
  Section Views, 473. *See also* section lines
    aligned sections, 480
    architectural sections, 484
    assembly sections, 479–480
    broken sections, 482

full sections, 477
half sections, 478
isometric sections, 484
offset sections, 481
overview, 473–476
removed sections, 483
revolved sections, 483
sectioning rigs, 482
slicing views and, 474
section views, 1179–1181
SE Isometric view, 1029
Select DWF File dialog box, 929
Select Image File dialog box, 952
selecting
  commands, 795
  modify commands, 183–184
  objects, tutorial, 144, 170–171, 173, 175, 177–181
  Quick Select dialog box and, 402–405
  solid modeling commands, 1040–1041
selection commands, 8
selection sets, 161
  crossing polygons, 166
  Crossing Window option, 164
  fences, 164
  object deletion, 165
  options, 161
  pickbox, 162
  pre-highlighting, 161
  QSELECT command, 168–170
  select previous, 166
  Window selection mode, 163
Select Reference File dialog box, 929
Select Template dialog box, 789
separating solids, Body editing, 1151
settings
  visual styles, 1035
Set Variable tool, Inquiry toolbar, 674
SHADEMODE command, 1030
shading, solid modeling, 1030
shadows
  materials and, 1218
shadows, lighting and, 1214
shape description
  Back view, 433
  Bottom view, 433
  Front view, 433

introduction, 433–436
Left Side view, 433
Right Side view, 433
Top view, 433
shapes
  dragging, 1107–1108
shapes, area, 674
  subtraction, 678
shelling, Body editing, 1152
shortcut menus, 13
  grips, 389–396, 398
Silhouette Edges setting, 1038
single line text, creating, 339
SLICE command, 1137, 1140
slices, solid modeling, 1136–1137, 1140
slicing views, 473
  Section Views and, 474
slots, dimensioning and, 578
snap
  menu, 96
  values, grids, 95–96
SNAP command, values, 95–96
Snap mode, status bar, 4–5
SNAP toggle, status bar
  Drafting Settings dialog box, 96
SOLDRAW command, 1171–1175
  section views, 1179–1181
  tutorial, 1188–1194
solid background, 1233
SOLIDEDIT command
  Clean option, 1153
  Copy Face option, 1149
  Delete Face option, 1148
  Extrude option, 1143
  imprinting, 1150
  Move option, 1144
  Offset Face option option, 1146
  Rotate Face option, 1145
  Separating Solids option, 1151
  Shell option, 1152
  solid model editing, 1141
  Taper Face option, 1147
solid fill hatch patterns, 489–490
solid modeling, 1003
  2D wireframes, 1031
  3D models, 1003–1004
  3D tutorial, 1080
  3D wireframes, 1031
  Boolean operations, 1050

chamfering, 1074–1075
Collar tutorial, 1084–1088
command selection, 1040–1041
Conceptual visual style, 1033
custom visual styles, 1033
edges, 1114–1115
editing, 1122
editing models, 1123–1131, 1141–1142
extrusions, existing faces, 1061
faces, 1114–1115
filleting, 1072
helix, 1068–1070
Hidden shade mode, 1031
interferences, 1132
lofting, 1065–1066, 1068
mass properties, 1076, 1132
overview, 1005–1007
polysolids, 1070–1072
pressing and pulling, 1115–1117
blocks and, 1118–1121
primitives, 1005
Realistic visual style, 1032
revolved solids, 1062
sectioning, 1133–1136
shading, 1030
slices, 1136–1137, 1140
sweeps, 1064
system variables, 1077–1079
terrain modeling, 1139
USC and, 1010
Vacuum cleaner attachment tutorial, 1089
viewports and, 1075
visual styles, 1035
visual styles and, 1035, 1037
solid primitives
Boolean operations, 1050
grips and, 1110–1111
cones, 1111
dynamic input and, 1111
editing and, 1111
grip location, 1110
solids
subtracting, 1052–1053
solids, 3D applications
moving, 1055
subtracting, 1056–1057
unioning, 1054
Solids Editing toolbar, 1141

body editing, 1141, 1150–1153
Boolean operations, 1050
edge editing, 1141
face editing, 1141, 1143–1149
Solids toolbar, 1041
solid unions, 1052
SOLVIEW command, 1171–1175, 1177–1178
Ortho option, 1175
section views, 1179–1181
tutorial, 1188–1194
SPACETRANS command, 746
spacing in grids, 94
spell checking, 347
SPELL command, 347
sphere
creating, 1046–1047
SPHERE command, 1046
SPLINE command, 262, 302
spline generation, 230
spotlighting, 1212
Staggered mode, dimensioning, 568
Standard toolbar, 3
plot icon, 770
Properties Palette, 399–402
Zoom tools, 143–144
start point, arc drawing, 264
states, layers, 120
status bar, 3–4
Clean Screen button, 7
Communication Center button, 7
Coordinate Display mode, 5
DUCS button, 5
Dynamic Input mode, 5
Dynamic User Coordinate System, 5
DYN toggle, 98–99
function key equivalents, 6
Grid mode, 5
GRID toggle, 96
Lineweight mode, 5
Model mode, 5
Object Snap mode, 5
Object Snap Tracking mode, 5
Ortho mode, 5
Osnap button, 5
Otrack button, 5
Paper mode, 5
Polar mode, 5

Snap mode, 5
SNAP toggle, 96
Status Bar Menu Controls button, 7
Toolbar/Window Positions button, 7
status of layers, 115
Status tool, Inquiry toolbar, 674
stretching
crossing windows and, 228
dimensions and, 228
dynamic blocks, 847–850
overview, 227
STRETCH mode
grips, 390
STRETCH modify command, 227
styles
dimension styles, 639–643
multilines, 285–286
Plot style tables, default, 725
point objects, 290
tables, 358–359, 373–375
text, 349–350
Styles toolbar, 3
subobjects, 1112
editing, 1113
SUBTRACT command, 1052–1053
holes, 1006
subtraction
area calculation and, 678
solids, 3D applications, 1056–1057
suffixes, dimensions, 633
sun icon, 103
surface models, 3D, 1004
Surveyor's Units, 88
SWEEP command, 1064
sweeping, 1064
SW Isometric view, 1029
symbols
dimensioning and, 589–590
multiline text symbols, 338
Symbols and Arrows tab, Modify Dimension Style: Mechanical dialog box, 618–619
symbols for hatch patterns, 493
system variables
attributes, 925
solid modeling, 1077–1079

T

TABLE command, 354, 358
tables
   blocks, 811
   calculations, 379–383
   calculations, tutorial, 379–383
   cells, merging, 356
   creating, 354, 358
   dynamic blocks, 857–860
   Excel spreadsheets and, 357–358
   fields, iinserting, 357
   grips and, 356
   moving, 356
   objects, tutorial, 373–375
   properties, matching, 422–423
   styles, 358–359, 373–375
Table Style dialog box, 358–359
Table View tab, Plot Style Table Editor dialog box, 781
tabs in indented text, 334–335
tags, attributes, 881
Tangent (Tan) Object Snap mode, 32
tangent points in circles, 278
tapering, face editing, 1147
template files, 131–133, 903–905
templates
   materials, 1223–1226
Temporary Tracking Point tool, 44, 46
terrain modeling, 1139
text
   adding, 330–332
   bold, 331
   bullets, 335
   commands, 329–330
   dimension text, 622–625
   editing, 342, 344
   fields, 351–353
   fractional, formatting, 332
   height, globally modifying, 344–345
   importing, 337
   indented, 334–335
   italic, 331
   justification mode, 333
   justified, globally modifying, 346
   linetypes, 109
   mirroring, 226
   Model Space, 746–748
   mtext, width, 336
   multiline, 334–335, 353
   Multiline Text Editor, 331
   multiline text symbols, 338
   multlline, 352
   numbering, 335
   objects, 338, 361–372
   plotted text size, 746–748
   properties, matching, 418
   single line, creating, 339
   spell checking, 347
   styles, 349–372
   tables, creating, 354, 358
   underlined, 331
Text Formatting toolbar, 331
   Insert Field button, 680
Text Style dialog box, 349–350
Text Style toolbar, 349–350
Text tab, Modify Dimension Style: Mechanical dialog box, 621, 625
Text toolbar, 329–330
   Convert distance between spaces, 746
TGA (Targa) images, background, 1233
Thaw layer state, 103
three-view drawings, 443
   tutorial, 451–458
through points, offsetting and, 203
thumbnails, viewing, 792
TIF (Tagged Image File Format) images
   background, 1233
tiled viewports, Model Space, 720
Time tool, Inquiry toolbar, 674
TOLERANCE attribute tag, 881
Tolerances tab, Modify Dimension Style: Mechanical dialog box, 635–636
tolerancing
   GDT, 587–589
   geometric tolerancing symbols, 587
toolbars, 9
   3D Navigation, 1026
   3D Orbit, 1026
   activating/deactivating, 9
   Block Editor, 829
   Dimension, 559
   docking, 10
   Draw, 3, 16, 261
   Draw Order, 958–959
   external reference commands, 928
   Inquiry, 673–674
   Insert, 799
   Layers, 3, 113
   Modeling, 1050
   Modify, 3, 183
   Object Snap, 23
   Properties, 3
   Refedit, 941
   reference, 928
   Render, 1199
   Solids, 1041
   Solids Editing, 1050, 1141
   Standard, 3
   Styles, 3
   Text, 329–330
   Text Formatting, 331
   Text Style, 349–350
   View, 1028–1029
   Viewports, 727, 987
Tool Palette, 12, 819–821
   creating, 822–823
   dashboard, 11
   materials library, 1214
Top view, 1029
   glass box, 435
   shape description and, 433
   tutorial, 448–450
TORUS command, 1047
Tracking mode, Object Snap, 35
tracking path, Polar Tracking, 36
transparency
   palettes, 824
   Tool Palette, 822
Tree View
   DesignCenter, 816
   External References palette, 945
Tree View pane layers, 118–119
TRIM command
   CIRCLE-TRR command and, 269
   toggle to/from EXTEND command, 214
TRIM command, 802
trimming
   blocks, 802
   chamfers and, 201

crossing box and, 207
cutting edges, 206–209
fillets, 197
fill patterns, 496
floor plan exercise, 210
hatch patterns, 496
TRIM modify command, 206, 208–210
tutorial, 250–251, 253
True Color tab, Select Color dialog box, 108
TTR (Tangent-Tangent-Radius) option, 691
tutorials
  architectural details, 994–999
  ARRAY command, 239, 241, 243–244
  Array dialog box, 245–247
  Attribute Extraction wizard, 920–924
  attributes, 885–890
  attribute value editing, 900–902
  auxiliary views, 532–534, 536, 538–541
  Block Attribute Manager dialog box, 917–918
  blocks, 861–863, 865–876
  dimensions, 595, 599, 601–602, 604, 646–649, 651–657, 659–660, 662–666
  Direct Distance mode, 62, 64, 66, 73–74
  DRAW command, 316, 318–320
  drawing layout, 754–756, 759, 762–764
  external references, 963–971
  Field function, 699–700
  Front view, 448–450
  geometric commands, 308–314
  hatch pattern property inheritance, 511, 513–516
  inquiry commands, 688, 690–694, 698
  layer creation, 134–135, 137–138
  LENGTHEN command, 239, 241, 243–244
  lighting, sunlight, 1243, 1245, 1248

  object selection, 144, 170–171, 173, 175, 177–181
  OFFSET modify command, 250–251, 253
  Polar Tracking, 62, 64, 66, 69–72, 77–79
  properties, 424–432
  redefining attributes, 908–912
  relative coordinates, 64, 66, 73, 74
  Right Side view, 448–450
  ROTATE3D command, 1154, 1159–1160
  rotating, 1154, 1159–1160
  sections, 505–509
  SOLDRAW command, 1188–1194
  solid modeling, 1080, 1084–1089
  SOLVIEW command, 1188–1194
  sunlight study, 1243, 1245, 1248
  table calculations, 379–383
  table objects, 373–375
  table styles, 373–375
  text objects, 361–372
  text styles, 361–372
  three-view drawing, 451–458
  Top view, 448–450
  TRIM modify command, 250–251, 253
  view arrangement, 973, 975–980
two-view drawings, 440, 442

U

UCS (User Coordinate System), 1011–1012
  3D Coordinate System icon, 1024–1025
  3D drawings, 1003
  modes, 1012
  solid modeling and, 1010
UCS command, 1010–1012
  3point option, 1014
  Face option, 1017
  New option, 1013
  object option, 1017
  options, 1012
  View option, 1018
  X/Y/Z rotation option, 1015
UCS dialog box, 1022–1023

UCSFOLLOW variable, 1025
UCSICON command, 720, 1024, 1031
UCS Icon dialog box, 1024
underlined text, 331
UNDO command, LINE command, 18
UNDO modify command, 237
Unidirectional Dimensioning System, 556
UNION command, 1005, 1052
unioning solids, 3D applications, 1054
Units Conversion tab, QuickCalc, 408
Unload option, External References palette, 948
user-defined points, xlines, 304–305
User Coordinate System, 3
  2D multiview drawings, 1172
  3DARRAY command, 1131
  icon, turning on/off, 720
  isometric views, 1182
  manipulating dynamically, 1019, 1021
  MIRROR3D command and, 1130

V

Vacuum cleaner attachment, tutorial, 1089
values
  architectural values, 88
  parameters, dynamic block creation, 835–837
  Polar Snap, 39
variables
  DIMASSOC, 636
  QuickCalc, 409
Variables tab, QuickCalc, 408
Vee1 variable, QuickCalc, 409
vee variable, QuickCalc, 409
Verify mode, attributes, 884
vertical text placement, dimension text, 623
View Manager dialog box, named views, 156–160
View modes, 1029
View option, UCS command, 1018
viewports
  creating, 975, 987–991

viewports *(continued)*
   Layout Mode and, 981–986
   locking, 730
   maximizing, 731
   properties, matching, 420, 422, 991, 993
   scaling, 728–729
   scaling images, 727
   solid modeling and, 1075
   tiled, Model Space, 720
Viewports dialog box, 721, 987, 1075
Viewports toolbar, 727, 987
views
   2D, FLATSHOT command, 1184–1186
   arranging, 973, 975–980
   auxiliary, 525–531, 1177–1178
   isometric views, 1182
   named, 156–160
   orthographic, 1175
   partial, 528
   relationships, 436
   Section Views, 473
   section views, 1179–1181
   slicing, 473
   tutorial, 448–458
View toolbar, 1028–1029
Visibility Normal attribute mode, 893
Visibility Off attribute mode, 893
Visibility On attribute mode, 893
visibility states, dynamic blocks, 853–856
Visibility States dialog box, 854
visible features, dimensions and, 554
visual Object Snaps. *See* grips
Visual Style Manager palette, 1035
visual styles
   custom, creating, 1033
   dashboard and, 1037
   Edge Jitter setting, 1038
   Edge Overhang setting, 1037
   Obscured Edges setting, 1039
   settings, 1035
   Silhouette Edges setting, 1038
Visual Styles dialog box, 1033
Visual Styles Manager palette, 1033
vpmax command, 731
VPORTS command, 1075

W

walking through models, 1238
   motion path animation, 1240–1242
WATTAGE attribute tag, 881
WBLOCK command, 798
WEDGE command, 1043
wedges
   creating, 1043–1044
wheel mouse, zooming/panning, 155
Window selection mode objects, 163
WIPEOUT command, 262, 303
wireframes
   3D models, 1003
   helix, 1070
   solid modeling, 1031
wizards
   Add-A-Plotter, 766
   Advanced, 94
   Attribute Extraction, 883, 913, 919
   Create Layout, 739–740
   drawing units, 93–94
   Publish to Web, 787
   Quick, 93–94
workspace
   AutoCAD Default XE AutoCAD Default workspace, 3
workspaces, 1
   3D Modeling, 1
   3D Workspace, 1
   AutoCAD Classic, 1
   AutoCAD Default, 1, 2
   choosing, 3
   customizing, 3
   Viewports, 1
Write Block dialog box, 798

X

X/Y/Z rotation option, UCS command, 1015
XBIND command, 938–940
XCLIP command, 946
XLINE command, 262, 304–305
xlines, 304–305
XR command, 928
XREF command
   binding external references, 943–944
Xref Manager dialog box
   binding external references, 944
X values, 47

Y

Y values, 47

Z

ZOOM-All, 148
ZOOM-Center, 148
ZOOM-Extents, 149
ZOOM-In, 153
ZOOM-Object, 151
ZOOM-Out, 153
ZOOM-Previous, 151
ZOOM-Scale, 152
ZOOM-Window, 146, 150
   PAN command and, 153–154
ZOOM command, 143–150
   realtime zoom, 146
   scaling viewport images, 729
Zoom tools, Standard toolbar, 143–144

LICENSE AGREEMENT FOR AUTODESK PRESS
A Thomson Learning Company

Educational Software/Data

You the customer, and Autodesk Press incur certain benefits, rights, and obligations to each other when you open this package and use the software/data it contains. BE SURE YOU READ THE LICENSE AGREEMENT CAREFULLY, SINCE BY USING THE SOFTWARE/DATA YOU INDICATE YOU HAVE READ, UNDERSTOOD, AND ACCEPTED THE TERMS OF THIS AGREEMENT.

Your rights:

1. You enjoy a non-exclusive license to use the enclosed software/data on a single microcomputer that is not part of a network or multi-machine system in consideration for payment of the required license fee, (which may be included in the purchase price of an accompanying print component), or receipt of this software/data, and your acceptance of the terms and conditions of this agreement.

2. You own the media on which the software/data is recorded, but you acknowledge that you do not own the software/data recorded on them. You also acknowledge that the software/data is furnished "as is," and contains copyrighted and/or proprietary and confidential information of Autodesk Press or its licensors.

3. If you do not accept the terms of this license agreement you may return the media within 30 days. However, you may not use the software during this period.

There are limitations on your rights:

1. You may not copy or print the software/data for any reason whatsoever, except to install it on a hard drive on a single microcomputer and to make one archival copy, unless copying or printing is expressly permitted in writing or statements recorded on the diskette(s).

2. You may not revise, translate, convert, disassemble or otherwise reverse engineer the software/data except that you may add to or rearrange any data recorded on the media as part of the normal use of the software/data.

3. You may not sell, license, lease, rent, loan, or otherwise distribute or network the software/data except that you may give the software/data to a student or and instructor for use at school or, temporarily at home.

Should you fail to abide by the Copyright Law of the United States as it applies to this software/data your license to use it will become invalid. You agree to erase or otherwise destroy the software/data immediately after receiving note of Autodesk Press' termination of this agreement for violation of its provisions.

Autodesk Press gives you a LIMITED WARRANTY covering the enclosed software/data. The LIMITED WARRANTY can be found in this product and/or the instructor's manual that accompanies it.

This license is the entire agreement between you and Autodesk Press interpreted and enforced under New York law.

Limited Warranty

Autodesk Press warrants to the original licensee/ purchaser of this copy of microcomputer software/ data and the media on which it is recorded that the media will be free from defects in material and workmanship for ninety (90) days from the date of original purchase. All implied warranties are limited in duration to this ninety (90) day period. THEREAFTER, ANY IMPLIED WARRANTIES, INCLUDING IMPLIED WARRANTIES OF MERCHANTABILITY AND FITNESS FOR A PARTICULAR PURPOSE ARE EXCLUDED. THIS WARRANTY IS IN LIEU OF ALL OTHER WARRANTIES, WHETHER ORAL OR WRITTEN, EXPRESSED OR IMPLIED.

If you believe the media is defective, please return it during the ninety day period to the address shown below. A defective diskette will be replaced without charge provided that it has not been subjected to misuse or damage.

This warranty does not extend to the software or information recorded on the media. The software and information are provided "AS IS." Any statements made about the utility of the software or information are not to be considered as express or implied warranties. Delmar will not be liable for incidental or consequential damages of any kind incurred by you, the consumer, or any other user.

Some states do not allow the exclusion or limitation of incidental or consequential damages, or limitations on the duration of implied warranties, so the above limitation or exclusion may not apply to you. This warranty gives you specific legal rights, and you may also have other rights which vary from state to state. Address all correspondence to:

AutodeskPressExecutive Woods5 Maxwell DriveClifton Park, NY 12065Albany, NY 12212-5015